"Janaki Ammal's was an exemplary life, and she has been lucky in her biographer. Historian of science, Savithri Preetha Nair, matches her subject's zest and energy, following her traces in far-flung archives in the United States, the United Kingdom and India. She closely tracks Janaki Ammal's relations with her scientific peers, and with her extended family (to whom she was very close). Her scientific research and achievements are narrated expertly, in language accessible to a lay audience but with no sacrifice as regards complexity and nuance. When published, this will be the best biography of an Indian scientist written thus far. In the authoritativeness of its research, and the sensitivity of its treatment, it far outdoes the existing biographies of male scientific icons such as C. V. Raman, Homi Bhabha, and Meghnad Saha."

—Ramachandra Guha in *The Telegraph*

"Savithri Preetha Nair's labour of love is truly inspiring, from the perspective of both biography-writing and writing history of science."

—Deepak Kumar, Historian of Science

"It is the definitive biography of an Indian woman botanist who made many notable contributions but who never received her due in her long career. With extensive archival and other research, Savithri Preetha Nair recreates, for the first time, the life and times of the pioneering E. K. Janaki Ammal and of her contemporaries. This is also a very fine contribution to the history of science in India in the twentieth century."

—Jairam Ramesh, MP, former Union Minister, and author

CHROMOSOME WOMAN, NOMAD SCIENTIST

This is the first in-depth and analytical biography of an Asian woman scientist—Edavaleth Kakkat Janaki Ammal (1897–1984). Using a wide range of archival sources, it presents a dazzling portrait of the twentieth century through the eyes of a pioneering Indian woman scientist, who was highly mobile, and a life that intersected with several significant historical events—the rise of Nazi Germany and World War II, the struggle for Indian Independence, the social relations of science movement, the Lysenko affair, the green revolution, the dawn of environmentalism and the protest movement against a proposed hydro-electric project in the Silent Valley in the 1970s and 1980s.

The volume brings into focus her work on mapping the origin and evolution of cultivated plants across space and time, to contribute to a grand history of human evolution, her works published in peer-reviewed Indian and international journals of science, as well as her co-authored work, *Chromosome Atlas of Cultivated Plants* (1945), considered a bible by practitioners of the discipline. It also looks at her correspondence with major personalities of the time, including political leaders like Jawaharlal Nehru, biologists like Cyril D. Darlington, J. B. S. Haldane and H. H. Bartlett, geographers like Carl Sauer and social activists like Hilda Seligman, who all played significant roles in shaping her world view and her science.

A story spanning over North America, Europe and Asia, this biography is a must-have for scholars and researchers of science and technology studies, gender studies, especially those studying women in the sciences, history and South Asian studies. It will also be a delight for the general reader.

Savithri Preetha Nair received her doctorate in 2003, from the School of Oriental and African Studies (SOAS), University of London, for her dissertation on the museum and the shaping of the sciences in colonial India. Nair's research interests include history of science, modernity and enlightenment at the turn of the nineteenth century, history and politics of collecting for science, sociology of knowledge, the public museum and women in science in colonial and post-colonial India. Among her publications is the co-authored (with Richard Axelby) *Science and the Changing Environment in India: A Guide to Sources in the India Office Records 1780–1920* (British Library, London, 2010), and the monograph, *Raja Serfoji II: Science, Medicine and Enlightenment in Tanjore, 1786–1832* (Routledge, 2012), besides several papers in peer-reviewed international journals and edited volumes. Nair is an independent scholar and divides her time between London and Kerala.

CHROMOSOME WOMAN, NOMAD SCIENTIST

E. K. Janaki Ammal, A Life 1897–1984

Savithri Preetha Nair

Routledge
Taylor & Francis Group

LONDON AND NEW YORK

First published 2023
by Routledge
4 Park Square, Milton Park, Abingdon, Oxon OX14 4RN

and by Routledge
605 Third Avenue, New York, NY 10158

Routledge is an imprint of the Taylor & Francis Group, an informa business

British Library Cataloguing-in-Publication Data
A catalogue record for this book is available from the British Library

ISBN: 978-1-032-03548-2 (hbk)
ISBN: 978-1-032-21168-8 (pbk)
ISBN: 978-1-003-26708-9 (ebk)

DOI: 10.4324/9781003267089

Typeset in Sabon
by Apex CoVantage, LLC

TO MY FATHER

CONTENTS

CONTENTS

CONTENTS

ILLUSTRATIONS

FOREWORD

Pnina Geraldine Abir-Am

This new, book-length, biography of South Asian cytogeneticist and scholar of the global origins and evolution of cultivated plants (including several pertinent disciplines, such as plant geography and ethnobotany) E.K. Janaki Ammal is a landmark contribution to the history of women in science, as well as to the history of biology. Several reasons justify this evaluation:

Though the history of women in science has been a most active and dynamic field for over four decades (the History of Science Society has been awarding prizes for 'outstanding research' on women in science, in book and article alternating formats, since 1987),[1] there are still relatively few book-length biographies of women scientists, especially biologists. Among the most remarkable such biographies to which this book can be compared, one finds Joy Harvey's on French evolutionary theorist Clemence Royer, 1830–1902[2]; Marilyn Bailey Ogilvie's on American ornithologist Margaret Morse Nice[3]; Sona Strabanova's on British bacteriologist Marjory Stephenson,[4] one of the first two women elected to the Royal Society in 1945; and of course several works on American cytogeneticist and sole 1983 Nobel laureate Barbara McClintock[5] (1900–1990). However, all these works are about women biologists from Europe and North America. By contrast, Savithri Preetha Nair's biography of E. K. Janaki Ammal (hereafter Janaki) introduces us to the largely unknown but most fascinating character of a woman cytogeneticist from South Asia. Born on the Southwestern coast of India, Janaki was a close contemporary of Barbara McClintock, as well as a citizen of the world. Indeed, Janaki took her doctorate in botany from the University of Michigan in 1931, just at the time McClintock published her early seminal papers. Janaki dreamt all her life to do her research at her alma mater and maintained life-long correspondence with the Chairman of her Department, H.H. Bartlett, and friend women geneticists Frieda Cobb Blanchard (known to us from the late Sylvia McGrath's essay on Blanchard's collaboration with her husband there)[6] besides the almost unknown Eileen Erlanson Macfarlane; Janaki's dream and desire to return to America did not materialise.

She revisited America once in the 1950s, as the only woman speaker at a landmark ecological conference at Princeton, and was given an honorary doctorate from her alma mater. However, race ('we Asians are not allowed to work in America; I only want to do research!') and gender, as well as the repressive spirit of the 1950s, prevented her from translating her feeling 'more American than British' into a permanent reality. Given what we know from the predicament of women scientists in America prior to the 1970s, McClintock included, it was probably better for Janaki to spend her life in India, except for the 1940s which she spent in the United Kingdom.

The importance of this biography to the history of women in science is not limited to its subject's unconsummated desire to conduct her research in America, but rather extends to Janaki's unusually resourceful and peripatetic life in science, moving between three continents and a considerable number of scientific institutions. All along, she managed to pursue her main interest in the origins and geography of cultivated plants by cytogenetic means, while retaining relative freedom from hierarchical and patriarchal institutions such as the government sponsored agricultural breeding stations.

From the viewpoint of a contribution to the history and philosophy of biology, this biography captures both the transdisciplinary nature of Janaki's practice, which thrived at the border zones between a wide range of fields and sub-fields, including cytogenetics, plant breeding, cytosystematics and ethnobotany; and her pursuit of a new mode of thinking that best suited her unstable, precarious research objects. The biography suggests that Janaki's efforts to distance herself from state science (practiced in the service of crop improvement and run by tyrannical directors who did not understand the importance of basic research) led her to develop her own practice of 'nomadic science,' a less stable and more precarious mode of knowing yet one that is more open to innovation and change than state science, which seeks to maintain the status quo. Ironically, post-colonial science which was supposed to be liberatory ended up adopting dominant models which discouraged change and innovation.

This is a very interesting argument which the biography grounds in the recent trend in science studies to focus on alternative modes of both being and thinking, modes that enable historical actors to have multiplicity of lives while privileging transitional objects of research such as polyploidy. Janaki is thus shown to have been first and foremost a cytogeneticist, master of fixating chromosomes, inducing chemically the formation of polyploidy, counting and systematising on the basis of new combinations of chromosomal sets, but also a 'field biologist, a plant geographer, a paleobotanist, an evolutionary systematist, and experimental breeder, an ethnobotanist, and not least a naturalist and a traveler-explorer' (Epilogue, p. 7).

This was a most remarkable achievement for a woman whose life in science spanned over half a century, from the late 1920s to the early 1980s, who published prolifically throughout that time, and who managed to acquire a measure of national and international recognition, such as membership in leading scientific societies (the Linnean Society, the Genetics Society and the Indian Academy of Sciences of which she was among its founders) even though the highest honours such as FRS or Padma Bhushan were reserved for those who laboured to maintain the status quo.

It is a great achievement of this biography to document in meticulous and extensive archival detail, having consulted circa 30 archives, roughly a third in each country of India, the United Kingdom and the United States, and complemented them by oral history. These archives were widely dispersed in each country, as a result of Janaki's intense peripatetic existence, which required that the historian too become the practitioner of 'nomadic' scholarship.

Janaki's various moves in space and time, while unpacking her immense range of research objects, tropical as well as temperate plants, plants of economic importance, most notably the sugar cane, ornamental plants, and plants used for subsistence by forest dwellers, the latter being part of a late-in-life interest in ethnobotany which coincided with the rise of environmentalist and conservation movements in the 1970s.

This biography inadvertently sheds new light on Barbara McClintock, the best known American woman cytogeneticist, sole Nobel Laureate (1983), contemporary of Janaki, and subject of several biographical works, by highlighting the key role of gender in trumping race and class, in the process of marginalising women scientists and depriving them of timely credit.

Much as with women scientists from Europe and America, this biography draws attention to progressive men scientists who helped Janaki prevail in her adamant quest for a life of research. It also raises the question as to whether such men may have been influenced by a few enlightened British scientists who opened their labs to women in the inter-war era, such as the biochemist F. G. Hopkins at Cambridge and the X-ray crystallographers W. H. Bragg and J. D. Bernal, in London. In India, as elsewhere, men of science largely favoured male students, which is why Janaki's trajectory can be safely described as an ingenious strategy to escape patriarchy in both science and society, during both colonial and post-colonial times.

The book is very enlightening on the intersection of Janaki's life with world history, whether during the blitz on London in World War II which made her feel as a 'world citizen,' but especially on the role of science in nation building in India, where she occupied a few key positions in science policy, such as director of the Botanical Survey of India, founding director of the Central Botanical Laboratory and member of advisory committees, in the decade after Independence. Her participation in the protests to stop a hydroelectric project in the Silent Valley in the 1970s illustrates how she managed to combine her roles as scientist seeking to map cytogenetically the deep forests, as an activist supporting the environmentalist movement and as a science policy figure serving on committees in search of solution for alternative sources of power. This part on whether the ancient forest will be saved, towards the end of the book, reads as a detective story, and serves as a superb end to a most remarkable life in science.

Another major contribution of the book pertains to using the biographee's life as a context for how science is created. On the one hand, it aims to show the pluralistic practice of modern genetics, as revolving not only around genes and model organisms but also around many ways of constructing knowledge and many networked actors. This is a story of science in the making, rather than the story of a seminal discover (as was the case, for example, with Barbara McClintock's discovery of 'jumping genes'). Though Janaki spent some of her time as an employee of state institutions such as agricultural experimental stations, doing plant hybridisation (mainly sugar cane) with an eye for improving crops, most of her time was spent on the move, either on solitary travels of exploration (all over India, from the Himalaya and Kashmir to Madras and Malabar) or in research on chromosomes, polyploidy and, especially their role in creating new, reproducible hybrids.

Along these lines, this biography combines the micro level of Janaki's intense interaction with her enormously diverse cytogenetical preparations, encompassing both tropical and temperate plants, and the special technical skills she developed in fixating them, with the macro level of interaction with institutions and individuals inhabiting the stable existence of state science.

The biography's argument that Janaki's ongoing mobility across interstices, borders and bylanes of science signified not only her ontology but also her epistemology, namely that she chose to become a peripatetic figure (the author uses the metaphor of 'nomad') which is embedded in literary and anthropological studies as a signifier of lacking the stability of a permanent abode, being on the move, to 'break free from the tyranny of hierarchy, patriarchy, and pseudo-science she encountered in the male bastions of state science such as crop breeding stations' (p. 10) but also to be able to contemplate the world via an archipelagic thinking, as a collection of interconnected islands rather than as closed continental forms.

The biography also has an important lesson for historians and others who seek to reconstruct the lives and unorthodox science of women scientists such as Janaki, namely to lobby for the collection and cataloguing of the Personal Papers of women scientists, so that future biographers be able to produce well-documented biographies. This work is a labour of love, and hence, it is disconcerting to learn that Janaki's incoming letters were not preserved; nor were her remarkable slides. Since the biography was pieced from letters and documents of hers found in the archives of her interlocutors, most notably her sponsors, colleagues, and friends and family in India, the United Kingdom and the United States, whose Papers did make it to institutional archives, especially if they were department chairmen, we may not find out how the intersectionality of gender, race, caste and age played out in her life-long meaningful relationships.

This biography should serve as an exemplar of the doability of recovering the lives and historical agencies of other women scientists from Asia and other regions which harbour such exciting biographees. Janaki's life, challenged by the intersecting constrains of gender and race in three continents, is inspirational for girls and women not only in South Asia but everywhere. As opportunities for practicing science become increasingly available to women, so we need dedicated historians, such as Savithri Preetha Nair, to make sense of the complex traces left by peripatetic women scientists in their countries of origins, as well as all over the entire world.

3–15–2022

Notes

1 hssonline.org/prizes/Rossiter prize
2 Harvey, Joy D. *'Almost a Man of Genius', Clemence Royer, Feminism, and 19th Century Science* (New Brunswick, NJ: Rutgers University Press, 1997).
3 Bailey Ogilvie, Marilyn. *For the Birds, American Ornithologist Margaret Morse Nice* (Norman, OK: University of Oklahoma Press, 2019).
4 Soňa Štrbáňová. *Holding Hands with Bacteria: The Life and Work of Marjory Stephenson* (Berlin and Heidelberg: Springer Verlag, 2016).
5 Keller, Evelyn Fox. *A Feeling for the Organism: The Life and Work of Barbara McClintock* (New York: Henry Holt, 1983); Kass, Lee B. Records and recollections: A new look at Barbara McClintock, Nobel Prize-Winning geneticist. *Genetics* 164 (2003): 1251–1260. www.ncbi.nlm.nih.gov/pmc/articles/PMC1462672/; or www.genetics.org/content/164/4/1251; Comfort, Nathaniel C. *The Tangled Field: Barbara McClintock's Search for the Patterns of Genetic Control* (Cambridge, MA: Harvard University Press, 2001).
6 See chapter 9 in Pycior, Helena M., Nancy G. Slack, and Pnina G. Abir-Am (eds.). *Creative Couples in the Sciences* (New Brunswick, NJ: Rutgers University Press, 1996).

ACKNOWLEDGEMENTS

One afternoon, in 2002, at the General Library of the Natural History Museum, London, which I had been visiting to consult materials for my doctoral dissertation, I was distracted by an old volume lying casually on the table that I was working on. Piqued by curiosity, I picked it up to discover that it was a list of members of the Eugenics Society (the Galton Institute) from its early years. Flipping through its pages, I came upon several familiar names: H. G. Wells, Havelock Ellis, J. M. Keynes, Julian Huxley, R. A. Fisher, Marie Stopes . . . and then in the most unanticipated of places, an Indian one, the only name I had spotted until that point. The year was 1932 but what baffled me most was that it was a woman, from South India evidently, but a name I had never heard before: 'E. K. Janaki Ammal, Lawley Road, Coimbatore.' Little did I know that this chance encounter would set me off on a humbling even if exhilarating journey, and one that would last so long, that it would end up consuming one-third of my lifetime, and by which time I had metamorphosed into a grandparent! I walked, scrambled and wandered in her trail, and for great stretches of time it was in darkness and isolation, until a tiny speck of light appeared at the end of the tunnel. I ambled towards it, and it grew bigger, but it was painfully slow striving, which I can only compare to trekking precariously up a steep mountain in a dreadful storm carrying an unwieldy load on one's back. It is that orb, a mere glimpse of which I had caught sight of, that I now offer to the readers in the form of this book. Nothing more. Nothing less.

I would like to see this biography as a triumph of independent scholarship against all odds. I could not have accomplished this alone however. The debt I owe—to people, plants and books—has grown into a great heap but I must first invoke those kindred spirits that pervade this volume, people who, when I began my journey, were with me, sharing their memories and whatever else they could, but are sadly no more: Janaki's nieces, Padma Padmanabhan and Shashi (a.k.a. Saraswathi, who sang the 'Koduvalli Blues,' so beautifully for me and my companion) and her nephews, E. K. Hari Krishnan, Ram Damodar, Muthukrishnan and more recently, Shyam Damodar. I also remember with gratitude, the late Oliyath Sayeed, who tirelessly showed me around his home town of Tellicherry, Malabar, Premnath Moorkoth who corresponded with me on 'Thiyas,' and Murkoth Ramunni, who spoke to me at length about Tellicherry, Laccadives and his close relationship with Janaki's adventurer brother, E. K. Madhavan. I am grateful to the late mycologist C. V. Subramanian, who readily met with me and presented me with a booklet on the University Botanical Laboratory. Justice Krishna Iyer, despite his advancing years, readily shared with me his memories as a tenant of the Edathil family, in Tellicherry; he recalled seeing Janaki Ammal, during the long vacations, out and about carrying baskets of plants collected from the neighbouring fields. I was utterly shocked to hear of the sudden demise of P. Lakshminarasimhan of the BSI, who I had been in regular correspondence with; his generosity and kindness will live with me. Another big loss has been that of Prof. M. K. Prasad. It was thanks to him, that I spoke on Janaki in Kerala (in June 2010) for the first ever time (for which thanks also to K. R. Lekha of the Kerala State Council for Science Technology and Environment, Government of Kerala), at the Shastra Bhavan, Trivandrum. Shortly after, thanks again to Prof. Prasad, I was invited to talk on Janaki at the School of Environmental Sciences, MG University, Kottayam. I pay homage also to the succulent specialist Gordon Rowley, who once trained under Janaki at Wisley, and Dorothy Blanchard, who as a young girl in Ann Arbor, interacted with her; both were generous in sharing with me their memories of her. It was very late that I learnt of the passing of Usha Zutshi, who so warmly received me in Jammu and took me to all of Janaki's favourite haunts; she introduced me to her fellow doctoral students (one of them B. L. Bradu, I regret is also no more), and sharing fond memories of her esteemed doctoral supervisor.

I can't thank Janaki's siblings and their children enough, some in particular, for preserving precious material—private correspondence and diaries—and hail their sense of history: Aarthi Ajit (and her loving family): how can anyone so young be so willing to share for knowledge-sake! Uma Ramachandran, the most senior among them, who has not only done more for this book than anyone else but also treated me as her own. Shamini Ramesh for her generosity. Geeta Doctor, for her words of encouragement and some Janaki-anecdotes. The late Shyam Damodar and Hyma for their candour and kindness. And Chithra, and her sister Lalitha, for sharing memories and rare photographs of their eminent aunt.

Janaki was a tireless traveller, and I had to match this at least to a modest extent, if I were to write her biography, and this required funds. I was fortunate to receive a small grant from the British Academy, in 2008, which enabled me to travel to the University of Michigan, to conduct research; I was at this time a postdoctoral scholar based at the School of Oriental and African Studies, University of London and the British Library, working on a completely different project. I am grateful to the British Academy, to Prof. David Arnold for recommending me to it and to the extremely helpful staff at the Bentley Historical Library at Ann Arbor, and my late cousin, Vinod and wife Asha, for hosting me during my stay in Michigan. Besides Dorothy Blanchard, I am grateful to Dorothea Colman of the Matthaei Botanic Gardens, Michigan, for the immense warmth with which she welcomed me at the Gardens and in tracing relevant material from the Garden-archives and *The Michiganensian*. It was also about this time that I would consult the substantial Darlington Papers at the Bodleian Library, Oxford. The five-decade long correspondence between Janaki and Darlington provided the chief leads and would serve as one of my chief founts of information.

For a great many years, work on the biography was carried out alongside my other researches, and without financial support; the result was that it was mostly put on the backburner, research being limited to focused reading, teaching myself basic genetics, hunting down sources (surprising myself at times) and nursing the ailing sapling of the *Magnolia kobus* 'Janaki Ammal' in my back garden in London. From 2007, I had also turned nomadic, taking my work (in progress) to various platforms, in the United Kingdom, United States and India. As part of a panel on twentieth-century science in India at the Annual Meeting of the British Society for the History of Science (BSHS), Manchester (28 June–1 July 2007), I gave a paper on 'The Impact of Soviet Biology on Indian Plant Genetics: E K Janaki Ammal and Her Contemporaries, 1930s–40s'; I thank my co-panelists Pratik Chakrabarti and Jahnavi Phalkey for this opportunity. This was followed by a presentation at the Annual Conference of the British Association of South Asian Scholars (BASAS), Leicester (26–28, March 2008).

A few sunny spots appeared in the firmament at about this time. I was awarded a New India Foundation Fellowship (2008–09), which was a great blessing. The Fellowship not only took care of my maintenance for a year but also funded my research travels in India (to Madras, Delhi, Calcutta, Lucknow and Jammu). I am most grateful to all those on the Board of the NIF, and to Ramachandra Guha in particular for this. Work on the biography received a great fillip with this grant, and I was able, in April 2010, to present something of the work in progress at the Indian Institute of Science, Bangalore, under the aegis of the Centre of Contemporary Studies (CCS), thanks to Prof. P. Balaram, then Director and Prof. Raghavendra Gadagkar, of the Centre for Ecological Studies and the CCS. Following this presentation, I was offered a year-long Visiting Fellowship at the CCS (which was mostly devoted to writing up my book on Tanjore however). It not only gave me the opportunity to reunite with a city I had grown up in, but living on campus, amidst its abundant nature—the Slender Lorises (making nights so live with their hoots), great trees and shrubs (including a beautiful *Magnolia grandiflora*), giant honeycombs and monstrous lianas—prepared me for Janaki, besides giving me limitless joy. I am ever thankful to Prof. Gadagkar for this. It was also here that I had the occasion to browse through the entire series of back volumes of the *Current Science* and the *Proceedings of the Indian Academy of Sciences*, groundwork so essential for the project. Thank you to Prof. Shankar Rao, for the tree walk and his volumes on the flora of the campus, which were an utter delight; thanks to Prof. Gadagkar, copies of these were gifted to me at the end of my Tanjore presentations; and also, to Ajay (Cadambi) for those endless conversations on music, art, history and queer-life.

More presentations followed. Thanks to Itty Abraham, I was invited to present a paper on Janaki at a conference on the history of science in India at the South Asia Institute, University of Austin, Texas, in May 2010. Soon after, I presented a paper, 'Doing Science Beyond the Nation: An Indian Woman Scientist in Britain in the War Years, 1939–1945,' at the *Making Britain: Visions of Home and Abroad 1870–1950 Conference* held in London, and organised by the British Library/University College, London/ Oxford University/Open University (13–14 September 2010). A couple of months later, I was back in India, as B. Venugopal, then Director of the National Museum of Natural History, Union Ministry of Environment and Forests, had invited me to deliver the key note lecture at a national seminar on Janaki Ammal, held in the town of her birth, Tellicherry. Over the next few years, more public presentations were made: a paper at a symposium, 'Profiles in Twentieth Century India,' organised by INSA on the occasion of its 77th anniversary meeting at Tezpur, Assam, in December 2011, thanks to Prof. Gadagkar; a lecture on the occasion of the Botanical Survey of India's 125th Anniversary Celebrations at Kolkata in February 2015, thanks to its Director Paramjit Singh and (the late) P. Lakshminarasimhan; and finally, in 2016, thanks to M. Sabu, Head of the Botany Department, University of Calicut, I was invited to contribute to the lecture series organised in memory of Janaki Ammal, at the 28th Kerala Science Congress, Calicut, in January 2016.

Between 2017 and 2020, I worked on more drafts, each one remarkably different from the previous, a reflection of the jumps I was making in my thinking. I had resisted publishing anything all this time, not even a journal article, because I knew that would be hasty and even foolish, for Janaki had continued to be elusive. New and important sources were unearthed as late as 2020, followed by more intellectual wandering, reflection and rewriting, until I felt that Janaki had forgiven me for this deed.

Among the institutions and their staff in the United States and United Kingdom, without which/whom this book would not have been possible, particularly in these pandemic times, I would like to thank: The Bentley Historical Library and the University of Michigan Herbarium; the Smithsonian Institution Archives and the US National Herbarium; the Bancroft Library, University of Berkeley; Huntington Library, San Marino, California; Mount Holyoke College Archives, Mass. (thank you, Debbie Richards); Oakridge Nuclear Laboratory, Tennessee (big thanks, David M. Whittaker); in Britain, the Bodleian Library, Modern Paper Collections (thank you very much, Colin Harris and the other staff); the John Innes Centre, Norwich (thanks a ton, Sarah Wilmot for all your help, warmth and kindness); the Kew Gardens Archives and Herbarium; the RHS Lindley Library; the RHS Wisley Collections; the British Library; the Wellcome Library; the Royal Botanic Garden, Edinburgh; and the Rothamsted Experimental Station, Harpenden. In India, I owe thanks to the Birbal Sahni Institute of Palaeo-botany, Lucknow; the Botanical Survey of India Herbarium and Library; the National Archives of India, New Delhi; the Regional Research Laboratory, Jammu; and finally, in Chennai, the Centre for Advanced Study in Botany, the Senate House Library, University of Madras; *The Hindu* newspaper archives, and the collections of the Women's Christian College.

Several scientists, historians and the 'interested' have contributed in various ways, ranging from words of encouragement, reading the manuscript, providing useful information or documents: I would like to thank Oliver Darlington, Clare Passingham and Oren S. Harman (in particular for his biography of Darlington); M. S. Swaminathan (Chennai), the late S. K. Jain (Lucknow), E. R. S. Talpasayi and Y. Ranga Reddy (Hyderabad), V. P. Prasad (BSI), Chandra Mohan Nautiyal (Birbal Sahni Institute, Lucknow), Profs. Mahesh Rangarajan, Deepak Kumar, Sudhir Chandra (Nainital) and K. S. Manilal, and Karunakaran (Shoranur); the Jammu group including the late B. L. Bradu and Usha Zutshi and Y. S. Bedi; the Coimbatore group of T. V. Sreenivasan, Z. Abraham and A. William Jebadhas; and the staff at the Centre for Advancement Study in Botany, University of Madras— N. Anand, Durairaj T. Kalaichelvan, P. Nagendra Prasad, besides Durairaj Rajiah and Mayakannan. I also thank Ram Guha, Prof. Deepak Kumar and Jairam Ramesh for reading the manuscript and for their reviews. Prof. Arjun Appadurai showed eagerness to read it despite his very tight schedule, but unfortunately could not make it in time for the book.

For their kind words of encouragement, I am indebted to Margaret Rossiter, whose work has been a great inspiration to generations of historians studying women's participation in science, to Lee Kass for her publications on Barbara McClintock, and to Pnina G. Abir-Ram; I cannot thank Pnina enough for her empathy, perceptive reading and so readily agreeing to write a foreword.

Every line in the manuscript was read with care by Rivka Israel, who played no small role in making it more readable. My thanks also to Aakash Chakrabarty and Brinda Sen of Routledge/Taylor and Francis for their kind support and effort and dedication in seeing this volume through the press.

There were those who continued to fan my flames, despite the great stretches of time taken to complete this book, to them many thanks and warm hugs: it was in the company of Ram that I began my journey in search of Janaki; Shivaji, kept a daily conversation going for years; Chithra and Venkitesh, for their companionship in the early years; Girija, for her deep concern and empathy; and as always to Dilip, for being there, every time. To

Ramesh, my long-suffering co-traveller, memory bank and sounding board of ideas, and to Gayathri, Shambu and little Siddharth, much love. Thanks also to my family, and my in-laws, for their unfailing support through the years, but it is to my dear father, who passed on to me his love for nature (among many fond childhood memories are visits with him to the Lal Bagh Gardens, and its famed glass house), science, history, art and good design, and his high regard for Janaki Ammal and the work I was doing to resurrect her, that I dedicate this book.

Kidangoor (Kottayam)/London, March 2022

E. K. JANAKI AMMAL: A TIMELINE

1897 November 4	Birth at Tellicherry
1905–15	At Sacred Heart High School, Tellicherry
1915	Joins Madras College for Women (later called Queen Mary's College) for Intermediate Course (FA)
1917–21	BA (Hons.) at Presidency College, Madras
1921	Appointed Professor of Natural Science at Women's Christian College (WCC), Madras
1923	Awarded Master of Arts (MA), University of Madras
1924–26	Master of Science (Research) as Barbour Scholar, University of Michigan, Ann Arbor, USA
1926	Returns to India and resumes teaching at WCC
1928–31	Barbour Fellow at the University of Michigan
1930	Made full member of Sigma Xi
1931	Awarded DSc for her dissertation 'Chromosome Studies in *Nicandra physalodes*', supervised by B. M. Davis; Researcher at John Innes (May–August); elected to Linnean Society as Fellow
1932	Research Fellow, Department of Botany, Presidency College, Madras
1932	Reader in Botany, Chairman of the Board of Examiners in Botany, Member of the Board of Studies for Natural Science and Academic Council, and Member of the College Council, University of Madras

1932–34	Professor of Natural Science, Maharaja's College of Science, Trivandrum, Travancore
1934 January 21	Meets Gandhi in Trivandrum
1934–39	Sugarcane cytologist at Sugarcane Breeding Station, Coimbatore
1935	Elected Honorary Secretary of Indian Botanical Society; in August, leaves for England representing India at the Imperial Botanical Congress at Cambridge and attends the Sixth International Botanical Congress in Amsterdam
1939	Leaves for Edinburgh to attend the Seventh International Genetics Congress; takes refuge at the Institute of Animal Genetics, Edinburgh
1940–43	Researcher at John Innes
1944–46	War time work and Researcher, Kew Gardens on subsistence allowance
1945	*Chromosome Atlas of Cultivated Plants* co-authored with C. D. Darlington
1946–51	Cytologist, Royal Horticultural Garden, Wisley
1947 July	Attends Sixth International Congress of Experimental Cytology at Stockholm, visits Uppsala and Lund
1948 November	Returns to India by flight after nine years in Britain; leaves for Simla and then Raxaul en route to Nepal
November–December	Plant collecting expedition in Nepal
1949 March	Returns to Wisley
November	Meets Nehru; presents him with copy of Darlington's 1948 Conway Memorial Lecture
1950 January	Director of Agriculture, Government of India; elected to membership of Royal Geographical Society
April	Returns to Wisley
September	Attends meeting of the British Association for the Advancement of Science, Birmingham
1951 August	Attends meeting of the British Association in Edinburgh
1951 September–1952 October	Researcher, Genetical Laboratory, Institut National Agronomique, Paris

1952–54	Special Duty Officer, Botanical Survey of India, Calcutta
1953	Re-elected to the Linnean Society as Fellow
1954	Director of Central Botanical Laboratory (CBL); CBL moved to Lucknow
1955 April	Leaves for Oak Ridge, USA, on four-week course on tracer atom techniques
June	Awarded Doctor of Laws, University of Michigan; participates in the Wenner Gren symposium at Princeton on 'Man's Role in Changing the Face of the Earth'
1956 March	Attends UNESCO symposium on study of tropical vegetation, and another on problems of humid tropical regions, at Kandy
1956–57	CBL moved to Allahabad
1957	Elected to the Indian National Science Academy
1958 December–1959 January	Attends 24th annual meeting of the Indian Academy of Sciences at Baroda and 46th session of the Indian Science Congress at New Delhi
1959	Elected president of the Indian Botanical Society; retires as CBL director; appointed Officer on Special Duty by CSIR to organise RRL Jorhat and put in charge of Department of Cytogenetics, RRL Jammu
1960	Receives the Indian Botanical Society's Birbal Sahni Medal; becomes life member of Institute of Tropical Ecology and Research
1961	Elected President of the Indian Society of Genetics and Plant Breeding
1962	Appointed Honorary Professor at the Jammu & Kashmir University
June	Goes on a secret high altitude agriculture mission to Ladakh organised by the Ministry of Defence; inspects work done at the Murtse farm on food farming meant for Indian troops
1964	Appointed Emeritus Scientist, RRL Jammu

1966	Presents paper at international symposium on The Impact of Mendelism organised by Indian Society of Genetics and Plant Breeding, Delhi
1969 July	Term as Emeritus Scientist at RRL, Jammu, ends; leaves J&K
1969 October	Advisor, Biology Unit of the Bhabha Atomic Research Centre, Trombay
1970 October	Professor Emeritus at the Centre for Advanced Study in Botany (CAS), University of Madras; returns to the Department where she began, after 50 years; resides at the Field Research Laboratory, Maduravoyal.
1971	Begins developing a medicinal/ethnobotanical garden at Shoranur, Malabar; Principal Investigator of the Indian National Science Academy (INSA) project, 'Ethnobotanical Studies of South Indian Tribes,' based at Maduravoyal.
1972 September	Chairs meeting of the Kerala State Committee on Science and Technology at Trivandrum for the first time, to discuss an 'approach paper' for the Science and Technology component of the state's Fifth Five Year Plan
November	Delivers the Second Silver Jubilee Lecture, 'Plants and Man,' at the Birbal Sahni Institute of Palaeobotany, Lucknow
1973 September	Writes to INSA protesting against proposal to terminate the ethnobotany project in February 1974
1975 September	Ethnobotany project contract renewed by the University Grants Commission
1976 April	Travels to Jammu to present paper, 'India's wealth in medicinal and aromatic plants: its exploitation and improvement,' at the symposium on Medicinal and Aromatic Plants
September	Botanical tour of the monazite sands of South Travancore
1977	Awarded Padma Shri; as member of the official working group for ethnobiological

July	studies in India, attends the first meeting to work out a national plan at New Delhi Janaki attends the second meeting of the official working group for ethnobiological studies in India
August	Janaki sends a booklet protesting against Silent Valley dam proposal to Darlington
December	Government of India's Department of Science and Technology sanctions three-year project to bring out an illustrated book on South Indian Medicinal Plants (with special reference to Ethnobotany)
1978 June	Kerala government commences work at Silent Valley despite protests from the scientific community; the Department of Science and Technology sanctions a sub-station at Ashok, Shoranur
December	Travels to Meerut to attend the first conference organised by the Indian Botanical Society and presents paper entitled 'Ethnobotany: Past and Present'
1979	Delivers the third M. O. P. Aiyengar Memorial Lecture entitled 'Ethnobotany: Past, Present and Future,' at the University of Madras Works on producing a modern version of the *Hortus Malabaricus*; plans to set up a cytology laboratory at Ashok, Shoranur
1980	Proposes a second project to the Indian government entitled Cytogenetic survey of the flora of the Silent Valley; seeks permission to visit the Silent Valley for her project; unsuccessful
1984 February	Death in Madras

PROLOGUE

The southwest monsoon was utterly devastating that year, battering the whole of South India. Travancore, Cochin and Malabar suffered terribly. The waters had risen to such incredible heights, the apocryphal story goes, that people crossing on boats could easily pick coconuts off trees. Seen drifting about the waters were carcasses of animals, furniture and articles of domestic use.

The year was 1924. The unabating rains and gusts of blustering winds caused much destruction also in the cantonment town of Tellicherry on Malabar Coast; even as the great deluge drowned most parts of the town, a 27-year-old Thiya woman, burning with desire to become an exceptional scientist, was preparing to set out on her maiden voyage across the oceans to America. It was as if she had walked on water, or the rivers had parted for her, for unsubdued by the floods and devastation, she surfaced miraculously in Madras ten days later. Her boat had departed already for the shores of New York, but she would find a berth on the next steamer, and reach her university in time for her course. In 1931, when she was awarded a doctorate by the University of Michigan, she was the first ever Indian woman to achieve this academic milestone in the botanical sciences. Her name was Edavaleth Kakkat Janaki Ammal (4 November 1897–7 February 1984), whose long and tempestuous life in science is the subject of this biography.

Janaki's birth into a progressive Thiya family of Tellicherry coincided with the dawn of the disciplines of cytology and genetics, in which she would specialise.[1] Using cytogenetics, agronomy, plant geography, geology, history and anthropology together with a sound knowledge of the cultural uses of plants, Janaki mapped the origin and evolution of cultivated plants across space and time, to contribute to a grand history of human evolution. Her concerns were moulded not only by contemporary ideas in cytology and genetics, especially evolutionary biology, but also by the urgent need to provide an antidote to the aggressive nationalist strategies adopted by Indian agronomists in the name of food security and progress following India's independence. Janaki is without doubt a worthy subject for biographical exploration, on several counts, but three reasons are central to the present undertaking.[2]

A Project of Recovery

Almost all published sources on the history of women in science have had as their primary focus Europe and North America, with Asia, Africa, Australia and South America hardly figuring in them. Wiping clean this selective clouding of history, this biography marks the beginning of a grand recovery project, of Indian women in science. The subject of this biography was a pioneer and the only one among those Indian women born prior to 1900 to make a successful career of science, and yet remains largely unknown; in the year she was awarded a doctorate, there were only one or two Indian men who had doctorates in botany, and none at all in cytogenetics, which goes to show how exceptional her achievement was even for this reason alone. It is hoped that this detailed reconstruction of her life and science will, at the least, enable people to know Janaki and her science better, and at best, write her back into the history of science. That she was not alone on this journey is what her story reveals; the biography offers a glimpse of the heterogeneous and unanticipated landscape that existed even in the early decades of the twentieth century, one that was inhabited by women of science (and men of course) across nationalities, who tended to form a creative counterculture, constellation or sorority, of an informal or formal kind, spanning continents, and co-producing scientific knowledge. Each of these women were path-breakers and made a success of their careers in their chosen fields of science.

Janaki was the first Indian woman to receive a doctorate in botany as noted earlier, but the completely unknown Miss Maneck Merwanji Mehta was the first to do so in any one of the sciences, a feat she accomplished six years before Janaki.[3] This biography exposes the names of a number of Indian women studying science at the University, or even pursuing careers in science in the first decades of the twentieth century,[4] and sheds light on the constraints, particularly in these early years that they faced in accessing education, resources and employment in science. It thereby also draws attention to those men of science who played important parts in altering the gender and caste ratios at University departments, even as early as the 1920s. Men of science like T. Ekambaram, M. O. P. Aiyengar and, later, T. S. Sadasivan of the University of Madras, for instance, self-consciously used their scientific and official authority as successive directors of the University Botanical Laboratory to enlist women in scientific research and inspire them to publish their work, to which they gladly lent their names as co-authors. In this respect, there was much in common between the UBL and the X-ray crystallography laboratories of the Braggs in London and Manchester, which had more women students than men engaged in research,[5] and the British Marxist physicist J. D. Bernal's laboratory in Cambridge,[6] which was open equally to men and women.[7] It was however in direct contrast to the laboratories of Nobel-physicist C. V. Raman at the Indian Institute of Science and of the Indian astrophysicist Meghnad Saha in Calcutta, with their preference for male students, to which some scholars in the recent years have drawn attention.[8]

A Twentieth-Century Life-World

A second reason that has propelled this extensive biography is that Janaki's story offers a defamiliarisation of the twentieth-century world. It reconstructs that world from the perspective of a pioneering and highly mobile Indian woman scientist, thereby renewing our perception of the socio-political and intellectual world of the times. Janaki's working life intersected with several significant historical events—the rise of Nazi Germany and World War II, the struggle for Indian Independence, the social relations of science movement, the Lysenko affair, the so-called green revolution, the dawn of environmentalism and the protest movement against a proposed hydro-electric project in the Silent Valley in the 1970s and 1980s, to highlight just a few. She corresponded with major personalities of the times, including political leaders like Jawaharlal Nehru, biologists like Cyril D. Darlington, J. B. S. Haldane and H. H. Bartlett, geographers like Carl Sauer, and social activists like Hilda Seligman, who all played significant roles in shaping her *weltanschauung* and her science. She was a participant in landmark scientific meetings and published prolifically in peer-reviewed Indian and international journals of science, most of which were founded during her lifetime, besides co-authoring a book, *Chromosome Atlas of Cultivated Plants* (1945), considered a bible by practitioners of the discipline. She would also be involved in the nation-building process, holding lofty public positions in her country of birth.

The understanding that science is shaped by the context in which it is created, or that science is as social as other kinds of knowledge, informs this biographical exploration. Janaki and cytogenetics were coevals (as were Barbara McClintock, Eileen Macfarlane and Irene Manton, to name three of the most eminent among early women cytogeneticists), and her science is a reflection of how the discipline changed its contours with the rise of the new Soviet State and its active encouragement of applied science (or state-led science), the advance of fascism and its abuse of science, and, eventually, World War II and the post-war reconstruction of the world.

Making Scientific Knowledge

This brings us to the third and perhaps the most important reason for this biography: to open a window into how science works.[9] When scientific 'discoveries' or scientific 'facts' are formally presented with a notion of objectivity, histories of actions or workings as well as the human touch behind these are eliminated, simply lost or rendered invisible in the narrative. In other words, they get black-boxed. To know how science works demands 'opening up' or 'unpacking' the black-box and making the 'messy' traces visible again.[10] Writing a scientific biography offers just such an opportunity. The biographee provides an axis about which to build a contextual history of how science is created locally, globally and also transnationally. Such a history takes on board the social, cultural and political contexts of science (the external social factors shaping science such as funding structures, institutions, political influences, peer reviews and bureaucracy, and the influence of caste, gender, class and race relations on the way scientific knowledge is shaped), the places and cultures of doing science (such as the laboratory and the field) and unabashedly 'the flesh and blood' or the human factors that fuel it—such as feelings, likes, dislikes, ambitions, ideas, ideologies, influences, friendships, career choices, idiosyncrasies, frustrations and joys—not just for the biographee but for all those integral to her social and professional worlds. In this sense, this is not just 'a' biography but a compound web of biographies, taking on a reticulate form (like a complex lace-like network) through which light shines bright on the system, the swarm of actors and the myriad ways in which scientific knowledge is constructed—in short, a global microhistory of a kind.[11] One of the things that this biography aims to demonstrate is that the history of modern genetics was not all about genes—or about Drosophila or Oenothera genetics—instead it was a pluralistic practice, involving a range of practitioners, aims, objects, techniques, styles and/or mixed cultures of doing science.[12]

Scientific biographies (one is here speaking of detailed reconstructions of lives in science based on archival material including oral histories, rather than popular accounts, short memoirs or journal articles) are a major lacuna in general, and in the Indian context in particular. Where such attempts have been made, the focus has chiefly been on the personal or the social, with little attention if at all to the content of their science or on science in action. This biography on the other hand undertakes the daunting task of 'following' Janaki's science—what she aimed to do, how she went about it, the strategies she adopted, what she accomplished and what she did not (her frustrations and failures)—situating it within a global history of plant genetics in the twentieth century. Her life cannot be written from within the 'Nobel-Prize' framework of finished science, for hers was not about making a seminal discovery of universal relevance (in cytogenetics, like a McClintock for instance),[13] but of producing a series of singular, small-scale and local discoveries, slowly and progressively, one concrete example after another—a necessarily slow and unfinished pursuit—, over a lifetime of wandering, which makes the task at once both thrilling and challenging to the biographer. Janaki's was a slow, small-scale, deep and thick or many-layered kind of science in the making. Worded differently, it alerts us to the fact that there are and have been other ways of doing science than those recognised by the Nobel Prize Committee, and that these deserve equal recognition if not more.

Also, at the centre of most accounts on Indian scientists are national institution-builders, or scientists of the state—invariably men, trained in the laboratory-based sciences of (nuclear) physics and chemistry,[14] representatives of the dominant model of doing science. On the other hand, probing the interstices, the borders, the folds or the dimly lit narrow bylanes of this heterogeneous landscape called science shines light on those like Janaki who chose to practise their science along the *holzwege*,[15] beyond the pathways determined and regulated by the state apparatus. Her science was about 'achieving personal perspectives while wandering, mostly, in the lonely byways of science.'[16] Following one's dreams across borders implied 'mobility,' which was not only a way of being but also of thinking. This biographical case-study shows that Janaki chose the liberating figure of the nomad as a way of being, and producing scientific knowledge, to break free of the tyranny of hierarchy, patriarchy and pseudo-science she encountered in the male bastions of state science such as crop breeding stations. She was a border-crosser in more ways than one, from her biological and social origins, to her nomadic subjectivity, her solitary travels of exploration and the nature of her scientific practice. Crossing borders and devising adventurous escapes made her life and science all the richer, and her story especially enthralling. Simply stated, this biography aims to reveal something of what it meant to live (dangerously at times) and make a career in the twentieth century as a nomad (woman) scientist.

If Janaki was a border-crosser, so were her objects of study. Way back in the nineteenth century, Jean Henri Fabre, the naturalist whom Charles Darwin referred to as 'that inimitable observer,' remarked that 'History celebrates the battlefields whereon we meet our death, but scorns to speak of the plowed fields whereby we thrive; it knows the name of the King's bastards, but cannot tell us the origin of wheat. That is the way of human folly.'[17] In a similar vein, or perhaps drawing on that, the feminist historian of science Londa Schiebinger stated that plants rarely 'figure in the grand narratives of war, peace, or even everyday life, despite their great relevance to mankind.'[18] This biography is no grand narrative, but is a story nevertheless, and has wandering plants at its centre, much as it has a nomad woman scientist at its heart. Alternating between the scientist's microcosm and the macro and public contexts of her scientific practice, this biographical narrative gradually unfolds to disclose, in parallel, the private and social lives of plants and their evolution and travels across landscapes in humans' trail—the story of the symbiosis between crops and humans, as articulated in Janaki's practice of science.

Janaki claimed multiple belongings as a nomad, the result of her deterritorialising wanderings (escapes) across myriad landscapes; she saw herself as a citizen of the world and as belonging to a transnational macrocosm of science, a view based on the belief that science knows no national, class or racial boundaries. This ontological position found epistemological translation in her science, as archipelagic thinking (as opposed to continental thinking) which, disrupting the notion of insularity, viewed the world as a collection of interconnected islands rather than as closed continental forms.[19] There has been a rising interest in recent years in thinking with the archipelago (enriched by the insights of Antonio Benítez Rojo, Derek Walcott and Edouard Glissant),[20] but the focus has largely been anthropocentric (on studies of human migration for instance), whereas in Janaki's science, we have a robust example of how archipelagic thinking provided a new ecological and transnational perspective to the cytogeneticist studying plant migration, speciation and evolution. Her larger interest was in mapping on a planetary scale, the migratory patterns or wanderings of plants as they occurred in the footsteps of humankind.

Finally, life-writing is a labour of love, demanding on the part of the biographer not only enormous mental resources and great stretches of solitary time to befriend and know intimately her biographee, but also objectivity and caution so as to avoid hagiography, the pitfall of most biographies. Further, in non-Western societies, especially for those individuals long gone, it is almost impossible to find a well-organised archive of private papers, or even a semblance of an archive that could be employed for the reconstruction of a life, posing a huge challenge to the biographer. On one occasion in the present case, material was recovered serendipitously from a garbage heap, and on another fortuitously rediscovered by the biographer, after years of lying buried, thanks to a typographical error in the catalogue of one of the world's finest university libraries. Corralling archival material (although locating them was no easy task, one can't complain here of a scarcity of sources) over several years, almost a decade in this case, and then sifting through this trembling mountain of abandoned workings to discover nuggets of gold has without doubt been painstaking, but also immensely rewarding, something of which I hope will be reflected on the pages of this volume—the first in-depth and archive-based study of the life of an Asian woman scientist.

Notes

1 Unless the context demands, in which case her formal name 'E. K. Janaki Ammal' will be used, this narrative will address her throughout by her first name, Janaki. It may be noted that 'Ammal' was only a polite title (like Esquire, for instance), and not her surname.
2 The earliest biographical accounts (in English) of E. K. Janaki Ammal were penned by two of her junior colleagues, one a forest geneticist (Kedharnath, 1988) and the other a respected mycologist (Subramanian, 2007). While Kedharnath's short memoir nicely sums up Janaki's chief scientific contributions and provides an almost exhaustive list of her publications, Subramaniam's, derived chiefly from the former with respect to her science, includes his personal memories of Janaki. For more recent studies, see Damodaran (2013, 2017); adopting an intersectional approach (the entanglement of gender with such social

divisions as caste and race), the two papers are based almost exclusively (but for a few items from private collections) on the cache of letters sent by Janaki to the British cytogeneticist C. D. Darlington, held at the Bodleian Library, University of Oxford. Damodaran's accounts however shy away from addressing Janaki's science in any depth.

3 Miss M. M. Mehta completed her DSc in biochemistry from the University of London in 1925.

4 It may be noted that their presence was not as rare as some scholars have claimed. See for instance, Sur, *Dispersed Radiance*, p. 34. and Damodaran (2013).

5 William H. Bragg and his son, Lawrence Bragg, who shared a Nobel Prize in Physics in 1915; Kathleen Lonsdale and John Desmond Bernal were two of William's best-known students.

6 Dorothy Crowfoot Hodgkin and Helen Megaw were two of the best-known of Bernal's female students.

7 Ferry, 'Telling Stories or Making History?' and *Dorothy Hodgkin: A Life*.

8 Sur, *Dispersed Radiance*, pp. 179–219.

9 This is hardly the place for a review of literature on the biographical genre or its place in the history of science but for a selection of reading material see Govoni and Franceschi (eds.), *Writing About Lives in Science*, and the special volume of the *Journal of the History of Biology* 44, no. 4 (2011), especially Söderqvist, T. 'The seven sisters: subgenres of "Bioi" of contemporary life scientists', pp. 633–650.

10 This outlook and the language reflect an obvious influence of the constructivist turn in the history of science discipline or Science and Technology Studies. Reading material on the subject is vast, but for a sample, see Biagioli and Riskin (eds.), *Nature Engaged*; Sismondo, *An Introduction to Science and Technology Studies*; Jan Golinski, *Making Natural Knowledge*; Hacking, *Social Construction of What?*; Pickering (ed.), *Science as Practice and Culture*; Latour, *Science in Action*.

11 The term global microhistory was first used by historian Tonio Andrade in his article, 'A Chinese Farmer, Two African Boys, and a Warlord: Towards a Global Microhistory'.

12 For a collection of essays written from within this perspective, see Campos and von Schwerin (eds.), *Making Mutations*.

13 For biographies of Barbara McClintock see Comfort, *The Tangled Field*; and Keller, *A Feeling for the Organism*.

14 For a sample see Chowdhury, *Growing the Tree of Science*; Shah, *Vikram Sarabhai*; Dasgupta, *Jagdish Chandra Bose and the Indian Response to Western Science*.

15 A reference to German philosopher, Heidegger's *Holzwege* (1950), literally meaning wood-paths, forest-ways, or off the beaten tracks.

16 Astrophysicist S. Chandrasekhar's speech at the Nobel Banquet, 10 December 1983. www.nobelprize.org/prizes/physics/1983/chandrasekhar/speech/

17 For a note on Jean-Henri Fabre see *Nature*, 96 (1915): 204–205.

18 Schiebinger, 'Following the Story', p. 49.

19 Since the 1950s, literary and cultural studies scholars, historians, sociologists have employed the concept of the archipelago to deconstruct linear narratives of historical and national development. For an introduction, see Pugh, 'Island movements: Thinking with the archipelago'.

20 For a recent collection of essays which explore archipelagic thinking across disciplines, including biology, see Martínez-San Miguel and Stephens (eds.), *Contemporary Archipelagic Thinking*.

1

TELLICHERRY

A Modern Thiya Family

The country is extremely pleasant . . . and the weather at this
season delightful. Indeed, the climate of Tellicherry is reckoned one of the finest in India, and the winds are generally
moderate, the sea breezes cool and refreshing. A constant
trade during the fair season, with vessels of all descriptions
from different parts of India, renders this settlement very lively
while the number of civil servants, with the garrison officers
and their families, beguile the rainy months in cheerful society
and domestic enjoyments.[1]

Travelling through Malabar in the late eighteenth century, James Forbes,
an East India Company official and amateur artist, was deeply taken in by
the extraordinary richness of the landscape in Tellicherry. Forbes found the
country pleasant for excursions and the weather at that season 'delightful.'
He observed: 'Indeed the climate of Tellicherry, in the latitude of 11° 47′
north is reckoned one of the finest in India.'[2] A salubrious and picturesque
town, it was 'situated upon a group of low wooden hills running down to
the sea, and protected by a natural breakwater of basalt rocks.'[3]

A traveller in the mid-nineteenth century described Tellicherry as 'a pretty
little straggling town on the sea-coast of Malabar, between the considerable
military cantonment of Cananore and the French settlement Mahe.' It was
divided into two parts, one the 'flat ground' constituting Tellicherry proper,
almost at level with the sea, and 'the high ground,' called Dharmadom. The
town consisted of about two hundred 'irregularly-built' European houses,
bazaars, the market place, an expansive prison 'built on a lofty bastion facing the sea, which includes the dens for criminals and the debtors' gaol,' a
lunatic asylum, the Zillah Court, and a chapel, besides a Catholic chapel
and a Protestant church. The burial grounds were 'situated on a high mound
nearly overhanging the sea.' Between the outer limits of the town and Dharmadom were a 'few straggling country-houses, and the court-house of the
now no longer existing judges of circuit who were three in number, besides
the registrar.' Dharmadom was itself located on the 'lofty cliffs and high
tableland' on the banks of a 'rapid and deep' stream, consisting 'of a few

DOI: 10.4324/9781003267089-1

scattered villages, occupied almost exclusively by native fishermen, and two immense mansions, more like palaces than private houses, and therefore the residence of two of the judges stationed at Tellicherry.'4 Near the sea-beach were extensive oyster-beds, and during low-tide, one could feast on oysters to one's heart's content. Tellicherry was famous for its pepper (as it remains today), most of which was dried on the spot for export. It was the headquarters of the pepper trade in Malabar. A great number were involved in the curing of coffee and ginger, again for export trade. Tellicherry was also well-known for its superior quality cinnamon, procured chiefly from the Randattara estate. Besides these, fruits, vegetables and poultry were always aplenty.

In 1844, the colonial government in India made knowledge of English a requisite for employment in government service. This new demand came as a blow to the vernacular-educated upper castes in Malabar under the Madras Presidency, chiefly Namboothiris and Nairs, who owing to caste inhibitions were less inclined to learning English. The Thiya community on the other hand, despite occupying the lower rungs of the caste ladder, embraced English education wholeheartedly. As a result, the turn of the twentieth century saw the burgeoning of a Thiya middle-class that was both English-educated and proficient in the vernacular and Sanskrit languages (the study of Sanskrit was by no means restricted to the upper castes as elsewhere), and cosmopolitan in outlook and taste. It was a colonial modernity marked by such activities as reading books in leisure time (the Malayalam novel was after all a product of these times), treating guests to a fusion cuisine of sorts, playing lawn tennis, badminton and cricket, indulging in exercise routines such as the one marketed by Eugen Sandow, resorting to a combination of Western and indigenous medicine to treat ailments, educating their children irrespective of gender, playing the piano, while also being connoisseurs of classical South Indian music and performances like Kathakali and Thullal, and very importantly keeping a daily journal, with entries on weather and books read, besides everyday happenings.

> The Thiyas have always been characterised by their persevering and enterprising habits. A large percentage of them are engaged in various agricultural pursuits, and some of the most profitable industries of Malabar have . . . been in their hands. They are exclusively engaged in making toddy and distilling arrack. Many of them are professional weavers, the Malabar *mundu* being a common kind of cloth made by them. The various industries connected with cocoanut cultivation are also successfully carried on by the Thiyas. They have among them good Sanskrit scholars, whose contributions have enriched the Malayalam literature; physicians well versed in Hindu systems of medicine; and well-known astrologers, who are also clever mathematicians. In British Malabar, they have made considerable progress in education. In recent years, there has

2

been gaining ground among the Thiyas a movement, which has for its object the social and material improvement of the community. Their leaders have very rightly given a prominent place to industry in their schemes of progress and reform. Organisations for the purpose of educating the members of the community on the importance of increased industrial efforts have been formed.[5]

Ideas of 'purity' and 'pollution' were simply 'antithetical to middle-class modes of life and work,' Andre Beteille has observed.[6] The Thiyas were mindful of being a cohesive and regulated community like the upper-caste Nairs and even claimed an equal status with them.[7] In fact, a contemporary Thiya writer and thinker Murkoth Kumaran criticised the Malabar Resident William Logan for his failure to highlight the Thiya social order in his *Malabar Manual*, an order that was as robust and socially evolved as that of the Nairs.[8] Thiya students leaving for higher studies in Madras, for instance, always sought the blessings of the elders in the community, who on their part ensured moral and financial support for the youngsters in their journeys towards progress. Graduates in Malabar were few in the late nineteenth century, but a good number of them were Thiyas; newly graduated Thiya boys were taken in procession through the town with nagaswaram and other musical accompaniments to give them a sense of having achieved an important goal. Employment in government service, which offered a sure means of liberation from the discriminating caste system, was an opportunity not to be missed.[9] By the late nineteenth century several Thiya gentlemen had attained high positions in government service or as independent professionals (chiefly as advocates); the names include Tahsildars Churyayi Canaren, Onden Kunhambu and Karayi Govindan; Deputy Collectors Uppota Kannan and Churyayi Kanaran; lawyers Potheri Kunhambu, Kottieth Ramunni, Churyayi Kunhi Kannan and Oyitti Krishnan; the Tellicherry Municipal Secretary Adiyeri Chathu; and then there was Justice Cheruvari Krishnan of the Madras High Court and the Sub-Judges, Panangadan Kannan and Edavaleth Kakkat Krishnan, the latter being Janaki's father.

Sub-Judge Edavaleth Kakkat Krishnan

E. K. Krishnan (1841–1907), on his maternal side, hailed from the ancient Thiya family of Oracheri at Chokli near Poyiloor in Tellicherry, a Municipality on the Malabar coast in the northern district of Cannanore (Kannur) of today's Kerala, renowned for its Sanskrit and Malayalam scholarship and expertise in medicine and astrology.[10] The oldest known member of the family was one Oracheri Kannan Vaidyar. He sired three sons, Kunhi Kannan, Chathappan and Othenan, and had two nephews, Kunhi Koran and Kunhi Chandan—the fivesome, owing to their scholarly reputation, were addressed collectively as the *Gurunathanmar*. The eldest, Kunhi Kannan, in his old age played *munshi* to German Basel missionary and lexicographer Dr Hermann Gundert (1814–93), teaching him Sanskrit and Malayalam. This association eventually led to the production in 1872 of the first ever Malayalam-English dictionary. In his *Tagebuch aus Malabar*, Gundert described 'Cugni Vei-dyen' (Kunhi Kannan Vaidyar) as an 'einfältiger Forscher' ('a simple-minded researcher,' in this case, teacher) who was helping him prepare a book on Malayalam grammar.[11] Besides the Oracheri *Gurunathanmar*, there were other well-known Thiya scholars in Tellicherry and its neighbourhood, men like Bapputty Gurukkal, Anandan Gurukkal, Canaren Vaidyar and Churu-kantan Vaidyar, who were all taught, incidentally, by so-called upper-caste teachers such as Kuttiappa Nambiar. That Gundert chose a man belonging to the Thiya community as his teacher rather than someone from the upper caste, tells us something not only about Gundert, the Protestant missionary, but also about Thiyas in general, who interacted with Westerners with ease, unconcerned about matters of pollution.

The second nephew of Oracheri Kannan Vaidyar, Kunhi Chandan (a cousin of Gundert's teacher Kunhi Kannan Vaidyar), chose Aroonda Neeli of Valiyayi as his partner; Neeli bore him two children, Kunhikutty and Cheeru. Aroonda Cheeru was given away in marriage to Kakkat Kungen Vaidyar, a renowned Ayurveda physician from Mayyazhi (or the French principality of Mahe); they had three children, daughters Neeli and Manik-kam, and an only son, Krishnan (the future Sub-Judge). Krishnan was born on the 6 June 1841, within six months of the death of his grand-uncle Kunhi Kannan Vaidyar, teacher to Gundert, who to the missionary's great regret had died 'ohne Christum bekanntzuhaben' ('without knowing Christ').[12] Contrary to the *marumakkathayam* (matrilineal) norms adopted by the Northern Thiyas, Krishnan took his father's family name, Kakkat, rather than his mother's, Aroonda, marking a break with tradition.

Krishnan completed his preliminary studies at Calicut in the Government Provincial School (established by the Zamorin in 1854), where he spent five years of his youth.[13] It was open to all classes of society and taught English, Malayalam and Sanskrit.[14] Krishnan was the recipient of the coveted Junior Conolly Scholarship, instituted in the name of the murdered Malabar Col-lector and Magistrate, Henry Valentine Conolly.[15] The Head Master of the School, Edmund Thompson, considered Krishnan 'the best English Scholar' among the natives who had studied at the institution, one who had 'read many works quite unconnected with the ordinary course of School study' and had 'obtained thereby a good deal of general information.'[16] In 1861, Krishnan entered government service as English Writer at the Civil Court, Tellicherry (Thalassery), and three years later was appointed Malayalam Translator at the Madras High Court. The Registrar Philip Percival Hutchins and the Acting Registrar Herbert Wigram of the Huzur Court of Madras cer-tified Krishnan as a highly intelligent and well-educated young man, and an excellent Malayalam Translator.[17] Krishnan passed the examinations for the office of 'Translator, Judicial Sheristadar and District Munsiff,' and in early 1869 was awarded a Bachelor of Law degree by the University of Madras.

Tellicherry's legal history might be traced back to 1802, when the first Zillah Court was established by the British. This institution was later upgraded as the Civil and Sessions Court and the Principal Sadir Amin's Court in 1845. In 1873, it was referred as the District and Sessions Court, and William Logan (1841–1914) was appointed Judge. Krishnan's tenure as District Munsiff of Tellicherry gave him ample opportunity to interact with his coeval Logan (both were born 1841), who would later be appointed Malabar Collector. Logan's *Malabar Manual* (1887) would occupy pride of place in Krishnan's library. Krishnan had been Munsiff at Tellicherry for some years, when he decided to compete for the superior post of Sub-Judge; in late 1883, he received a telegram notifying him of his elevation to the position of Sub-Judge at Calicut. The District Judge of Malabar, John William Reid, in his testimonial stated that Krishnan had 'sustained up to this the high character for integrity,' an opinion Reid had entertained of him when he had first given him a testimonial in August 1875, in his application for the post of Munsiff. Reid had then hoped that Krishnan's successes would act as an incentive to those natives who ceased to pursue studies quite early in their lives. In late 1892, Krishnan was posted as Sub-Judge in Palghat, where he would remain until his retirement in 1896. At the end of his official tenure as Sub-Judge, Krishnan would re-enter government service as Deputy Collector of Tellicherry, eventually retiring in 1901. From then on until his death in 1907 Krishnan would serve as Chairman of the Tellicherry Municipality.[18]

Sub-Judge E. K. Krishnan wielded immense power and influence over the Thiya community in Northern Malabar, given his official position, sense of judgement, and vast and eclectic erudition. He enjoyed a wide network of friends and acquaintances from the top rungs of the social ladder, a veritable 'Who's Who' of the region, including highly placed Indians, colonial officials and other Europeans of the station. He refused to be part of any caste or religious organisation, and thought of himself an agnostic. Krishnan was the epitome of the newly emerging indigenous reading public in Malabar; he was a voracious reader of diverse literature, from fiction and poetry to nonfiction, history, government manuals and legal tracts, in English, Malayalam and Sanskrit. At his house, there was a designated space for a library, complete with revolving book stands and walls lined with wooden book shelves, which were fumigated regularly to keep out pests and mould so widespread in the humid tropics. An avid gardener, with a partiality for hibiscus and mangoes, and amateur ornithologist, Krishnan published two books on the birds of Malabar (one of these exclusively devoted to the birds of Tellicherry)[19] as if in continuation with the Scottish surgeon T. C. Jerdon's researches in the region at the turn of the nineteenth century.[20]

In 1867, while employed as Malayalam Translator at the Madras High Court, Krishnan married Kalyani, daughter of Poovadan Kelu of the

Kuruntharatta House. The marriage was a small affair, having aroused much displeasure among his relatives, especially on his paternal side, for they had hoped he would marry one of his cousins, as per custom. Kalyani however died young, leaving Krishnan with two little ones to fend for, a girl called Sharada and a boy, Govindan, a dire situation which forced him to marry again. By this time, Krishnan had also taken charge of the Edavaleth House, Mahe, his paternal *tharavad*, which had been in the hands of strangers and distant relations. The bride this time was Devaki Amma, a young girl of mixed-race parentage, more than twenty years his junior, of Kuruvey House in the old part of Tellicherry town.

Kuruvey Devaki Amma

On 18 March 1895, the *Madras Mail* carried an obituary notice of Lt. Col. John Child Hannyngton (1835–95), who had died at the age of fifty-nine on 25 February at Lewisham, London. He was a victim of the influenza epidemic that had peaked in England and was spreading at an alarming rate owing to 'a long spell of severe weather' lasting sixteen weeks.[21] The news would bring great grief to Sub-Judge E. K. Krishnan, who recorded the event faithfully, but simply, in his diary: 'I well remember being photographed by this genial European in 1862. He kept up a correspondence with me.'[22] Krishnan was a junior Court official in Madras during Hannyngton's five-year sojourn in Tellicherry, but they had had several opportunities to interact both within judicial circles and otherwise.[23] The two men shared a professional and intellectual bond but also a certain guarded intimacy, not ordinarily expected of an Indian subordinate and a British superior, for Hannyngton was also father-in-law to Krishnan, albeit publicly unacknowledged.

In the nineteenth century it was not uncommon for Thiya women to enter into liaisons with low-ranking British men in the military or civil service; in these times this was not thought of as 'improper' or leading to a 'loss of caste.' Thurston noted:

> the marumakkatāyam system (inheritance through the female line), which obtains in North Malabar, has favoured temporary connections between European men and Tiyan women, the children belonging to the mother's tarvad. Children bred under these conditions, European influence continuing, are often as fair as Europeans. It is recorded in the Report of the Malabar Marriage Commission, 1894, that 'In the early days of British rule, the Tiyan women incurred no social disgrace by consorting with Europeans, and up to the last generation, if the Sudra girl could boast of her Brahman lover, the Tiyan girl could show more substantial benefits from her alliance with a white man of the ruling race. Happily, the progress of education, and the growth of a wholesome public opinion, have made shameful

the position of a European's concubine, and both races have thus been saved from a mode of life equally demoralising to each.'[24]

The incidence of such liaisons was particularly conspicuous in cantonment towns like Tellicherry, observed William Logan in his *Malabar Manual* (1887):

> The head-quarters of the [Thiya] caste may be said to lie at and round the ancient European settlements of the French at Mahé and of the English at Tellicherry. The women are not as a rule excommunicated if they live with Europeans, and the consequence is that there has been among them a large admixture of European blood, and the caste itself has been materially raised in the social scale. In appearance some of the women are almost as fair as Europeans, and it may be said in a general way that to a European eye the best favoured men and women to be found in the district are the inhabitants of ancient Kadattunad, Iruvalinad, and Kottayam, of whom a large proportion belong to the Tiyan or planting community.[25]

Our story begins with one such Thiya girl, Kunhi Kurumbi (1845–1918) of the Kuruvey family, except that her liaison was not with the rank and file but with the Judge of the Court of Small Causes at Tellicherry, none other than J. C. Hannyngton. The Irish-born Hannyngton was the son of Major General John Caulfield Hannyngton, inventor of the Hannyngton Slide Rule used in astronomical computations and Fellow of the Institute of Actuaries. He joined the Madras Civil Service in 1857, the career ending in 1892, a major part of which was spent on the west coast of South India. In 1878, he acted as the Resident of Travancore State; a short stint in early 1879 as District and Sessions Judge in Salem was followed by an assignment in December 1880 as arbitrator of boundary disputes between the States of Travancore and Cochin. In 1881, he was once again appointed Acting Resident, before leaving on special duty and accompanying the Maharaja of Travancore on his tours to Calcutta, Banaras and Madras. His official post was confirmed in 1882 and from then on until his retirement, save for short intervals, Hannyngton served as the Resident of Travancore and Cochin States.[26] A popular personality, he was considered a Sanskrit scholar of repute, a keen sportsman and 'a capital shot not only with gun and rifle but with camera and lens.'[27] He was also passionately interested in natural history, especially botany; in the 1880s, for instance, while Resident of Travancore, he sent a collection of 'orchid roots' from the Western Ghats, growing at an elevation of 3,000 feet, to the Director of Kew Gardens, London.[28]

Young Hannyngton's liaison with Kuruvey Kunhi Kurumbi led to the birth of two girls at the Kuruvey House: Devaki (1864–1940) and another of uncertain name, perhaps Matha (1865–89). Hannyngton would not publicly acknowledge the girls. The heart-wrenching story goes that within a couple of weeks of Matha's birth, the baby was forcibly taken away to Madras for adoption. The elder girl Devaki escaped a similar fate because she clung desperately to her mother, refusing to be separated. The adoption plan had been masterminded by Hannyngton, then Sessions Judge at Salem. On 24 March 1866, immediately after Matha's adoption, he married a nineteen-year-old English girl, Laura Elizabeth Onslow.[29]

Perhaps it was also Hannyngton who suggested to Krishnan, when the latter was Munsiff at the Tellicherry Court, to marry his elder daughter, Devaki; he had high regards for Krishnan, knew he could keep a secret and take good care of his daughter. As for Krishnan, he was more than pleased with the suggestion, and sometime after October 1879, began a *sambandham* with fifteen-year-old Devaki; she was twenty-three years his junior, but the age difference was hardly surprising for the times. In 1890–91, Krishnan built a grand *padinjitta* (a west-facing house) called Edathil (or Edam) in Chetamkunnu, Tellicherry, not far from the Courts and the sea, and in close proximity to Kalathil House, the house he had built for his first wife, Kalyani and their two children, E. K. Sharada and E. K. Govindan (1875–1944). Devaki Amma (called Devi Amma, and referred to as Mrs K in Krishnan's diaries) would bear him thirteen children in all (seven boys and six girls); the initials in their names stood for Edavaleth Kakkat (taking after the paternal family name, rather than the maternal, the latter being the case in matrilineal Thiya families): E. K. Damodaran (1879–1904), E. K. Lekshmi (1881–1952), E. K. Raghavan (1882–1973), E. K. Vasudevan (1885–1936), E. K. Krishnan (also known as Kunhi Krishnan or Kittu) (1887–1958), E. K. Cousalya (1889–1971), E. K. Parvathi (1891–1984), E. K. Sumithra (1892–1972), E. K. Padmanabhan (1894–1974), E. K. Janaki (1897–1984), E. K. Madhavan (1899–1967), and the twins, E. K. Devayani (1902–80) and E. K. Varadan (1902–59).

Janaki's Early Years

Krishnan and Devi Amma had their marriage registered on 13 July 1897 before the Registrar of Marriage, Tellicherry, as per the 4 November 1896 Act of Madras; she had by this time given birth to nine of their children. The year ended for them with the birth of yet another, a girl named Janaki born on 4 November 1897. In early 1898, only a few months after her birth, the colonial government conferred Krishnan with the title Diwan Bahadur. The birth certificate issued by the Tellicherry Municipality recorded Janaki's date of birth as 5 November, a discrepancy that might have arisen from a conversion of the date from the lunar calendar to the Gregorian; however, the date Janaki herself cited in official documents was 4 November 1897.

Little Janaki was fondly called Jani. An entry from Krishnan's diary of mid-1899, records: 'Baby Jani is [a] . . . pet in the family. She is growing a great beauty . . . Jani is very amusing & eats betel nut freely.' Krishnan would frequently take little Jani on drives in his four-wheeled carriage, which included dropping off her elder sisters at the local convent school for girls. On one occasion, on 18 February 1901, brother Kittu after seeing his sisters off at school took three-year-old Jani, who had suddenly taken ill, to the hospital on his *jutka* drawn by two bullocks. A couple of months later, on 16 April 1901, her ears were pierced in a *kathu-kutthal* ceremony as per custom. That same month, Krishnan would receive the good news of E. K. Damodaran, his eldest son with Devi Amma, having passed the MB&CM examination of the Madras University, the first one to do so among the Thiyas. Krishnan, the proud father, attended a dinner party hosted by V. Karunakara Menon (a lawyer and honorary Managing Director of the Cosmopolitan Club, Calicut) at his house in the company of friends, District Munsiff Mundappa Bangera (with whom he went on drives too), Adiyeri Chathu, Kottieth Ramunni *vakil* and Mangat Gopala Menon, the new Municipal Chairman. The friends had a brief chat before they sat down to a dinner 'consisting of some of the nicest dishes of Malabar prepared by Menon's servant, including *sambar, koottukari, avial, mangakari, rasam* followed by *payasam.*' Karunakara Menon and Gopala Menon would in turn visit Krishnan; this time it was a grand Thiya fare, comprising mutton soup, fish moilee, mutton cutlets, duck roast and rolong *pilao* (a spicy *pilao* made of semolina or *sooji*),[30] and 'fine old whiskey to wash it down well.' On this occasion son Damodaran was introduced to them as a 'private practitioner.' The children including Janaki and her toddler brother Madhavan were kept entertained at such times by being driven around the house in a hand-cart.

On 6 June 1903, Krishnan turned sixty-two: 'I enjoy fine health with no organic disease. All my children are with me,' he recorded in his diary, in obvious contentment. That very month, a private tutor, Kunger *kanisan* was hired to teach the six-year-old Janaki the Malayalam alphabet.

The following year, in his journal dated 28 April 1904, Krishnan noted, 'This is an eventful day in my family. Sumithra and Janaki [now almost seven] and their cousin Narayani daughter of Rugmani had the Talis tied round their neck.' Within the *marumakkathayam* system, every Thiya girl of North Malabar was to undergo this *rite de passage* called *tali-kettu-kalyanam*, wherein a *tali* ('a small, thin, gold, neck ornament, with a gold bead on either side of it, and attached to a string made of cotton-thread') would be tied around her neck by her aunt, before she attained puberty so as not to 'lose caste.'[31] Subsequent to this ceremony, the girl would be formally addressed with the honorific title, 'Amma' or 'Ammal'; it appears that only Janaki among the EK sisters chose the title 'Ammal' for herself. The ceremony, being an expensive affair, was usually conducted for several girls of the family together, by they had to be an odd number of years old (5, 7, 9, or 11); postponement beyond the age of 11 years was not considered appropriate.[32] An auspicious day and hour for tying the *tali* would be fixed beforehand in consultation with the *kanisan*, or *kaniyan*, the astrologer. Summing up the evidence laid before the Malabar Marriage Commission of 1891, Thurston wrote about the ceremony thus: 'Of those who gave evidence before the Malabar Marriage Commission, some thought the tali-kettu was a marriage, some not. Others called it a mock marriage, a formal marriage, a sham marriage, fictitious marriage, a marriage sacrament, the preliminary part of marriage, a meaningless ceremony, an empty form, a ridiculous farce, an incongruous custom, a waste of money, and a device for becoming involved in debt.'[33] That Krishnan supported such customary practices despite his progressive outlook comes as a surprise, but perhaps for the elite Edathil girls going through the ceremony was crucial, to rid them somewhat of the social stigma, or the 'white-stain' on their maternal side that tainted the reputation of the family.

Sumithra, Janaki and Narayani, dressed up in *thangakasavu* (gold zari) bought at the Cannanore bazaar from 'Camp' Rameshwara Aiyer's shop, and accompanied by five or six women, were taken in procession to the Keloth family accompanied by music, from thence to the Cannanore Road by Komachankandi Lane and then back to Edam through the old Pallikunnu Road. The ceremony lasted four days and at its close the girl had the freedom to dispense with her *thali* if she so wished. While Sumithra and Narayani would be married off to Thiya gentlemen a few years later, Janaki chose to remain unmarried and would officially be addressed as E. K. Janaki Ammal.

At School

With E. K. Krishnan's death in 1907, a great cloud of gloom descended over Edam. The grand life had come to an end and the family had fallen on bad times; Devi Amma was inconsolable, for her eldest son Damodaran, the doctor, had died tragically in 1904 of the plague, and the younger children had got nowhere in life. The eldest of her surviving sons, E. K. Raghavan, the family's sole bread-winner, was residing at Gyogin in the Insein Township, Northern Rangoon. A responsible and loving son, Raghavan would send monthly remittances for the upkeep of Edam, and the education of his younger siblings (Madhavan, Janaki, Varadan and Devayani), but he had only started off, and his earnings were far from sufficient. By then Janaki had joined her sisters Parvathi and Sumithra at the Sacred Heart High school in Tellicherry, but drastic changes were inevitable at Edam: Krishnan's carriage was to be sold off to the Munsiff residing at the Kalathil House. Three of his guns, which had been kept at the Taluk Cutchery for renewal of licence, had to be disposed of. The cook Raman had already quit and his wife was expected to follow suit as Devi Amma was no longer able to employ them. Some of Krishnan's precious volumes such as the *Encyclopaedia Britannica* were offered on sale.

The nuns at school encouraged the girls to pursue higher education rather than view marriage as the end. They taught them the importance of maintaining daily journals (unfortunately Janaki did not keep one) and communicating clearly and confidently, in both English and Malayalam. Sometimes they visited the girls at home to check on their welfare. Besides studies, the girls took an active part in extra-curricular activities, such as acting, music, needlework and embroidery. The EK girls would have fond memories of school, which would prepare them in several ways to become formidable women, each in their own way. The girls all read English novels and books on natural history borrowed from their erudite father's personal library. They knew their plants and birds well and occasionally enjoyed angling.

In a few years' time, Raghavan's younger brother, Vasudevan would find a job with the police department, and Kittu with the forest service. Kittu had completed his BA from the Presidency College, Madras (specialising in botany) in 1909–10, in a commendable second class; before being sent to Dehra Dun on the usual three-year course, a probationer of the forest department had to spend six months as an Attached Officer in a District Forest Department. Accordingly, in early 1911, Kittu left for Ootacamund (Ooty), soon to be joined by Devi Amma and the younger children, who were on summer vacation. 'Though it is very cold the country is delightful,' fourteen-year-old Janaki wrote from Ooty to brother Raghavan in Burma. The family visited the Botanic Gardens, which they all loved. She also wrote about the hail storm they witnessed there: 'Some two days ago hail as large as marbles fell here. We all ran out to pick them. It was really lovely.' The cold was somewhat extreme however, and Devi Amma was desperate to return to Tellicherry. Janaki's letter to Raghavan ended with a request for an instrument box to be used at school, containing a compass, divider, ruler, protractors and a pair of set-squares. 'They can be got at Spencer & Co.,' she added very helpfully.[34]

In 1913, Janaki was in the Fifth Form. Her steadfastness, industrious nature and keen eye for detail were already evident. A descendant recalls a telling moment from her early years at school. When the visiting School Inspector asked the class to identify the plant displayed before them, a youthful Janaki without a moment's hesitation answered, much to the Inspector's delight, that it was a *Gloriosa superba*, the very plant that had captured James Forbes' attention when he visited these parts in the late eighteenth century.[35] Plants would become the centre of her life.

Janaki's elder sister Sumithra was a great support to Devi Amma; she was extremely resourceful and industrious, even if somewhat of a disciplinarian. Every day, she would toil hard to ready her younger siblings for school. On some days, young Janaki assisted her elder sisters in pounding

rice and preparing breakfast for several members of the household.[36] Sumithra would however make time to read English novels, the likes of Allen
Raine's *The Queen of the Rushes* (1906)[37] and Dickens' *David Copperfield*.
Interested in the English language, she would miss no opportunity to teach
herself new idioms and words, and help her younger siblings with their
homework, especially Janaki. In 1913, we find her reading novels such as
Destiny by Alice and Claude Askew (1911) and Charles Gravis' romantic
melodrama, *My Love, Kitty* (1911) as well as Malayalam ones such as
Murkoth Kumaran's *Vasumathi*, a novel based on Thiya life, which she
found, 'interesting' but written in 'very difficult Malayalam.' One entry in
her diary records that she had spent the 'whole morning in copying English
notes for Jani—idioms & meanings from Sr Letitia's notebook,' which had
turned out to be 'instructive' for her. Sumithra observed wistfully and not
without a hint of envy, that Janaki was improving her English 'by leaps &
bounds & has adopted a flowery style of language from Sr Letitia,' while her
own English, especially pronunciation, was 'being slowly forgotten.' Janaki
had in fact received a long letter 'full of praise' from brother Raghavan for
her facility with the English language. Referring to herself, Sumithra would
remark with regret: 'What a shame to be master of no language!' It was
indeed from Janaki, who had learnt it at school, that Sumithra learnt the
Scottish ballad 'Jessie's Dream' or 'The Relief of Lucknow' adapted for the
pianoforte.[38]

In the absence of Sumithra, who would occasionally go to Tayilekkandi
(aunt Kalyani's house, in the old part of town, where youngest sister Devayani resided),[39] it fell on Janaki to prepare her younger siblings and nieces
(daughters of elder sisters Lekshmi and Cousalya) for school. The girls left
for the convent every weekday at eight in the morning 'at breakneck speed
under Jani's orders.' The older girls including Janaki wore saris, and walked
to school, while the younger ones went by *jutka*. Occasionally, an ill little
Devayani would be brought back home by Janaki in a *jutka*; the girl was
prone to vomiting and weakness. Tired and over-worked, Janaki would at
times sulk and go to bed without dinner, especially after a quarrel with the
ever dominating Sumithra, and at others stay back at the convent. Sumithra, who enjoyed needle and crochet work, stitched 'nainsook' (fine, soft
cotton) blouses for herself, and for Janaki, soft muslin and chintz blouses,
besides laced handkerchiefs. She would also make *ravikkas* (blouses) for
Devi Amma, who was addressed as 'Ma' by her children. Sumithra would
teach Janaki to make pillow-lace, a kind of hand-made lace worked on a
pillow using threads wound around bobbins.

The struggle for money led to frequent quarrels at Edam, leaving Devi
Amma miserable and anxious. Sumithra would write: 'Lamentations &
noise. Edam is disgraced!. . . No peace or money in the house & everyone is
miserable.' She blamed herself for 'keeping anger to heart.' One day, Janaki
made a scene before starting for school. She had purchased a 'Mangalorean

sari' without her mother's permission. When Devi Amma lost her temper and refused to pay up, sisters Parvathi and Sumithra were forced to intervene. They lent Janaki the Rs 3 that she so desperately needed to repay her loan. 'Jani is too independent and has no fear of debts. This is a lesson,' Sumithra would comment. On the other hand, she greatly admired Janaki for her 'perseverance' and generosity.' For instance, she would religiously take an oil bath every morning in the auspicious month of *karkadakam*, which none except her father did. Even if she was fierce and demanding, 'Jani [was] very considerate about [Sumithra's] needs,' Sumithra observed. 'I wish I had some of the admirable qualities that Jani possesses—unselfishness foremost,' she would add. And on another occasion, she would remark, 'Jani cheered me. She is a treasure.'

In February 1914, Devi Amma attended the Parents' Day at the convent, with her daughters Lekshmi, Cousalya, Parvathi and the younger girls, Janaki and Devayani, and nieces Tara and Leela (Cousalya's daughters), who were beautifully turned-out. Sumithra remained at home to take care of Lekshmi's young daughters and to await the return of the boys from school. Incidentally, Shantha (Lekshmi's elder daughter) assigned nicknames for her aunts and the younger ones followed suit—Sumithra became 'Chummai,' Cousalya, 'Kuchamana' and Janaki, 'Nachi' and later 'Nachi Amma.' A grand entertainment had been organised at the convent, where the District Judge Roberts was chief guest. Janaki, dressed in a 'gala' style for the occasion, was expected to recite a few lines of English poetry and then quickly change her clothes to appear as a house maid in the school-play. Raghavan who had arrived in time from Burma to attend the function hailed it a grand success and when the family reunited at Edam late in the evening, 'Convent news figured prominently in the conversations.' In the days following, they would indulge in moonlit strolls and extended music and poetry sessions, but the dreaded final examinations were fast approaching. Unfortunately, Janaki sprained an ankle, but only after the exams were out of her way. Accidents of this nature would recur in her life. That year, all the girls cleared their examinations.

With the convent now closed for summer vacation, Janaki began indulging in what she loved most: reading. A frustrated Sumithra commented: 'Very noisy days. Boys & girls school closed so they do nothing but make mischief. Jani spends the day like a grand madam reading. Education is a bar to helping at home.' That vacation, brother Vasudevan invited Janaki to Palghat to keep him company, as his wife, presently at Edam, was intending to spend a few days at her own house, Ambalavattom. Janaki however discovered, to her utter distress, that she had no good blouses or jackets to carry with her to Palghat. She begged Sumithra to 'sell' her one of her new 'nainsook' blouses, to which the latter reluctantly yielded. Over the next couple of days, Janaki would be busy altering the blouse to her size using sister-in-law Yeshoda's tailoring machine, fixing hooks and stitching borders for it. Sumithra was often annoyed by Janaki's 'untidiness' and at times the two quarrelled fiercely, but they were also inseparable. When Janaki returned from Palghat two weeks later, the two chatted late into the night, catching up on Palghat news. Part of the summer vacation would also be spent with forester brother Kittu in Ootacamund, whom Janaki would help in making botanical illustrations.

When the results of the school-leaving examination were announced in April 1915, there was much elation at Edam; Janaki had passed with

Figure 1.1 Sub-Judge Edavaleth Kakkat Krishnan (1841–1907) and Kuruvey Devaki Amma (1864–1940), c. 1890.

Source: Courtesy of the EK family.

flying colours. Girls and boys from Malabar usually went to Mangalore for their intermediate education. In fact, Janaki's friend and schoolmate, Amy Saldanha, daughter of C. T. Saldanha of the Indian Medical Service posted at Tellicherry, tried to persuade her to join the Government College, Mangalore, but she would choose to apply to the newly established Madras College for Women (renamed Queen Mary's College in 1918) in Mylapore for the FA course (First Examination in the Arts), perhaps on the advice of her brother Kittu, who had been appointed probationary 'EAC of Forests' (Extra Assistant Conservator of Forests). Founded as a junior college in July 1914, the Madras College for Women was controlled by the Director of Public Instruction and managed by the local government; it was the city's first women's college. Besides Kittu, Janaki's elder brothers Raghavan and Vasudevan would also agree to support her education in Madras.

Figure 1.2 John Child Hannyngton (1835–95).
Source: Courtesy of the EK family.

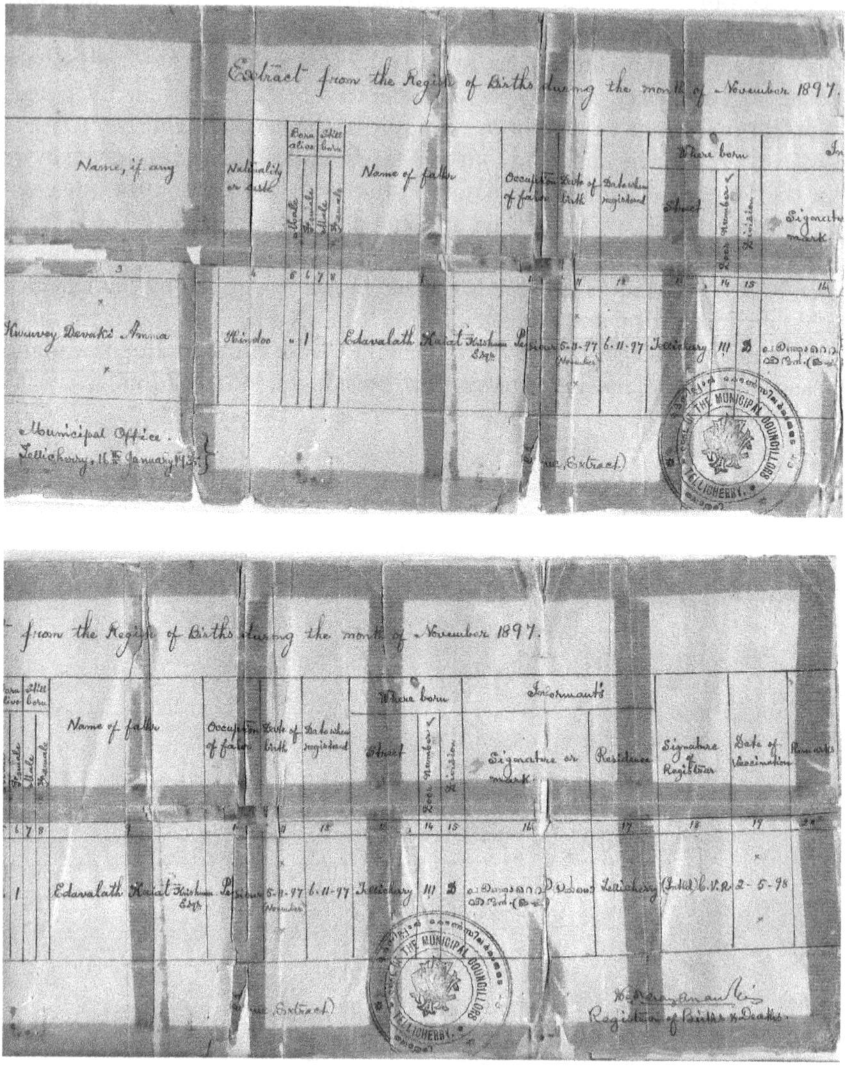

Extract from the Register of Births during the month of November 1897.

Figure 1.3a-b Janaki's Birth Certificate (copy issued by Tellicherry Municipality in 1935), with names of parents, date of birth and date of vaccination.

Notes

1 Forbes, *Oriental Memoirs*, Vol. 2, p. 456. Forbes travelled across India between 1765 and 1784.

2 Ibid, p. 181.

3 Innes and Evans, *Madras District Gazetteers*, Vol. 1.

4 'The Home Friend, A Weekly Miscellany of Amusement etc.' http://malabardays.blogspot.co.uk/2010/03/tellicherry-in-1850.html

5 Edgar Thurston, Superintendent of the Madras Government Museum, quoting from an article published in the *Indian Review* (October, 1906) in Thurston, *Castes and Tribes of South India*, Vol. 7, pp. 115–116.

6 Beteille, 'The Social Character of the Indian Middle Class', p. 83.

7 Mundon, 'Renaissance and Social Change in Malabar', p. 166.

8 Ibid., p. 167, f9.

9 Kurup, *Modern Kerala*, pp. 84–86.

10 Mundon, 'Renaissance and Social Change in Malabar', p. 175.

11 *Malayalabhaasha Vyakaranam* (1859); Gundert paid his researcher-munshi Kunhi Kannan Vaidyar a sum of Rs 10 per month. See Gundert, *Tagebuchaus Malabar*, pp. 62, 64.

12 Ibid., p. 75.

13 When Hermann Gundert founded a primary school in Tellicherry, among the students who opted to study there a great number belonged to the Thiya community; Gundert taught them science, geography, English, Malayalam and the Bible. See *Report of the Basel Mission*, 1840, 1841 and 1842; about the same time as the school in Tellicherry, a primary school was begun at Bernassery near Cannanore (1842) and a few years later at Kallayi near Calicut (1848). These schools played a major role in providing an education that was modern and the opportunity to mix with students from a range of social backgrounds. Several of the Basel Evangelical Mission primary schools were later upgraded as middle-schools and high-schools. Kurup, *Modern Kerala*, pp. 84–85.

14 In 1879, this institution went on to become a second-grade college (later called the Calicut Government College).

15 H. V. Conolly had been brutally murdered only a couple of years earlier, in 1855, by four Mappila convicts averse to him and his intentions.

16 Testimonial provided by Edmund Thompson, private collection.

17 Private collection.

18 The Tellicherry Municipality was constituted in 1866. The right of election of the Municipal Chairman having been withdrawn in 1898 'owing to the unsatisfactory state of municipal affairs', it then fell on the Government to nominate the right person for the post.

19 Rao, The *Indian Biographical Dictionary*. Unfortunately, neither of these books has been traceable.

20 Thomas Caverhill Jerdon made Tellicherry his home between 1847 and 1851. Appointed Civil Surgeon of the station, Jerdon, a keen naturalist, missed no opportunity in exploring nature around him; he is believed to have kept a pet otter at his house in Tellicherry. Jerdon studied the many species of ants in Malabar, and discovered at Tellicherry the *Harpegnathos saltator*, referred to as the Jerdon's jumping ant. A pioneering ornithologist, Jerdon described several species of Indian birds hitherto unknown to the scientific world, including many Malabar ones, and regularly corresponded with that other well-known figure of Indian ornithology and politics, Allan Octavian Hume, while stationed at Tellicherry. See Jerdon, *Birds of India*; among his other publications is the beautiful *Illustrations of Indian Ornithology* (1847).

21 Influenza-related mortality figures for London in this year were among the highest in a decade (1890–1900). *Evening Post*, 23 February 1895, p. 1.

22 Sub-Judge E. K. Krishnan's diaries, private collection. Unless otherwise stated, the chief source for this chapter are these diaries.

23 J. C. Hannyngton's career began as a writer in the service of the East India Company in 1857. In 1859, he was sent as Assistant to the Collector/Magistrate, Trichinopoly (Tiruchirappalli); he moved to Malabar in 1861, where he spent

a period of ten years. Hannyngton was appointed Acting Judge of the Court of Small Causes at Tellicherry in 1866, and later given the additional charge of Acting Collector and Magistrate of Malabar. In April 1867, he was returned to Tellicherry as Judge of the Court of Small Causes. The next year saw him act as Judge of the Civil and Sessions Court of both Tellicherry and Calicut. He would hold the position until 1871, before moving on to Salem and Guntur. On retirement from the Madras Civil Service, he moved to the Travancore service where he served four stints as Resident of Travancore and Cochin: 20 February 1878– March 1879, 1 April 1881–May 1883, 15 August 1884–July 1887 and 7 October 1888–July 1890. For a brief note on J. C. Hannyngton's career in India, see the *Asylum Press Almanac*, Madras, 1892, p. 119.

24 Thurston, *Castes and Tribes*, Vol. 7, p. 36.

25 Logan, *Malabar Manual*, Vol. 1, p. 143.

26 During this ten-year tenure, he had been signatory to the contentious Periyar Lease Deed of 1886 between the State of Travancore and the Government of India, whereby it was agreed that the Madras Presidency had the right to divert waters below the 155-foot contour over a lease period of 999 years.

27 *Madras Mail*, 18 March 1895.

28 Library and Archives at Royal Botanic Gardens, Kew (RBG Kew, Library & Archives): Director's Correspondence 157/428a, letter dated 26 June 1883.

29 Hannyngton's wife, Laura Elizabeth Onslow, would bear him five children from what we know; of these, at least two sons, John Arthur Hannyngton and Patrick Hannyngton, were born in Tellicherry, in 1868 and 1871 respectively. William Onslow was born at Salem in 1874 and Frank Hannyngton a year after. Their sister Agnes Bernice Hannyngton was born in Trivandrum in 1878.

30 *Hobson-Jobson* (1903) explains that 'rolong' was a corruption of the Portuguese 'roláo' or 'raláo' meaning semolina or *sooji*, p. 767.

31 See Moore, *Malabar Law & Custom*, p. 73, wherein he cites Justice Muttusami Aiyar's (President of the Malabar Marriage Commission of 1891) statement: 'As a religious ceremony it [tali-kettu-kalyanam] is taken to give the girl a marriageable status and in North Malabar she is addressed afterwards as Amma or lady'.

32 For a detailed description of the *tali-kettu-kalyanam* followed by the Thiyas of North Malabar in the early twentieth century, see 'Marumakkathayam Marriage Commission: Answers to Interrogatories by Onden Ramen, Sheristadar, Chirakkal Taluk' in *Report of the Malabar Marriage Commission* 1891, Appendix 3, part 6, pp. 2–5; also see C. K. Revathi Amma, *Sahasrapoornima*.

33 Thurston, *Some Marriage Customs in South India*, p. 178.

34 E. K. Janaki to E. K. Raghavan, letter dated Ootacamund, 10 May 1911, private collection.

35 See, Forbes, *Oriental Memoirs*, Vol. 2, p. 181.

36 Unless otherwise cited, the other important archival source for this chapter is E. K. Sumithra's diaries, private collection.

37 Allen Raine was the pseudonym of the Welsh novelist Anne Adalisa Beynon Puddicombe (1836–1908).

38 This song is a reference to the Siege of Lucknow (1857), which was one of a series of sieges and battles collectively called (by the British) 'the Indian Mutiny' of 1857–58. It was composed and arranged in the late nineteenth century for the pianoforte by John Blockley. For more on the ballad see Llewellyn-Jones, *The Great Uprising in India*, pp. 23–24.

39 Devi Amma's half-sister Kalyani (born to Kunhi Kurumbi and a Nair gentleman) had no children; she acted as a foster mother to Devayani (E. K. Varadan's twin), the youngest in the EK family.

2

MADRAS I

Science and Politics in a Cosmopolitan City

I am glad to tell you that I had the luck to listen to Mr Ghandi's lecture the other day . . . So plain an Indian. One cannot but be impressed at his simplicity. His eloquence is like his external self. So simple yet so powerful in its effect. I think I could fall at his feet and worship him. Listening to his thrilling words on Social Reform I could not but be moved a little with feelings both of Patriotism and self sacrifice . . . I have a great mind to give up everything and devote to my life to the service of the mother country . . . but I am sure such a life, as a young unmarried woman will be hard as well as dangerous.
—E. K. Janaki Ammal (1916)[1]

On 30 June 1915, seventeen-year-old Janaki boarded a train for Madras with five other students (and her mother as far as Tirur), to join the city's College for Women. With so much company, she found the journey 'far from wearisome.' Moreover, her brother Raghavan's friend from his school days, A. K. Govindan, had promised to receive her at Madras Central railway station. When the train pulled into the station, Janaki was met by a cheerful Govindan and his young wife, who took her to their house, fed her well and made sure she rested before enrolling for the FA course (1915–17) the following morning. The college provided accommodation for non-residential students, but as the hostel building was not yet complete, students were put up at the Capper House, a garden house on the Marina belonging to Col. Francis Capper: 'It is a fine building close to the Cathedral of St. Thomas' (the area called San Thome near the Marina in Mylapore), she wrote to brother Raghavan, after she had moved in. They were in all fourteen girls, ten of whom were Malayalis, 'mainly Nairs,' she noted. Besides her, there were three other Thiya girls: the 'two Miss Palpus' [daughters of social reformer Dr Padmanabhan Palpu or his elder brother Velayudhan Palpu] and 'N. K. Narayani, sister of the dresser employed somewhere in Burmah.'[2] The College Principal Miss Dorothy de la Hey and a teacher (probably Miss Philips) resided with the girls at Capper House.[3] Incidentally, the first 'Hindu' lady

DOI: 10.4324/9781003267089-2

to be appointed at the Madras College for Women was Miss Kamakoti Natarajan; a graduate of the Bombay University, and daughter of the editor of the *Indian Social Reformer*, Kamakashi Natarajan, she taught history.

In her letter, Janaki described her room and the mess at Capper House: 'On the whole it is rather comfortable. Each room occupies two and contains two cots, two chairs, two tables and a mirror and bureau. We all eat together the food cooked and served by a Brahmin. Sitting on the floor with leaves in front of us we cut a very strange scene especially to the European ladies who have never had a chance of seeing a Malayali mess. We get nothing but vegetable together with ghee, sour milk and *rasam*.' At first, Janaki did not care much for the curries and was even homesick, but with time she settled down well to it. The mess charges came to a total of Rs 16 per month, but this did not worry her much because she was hopeful of winning a scholarship. The students were however expected to pay a security deposit of Rs 25, and it was this that was making her anxious: 'this sum will be returned . . . on leaving college. I have not yet paid the amount. I don't know what to do. The cost of books too comes up to a good amount. This month being the first will be very expensive. I have not got anything from Kittuattan [brother Kittu] or Vaston [brother Vasudevan]. I hope you have sent me something. You can more or less guess the amount of expense when joining a college. So please send me something more. From next month forward you will have to pay just a third of the messing & a small pocket money. I go to Devu's [perhaps a college-mate] house on Sundays and there at least a nice meal awaits me. Her house is quite close so that I even run up to it at times when water is scarce and I want to bathe. . . . How are you doing there? I'll be waiting for your M.O. and letter,' she wrote to Raghavan.[4]

First Examination in the Arts

At the Madras College for Women, Janaki had opted for 'Group II,' which included Natural Science (Botany and Zoology) and Physical Science (Chemistry and Physics), besides English, but because there were no professors at the College to teach the science group, students had to go to the Presidency College, also on the Marina, to attend lectures. Government 'rickshaws' (*jutkas*) transported them between the Colleges free of charge. In 1915, for her intermediate examination in botany, Janaki studied P. F. Fyson's *A Botany for India* (1912), David Thoday's *Botany for Senior Students* (1915) and K. Rangachari's *Manual of Elementary Botany for India* (1916). Raghavan had sent her a collection of Shakespeare's plays when she enrolled for the course and within months she had already read several plays from it: 'At present I am with *Hamlet*. I read each play twice. Then only am I able to digest and relish Shakespeare,' she wrote to Raghavan.[5] Janaki would miss no opportunity to quote from the bard in her letters.

In the first year of study, the Madras University English curriculum for the intermediate examinations included a detailed study of Shakespeare's *Henry V* besides Milton's *Paradise Lost* (Book II), Coleridge's 'The Ancient Mariner,' Matthew Arnold's 'Baldur Dead' and 'The Forsaken Merman' and Boswell's *Life of Johnson*, 1763 to 1767 (Blackie's English Texts). In addition, students were also expected to study as 'non-detail,' Walter Scott's *Quentin Durward* (1826), William Morris' *Sigurd the Volsung* (1876), A. J. Church's *Henry V* (English Men of Action, 1891) and W. W. Skeat's *The Past at Our Doors* or *The Old in the New Around Us* (1912).[6] Over the second year, the students would study Shakespeare's *Julius Caesar* besides poetry by Keats and Tennyson, and some more Milton and Boswell.

Edam's Saviour

Janaki was a good correspondent, writing regularly to her mother, sisters and brothers, and always in English, with occasional words in Malayalam thrown in for emphasis or effect. She would remain a responsible sibling, even trying her hand at matchmaking on behalf of her unmarried elder siblings. She was mindful of the social class she belonged to as the daughter of late Diwan Bahadur Sub-Judge E. K. Krishnan but was also deeply tormented by the social stigma her family suffered on account of the 'white stain' on the maternal side. A few months into her FA course, Janaki wrote to brother Raghavan:

> what if you marry my room-mate and class-mate Narayani, the sister of the dresser in Burmah? You saw her photo, what do you think of her? Not ugly, is she? And the most educated girl you can find. She is up to date and can move freely in any society. I think she is a fit sister-in-law. Only about her people, they are, rather, they were insignificant people. Her father was *masalchee* or some personage like that but she is of an unstained *tharavad* though it is a very low one in the sense of the world. I am sure her people will jump at the idea of her marrying an E. K ... do you think Parvathiattathi [elder sister Parvathi] can be married to her brother?[7]

Janaki would spend her Christmas vacation of 1915 at the college. This was the first time ever she was away from home during the holidays, but had to save to make ends meet; her roommate Narayani on the other hand was lucky to go home to Tellicherry. Devi Amma would send a parcel containing pickles and bananas through Narayani; the other girls brought back sweets from their homes. In a letter, Janaki described to Raghavan the 'large ripe papayas,' she had recently relished, each 'like a pumpkin,' the seeds of which she had saved to take back home. Her passion for gardening, something she had inherited from her father, was already evident, so much so that despite limited means and far from robust health, she had begun to garden at Capper House.

> I suppose you have a garden of your own there. I have purchased half a dozen flower pots to do some gardening during my leisure hours. I have placed them just in front of my room. I am owner of two pots of chrysanthemum (white) and two chilli plants. I have also sown two seeds of the garden palm. If you have any seeds can you send me a little of each! I think you can just drop a few of them in your letter each time you write to me. I believe it is not wrong to do so? [I want to make] a pretty garden in front of the old building. It is [very] small and crowded. At present balsams. Zinnias and Holly Hocks form the majority. Chrysanthemums and lilies are also seen flowering from time to time. It is a pity the space for gardening is so limited. I have planted vegetables like *thovara* [written in Malayalam] and chillies near our kitchen. They are growing quite well.[8]

Despite having remained at the College during vacation, Janaki and her hostel-mates would have no reason to complain: 'We had plenty of outings,' she reported to brother Raghavan, for the Irish-born Miss G. C. McCormick, one of the lecturers, had ensured their vacation would be an enjoyable one. Sometimes McCormick would go 'out for a drink or a shopping' with her. In late December, Mrs Bedford, associated with the National Indian Association, and the Ladies' Recreation Club, Madras, and a well-wisher, took the students to People's Park to see the grand Madras Exhibition (December 1915–January 1916) inaugurated by Lord Pentland (John Sinclair, Governor of Madras), a radical liberal who had supported women's suffrage as Secretary of Scotland and was a popular figure in Madras during his tenure as Governor (1912–19). He took an abiding interest in urban planning, in the development of local industries and in Hindu culture.[9] Only a year before he had facilitated mathematician Srinivasa Ramanujan's journey to England, for the latter's legendary meeting with the Cambridge mathematician G. H. Hardy. However, he would also be disliked for his attempts at crushing the Home Rule League and arresting Annie Besant and other leaders of the movement, in June 1917.

The Madras Exhibition included a medical section, curated by the Surgeon-General of the Madras Presidency, Dr W. B. Bannerman. Bannerman, incidentally, was President of the Indian Science Congress held in Madras in January that year, six months before Janaki arrived in Madras. When the Women's College group progressed to the medical exhibition venue, the Governor was present, as a result of which they 'were saved from the push and knocks of the crowd,' and moreover 'His Excellency was kind enough to ask Dr. Bannerman to explain' to them 'the various photos of diseases in India' exhibited.[10]

Keen on seeing new lands, Janaki much wished she could visit Burma, where her brother lived and worked. The eighteen-year-old wrote to Raghavan telling him how much she longed to go on a 'voyage': 'I am sure it will be very easy for me to [cross] the Bay to you . . . to spend a . . . vacation . . . but it all lies in your [enticing?] a girl to wife. I hear that Telly [Tellicherry] is teeming with [men looking for] girls. Don't you think it will be very good of you to add one to the number. All the pretty girls are being picked away one by one. You had better try your hand soon.' She also spoke of how much elder brother Padmanabhan also had 'a great fascination for Burma,' but of late had been 'getting attacks of asthma often.' Despite bad health he would 'imprudently' go to play cricket matches, she complained, and as a result 'study was out of question.'[11]

Janaki was still awaiting the books she had requested Raghavan to send her and wondered whether the delay was because they were unavailable in Rangoon. She had received her monthly remittance of Rs 5 from him in December, but needed more the following month having incurred some expenses during the vacation: 'Can you send me Rs 7 instead of Rs 5 this month Ragton [Raguattan]? I had some extra expenses in the shape of carriage hire and [Christmas] cards last month, besides our food money has come a little more owing to the lessening of numbers during the vacation.' Most of the results of the half-yearly examination were out and she had done quite well, but was dreading her Physics results. 'I don't expect good marks in it,' she would remark. Janaki was happy in Madras, finding it very pleasant at this time of the year, but found that she was turning into 'a regular negro.' 'Some say it is due to the salt air but have we no salt air in Telly?' she quizzed her brother. 'How did you spend Xmas and New year? . . . Please give me a long letter. I am anxiously waiting to hear from you. Or do you mean to give me a surprise visit. That will be much better,' she added.[12] Janaki wrote to him again in February that year: 'Do you think you can send Rs 10 next month Raguattan? I am rather hard up without coins in hand to buy even the bare necessities of life. My stationery, soap etc have not been refilled this year and I had to buy some more books this year. You can understand what all an Intermediate student will require. I have not even a pie for pocket money.'[13]

Seeing Gandhi

Incidentally, the most popular soaps in Madras at this time were imported English ones like Wright's Coal Tar ('The Soap for India, Good for prickly heat, wards off insect bites'), Pears ('Good Morning! Have you used Pears Soap?') and later Palmolive. In just a few years' time, the Calicut Kerala Soap Institute, in the heart of Calicut on the land originally owned by Rarachan Moopen of Kallingal Madom, a friend of Janaki's late father, would market its Washwell and Vegetol soaps ('genuine soaps' with no 'fillers') as *swadeshi* products, made of locally available vegetable oils under the able direction of chemist Dr Ambat Keshava Menon. Gandhi would speak on the meaning of *swadeshi* at a Missionary Conference in Madras on 14 February 1916, besides addressing the annual meeting of the Social Service League at Ranade Hall on Brodie's Road a couple of days later. He was keen that his message reach the college students of the city and for this purpose visited several institutions including the Victoria Students Hostel at Chepauk, YMCA Hall on the Esplanade and Anderson Hall (as part of the official inauguration of the Madras Christian College's Debating Society). He explained to the large gathering of students, the aims of the Satyagraha Ashram which was opposed to the accumulation of wealth and the need to take the *Swadeshi* vow. An opportunity to listen to Gandhi in Madras was unmissable. It was perhaps at the YMCA Hall that Janaki saw Gandhi for the first time ever; it made a lasting impression on her, as is clear from her letter to brother Raghavan, quoted in the epigram to this chapter, in which Janaki wrote about her wishes for the future, which included dedicating her life to social service:

> I often think over what he said that day. I have a great mind to give up everything and devote my life to the service of the mother country why not join the Servants of India Society. It is doing so much good to our land. I think it is the best way I can devote my life to a good end but I am sure such a life, as a young unmarried woman will be hard as well as dangerous. I must pass my BA and then I must think of what to do. I often wish I were old, an old maid so that I could do and go anywhere. At other times, I think of going in for medicine. That is what India needs most. I must consider before I take to some conclusion.[14]

The lesson in simplicity Janaki would instantly imbibe; in a few years, her style of dressing would undergo a radical transformation and she would begin wearing light or pale-yellow saris, like a renunciate or even a Buddhist. She would also begin experiments in vegetarianism.

Entertaining Diversions

Like her father and brothers, Janaki enjoyed playing tennis (a popular sport at these times) and being physically active in general. Janaki had become a tolerably good tennis player within months, 'improving with long strides' but was unhappy as she did not own a racquet and had to borrow one each time. Not one to lose an opportunity to quote from Shakespeare, she remarked: 'I want to play and Shakespeare says, "Neither a borrower nor a lender be" (quoting Polonius's advice to his son, Laertes in *Hamlet*, Act I, Scene III).' She chided brother Raghavan, passionate about cricket like all true-bred Tellicherry boys, but who had now begun to find it physically challenging: 'Why do you stick to cricket when you see it does not keep well with your health? Pappuattan [brother Padmanabhan] is as crazy after it as you are. And he seems to get on alright in spite of his asthmatic body. He writes he has been scoring a good deal since late.'[15]

An interest in travel and adventure would become a major driving force behind Janaki's choices in life, including her scientific practice, but more on this later. Both her roommate Narayani and herself had brothers living in Burma and so it is hardly surprising that she kept track of the Burmese festival calendar. She wrote to brother Raghavan:

> You say the Burmese festivals will soon begin, Yes, I remember you sending us pressed flowers some time near Vishu [April]. I am sure it is that festival that is going to come off. What about the water festival, Is it over? One of our professors has been to Burmah. She speaks of the land very favourably. How I wish to see it. I think I will surprise you sometime during my stay in Madras. I have a classmate a girl named Janaki, niece of the Peria Samy Pillai . . . who has many of her people there. What do you think of the idea?[16]

Janaki was doing well at college, but was somewhat annoyed that they had classes throughout the week: 'Had it not been for the sea in front of us, I am sure our life with all its company would be a bit monotonous,' she remarked.[17] This love for the ocean, and curiosity about the world beyond, would remain undiminished throughout her life. She would not miss the opportunity to play Shylock in the *Merchant of Venice* (Second Part) in the month of February (1917) at Snowdon, Adyar, the residence of Mrs Bedford. Janaki's roommate Narayani would play Antonio. It was Miss McCormick and Miss Phillips who had trained the students in their parts. The *Indian Ladies Magazine* reported: 'It was a pretty sight to see many girl students- most of them Indians, the future hope of our India, trying to enter into the spirit of the immortal Shakespeare, and speaking his grand and gracious words with great understanding and enjoyment.'[18]

A Bachelor's in Botany

Janaki passed her FA in the summer of 1917. Moved by Gandhi's call to serve society and the frenzied social reform drive in Madras, she had toyed briefly with the idea of becoming a doctor like her late brother Damodaran; women like Ayyathan Janaki Ammal from Janaki's neighbourhood of Chetamkunnu, and V. V. Janaki of Calicut, had qualified as doctors at least a decade earlier.[19] It was also in 1917 that Sarojini Naidu and Margaret Cousins (founder of the Women's Indian Association, Madras, established to serve as a platform for women to influence government policy) led a group of prominent Indian women across the country to demand that women's suffrage be included in the nascent Franchise Bill being developed by the Government of India.[20] Janaki abandoned the plan however in favour of botany and enrolled herself in 1917 for a BA (Hons) degree at the prestigious Presidency College on the Marina, just a short distance away from the Madras College for Women, where she resided.[21]. She was determined to become a scientist rather than study medicine, which incidentally her roommate N. K. Narayani would pursue.[22] It might be noted that Janaki's decision to become a scientist was a most unusual one for the times, when the only career options for educated women/graduates, invariably pulled into the vortex of social reform, were that of an educator, doctor or social worker, or later nurse.

There were six first-grade colleges affiliated to the University at this time: Government Presidency College, Government Muhammadan College, Madras Christian College, Pachaiyappa's College, Madras College for Women and Women's Christian College. For the three-year honours degree, one enrolled at the Presidency College, which offered courses in English Language and Literature, History and Economics, Mental and Moral Science, Sanskrit, Mathematics, Physics, Chemistry and Natural Science. The only other institution to offer these courses was the Madras Christian College, but it was located on the outskirts of the city. For a degree in Natural Science, students were examined in two Parts: Part I dealt with English language and literature, chiefly two Shakespeare plays and some prose and poetry from the sixteenth to eighteenth centuries, while Part II covered the main subject of study and a subsidiary one. Janaki chose Botany for her main and Geology as her subsidiary subject. For Botany main, students were taught the general morphology and physiology of plants and the peculiarities of form or structure depending on habit or habitat; the systematic position and relationship of the chief flowering plants and ferns of India and in general of flowerless plants, especially those of economic importance; general palaeobotany especially with reference to the relationship of modern plants; and lastly the phenomena of heredity, and selection, natural and artificial. All these aspects would come into play, especially the last, and flowering plants in particular, in Janaki's researches in the years to come. In December 1919, she bought herself Charles Darwin's *The Various Contrivances by which Orchids are Fertilised by Insects* (1877). The student was also required to submit for examination her laboratory notebooks containing drawings relating to the practical work conducted during the period of study. Janaki had enough experience in this department, having made several botanical illustrations for her forester brother Kittu.

The honours students were expected to own a copy of Eduard Strasburger's *Textbook of Botany* (1894; 1908 edition), Ludwig Jost's *Lectures on*

Plant Physiology (1907), Coulter, Barnes and Cowle's *Textbook of Botany* (1912) and John C. Willis' *A Dictionary of Flowering Plants and Ferns* (1919). Strasburger's book provided students like Janaki with an introduction to cytology, histology, organography and the theory of descent and the origin of new species; in another book, *On Cell Formation and Cell Division* (1876), Strasburger had explained the basic principles of mitosis. The third chapter in Jost's book also dealt with heredity and variation, but only to a modest extent. At the library of the Presidency College, which students like Janaki made good use of, were scientific journals such as the *New Phytologist, Annals of Botany, Curtis' Botanical Magazine, Memoirs of the Department of Agriculture in India, Journal of Indian Botany, Philippines Journal of Science* and *Journal of Genetics*.

Ill Health

Raghavan would receive a long letter from Janaki, only a few months into her BA (Hons) course, filled with news on life at college. Congratulating him in advance 'at the prospect of becoming a father,' she wrote, 'I do hope you will keep Sil [sister-in-law] from all harm and bless you both with a little son.' Janaki had been unwell and in fact was suffering from bronchitis and running a fever when she left Tellicherry for Madras some months earlier. The Anglo-Indian temporary lecturer, Miss L. H. Philips, had taken her to the tuberculosis expert, Dr Kesava Pai for consultation[23]; on examination of Janaki's chest, Pai found her lungs to be quite alright but felt she needed to improve her general health. Heeding Dr Pai's advice, Janaki began taking 'some wines' regularly.[24] Her mother was at this time at Vellore, not far from Madras, visiting son Kittu (who would marry Kanoth Yeshoda, daughter of Rao Sahib Kanoth Chandan of Talap, Cannanore, on 30 December that year), in the company of daughter Sumithra and asthmatic son Padmanabhan. In fact, Devi Amma had taken a loan for the sole purpose of having Padmanabhan treated by the eminent doctor, Rao Bahadur Lakshmana Perumal Pillai of Madras.

During the Michaelmas term, Janaki managed to visit Vellore to be with her mother and brother: 'Mother is already tired of the place. Kittuattan is out in camp 20 days in the month and then life is so very dull there. They live in one of the Tamilian houses that face a street with not a square inch of compound. Mother has very little to keep her occupied and the constant sight of Papton [Padmanabhan] suffering makes her very desperate,' she reported to Raghavan. While she was in Vellore, Miss Philips paid them a 'flying visit' to bid goodbye, as she was to leave Madras by the end of the year to enter a nunnery at Ernakulam. Janaki had done her best to dissuade her, but Miss Philips, was 'determined to bury her talents and her goodness within convent walls.' 'I hear we are going to get an Indian in her place,' Janaki added.[25] There was already 'a Syrian Christian on the staff,' she wrote, 'one Miss Joseph [perhaps Theresa Joseph, who taught economics] from the Trivandrum College,' and the 'Jewish Professor,' who had lately left to practice Law, 'as some were saying—you see she is a BA BL.'[26]

A Matter of Dignity

Brother Raghavan had been supporting Janaki's education, but being self-reliant by nature, she had begun looking out for alternative sources of funding. At the start of the academic year (1917), she had applied to the Sri Narayana Dharma Paripalana Yogam (a caste organisation of the Ezhavas/Thiyas, founded in 1903 with the blessings of the social reformer Sri Narayana Guru) for a scholarship, but E. K. Govindan (her half-brother, the future Diwan of Pudukottai) was very offended by this; he wrote her a stern letter saying that her action had displeased everyone in the family and had brought 'discredit to Father's name.' He had suffered humiliation, he claimed, having been subjected to an examination by the Secretary of the Yogam 'regarding the financial difficulties of the E.K.' family and being asked 'whether she hadn't a brother who is a district forest officer and the like.'[27] Her hopes for a scholarship were thus rudely dashed to the ground, but fortunately for her, Principal Dorothy de La Hey recommended her name for an extension of a previously held scholarship, worth Rs 9 a month.

Since the doctor had prescribed a period of rejuvenation, Janaki stayed away from the tennis and badminton courts, but this did not stop her from tracking the sports events at the College. Janaki exclaimed in a letter to Raghavan: 'Our Games Club is making a name for itself. We won a Shield the other day for Badminton . . . hoping to get a cup for the same very soon. Are you having any cricket matches now?' She had also been keenly following the political developments in the country; Annie Besant, the ageing theosophist and organiser of the Home Rule movement (advocating self-government within the British Empire for all of India), had been sworn in as President at the meeting of the Indian National Congress at Calcutta upon release from prison for her political activities. Besant had begun to organise Home Rule Leagues across the country, with the first in the city of Poona (April 1916), followed by Madras in September that year. A sceptical Janaki queried, keen on knowing what Raghavan thought of all this: 'What do you think of Mrs Besant's release? There was great rejoicing here. I believe the old Lady [made] a grand time of it' and is now paying visits to the 'great cities of India.' She was obviously not too impressed by Annie Besant or the idea of Home Rule for India. Making light of the whole affair, she commented: 'We have a few "home rulers" in the College. It is a common question now: Are you a home ruler? A friend of mine asked me the same question. I replied not yet, as I have not come in the possession of my home!'[28]

Becomes a Graduate

At the convocation in 1921, Janaki was awarded a BA (Hons) degree, one among the twenty-odd women who had read for an honours degree at the Presidency College[29]; she was placed in the third class, but this was not bad at all, when a second was not common and a first extremely rare. Among the women who received an honours degree at the same time as Janaki were Checha T. George from Travancore, and M. Lakshmi Ammal and K. Shanti Bai, from Madras.[30] Upon payment of a fee of Rs 25 an honours graduate of the Madras University could proceed to the MA degree without a further examination, but only after two years from the date of passing the BA. So, when the recently graduated Janaki received an offer to teach botany at the Women's Christian College (WCC), she gladly accepted.

Professor of Natural Science

The Women's Christian College was founded in July 1915 as a first-grade international and interdenominational missionary college, under the control of a Council composed of representatives of several missionary societies in Britain, Canada and the United States;[31] from 1920, the college became a sister college of Mount Holyoke College, South Hadley, Massachusetts. Structured on the lines of a liberal arts college (like Mount Holyoke), the WCC played a major role in promoting the teaching of science at college level among young women in Madras in the early twentieth century. Several of the science teachers at the WCC were products of Mount Holyoke, which enjoyed the reputation of being a major institution of science education for women in America as early as 1837, when it functioned as a seminary; the women scholars were all single and deeply religious then. In 1938, Mount Holyoke College was the third largest employer of women faculty members in science in America.[32]

Situated in its own grounds on College Road, Nungambakkam in Madras, the WCC provided residential accommodation for term professors and about a hundred students within the campus. Miss Eleanor McDougall was the Principal of the institution and Professor of English, Ancient History and Latin. Miss Edith Marion Coon was the Vice-Principal and Professor of Physics, from the time she arrived in India in 1916. In 1921, Janaki assumed her post of Professor of Natural Science (Botany) at the institution, a task she would share with physiologist Eleanor Dewey Mason (1898–1993), who taught Zoology. A postgraduate from Wellesley College, Massachusetts,

Mason was born in Tura (Assam) to Baptist missionary parents. Among the other faculty members were Miss E. T. Stevens, Miss A. L. Jackson and Miss Somakumari Seneviratne (from Colombo, a Christian convert, who went to Girton College, Cambridge), all Professors of English.[33] From 1921, Miss Checha George taught History and Economics, as would Professor Miss D. E. Hitchcock. The WCC was affiliated to the Madras University in Group I (Mathematics), Group III (Natural Science, Botany with Geology), Group IV (Philosophy) and Group V (History and Economics), of the BA degree course. The teaching of scripture was an integral part of the course. Besides, the college supported a number of student clubs such as the Historical Society, Star club (astronomy), Glee club (musical or choir group), Art club and Modern Poetry club, and with the joining of Janaki and Mason, also a Natural History Association. In addition, the WCC made adequate provision for games and sports including netball, badminton, volleyball and tennis, the last being a particular favourite of Janaki'.[34] Miss George, Miss Mason and Miss Hitchcock contributed to raising the level of athletics at the College, by breathing new life into the drill classes.

In 1921, when Janaki joined the teaching staff, there were in all 130 students at the institution. Nineteen resided at the college, of which eight spoke Malayalam, ten Tamil and one Telugu; all were bare-footed and wore saris, which in the case of Syrian Christians were invariably white. In early November 1922, a year after Janaki began teaching, the Nobel Prize–winning (1913) poet-laureate Rabindranath Tagore visited the college, triggering a great wave of excitement among students and teachers alike.[35] Exactly a year later, the young poet Harindranath Chattopadhyay (1898–1990), a contemporary of Janaki, visited; the poet had just returned from England. One afternoon, he read aloud some of his poems 'to an audience enthralled by his wonderful elocution and picturesque appearance.' He was the younger brother of the poet Sarojini Naidu and husband of the once widowed Kamaladevi (1903–88), who had studied at the Queen Mary's College (formerly, Madras College for Women). It was at the age of twenty that Kamala had married Harin and travelled with him to England, where she would pursue a diploma in Sociology at the Bedford College for Women affiliated to the University of London.

A Thiya Woman Botanist

Kamaladevi's college mate at Bedford College was a Thiya woman science graduate from Cannanore town, not far from Tellicherry, a government scholar, who in 1922 had enrolled for a second degree (BSc) in Natural Science (Botany).[36] Her name was Cheruvari Kottieth Kausalya (1890–1965), whose uncle on the maternal side was none other than Justice Cheruvari Krishnan. Kausalya's passport issued in Madras, in 1932, indicates that she was small-made, and had dark brown eyes and black hair.[37] Like Janaki, C. K. Kausalya was of mixed-race ancestry on her mother's side, but hers went back two generations (that is both her mother and grandmother were of mixed parentage). Kausalya's father was a very wealthy business man named Kottieth Choyi, known locally as Choyi 'Butler,' the proprietor of the fashionable Hotel Esplanade at Cannanore by the sea.[38]

The Bedford College had been founded in London in 1849 as an institute of higher education for women, the first one of its kind on Britain. In 1900, the college became a constituent school of the University of London. When Kausalya attended the college in 1922–24, Dame Margaret Janson Tuke was Principal. Incidentally, that same year, Miss Theresa Joseph of the Madras University[39] had won a government scholarship to study at the London School of Economics, and Miss A. Pitchamuthu, for medicine, at the University College Medical School, London. At least seven years older to Janaki, Kausalya had completed her BA in 1910 (convocation in 1911) from the Presidency College in the second division, perhaps the first Malayali science graduate among women or at least the first to graduate in botany. Her classmates at the Presidency included such high-achieving women as Rishiyur S. Subbalakshmi, the future 'Sister' Subbalakshmi (a recipient of the Grigg Memorial medal)[40] and Hilda M. Lazarus (awarded the Anna Isabella Subramanyan scholarship for Indian Christians in the field of medicine). In 1913, Kausalya obtained a Licentiate in Teaching qualification, following which she joined the Government Girls' High School, Cannanore as an Assistant. It was after her return from England in 1924 that she had been appointed Professor in Natural Science at Queen Mary's College (QMC); the small museum devoted to natural science (comprising six cases of zoological specimens, seven of botanical specimens, a multi-leaf frame of herbarium sheets and two cases of physiological exhibits), set up only a year previously in a room in the college premises, would come under her charge.[41] It was only natural that Kausalya and Janaki became close friends; after all, their families knew each other and they both taught botany, at the two women's colleges of Madras.[42]

Wins a Barbour Scholarship

By late 1923, Janaki had received her MA degree from the Madras University, just as the new science block at the WCC had grown to be one storey high, and a few classes had begun to be held there, taught by herself, Miss Coon and Miss Mason. The building had been funded by a joint missionary committee, which distributed funds raised in America in 1923 (referred to as the three-million-dollar drive) for the seven Woman's Union Christian Colleges in the Orient.[43] The building would take several more months to complete, and Janaki would miss the inauguration as she had been awarded the prestigious Barbour scholarship to do a Master's at the University of Michigan. In March 1924, a grand farewell meeting was organised at the college for three of the staff: Janaki, who would be leaving for Michigan in a few months, the missionary Catherine Justin who was returning to Kansas after spending three months at the college as chemistry lecturer, and Miss Jackson who was returning to England for a year of rest and recovery. Janaki's work would be shared between an Assistant Professor of the Madras Christian College and a young graduate, Miss Sugirtham Swamidas, an alumnus of the WCC. Incidentally, in July that year, the brilliant Miss Mariam Oommen, who passed the BA (Hons) examinations in the first division from the Presidency College, would be appointed temporary lecturer in chemistry.[44]

The Deluge

The year 1924 was an eventful one in more ways than one, serving as a major marker in the social history of Kerala. It was the year of the great deluge in South India, and the year that launched the historic struggle referred to as the Vaikom Satyagraha, against caste-based discrimination, which limited access to public roads leading to the Vaikom Mahadeva temple near Kottayam. And for Janaki personally, the year she would leave for America for higher studies as a Barbour Scholar.

The *edavapathi* or southwest monsoon was unprecedented and utterly devastating that year, battering the whole of South India. Travancore, Cochin and Malabar suffered terribly. From 3 July 1924, the newspapers were filled with news of heavy rains, landslips and breaches along the railway lines. A fortnight later, on 18 July, the papers reported that the west coast (Calicut and Mangalore) had been totally cut off from the rest of South India. The traffic manager of the South Indian Railway issued orders to stop booking passengers or goods beyond Olavakot station (Palghat) on the broad gauge. It was feared that reservations would not resume at least for a fortnight.[45] On 21 July, it was reported that the Shoranur-Cochin railway was entirely washed away, and communication would only be after two months; serious breaches had occurred between Shoranur and Pattambi. Extensive destruction of standing crops and houses was reported from all corners of the region. People residing in the lowlands had been rendered homeless.

Incessant rains and gusty winds also caused much destruction in Tellicherry, and the interiors had been cut off from all communication. Despite the terrible misery all around, the news of Janaki's scholarship was enough to bring joy and excitement to Edam. In fact, Janaki had been at home in Tellicherry since May that year (1924), her college having closed for vacation. Her initial plan was to set out for Madras in late July, take proper leave of friends and colleagues at the WCC, and then depart for New York in early August. However, tempted to spend more time with her family, she delayed her journey by a few days; this would prove imprudent, for the great downpour commenced and the railway network connecting Malabar and Madras became dysfunctional. Janaki was nevertheless rock-solid in her determination to make it to Madras before the ship departed. At the WCC, Principal Eleanor McDougall and her colleagues were deeply worried for Janaki when news reached them of the floods in Malabar. When she appeared before them late evening, ten days behind schedule, with a bright smile on her face, it felt nothing less than a miracle. Principal McDougall did not fail to record the incident in her journal:

> Our botanist left home one day too late, and on entering the station to take her ticket to Madras was informed that in the previous night the flood had carried away the great railway bridge at Shoranur. She made her way however with great courage and resource through the flooded country, accompanied by a young nephew and armed with a heavy stick. We of course could hear nothing of her movements and were very anxious, but 10 days after her expected date she quietly walked in just before dinner.[46]

How exactly she managed this is unknown, but her amazing will-power, courage and resourcefulness in the face of danger and seemingly impossible situations such as this would be demonstrated several times over in her life. The ship she was to sail on had already departed for New York, but she was very fortunate to find a berth on the following steamer and reach Michigan in time for her course.

Figure 2.1 Queen Mary's College group, Madras, 1915. Janaki seated on chair, second from left.

Source: Courtesy of the EK family.

33

Figure 2.2 The EK sisters, at Edathil. Parvathi (seated), Sumithra (standing right) and Janaki (kneeling). April 1916.

Source: Courtesy of the EK family.

Figure 2.3 The EK brothers, at Edathil. Clockwise: Kittu, Raghavan, Varadan, Vasudevan, Padmanabhan and Madhavan (seated on floor). December 1916.

Source: Courtesy of the EK family.

Figure 2.4 The EK sisters with Devi Amma, at Edathil. Clockwise: Janaki, Parvathi, Lekshmi, Cousalya, Sumithra and Devayani (seated at Devi Amma's feet). December 1916.

Source: Courtesy of the EK family.

Figure 2.5 Janaki and Checha T. George, c. 1923.
Source: Courtesy of the Mount Holyoke College Archives and Special Collections.

Notes

1 E. K. Janaki to E. K. Raghavan, letter dated Madras, 20 February 1916, private collection.
2 E. K. Janaki to E. K. Raghavan, letter dated 6 July 1915, Madras College for Women, Mylapore, private collection.
3 Capper House was named after Colonel Capper, who built it in the late eighteenth century as a private residence, when he returned from the command of the Madras Artillery at St Thomas Mount. The House, according to a woman writer in the early part of the twentieth century, was 'buried in a casuarina grove . . . a fine building, pillared with polished chunam columns that look like marble'. Penny, *On the Coromandel Coast*, p. 25.
4 E. K. Janaki to E. K. Raghavan, letter dated 6 July 1915.
5 Ibid.
6 Among the textbooks used were J. H. Fowler's *Nineteenth Century Prose* (1897) and J. A. Froude's *Short Studies on Great Subjects* (1867–82). Information on the curriculum and the prescribed reading lists have been gathered from the University of Madras Calendars of the respective years, in the collection of the Madras University Library, Chennai, and the British Library, London.
7 E. K. Janaki to E. K. Raghavan, letter dated 6 July 1915.
8 E. K. Janaki to E. K. Raghavan, letter dated 4 January 1916, Madras College for Women, Mylapore, private collection.
9 Lord Pentland's fascination for the work of fellow Scotsman Patrick Geddes, architect and urban-planner, brought the latter to Madras in late 1914; Geddes delivered a talk on cost-effective town-planning and sanitation (accompanied by detailed illustrations and maps) at the Cities and Town Planning Exhibition organised as part of the Madras Exhibition (1915–16).
10 E. K. Janaki to E. K. Raghavan, letter dated 4 January 1916.
11 Ibid.
12 Ibid.
13 E. K. Janaki to E. K. Raghavan, letter dated 20 February 1916, Madras College for Women, Mylapore, private collection.
14 Ibid.
15 Ibid.
16 The water festival is celebrated on the second day of the Thingyan, the Burmese New Year festival occurring in the middle of April; a Buddhist festival, it is celebrated over four or five days ending in the New Year and involved the sprinkling of water from a silver bowl to wash away sins.
17 E. K. Janaki to E. K. Raghavan, letter dated 20 February 1916, Madras College for Women, Mylapore, private collection.
18 *The Indian Ladies Magazine* 16, no. 4 (February 1917), p. 130.
19 A Thiya woman, Ayyathan Janaki Ammal earned her a Licentiate in Medicine and Surgery qualification from the Madras University in 1907; she was the youngest sister of Ayyathan Gopalan, who was the first LMS from Malabar and founder of the first Brahmo Samaj branch in Malabar (in Calicut) in 1898. V. V. Janaki (belonging to the Dheevara/fishing community) from Calicut, qualified two years later, in 1909. In 1927, V. V. Janaki joined the Women's and Children's Hospital, Calicut as 'Lady Assistant Surgeon'.
20 The delegation included Annie Besant, Parvathi Ammal, Mrs Guruswamy Chetty, Nalinibai Dalvi, Dorothy Jinarajadasa, Dr Nagutai Joshi (nee Rani Rajwade), Kamalabai Kibe, Mrs Z. Lazarus, Begum Hasrat Mohani, Saralabai Naik, Sreerangammal and Herabai Tata. Full suffrage for women was introduced by

the Indian Constitution only in 1949. For a history of the suffragette campaign in India, see Geraldine Forbes (2004), pp. 91–120.

21 The Presidency College was established in 1841 under the name 'The High School of the Madras University'. In 1853 collegiate classes were started at the institution and two years later, when the control of the institution was transferred to the newly appointed Director of Public Instruction, the college was duly constituted and a principal and professors appointed. It was then that the institution was renamed Government Presidency College. A Bachelor of Arts (BA) degree was awarded irrespective of the subject of graduation, whether in the sciences or the arts.

22 Narayani would however only complete her Licentiate in Medicine and Surgery in 1928. She would later marry Lt. Gen. Benegal Mukund Rao, from a Konkani Saraswat Brahman family, and come to be known as Dr Narayani Rao. Incidentally, in the late 1920s, Narayani would donate to the Free Hospital, Tellicherry, run by the Guild of Service.

23 M. Kesava Pai (1879–1965), conferred with the title 'Rao Bahadur' in 1932, was a brilliant surgeon and bacteriologist, based initially at the Pasteur Institute of South India and later the Government Tuberculosis Hospital and the King's Institute, Madras. He is best known for his pathbreaking paper in mathematical biology, co-authored with Anderson G. McKendrik, and relevant particularly today, titled 'The rate of multiplication of micro-organisms: A mathematical study'. Pai was also a social reformer, the co-founder of Mahila Sevashram, Mangalore, for the upliftment of Saraswat Brahman widows and destitute women,

24 E. K. Janaki to E. K. Raghavan, letter dated 7 October 1917, Queen Mary's College, private collection.

25 Ibid.; it appears however that Miss Philips after all returned to Madras to become a lecturer at the Lady Wellington Training College for Women.

26 Ibid.

27 Ibid.

28 Ibid.

29 The total number of students studying for an honours degree (in all the subjects put together) at the Presidency College in 1921 totalled 330, of which 20 were women. Indian Science Congress, *Madras Handbook 1922*, p. 61.

30 There were others like Edith Britto, Monica Fernandez and Acca W. Oommen from Trivandrum, Saramma M. Korah from Tiruvella, K. Ammukutti Ammal from Cochin, C. Chinammu Ammal from Palghat and K. Parukutti, who received regular BA degrees (as against the honours degree) that year.

31 These included the Church Missionary Society, London Missionary Society, Church of England Zenana Mission Society, Wesleyan Methodist Missionary Society, United Free Church of Scotland Mission, American Arcot Mission, American Baptist Telugu Mission, American Madura Mission, Methodist Episcopal Mission, Canadian Presbyterian Mission and American Lutheran Mission, Guntur.

32 Levine, *Defining Women's Scientific Enterprise*, p. 134.

33 Between 1919 and 1920, one Miss Elizabeth Zachariah from Travancore was employed as a visiting lecturer in English at the WCC.

34 For an early history of the WCC see McDougall, *A Missionary College at Madras*, 1926.

35 Tagore had left Ceylon for Travancore on 8 November 1922, where he was to be a State Guest, and it was from there that he had travelled to Madras. He would visit Madras again in early July 1929; this time he would spend a few hours at the house of Madhavan Arathil Candeth, Professor of History, Presidency College.

36 BL: 'The Women's Christian College, Madras, Principal's Journal', November 1923, p. 15.
37 BL, IOR: L/P & J/11/1/711; the passport tells us that she was born on 15 July 1890.
38 He was well-known for his signature 'Choyi's Pudding', a dessert made of mashed ripe bananas, mixed with sugar, clarified butter and crushed pappadams, and served at his seaside hotel near the Fort Maidan, Cannanore.
39 In 1930, Miss Theresa Joseph was extended a fellowship by the Barbour Committee; she joined the University of Michigan as a Barbour Fellow in 1931–32.
40 One of R. S. Subbalakshmi's nieces was Lalitha Doraiswamy, C. V. Raman's student at IISc, Bangalore, who would later marry the astrophysicist S. Chandrasekhar (a nephew of Raman).
41 Markham and Hargreaves, *The Museums of India*, p. 181.
42 Among the women who graduated in 1923 from the Madras University were Liza Jacob (Quilon [Kollam]) and N. K. Lakshmikutty Amma (Cranganore [Kodungallur]); while T. K. Sara (Kottayam) completed a BA (Hons) degree from the Queen Mary's College. That same year, two girls from Malabar, Oyitti Kunnathidathil Savithri Ammal and Upottu E. Sumithra, graduated with honours from the WCC, as did three Travancore girls, P. Anna Varkki of Kottayam and Rachel Thomas nee Jacob and Annamma Alexander of Tiruvalla.
43 These included the Woman's Christian College, Tokyo, Japan; Yenching College, Peking University, Ginling College, Nanking and the Women's Department of Medicine of Shantung Christian University, Tsinanfu, China; in India, the Women's Christian College, Madras, the Isabella Thoburn College, Lucknow, and the Missionary Medical School for Women, Vellore.
44 Coon, 'The Women's Christian College'. Also see Levine, *Defining Women's Scientific Enterprise*.
45 *Madras Mail*, 18 July 1924.
46 BL: 'The Women's Christian College, Madras, Principal's Journal', September 1924, p. 10.

3

MICHIGAN I

First Lessons in Internationalism

> I am teaching in a small Women's College which is very much indebted to America for its existence. Our new Science building is the result of your country's generosity and due to that we are the only Women's College that makes it possible for a girl to take a science degree without going to men's colleges.
> —E. K. Janaki Ammal (1928)[1]

In the early twentieth century, Levi Levis Barbour, a graduate of the University of Michigan, made generous gifts to his alma mater in the interest of women. These included a property in Detroit, which led to the setting up of a gymnasium in his name, and the Betsy Barbour House. On 22 June 1917, Barbour presented the University with a fund of $50,000 to found scholarships for young women from 'Oriental' countries. The chief purpose of the scholarships was to 'bring girls from the Orient, give them an Occidental education and let them take back whatever they find good and assimilate the blessings among the peoples from which they come.'[2] The idea was not to 'replace the Oriental pattern by an American model, but to supplement the basic and permanent native characteristics with Occidental values to form a new type of world citizen, willing and able to serve a broader internationalism.' To make the scholarships known in the East and to develop a method of attracting applications and making selections, letters were sent to Michigan alumni, government officials and women's colleges in the East, including the WCC. The response was 'phenomenal.'[3]

The amount allocated to each appointee was not considered sufficient to meet all the expenses of a scholar; holders of the scholarship were expected to pay their own travel expenses, pay University fees and meet all their personal expenses. With economy, however, the scholarship was enough to take care of fees as well as maintenance. Advisory Barbour Scholarship committees, whose chief function was to scrutinise applications and make recommendations, were appointed in several countries, including India.[4] The chief factors considered by the selection committee included the 'character' of

DOI: 10.4324/9781003267089-3

the candidate, her scholastic achievement, suitability for University work, marked ability in a special field of study and very importantly 'her desire to return to her own country for service after suitable preparations shall have been made.' In India, in the 1920s, the Chairman of the committee was Miss Martha Downey; members included Miss Eleanor McDougall, Principal of the WCC, Miss Dorothy de le Hey, Principal of the Government College for Women (QMC), the American zoologist Miss Eleanor D. Mason of the WCC teaching staff, and the Rev. W. Meston, Professor in the Men's Christian College, Madras. Two trips were made by the Secretary of the Barbour Scholarship Committee to the relevant countries to meet with advisory committees, personal advisors, heads of women's and co-educational colleges and again later to interview prospective applicants.

Almost all Barbour scholars from India have had as their alma mater one or the other of these institutions: the Allahabad University, the Crosthwaite College for Women (Allahabad), the Isabella Thoburn College (Lucknow), the Madras University or the WCC. Most were in positions of authority themselves, as professors and/or academic administrators. On 13 March 1924, *The Michigan Daily*, the University-based newspaper, announced that two Japanese, four Chinese, one Korean and two Indian women had won the Barbour Scholarship worth an annual grant of $800; the Indians were E. K. Janaki Ammal and Mainabai Wasdeorao Shahane.[5] Credit for being the first Indian Barbour Scholar (in 1920) however went to Ashalatika Haldar, a philosopher by training and member of the teaching staff at the University of Allahabad. A year after Haldar, in 1921, another Calcuttan, Probhabati Dasgupta, daughter of Sub-Judge Rai Tarruk Chandra Bahadur won the prestigious scholarship; Dasgupta later became a trade unionist, organising jute workers of Bengal, in 1929.[6] The *Madras Mail* announced under the title 'Indian Women in America—Opportunity for Study' that the Levi Barbour scholarship for 1924–25 had been awarded to 'E. K. Janaki Ammal, Lecturer in Botany in the WCC, Madras.'[7] The Secretary of the Barbour Scholarship Committee had the authority to make an additional grant, and where necessary, to increase the value of the stipend by waiving the Special University fees. Accordingly, in September 1925, Janaki was granted an extra fund of $68.50.

Graduate Studies in Michigan

After a long and tedious journey by ship via England, Janaki reached New York (landing at Ellis Island) in early October 1924. She had travelled in a 'bunk cabin below, in a slow boat across the sea, [was] feeling miserable, lonely & frightened to land in a strange land with none of her own people for solace, only to be collected by two strange women of the YWCA at Ellis Island, who gave her the impression that they were the saviours of a lost body & soul!' On waking up the first morning in America, she looked out of her window 'to find the whole inner courtyard of the YWCA full of nude females,'[8] recalled nephew Hari Krishnan, to whom she often recounted stories from her life. We know that Janaki had broken journey for a few days in England,[9] but we have no clue as to where she stayed or what she did there. From New York she made her way on her own by train to Ann Arbor. 'Oriental Women' were usually housed in the university residential buildings reserved for women, and the League House would sometimes be used to provide temporary accommodation. Janaki was however lodged in the Martha Cook Dormitory, the largest women's dormitory on campus. That Janaki was a curious sight on the campus is hardly surprising. *The Michigan Daily* reported that although students 'come to Michigan from all parts of the world,'

> it is decidedly unusual to see on the campus the native costume of the Hindu woman. E. K. Janaki, grad., comes to the University from Malabar in South India, to do special work in the natural sciences. Miss Janaki holds two degrees from Madras Christian College, Madras, South India. She says college work in India and at Michigan cannot be compared because it is very different but she comes to Michigan for the wide opportunity it offers to do research work in her chosen line of study.[10]

On 1 April 1925, H. H. Bartlett, Chairman of the Botany Department, delivered the Presidential Address of the Michigan Academy of Science, Arts and Letters, which had just opened in the Natural Science Building on the campus, on the subject of colonial botany; Janaki surely would have attended the function. Harley Harry Bartlett (1886–1961) was born in Anaconda, Montana, and graduated from Harvard with a chemistry degree and worked as a chemical biologist for the US Department of Agriculture in Washington, where he became interested in the work of Dutch botanist Hugo de Vries on evolution and began research on the genetics of the evening primrose (genus *Oenothera*). de Vries had argued that species originate through sudden, spontaneous mutations (evolution by sudden leaps), of course under the assumption that mutants were pure species and not hybrids.

Janaki was deeply inspired by Bartlett, particularly, his sense of adventure and his synthetic approach to science. A pioneer in plant genetics, the unmarried Bartlett chaired the Botany Department at the University for several years. He was an able administrator, hugely interested in the botany (and culture) of the tropics, and would undertake botanical expeditions to Formosa, Sumatra, Mexico, British Honduras, Guatemala and the Philippines, Panama and South America during his lifetime. Bartlett was also passionate about music, Batak and Malayan ethnography and linguistics.[11]

In the summer that year, several members of the Department of Botany were away on research work, during which time graduate students like

Janaki focused on their own experiments and microscopic work. While B. M. Davis spent the summer doing experimental work on the American evening primrose at the John Innes Horticultural Institution (hereafter John Innes) in England, Sterling H. Emerson, Instructor in Botany, was travelling in Sweden, Denmark and Holland to undertake a 'special study of genetics and histology'; Mrs Eileen Erlanson, holder of the Cole fellowship in Botany was continuing her study on the American wild rose, in Brussels and at the Kew Gardens in London, while her husband Carl Erlanson was on a geological expedition in Texas; J. H. Ehlers and Carl D. La Rue were teaching at a camp at Douglas Lake in East Tennessee, the biological station of the University of Michigan. As for E. G. Anderson, he was with E. E. Dale on a collecting excursion in Southern Illinois.[12]

Academic clubs such as the University Women's Research Club, the Botanical Journal Club and the Michigan Dames held regular meetings, some of which Janaki attended. Botanical seminars were also periodically held on campus; graduate students were expected to religiously attend these. Only a few months into her course, Janaki was invited to speak at a meeting of the Michigan Dames:[13] 'Edavaleth Janaki, grad., of Malabar, India gave an interesting talk Tuesday night at Wesley hall, at a meeting of the Michigan Dames. Miss Janaki spoke of her home, and contrasted it with other countries.'[14] A few months later, in early October, the 'Oriental women' were entertained by the Dean of the University at 923, Olivia Avenue, on campus; one of the invitees was 'Miss E. K. Janaki Ammal of Martha Cook Building.'[15] The University Women's Research Club organised periodic meetings at half past seven in the evening in the Natural Science Building, which housed the botany department among others; on 18 January 1926, at one such meeting Janaki delivered a talk curiously titled 'The Racial History of the Hindus.'[16] We do not know what exactly she spoke on this occasion, but we can safely surmise she did not allude to her own mixed race origins. It is more likely that she introduced her American listeners to the Hindu caste system and its relation to race from within a colonial worldview. That she chose this topic at all reveals her early, and later abiding academic interest in eugenics and human genetics. It must be remembered that eugenics during these times was somewhat popular in America, despite the forced sterilizations of the 'feeble-minded' in the country, but with the rise of Nazi Germany in the early 1930s, its undeniable diabolic side could no longer be ignored. It was not just the right-wingers who were attracted to eugenics however; for in England, Fabian Socialists like Sidney and Beatrice Webb, Harold Laski and J. M. Keynes were keen advocates of the movement, but more on this later.

During Janaki's time on campus, there were such social and cultural clubs as the Hindustan Club and Cosmopolitan Club, which organised gatherings on a regular basis. During the first semester, Janaki was joined by another Indian graduate, Achy Iype of Travancore, a former student of the WCC who had finished medical school at Chicago, and was now studying for an MD at Michigan; Iype would intern at the Women's Hospital, Philadelphia, before returning to India and becoming a medical missionary at the Kottayam General Hospital.[17] Janaki and Iype would become active members of the Hindustan Club of the University of Michigan.

By the time she left Michigan in April 1926, Janaki proved herself to be an 'excellent student,' with thirty-five hours of graduate credit, including twenty-six of A grade (she had Botany as her main subject, and Zoology as subsidiary). On the basis of her outstanding performance, she was granted $100 on the recommendation of Professor Bartlett, to enable her to cultivate eggplants after her return to India 'in continuance of her research programme to obtain material for her thesis.'[18] The understanding was that Janaki would return to America within a short period of this to finish her thesis and complete the requirements for the degree of Doctor of Philosophy.[19] In late April, just before her departure for India, Janaki was elected to the associate membership of the Michigan chapter of Sigma Xi (founded in 1886 at the Cornell University), a national honorary society for the promotion of research in both pure and applied sciences.[20]

Applicants for a Barbour scholarship from India in 1926–27 included Miss Aley Checha Kuriyan, a graduate from the WCC, whose special field of study was Education, and Mary B. Dagmar da Costa, a BA from QMC, wishing to specialise in English Language and Literature. Aley C. Kuriyan was employed at this time as a teacher in a Syrian Church School in Tiruvella, in erstwhile Travancore. As Professor at the WCC, and a Barbour scholar herself, Janaki was consulted on Miss Kuriyan's suitability for the scholarship; she was more than happy to write out a recommendation in Kuriyan's favour. Janaki would highlight Kuriyan's privileged upbringing (not unlike her own), which made her an ideal candidate for the award: 'Miss Kuriyan comes from a very cultured Syrian Christian family in Travancore and she has therefore the background of broadmindedness and tolerance which will make her stay in the West very profitable both for herself and those who will come into contact with her . . . I feel quite confident she will be a great asset to any University abroad both socially and academically.'[21]

Teaching, Research and Social Life

On return to Madras in late April 1926 with a Master of Science (Research) degree from the University of Michigan, Janaki resumed teaching at the WCC. The biology students numbered five that year and they were elated to have her back. She had just purchased T. H. Morgan's *The Physical Basis of Heredity* (1919), which would be her constant companion for some time to come. It is possible that Janaki incorporated some of Morgan's radical ideas in her lectures, and brought Mendel and Darwin to their attention

45

for the first time. The WCC had inaugurated the new science block in the late 1920s (construction had begun in 1923), complete with laboratories to attract girls to experimental science.

The staff of the WCC now had a new motor car for their use, made possible by the generous benefaction of the Mount Holyoke College. It was a Chevrolet, like the one owned by Principal McDougall, but only younger; the two cars would henceforth be known as Jack and Jill. The 'joint ownership and control of Jill' was thought to be 'an interesting experiment.' The College also kept an old horse called Atlanta, a 'faithful animal who coldly repels all demonstration of affection but serves . . . loyally.'[22]

On her return, Janaki was elected (along with C. K. Kausalya) to the Madras University's Board of Studies for Natural Science and the Academic Council (1926–28); members of the Board included Janaki's teachers T. Ekambaram and P. F. Fyson, besides M. O. P. Aiyengar and M. S. Sabhesan, James Pryde (Principal of the Maharaja's College of Science, Trivandrum), M. A. Sampathkumaran and S. Sundararaman (both of the Bangalore University); and C. Tadulinga Mudaliar, T. S. Venkatraman and Diwan Bahadur K. Rangachari of the Agricultural College, Coimbatore. Janaki's colleague, Eleanor Mason, Professor of Zoology at the WCC, was also elected to the Board of Studies; she had obtained an MA degree from the Wellesley College, Massachusetts, and was keen on including physiology in the syllabus.[23]

At this time in Madras, hardly any of the staff in the departments of Zoology and Botany were trained or even familiar with the fields of cytology and genetics. In a letter to Aby Howe Turner, Professor of Physiology at the Mount Holyoke College (who had only recently received her doctorate from Harvard, and given an evolutionary and evangelical twist to the science),[24] Mason remarked: 'The Board [of Studies] is too funny! There are 12 members. I am the only woman on it and distinctly the youngest of the crowd. There is one other European, an Englishman [F. H. Gravely] who is in charge of the Government Museum. The whole crowd are through and through systematists, and I sit on the side lines and groan! I wish I were an authority on anything! I do think I know more of the genetics than any of them but that subject simply does not exist for them.'[25]

Janaki would nurse a similar resentment, for none among the Board of Studies for Botany were cytologists or geneticists, except herself; they were all taxonomists of the classical kind. The research-oriented Mason indicated in her letter that there were plans afloat to establish in Madras an 'Institute for Scientific Research' under the management of the University, which would offer research opportunities for Honours and graduate students and members of staff, if they wished to use it. Initially, the focus was to be on research in marine zoology, Madras being on the coast. Mason was however not too optimistic: 'At the rate things move in India, this will

46

probably be ready when I am gray-haired, but at least it is being talked about!'[26]

Sometime in September, Janaki was twice down with influenza; the second and more severe attack occurred just after her return to Madras from a botanising excursion in the Godavari region in the company of her forester brother. She had by this time begun to feel 'rather discouraged,' for she felt her college was not 'particularly empathetic' towards her. 'I would appreciate a little coaxing and gentle words,' she would remark. 'She missed the American mode of life and her congenial friends' and found it challenging to continue as a teacher anymore. Janaki felt that Michigan had 'made her younger' so much so that she found it 'rather hard to be "good and nice and old".' The 'mystical side of herself' had vanished, and she had even begun to consider 'religiousness as an abnormal state of mind': 'Hindu ceremonies do not appeal to me and Hindu philosophy is so bold it sometimes awes me.' The West had made her 'more individualistic,' Janaki believed. If America was 'a new world for [her],' India was now 'even more of a New World. Altogether it is a muddle. I am neither East nor West—I often laugh at myself,' she would comment in a moment of self-reflection. The only joy in life was her research, but she felt 'very much alone since there [was] nobody to discuss it with.'[27]

Among the new students at the WCC that year (1926) was a sister of the Maharaja of Travancore and a princess from the palace of the Nawab of Arcot, Rahamathunissa Begum Sahiba. Miss Mariam Oommen, the chemistry lecturer, also taught the 'Beginner's Scripture' class. When the new science building was completed towards the end of 1926, it was Miss Oommen with the assistance of Miss Aley George and Miss Thaniammal Joseph who worked hard to set up the laboratories. Soon after, Oommen would leave for England to pursue an advanced course at the Imperial College of Science. At the WCC, progress had also been made in teaching Physical Drill, thanks to the arrival of Mrs Marie Buck, wife of Harry Crowe Buck, the YMCA Physical Director, who had returned from America.[28]

Every summer, members of the WCC residential staff would go on holiday, usually to the hill stations of Kodaikanal and Kotagiri, for rejuvenation, trekking and plant hunting. During the Christmas vacation of 1926, a few went on a trip along the West coast in a 'dilapidated and maltreated' Ford bus; from Mysore, they travelled westward, passing through pleasing scenery and then they climbed higher into the Western Ghats and were soon surrounded by thick forest on both sides, with tall trees and giant ferns. Janaki, who had returned from Michigan in April that year, joined them on the last leg of their journey; their destination was her house, Edam, in Chetamkunnu, Tellicherry. Dorothy E. Williams, the new Zoology Professor at the WCC, also on the tour, described Tellicherry as a 'typical West Coast city being more like a large spread out village with each house more or less of an ancestral estate with its own compound, the house often being obscured by coconut palms and plantain trees,' while Janaki's house was

> a characteristic Hindu establishment delightfully located among palms, rice fields and with even pepper plants and coffee trees growing in their own compound. The house itself is, according to the usual Hindu plan, built around a middle room, which is a ceremonial center where weddings are celebrated and funeral rites observed, a room on the north and south respectively, and a long east room into which the other three opens where the ordinary family gatherings for meals and social intercourse are held. This last leads on to an extensive verandah where at dusk the charming night lamp is lighted around which the children of the household gather for their song of evening worship.[29]

On their first evening at Edam, Janaki arranged for a 'devil-casting' performance as an anthropological curiosity. In fact, she offered herself as the subject, 'even though she was as amused at the whole affair' much as they were. Janaki sat at the edge of a 'magic circle' of ashes 'with her feet and hands pointed outwards so that the evil might readily flow outwards from her fingers and toes.' The *velichapadu* (sorcerer) solemnly performed the rites of 'extracting the evil from his subject and transferring it to the little pink effigy of rice and colouring matter containing a hair, a toe-nail and a finger-nail of Janaki!' Her love for anthropology, like that of her father, would stay with her throughout life, even becoming part of her research methodology in plant cytogenetics.

With Tellicherry as their headquarters, the WCC group travelled around, including a short trip to Mangalore. On their return to Janaki's house, 'they tarried long enough to enjoy [the] Christmas dinner, a typical Malayalam meal' which, draped in saris and seated on the floor, they ate off plantain leaves, with their fingers.[30] After a most enjoyable time at Edam, the group headed to Kuttanad in central Travancore, where they travelled on a *vallam* (country boat) to visit the house of another colleague, the chemistry instructor Aley George.

During the whole of 1927, besides teaching at the WCC, Janaki was busy making a collection of *Nicandra physalodes* chiefly from the Wayanad plateau, in preparation for her doctoral research. She felt her work was not extensive enough; she had got the F2 seeds ready for planting, but several were not pure strains, and so feared she would have to repeat the process. Janaki began desperately scoping for more options to do her doctoral research, as she was unsure her application to the University of Michigan for a fellowship would receive a positive response, despite it being strongly recommended by her mentor at the Department of Botany,

Bartlett. Amiable by disposition, an effective communicator and a highly focused individual, Janaki had been in correspondence with Bartlett ever since her return to India in 1926. She however became anxious when a letter posted to him at his temporary address in Formosa (Taiwan) through the American Express Co. was returned, although she was aware he was on a long collecting tour in Southeast Asia. Fortunately for her, the England-born cytobotanist Eileen Jessie Whitehead Erlanson (1899–2002),[31] her friend at Ann Arbor, updated her with news of the department, and of Bartlett in particular, which relieved her somewhat.[32] Eileen was at this time wrapping up her doctoral research on the genetics of the *Rosa* under Bartlett's guidance. In 1928, when she earned her doctorate, she was one of the earliest women anywhere in the world to achieve this milestone in plant genetics (McClintock had been awarded a doctorate only a year before), and surely the only one on the *Rosa*, and yet has not received the recognition she deserves, what with a mere mention in the *American Men & Women of Science*.

At Tellicherry

Janaki was at Tellicherry on vacation in mid-June, when she heard from Bartlett; her joy knew no bounds when she discovered *Solanum* seeds enclosed in the post. She had herself made 'quite a large collection of egg plants as well as local species of *Solanum*' and intended 'to do more cytological work.' In her reply, she let out her plan to 'bring out a Flora of Malabar' (towards which she had been collecting in and around Tellicherry), but the onset of monsoon was making the preservation of specimens a big challenge.

Meanwhile, Janaki heard from the Local Barbour Scholarships Committee that Carl Rufus, head of the Barbour Scholarships Committee, University of Michigan, wished to meet with her during his impending visit to Madras; the aim was to assess the progress she had made in her research. Janaki wondered if it had to do 'with the grant of 100 dollars' they had given her the previous year; she had not yet applied for a grant renewal. She longed to return to Ann Arbor 'someday,' but sensed her college (the WCC) was not 'very keen about it.' The WCC was however a 'jolly place to work,' she quickly clarified to Bartlett: 'We have very well-equipped laboratories and I am very happy in them.'[33] That year (1927), she would visit brother Raghavan in Burma; for her, it would be the realisation of a long-cherished dream.

Corresponding with Mary Agnes Chase

Incidentally, by this time, Janaki had begun a correspondence with the American grass expert, Mary Agnes Chase (1869–1963). We know she had started collecting and studying the grasses of North Malabar diligently even before she had left for America the first time (1924). She had however become frustrated 'with no facilities for identification and very little encouragement from S. Indian Botanists.'[34] Grasses were indeed difficult to know, despite man having used them for thousands of years. When she mentioned this to Bartlett during her first stint in Michigan, he introduced her to Chase, by way of a letter. Janaki was unable on that occasion to travel to Washington to meet Chase but on her return to Madras (in 1926) was met with a letter from the expert, offering her services in support of her interest in grasses. As an encouragement, Chase would send Janaki an autographed copy of her book, *The First Book of Grasses, the Structure of Grasses Explained for Beginners* (1922). Despite her lack of institutional power owing to her gender or even a doctorate for that matter, Chase had built a network of women agrostologists and serious grass collectors across the world, several of who she mentored.[35] Janaki was one of them, and perhaps the only one in South Asia to have kept up a regular correspondence with her throughout the 1930s.

When Chase's book reached her in Madras 'for a few days [her] brain was very active devising ways and means of finding more time to devote to [her grass] work.' Deeply touched by Chase's gesture, Janaki wrote to her:

> I feel very highly honoured by the gift especially after reading Prof. Bartlett's introductory letter. I wished I had met you when I was in America. I hope I shall have that privilege some time in the near future. Your promise through Prof. Bartlett to identify any grass I may send you encourages me to take up once again my collecting hobby. . . . Your book and promise to help have given me a new impetus.

Janaki also enlightened Chase about her college (the WCC, where she was teaching):

> I am teaching in a small Women's College which is very much indebted to America for its existence. Our new Science building is the result of your country's generosity and due to that we are the only Women's College that makes it possible for a girl to take a science degree without going to men's colleges.[36]

Born in Iroquois County, Illinois, Chase specialised in the study of American grasses and conducted extensive fieldwork in South America at a time when few women dared to, sometimes even funding her own research trips. In 1903, she joined the United States Department of Agriculture as a botanical illustrator and then became Scientific Assistant in Systematic Agrostology. Chase collaborated closely with Albert Spear Hitchcock (1865–1935) publishing *The North American Species of Panicum* in 1910. In 1925, she became Associate Botanist and, after Hitchcock's death in 1935, was appointed Principal Botanist in charge of Systematic Agrostology and Custodian of the Section of Grasses (Division of Plants), United States National Museum of Natural History of the Smithsonian. In Chase, Janaki found an inspiring role model, of a world-class botanist and an intrepid explorer, who went on solitary collecting expeditions, climbing treacherous mountains in search of grasses, and was fearless in speaking her mind; Chase was a suffragette, a socialist and a pacifist, and a lover of all things Brazil, a country she would frequent throughout her life.[37] Janaki would continue to correspond with Chase until the latter's retirement from the US Department of Agriculture in 1939.

Due to 'scarcity of water supply' in Madras Janaki had not dared to 'sow [her] precious F2 seeds of *Solanum*' [sent by Bartlett in mid-1927] but feared for them as eggplant seeds did not survive very long: 'until I find they are alright, I shall have no peace,' she wrote to Bartlett. 'We are just getting into our hot weather. Schools and colleges close for the summer at the end of this month and Madras will be unbearably hot and dry till August. I shall therefore spend the summer collecting in Malabar. I did quite a considerable amount last summer but most of the specimens are lying unidentified and unmounted. Sometime I shall have to have a year off and work on my collections.' Reiterating her aim to produce a 'Flora of N. Malabar,' she told Bartlett she was very glad she now had Chase to help identify her grasses.[38] Janaki would group plants separately under the headings, 'Flora of Chingleput District' (being stationed in Madras, she would go out collecting in its suburbs and beyond, including Gingee or Senji in the erstwhile South Arcot District) and a 'Flora of North Malabar' (in the region where she spent her vacations, at Tellicherry and Shoranur in particular), which would together form her 'Flora of South India,' and this included grasses.

Too Many Distractions

Janaki however complained that she was 'being pulled more and more away from quiet work'; this would become a constant refrain in her life. As a member of the Board of Studies, she had to take on the extra tasks the university periodically assigned her, including serving as examination invigilator, besides she was in the thick of 'many social reform movements' that were appealing for support. There were times when Janaki felt 'particularly miserable' when she could not 'get to a corner and examine' her *Solanum* slides in quiet. Moreover, she was hoping to read a paper at the next meeting of the Indian Science Congress (1929) in Madras, which was to have physicist C. V. Raman as President.

'Some day I want to get back to Ann Arbor. I wonder if I shall,' Janaki remarked to Bartlett, yet again. Eileen had been giving her news of the department but to her dismay had not sent her a copy of Bartlett's 'letter to the Journal Club from Sumatra,' which she much wanted to read; it was on his second expedition to this Indonesian island (1927) that he had fallen in love with the culture and language of the Batak of Asahan. She had fervently hoped Bartlett would visit India on his return journey, to meet with some of the Indian tribes, but this was not to be; in fact he would never visit India in his lifetime. 'I hope you are coming some day— and please come before India gets too civilised. I found all the Kurumbas and Todas going about well-dressed—and I was so disturbed to see them get less interesting,' she wrote. In fact, at a 'social reform meeting' Janaki had spoken against marriages between castes and tribes, 'on the plea it would spoil the homogeneity [purity] of the races,' so much so that she was attacked by some for wanting to reduce India to 'an anthropological museum.' She commented to Bartlett, 'Missionaries are making all our men into good Christians—please come before they get unrecognisable. Of course, with your head-measuring you may be able to sort the types.'[39] It may be pertinent to note that Eileen on one occasion remarked on Janaki's 'anti-miscegenist ideas.'[40]

Some Family Time

In February 1928, Janaki's forester brother Kittu, an EAC of Forests attached to the Wayanad division, was placed temporarily in charge of the Sultan's Battery. On the 5th of that month, younger brother E. K. Madhavan employed with the Madras Fisheries Department was married to Kanthi, daughter of Rao Bahadur Panangadan Raman, who had been their father's (Sub-Judge E. K. Krishnan) good friend and neighbour. All members of the family attended except Janaki and younger brother Varadan, but she would pleasantly surprise them all by turning up on the following day. In fact, she had spent the whole of her Christmas vacation at Edam, and only returned to Madras a few weeks ago, after a spot of intense collecting in early January, at Tellicherry and Beypore; the plants collected included a *Euphorbia thymifolia* and *Euphorbia hirta*. A family photo on the wedding day, taken against the lovely green wall at Edam (covered by the *Thunbergia grandiflora* planted by Sub-Judge E. K. Krishnan), a constant feature of all Edathil family pictures, was a particularly cheerful one, as the lot had laughed uncontrollably, amused by the mute photographer's actions.

Applies for Government Scholarship

By early March that year (1928), unsure of what lay ahead, Janaki decided to turn in an application for a Government of Madras Scholarship, instituted for pursuing an advanced course of study in England; two scholarships were being offered, one each for Botany and Zoology, for a period of two years and with the prospect of extension for a third year if the work was found 'satisfactory.' For courses at Oxford or Cambridge, the scholar was to receive £300 plus a 'cost of living bonus' to the tune of £45, while for other Universities, it was £250 plus a sum of £40. In addition, women scholars were provided with second-class passage to and from England. On account of there being only a 'few women in S. India who would apply for it,' Janaki imagined she stood 'some chance of getting it,' that is if they did 'not decide to give it to a man'! She requested Bartlett to send her a 'certificate of capacity for research' but only if he believed she deserved one.

> I want very much to continue my work in Genetics & Cytology either at the Royal College of Science [Imperial College] or Cambridge. Please also give me some advice as to whom you would like me to work under . . . in case I am fortunate enough to get the scholarship. England is but one step to America and I cannot help but dream a few dreams.[41]

America would remain her dream destination, if ever there was one.

Collecting in Ganjam

A month later, in early April (1928), Janaki travelled alone to Chatrapur in the Ganjam district of Orissa, where forester brother Kittu was now posted. Both, botanists, they would wander tirelessly in the Sal forests of the district looking for plants. One day they drove up to Gopalpur and on another, to a fishing village; on several days they collected plants in the Reserve Forest. Kittu was passionate about orchids. His eldest son Hari had now taken Janaki's place in indexing the plants collected by him. On her return to Madras, Janaki wrote to Bartlett about the Ganjam visit, a letter that reveals once again her growing interest in anthropology:

> I have just come back after camping a few days with my brother in the Sal forests of Ganjam. Both the flora and the people are most interesting. I had a good opportunity of studying the Khonds at close quarters. There [are] very distinct Mongolian characteristics in some of the tribes and I am bringing (to wear in the [botanic] Garden) a hat they wear which is very Chinese.[42]

It was as if she was almost sure she would make it to America soon.

Back in Madras, she would do more collecting; she had already collected some Euphorbiaceae in the month of March. Some of these including the flowering plants, *Tragia involucrata* (the Indian stinging nettle or *choriyanam* in Malayalam) from Vandalur, and the succulent, *Trianthema triquetrum* collected from the Madras Beach, she would eventually present to the University of Michigan Herbarium, as part of her 'Flora of South India.' She would sometimes record her name in the label on the herbarium sheet as 'E. K. Janaki Ammal' and at other times simply as 'E. K. Janaki,' and when the desiccated specimen contained seeds, care would be taken to place that part of the plant within a transparent pouch, which would then be hemmed to the sheet.

Awarded a Barbour Fellowship

In fact, it was while Janaki was in Chatrapur that the grant of fellowships 'for Eastern Women of Noted Achievement' was announced by the University of Michigan. The three appointees were Miss Lucy Wang, Dean of Hwa Nan College, Foochow, Miss Sugi Mibai of Kobe College, Japan and 'Miss E. K. Janaki of Women's Christian College, Madras.' The Michigan newspaper reported:

> Miss Janaki took her Bachelor's degree with honours at the Presidency College, Madras, India. She came here as a Barbour Scholar from 1924–26, residing at the Martha Cook dormitory while here. She received her A. M. in 1925 but since she was on a leave of absence, she felt bound to return to her teaching position before completing her doctorate. Since 1926, she has been teaching and studying botany. 'Keen interest is felt in her acceptance of the fellowship.'[43]

A group of Barbour Fellowships had been instituted just that year (1928) 'to be awarded upon invitation to Oriental Women of noteworthy achievement.'[44] They were worth a lot more than the Barbour scholarships and were intended to provide for a year's leave of absence with an opportunity to use the university classroom, libraries and laboratories for research. Janaki was overjoyed. On 23 April 1928, when Kittu received the news that she had been awarded a Barbour Fellowship to the tune of $2000 to complete a doctoral thesis at the University of Michigan, he would promptly send her Rs 32 for the purchase of saris and other requirements, ahead of her travel.[45] In June, she would be granted a $200 advance to help defray expenses in preparation for the trip to America.

Janaki lost no time in writing to Bartlett to thank him 'most of all for this great opportunity to return to Ann Arbor.' She could hardly believe that she would be back in Michigan in a matter of just months. 'Please let me know if there is anything botanical or otherwise that you or any of the department wish me to bring,' she wrote. Janaki was planning a contribution to the University Herbarium of 'a small collection of S. Indian plants . . . and also seeds.' The Herbarium was one of the four main divisions which constituted the institution called the University Museums and was established in 1921. It was located in the north wing of the fourth floor of the Museum building on Washtenaw Avenue and was well equipped for all aspects of plant research; C. H. Kauffman was director of the Herbarium, and J. H. Ehlers the Curator. In 1929, a collection of specimens collected by Bartlett from Sumatra would find a place at the Herbarium. 'I wish I could bring some live plants,' she wrote to Bartlett, but this was impossible owing to the severe restrictions imposed by the United States. As she was to leave India soon, Janaki decided to 'roam around Malabar' rather than go up to Sampathkumaran's laboratory in Bangalore to work on her *Solanum* slides, as earlier planned.[46]

Hears from Bartlett

Early that month, she would receive Bartlett's response, which assured her that it was her own merit that had attracted the Fellowship: 'I cannot take any credit for securing the appointment for you, since that was done at the initiative of the Barbour Scholarship Board and I did not even know that it had been done until it was announced in the paper. I had of course been asked if I really wanted you to come back and thought you were eligible for additional assistance, however, that anything so good in the way of an appointment was at all possible, I congratulate you heartily.'

Bartlett discouraged Janaki from bringing 'any live plants to the Botanical Garden on account of the quarantine restrictions' and suggested instead 'to confine' herself to seeds, which could be carried 'freely and without difficulty.' 'As for Herbarium specimens you know we have nothing from India and would be delighted with anything you could bring, even common species . . . send them by mail . . . we could reimburse you for the postage,' he advised. Bartlett was keen to learn more about Janaki's Ganjam explorations; he wrote to her, much like the ideal mentor he was: 'If you made any ethnobotanical notes during your work in the Sal Forest of Ganjam you should bring them with you, together with any photographs you may have taken,' in order to work them up into 'a little paper for the Anthropological Section' of the *Michigan Academy of Science, Arts and Letters*. 'So perhaps you will branch out a little bit into Anthropology as well as botany,' Bartlett suggested.[47]

A New Botany Professor at the WCC

Meanwhile at the WCC, Janaki's position of Professor of Botany was occupied by the much older Alma Gracey Stokey (1877–1968) from Mount Holyoke. Stokey had obtained her doctorate on fern gametophytes from the University of Chicago, under the supervision of John M. Coulter and W. J. G. Land in the first decade of the twentieth century. It was when she

was about to conclude her research on the *Cyathea* that she had heard of
WCC's hunt for a competent teacher to reorganise the botany department.
Losing no time, she offered herself up for the post and set out for Madras on
a two-year leave of absence. Stokey showed great enthusiasm for teaching,
was a tireless plant collector, had a wonderful sense of humour and nurtured
varied interests, including classical music, and while in Madras never failed
to attend concerts. In the late 1930s (1936–37), she would return to Madras
to teach for a year; on her way back to America, she would break journey
at Jakarta and collect extensively; she had hit sixty by this time. From the
Dutch Buitenzorg (also called Bogor, near the volcanic Mount Salak in the
south of Jakarta, a region renowned for its biological diversity), she would
write in July 1937, a letter that would demonstrate the extent of her passion
for botanical field exploration even in her advancing years, and an outlook
that would resonate with Janaki herself: 'It was a most enjoyable excursion. I contrasted it with the expedition in Hakgala [Nuwara Eliya], Ceylon,
where my attendants were more concerned with keeping me intact than in
finding ferns . . . no one seemed unduly concerned [on Mount Salak] about
my skin or bones and did not help me unnecessarily. I would rather be a
Botanist than a Lady,' wrote Stokey.[48] Besides Stokey, on the botany faculty
of the WCC during the time Janaki was away in Michigan, was the future
Barbour Scholar, Maria Francisca Theresa Swamikannu (later Thivy),
daughter of a noted Madras lawyer. Francisca (1904–89) joined the Department as lecturer in 1931.[49]

Leaves for America a Second Time

Janaki intended to leave for Ann Arbor within a month. She planned to
break journey in England (and Switzerland) and much wished friend Eileen
'would come searching for wild roses on the continent!', in which case, they
could travel to Michigan together. Eileen was at this time a National Fellow
in Botany. Janaki's journey would however be delayed by a month or so. In
late August, she would board the Cunard Line's *RMS Berengaria*, one of
the most popular liners of the time, 'the one on which the Prince of Wales
went to America,' she would enlighten brother Raghavan. Once the vessel
left Bombay, it called at Port Sudan, Suez, Port Said, Gibraltar, Liverpool,
Southampton, Queenstown, Galway and Boston, to finally reach New York,
about two months later. From the ship, Janaki wrote letters to her family,
describing life on board and of her plant-collecting adventures in some port
surrounds. In a long letter to Raghavan, she wrote: 'Here I am on the Atlantic. In about 6 days I shall be at the end of my long trip. It has been very nice
up to this except for a brief spell of fever I had after the boat left Naples.
I must have caught the "flu" that was raging along the Mediterranean.' She
had to disembark at Galway (Ireland), and take another boat to England to

reach 'the lovely district of Devonshire,' where her friend Doris Mary Hold-rup (recipient of a Frances E. Riggs Fellowship at the University of Michigan) was spending time with her family. They had excellent weather throughout, and Janaki found the seaside town of Torquay 'beautiful.' With Doris, she even went up to London and 'spent a few very enjoyable days' there.

To her utter joy, Janaki found letters from her family awaiting her at Holdrup's house. She was especially happy to see a couple from her mother, who had been unwell when she bid her goodbye. Janaki had begged brother Raghavan to allow sister Sumithra to remain at Edam (and not leave for Rangoon where her husband lived), until Janaki returned to India. She had then remarked:

> Edam would be rack and ruin without Sumthi [Sumithra] to take care of it. . . . I do hope all the brothers appreciate her work for the house . . . her methods might sometime clash with your ideas but in the long run you will realise they are for the best. With just the minimum as far as money is concerned Sumthi is keeping alive the position & status that was ours as the children of Diwan Bahadur E. K. Krishnan. We have somehow—with the Grace of God kept it in spite of many falls . . . and really when you look back at it all you will realise it was mostly due to our great mother—and to Sumthi's strength of character. I have contributed a little money towards the education of the boys & girls—and that was all. My hope for Edam is that some day we shall be looked upon as the most enlightened and grandest of Thiyya *tharavads*—upholding the best in the tradition of our land.[50]

Reaches Ann Arbor

On 8 September 1928, Janaki resumed her journey to New York, from Southampton, 'comfortably settled in an II class berth cabin with Barbour scholar Miss Kuriyan [Aley Checha Kuriyan], the Travancore girl and a former student of the Women's Christian College.' We might recall that it was Janaki who had recommended Kuriyan to the Barbour committee. There were a number of other Indian students on the boat as well; one of them, Janaki informed her brother, was a boy who had studied with her for an Honours degree in Botany at the Presidency College. More than a month later, on 15 October, the ship touched New York, from whither Janaki travelled by train to Ann Arbor. At the University, she would reside once again at the Martha Cook Dormitory. The Social Director of the dormitory, Elva M. Forncrook, with whom Janaki enjoyed a congenial relationship, would soon be replaced by Miss Ethel G. Dawbarn. The dormitory regularly published

The Martha Cook Annual, carrying humorous quotes from or on its members, such as this one on Janaki:

> Janaki (on receiving her box of Old Golds): 'They say that smoking injures the germplasm but I am not going to use mine anyway.'
> 'Just what do you mean Janaki?'[51]

She expected to take two years to complete her doctoral thesis; the Barbour Committee after some deliberation agreed to extend her fellowship until 1929–30.[52]

Edam as Her Anchor

From Ann Arbor, early in 1929, Janaki sent her eldest niece E. U. Shantha a copy of H. G. Wells' *The Outline of History: Being a Plain History of Life and Mankind* (1920), a book which would have considerable influence on the teaching of history at higher educational institutions in the West. Forester brother Kittu was at this time at Edam, making a collection of plants for Janaki; his nieces would help him organise the specimens for despatch to Michigan, while elder son Hari would help with the indexing of their botanical names. Janaki's family, it might be noted, rather than impede, was a great resource for her science; botany was a familial enterprise as far as Edam was concerned. Her siblings, nieces and nephews would contribute enormously by collecting, preserving, labelling and despatching plant material to her across the world or wherever it was that she needed them; some of them would also join her on collecting excursions and fieldwork, which were sometimes nothing less than adventurous. As we journey through Janaki's life, we will also obtain a good measure of the lives of her interesting siblings, who were united in their love for nature, the outdoors and literature. Janaki's voluminous correspondence with her family (she however did not keep a daily journal) provides us with many insights into her family, which might seem inconsequential at first, but are key to construct a more rounded picture of her life, which had Edam at its core. Although privileged, the family suffered, it must be noted, from the social stigma of being 'white-stained.' It was a dark secret, of immense interest to Janaki as a geneticist, but a fact to be closeted from the younger generations of the E. K. family even; sometimes coming to terms with the hybrid nature of her family's origins posed a much bigger challenge to her, than the complex botanical puzzles, which her science attempted to resolve.

In late April that year, Kittu would hear from Janaki; she had enclosed a newspaper cutting in which a group photograph of the Barbour scholars residing in the campus had been published. The group included the Indians Achy Iype, Sharkeshwari Agha and herself. Janaki was at this time just about to leave for West Virginia, located in the Appalachian region, perhaps on a holiday with Eileen, who was now teaching at the Kent State College in Ohio.[53] In fact, by this time, the prolific Eileen had published a *Flora of the Peninsula of Virginia* (1924).

Campus Life

In Ann Arbor, a banquet was held in aid of the Barbour Scholars and Fellows at the new Michigan League Building on 22 May 1929, which included an after-dinner programme where Janaki, Sugi Mibai and Lai-wing Fung responded respectively on behalf of the women of India, Japan and China. Janaki spoke on Barbour Scholars in India.[54] An active member of the Cosmopolitan Club and the Hindustan Club on campus,[55] she would imbibe her first lessons in internationalism as an Oriental Barbour Fellow, fully embracing the idea of 'world citizen' which the Fellowship epitomised. It was a worldview to which she would tenaciously hold on until the end of her life. On no public occasion would she identify fanatically with a caste, race or nation. In fact, in a letter to her elder sister Parvathi, Janaki would be vocal about the need to think beyond the Hindustan Club and form an All-Asian sorority (a society of female students on University campuses); she had after all met several Chinese and Japanese Barbour scholars on campus, and was convinced they shared similar cultures, concerns and ambitions:

> I realize what a lot Asia has in common. You know, I have started an organization that is going to link the University Women of Asia— we are still in an embryonic state—but I am getting new members for all quarters. It is my dream to send some Indian girls to study in China and Japan and have girls from these countries to come to our country. I have just had an invitation from a college in China to teach botany. Of course, all this means money, but I feel that it will come somehow.[56]

Unfortunately, nothing really came out of this, but without doubt, this world-view would find expression in her approach to cytotaxonomy, but again more on this later.

Doctorate in Botany

Between 1910 and 1912, Bradley Moore Davis (1871–1957), Janaki's doctoral supervisor, had demonstrated that the *Oenothera* mutations actually followed Mendelian laws of inheritance. It would later be discovered that what de Vries witnessed were not mutations after all, but only recombinations of existing genetic materials; in other words they were hybrids, and not species.[57] In the early twentieth century there was much confusion over the identity and meaning of the terms 'hybrid' and 'mutant' and biologists were extremely guarded about making hasty claims.[58] When Bartlett joined the University of Michigan as an assistant professor in 1915, he began to plant *Oenothera* at the newly established botanic garden for purposes of research. The garden provided the 'best facility in the country for work in genetics and plant breeding.'[59] He was assisted in the management of the garden by the pioneering animal geneticist Frieda Cobb Blanchard (1889–1977), who had completed her doctorate in 1920 and married herpetologist Frank Blanchard. Unfortunately, Frank would meet with premature death in September 1937, leaving her alone to raise their three children.[60] Frieda was the first to demonstrate Mendelian inheritance in reptiles. Together with Bartlett, whose graduate student she was, Frieda developed the garden as a major centre for research in *Oenothera* genetics. Janaki however did not choose to work on the *Oenothera* despite the genus being the focus of both her doctoral supervisor B. M. Davis and her mentor, Bartlett; the two men were the most committed supporters of the mutation theory in the early twentieth century, along with R. R. Gates and Theodor J. Stomps. Incidentally, Chandrakant Ganapatrao Kulkarni, one of Janaki's Indian batchmates, unlike her, worked on the *Oenothera* under Davis.[61]

Garden and Laboratory

Janaki's subject of research was the *Nicandra*. The *Nicandra physalodes* of the nightshade family was the only species within the genus *Nicandra*; the plant was native to Western South America and known by the common name, Apple of Peru. Her doctoral research chiefly involved fieldwork at the University Botanic Gardens and microscopic study at the department laboratory. A few lines on the history of the Gardens will be useful information. It was situated on campus on Packard Street, in the heart of Ann Arbor's south side, where today's Graduate Library of the University of Michigan stands.

It was Asa Gray (1810–88), the great botanist associated with the University from its inception, who procured books for its first library, organised the original campus and transformed the twenty acres on the east into a botanical garden. The Dean, chemist Julius O. Schlotterbeck (1865–1917) of the College of Pharmacy, later laid out a garden on the campus near the library, where he grew medicinal plants to aid his teaching. This garden evolved to become a fifty-two-acre botanic garden, off Iroquois, playing a very important role in University life.[62] Its first director was Prof. Henry A. Gleeson (1914–19). Plants were grown for research and teaching chiefly, but also provided ornamentals for University functions. The actual work of raising plants was done by excellent gardeners. The garden closest to the greenhouse was a large oval, with horizontal beds divided according to family and genus, referred to as 'the graveyard' because of the arrangement. The land beyond 'the graveyard' was set apart for specific research projects such as Eileen Erlanson's wild roses (the University Gardens were the only ones in which 'the wild origin of every variety of rose [was] known'), Janaki's *Nicandra*, Kenneth Jones' ragweed, E. G. Anderson's Indian corn, E. E. Dale's red peppers and Felix Gustafson's tomato plants. Frank Blanchard's snakes were kept in cement-lined pits, where some plants were wintered.[63]

When Bartlett went on plant expeditions, which he often did, Frieda Blanchard would become the chief administrator of the garden, maintaining facilities and an atmosphere conducive to scientific research. Janaki adored the Blanchard children; she especially doted on the girls Dorothy and Grace and on one occasion had little dresses made out of a sari for them. Interestingly, Janaki insisted on wearing a sari (with boots) even while on fieldwork. Dorothy Blanchard recalls her mother asking her father one day to take Janaki along on fieldwork, to which Frank Blanchard responded: 'if she dressed appropriately.' Frieda accordingly provided her 'with suitable clothes—knickers, skirt etc' but Janaki refused to wear them because 'she felt naked' in them[64] Biology students and teaching staff invariably spent summers at the biological station of the University of Michigan at Douglas Lake, a scientific and social camping life, which they much looked forward to. Janaki's work was centred around the University Botanic Gardens, where her *Nicandra* fields were, but in 1930, she decided to spend a week at the camp, taking in all the natural beauty of the Michigan peninsula. She described the geography of the place (where she had the occasion to witness the Aurora Borealis), and life at camp, in a letter to elder sister Parvathi, her interest in anthropology evident again:

This is a lovely place—my back yard is the lake shore and I just have to run in for a bath when I feel inclined. There are small wooden houses scattered about the place with just the necessary camp furniture—for both students and staff. Most of the professors bring their family. So this is a very interesting sociological colony. I have a guesthouse all to myself and am having a real rest . . . We all have a common dining room. All the classes are in the form of excursions and it is like one grand picnic. There are motorboats and trucks to take the students about. I think only Americans can plan things on such a scale. There are only about 150 students. But when you consider that we are cut away from civilisation, you can understand how difficult it must be to keep everything going . . . On all sides are lovely pine forests with Red Indian villages scattered around. The American Indians are fast dying out. They make bags and baskets and bring them for sale here. It is rather cold up here. I saw the Aurora Borealis yesterday night. It was a gorgeous sight to see the whole sky lit up with beams of light.[65]

That same year, Janaki would be elected a full member (advanced from associate member) of Sigma Xi, as well as of the American Association for the Advancement of Science and the Genetic Society of America.[66]

Awarded a DSc

On 18 December 1930, it was decided by the Barbour Scholarship Committee that Janaki's stipend be extended by a month at a time, until she completed her work, but not to continue it beyond June 1931.[67] Janaki would however submit her dissertation, 'Chromosome Studies in *Nicandra physalodes*,' well within the time limit set by the University for her. Her dissertation would be accepted unconditionally and the University would decide to award her a DSc degree; she had now officially become Dr Edavaleth Kakkat Janaki Ammal. In her dissertation, Janaki expressed her greatest appreciation for Bartlett and Frieda Blanchard, for the assistance given to her by the staff at the University Botanic Gardens, and for Bradley Davis, her supervisor, for providing helpful criticism throughout the course of the work.[68] The Committee's Minutes of 26 March 1931 recorded that Janaki had completed the requirements for the degree of DSc and would soon be returning to India.

Janaki would leave Michigan having made life-long friends, in particular Bartlett, Frieda Blanchard, and of course, Eileen Erlanson. Carl D. La Rue of the botany faculty gifted her several books, including P. R. White's *A Handbook of Plant Tissue Culture*, V. R. Gardner et al.'s *The Fundamentals of Fruit Production* and O. W. Barrett's *Tropical Crops*. On the eve of her departure, a friend Martha gave Janaki a copy of Walt Whitman's *Leaves of Grass*, inscribed: 'To E. K. Janaki, DSc, with Fondest Aloha, Martha.' Lois Ehlers, wife of John H. Ehlers, Curator of the Herbarium, in charge of the flowering plants, presented her with *The Travels of William Bartram* (New York, 1928), signed by her thus: 'To Janaki, so she will remember all about Asa Gray!'[69] Among the books Janaki purchased for herself at Ann Arbor was Neltje Blanchan's *Nature's Garden* (1900), whose writings were known for their mix of scientific interest and poetic articulation.

For the family at home, letters from Janaki had become few and far between; on 15 January 1931, Kittu, Janaki's elder brother eventually heard from her and a couple of months later, on 19 March, sister Sumithra received the happy news of her being awarded a doctorate, followed by a letter enclosing photographs, and announcing her departure from America. Janaki was the only Indian woman who had earned a doctorate in botany, and only the second (M. M. Mehta being the first) to have achieved this academic milestone in any one of the sciences thus far, but ironically neither her family nor Janaki herself were fully conscious of the enormousness of her achievement.

Battles Rejection and Triumphs

Early in 1930, Janaki had requested the Committee to permit her to spend two months (August–September 1930) in absentia, in order to visit botanical institutions (including the Smithsonian in Washington, where Chase was based) and attending the Fifth International Botanical Congress at Cambridge, England (16–23 August). Her application was strongly supported by Bartlett, but the Committee voted against it.[70] Incidentally, it was at this Cambridge Congress, that the Russian botanist Nikolai Vavilov would speak of his explorations in Afghanistan and W. Burns, L. B. Kulkarni and S. R. Godbole from India would discuss their work on xerophytic Indian grasslands (later published in the *Journal of Ecology*).[71] The young George Ledyard Stebbins, only in the final year of his doctoral programme at this time, would also attend the Congress, where he would meet fellow American Edgar Anderson (at this time a visiting fellow at the John Innes, working on the genetics of hybrid populations of the *Iris*) and the British cytologist Cyril D. Darlington, whose insightful evolutionary approach had impressed him immensely.

Unwilling to give up, Janaki put together her sparse savings and set out bravely on a solitary American tour of scientific institutions of botanical interest. Her first halt was Washington, where she was met with the ecologist Carl Erlanson (married to the British rose geneticist Eileen Whitehead, her good friend), attached to the US Department of Agriculture, who showed her around the city, besides the Smithsonian. Janaki and Erlanson spent most of one morning at the Smithsonian, where she was formally introduced to the 'very friendly' agrostologists Albert Hitchcock and Mary Agnes Chase. She was particularly thrilled to see Chase with whom she had been corresponding since 1926, but had never met before. From Washington, Janaki travelled to Boston (Massachusetts) to see the Arnold Arboretum of the Harvard University, the oldest such public institution in North America and a world centre for the study of plants. It is not known whether she travelled alone to Boston, but she resided at 5 Gibson Terrace, Cambridge. 'The Arnold arboretum is gorgeous with *Prunus* in flower,' Janaki remarked to Frieda Blanchard in a letter written after her return to India.[72] She was however greatly disappointed that she was unable to visit Asa Gray's herbarium at the Harvard University. Asa Gray, considered one of the foremost American botanists of the nineteenth century, had been Director of the Harvard Botanical Garden between 1842 and 1873; we also know he was responsible for the birth of the University of Michigan Botanic Gardens. His classic, *Manual of the Botany of the Northern United States*, referred simply to as *Gray's Manual*, saw several editions.

Figure 3.1 Janaki and friend Doris Mary Holdrup of Devon. University of Michigan, Ann Arbor. 1925.

Source: Courtesy of the EK family.

Figure 3.2 Janaki in the *Nicandra* fields, University Botanic Gardens, Ann Arbor. August 1925.

Source: Courtesy of the EK family.

Figure 3.3 Janaki with Barbour Fellows Lucy Wong and Sugi Mibai, University of Michigan, Ann Arbor. c. 1928.

Source: Courtesy of the Bentley Historical Library, University of Michigan.

Figure 3.4 Barbour Scholars from India: Clockwise from Janaki, Achy Iype, Shakeshwari Agha and Premola Shahane. c. 1929.

Source: Courtesy of the EK family.

Figure 3.5 Picnic at Old Botanic Garden, Ann Arbor. 8 June 1929. Janaki (left extreme), Eileen Erlanson and Frieda Blanchard among others.

Source: Courtesy the Matthaei Botanic Gardens, University of Michigan.

Figure 3.6 Harley Harris Bartlett (1886–1960).

Source: Wikipedia Commons.

Figure 3.7 Mary Agnes Chase (1869–1963).

Source: Wikipedia Commons.

Notes

1 Smithsonian Institution Archives, Washington: Record Unit 229, United States National Museum, Division of Grasses, Box 4, E. K. Janaki Ammal to Mary Agnes Chase, letter dated 21 Feburary 1928.
2 Bentley Historical Library, University of Michigan (BHL, UoM): Barbour Scholarships for Oriental Women Papers 1918–69. Carl Rufus, 'Twenty-five years of the Barbour scholarships', p. 15.
3 Ibid., p. 18.
4 The office in India was situated at 134, Corporation Street, Calcutta.
5 The report missed the name of Shahane however. See *The Michigan Daily* (*TMD* from now on), 13 March 1924, and 2 October 1924 (announcing the arrival of new students from the Orient on the campus).
6 She was the sister of the revolutionary leader, Khagen Dasgupta, who studied chemistry at Stanford University in 1910, and returned to India to start Calcutta Chemicals (1916), which produced soaps, including the neem-based Margo, a quintessentially Indian/*swadeshi* product.
7 *The Madras Mail*, 4 July 1924.

68

8 Nine years before Janaki visited, in 1915, the YWCA had bought the 610 Lexington Avenue building in New York city and opened the first public swimming pool in the state.
9 BL: 'The Women's Christian College, Madras, Principal's Journal', September 1924, 10.
10 *TMD*, 12 October 1924; the report was incorrect in claiming that she obtained 'two degrees from the Madras Christian College', when it should have been the Presidency College, Madras.
11 See Michener and Reznicek (2017).
12 *TMD*, 26 September 1925 and 1 July 1925.
13 The Michigan Dames was affiliated to the National Association of University Dames and was formed in 1921, the same year that the Faculty Women's Club was started by a small group of women. The Dames were usually married women who were University students or wives of University students. http://umich.edu/~fwc/FWC_Website/STORIES_files/Michigan%20Dames.pdf accessed 13 August 2020.
14 *TMD*, 19 February 1925.
15 *TMD*, 8 October 1925.
16 *TMD*, 16 January 1926.
17 In 1928–29, Premola Shahane from Poona was studying medicine at the Women's Medical College, Philadelphia. Anandibai Joshee (1886), Sophia Johnson (1888, she was of Indian and Scottish descent), Gurubai Karmarkar (1893), Dora Chatterjee (1901), Ethel Maya Das (1908), Chumpa Sunthankar (1910), Tugabai Mary S. Kukde (1911) and Achy Iype were some of the earliest among Indian women to earn American medical degrees. In 1927, Shahane also appears to have spent time at Chicago, about the same time as Iype. All these women were connected to missionary networks; Johnson and Karmarkar for instance were affiliated to protestant American missionary societies. See Pripas-Kapit, 'Educating Women Physicians of the World', especially pp. 18–73 and 107–153.
18 BHL, UoM: Barbour Scholarships for Oriental Women Papers 1918–69, Minutes, 28 April 1926.
19 Ibid.
20 *TMD*, 29 April 1926.
21 Ibid., 26 February 1926.
22 BL: 'The Women's Christian College, Madras, Principal's Journal', September 1924, p. 15.
23 In 1934, Mason received a doctorate in physiology from the Radcliffe College of the Harvard Medical School.
24 Levine, *Defining Women's Scientific Enterprise*, p. 131.
25 In 1933, the Board of Studies for Natural Science would include Miss Anna K. Joshua, the author of *Microbiology*, a much-used textbook in college courses dealing with food science, nutrition or dietics, besides the popular, *Text-book of Botany* (1948), co-authored with her sister Susan George Pulimood. Daughters of the Travancore Judge, K. C. Joshua, the sisters spent a major part of their careers teaching in Ceylon (Sri Lanka).
26 Mt Holyoke College, MA: Eleanor Mason Letters, letter dated Kotagiri, Nilgiri Hills, South India, 21 April 1925, www.mtholyoke.edu/~dalbino/letters/text/mason18.html, accessed 12 April 2019. The Board members were actually 11 in number and included K. Ramunni Menon, D. W. Devanesan, R. Gopala Aiyar, F. H. Gravely of the Madras Museum, K. Karunakaran Nair, C. Lakshminarayanan, C. R. Narayana Rao, K. S. Padmanabha Aiyar, Y. Ramachandra Rao and B. Sundararaj, besides Miss Mason.

27 BHL, UoM: M. U. Botanical Gardens (Correspondence series), Box 12, Eileen Erlanson to H. H. Bartlett, letter dated 27 November 1926.

28 H. C. Buck (1884–1943) began the first school of physical education in India at the Madras YMCA in Nandanam; the first Indian contingent for the Olympics was selected from this school and in 1928, Buck founded the first Indian sports journal, called *Vyāyām*.

29 www.mtholyoke.edu/~dalbino/letters/text/dew07b.html

30 Ibid.

31 Eileen was married three times: to botanists Earl J. Grimes and Carl O. Erlanson and a third time, to an engineer, James B. Macfarlane in the 1930s. She had no children of her own.

32 BHL, UoM: Harley Harris Bartlett Papers 1909–1960, Box 1, E. K. Janaki Ammal to H. H. Bartlett, letter dated 19 June 1927.

33 Ibid.

34 Smithsonian Institution Archives, Washington: Record Unit 229, United States National Museum, Division of Grasses, Box 4, E. K. Janaki Ammal to Mary Agnes Chase, letter dated 21 February 1928.

35 Henson, ' "What holds the earth together": Agnes Chase and American agrostology'. Also see her 'Invading arcadia: Women scientists in the field in Latin America, 1900–1950'.

36 Ibid.

37 Chase was jailed twice and force-fed for holding a protest before the White House for women's right to vote.

38 BHL, UoM: University Herbarium Records (University of Michigan) H. H. Bartlett series, Box 11, E. K. Janaki Ammal to H. H. Bartlett, letter dated 29 February 1928.

39 Ibid.

40 BHL, UoM: M. U. Botanical Gardens (Correspondence series), Box 12, Eileen W. Erlanson to H. H. Bartlett, letter dated 30 May 1934.

41 BHL, UoM: University Herbarium Records (University of Michigan) H. H. Bartlett series, Box 11, E. K. Janaki Ammal to H. H. Bartlett, letter dated 8 March 1928.

42 Ibid.

43 *TMD*, 6 April 1928.

44 Carl Rufus, 'Twenty-five years of the Barbour scholarships', p. 21.

45 E. K. Krishnan Jr's Diaries, private collection.

46 BHL, UoM: University Herbarium Records (University of Michigan) H. H. Bartlett series, Box 11, E. K. Janaki Ammal to H. H. Bartlett, letter dated 25 April 1928. Sampathkumaran had established the Botany Department of the Central College in Bangalore city (affiliated to the Mysore University at this time) in 1919, which became one of the foremost centres of research in India for plant morphology. In the early 1930s, along with W. Dudgeon of Allahabad, Sampathkumaran initiated the study of angiosperm embryology (incidentally, both botanists had earned their doctorates under C. J. Chamberlin of the University of Chicago).

47 Ibid., H. H. Bartlett to E. K. Janaki Ammal, letter dated 5 June 1928.

48 Atkinson, 'Alma Gracey Stokey', p. 150.

49 Francisca Swamikannu completed a BA in Botany in 1926, followed by a Licentiate in Teaching two years later, both from the University of Madras. She was the daughter of Diwan Bahadur Lewis Dominic Swamikannu Pillai, lawyer, astronomer and author of the much-consulted book *Indian Chronology (Solar, Lunar and Planetary): a practical guide to the interpretation and verification of tithis, nakshatras, horoscopes, and other Indian time-records, B.C. 1 to A.D.*

2000 (published in 1911). He was the first elected president of the Madras Legislative Council in 1925. In August 1929, Francisca married Edward Thivy (son of Louis Thivy, a retired Station Master residing in Kuala Kangsar, Malaysia and in 1929, the first Indian to be elected to the Perak State Council) at the St Theresa's Church, Nungambakam.

50 E. K. Janaki to E. K. Raghavan, letter dated 9 September 1928, 'On Board the CUNARD R. M. S. Berengaria'.
51 Vol. XIV, June 1929.
52 BHL, UoM: Barbour Scholarships, Minutes dated 14 May 1929.
53 E. K. Krishnan Jr's Diaries, private collection.
54 *TMD*, 23 May 1929.
55 Besides Janaki, the Indian members of the Hindustan Club from this time included Ambalavattath K. Sukumaran, Bhagat Ram, Bahwanth S. Sindhu, Miss P. Shahane, Miss Sharkeshwari Agha, P. S. Dhariwal, A. S. Dhillon, C. S. Gill, Madhusudan Mozumdar and Nirpendra N. Roy.
56 E. K. Janaki Ammal to E. K. Parvathi (1930), an extract of the letter quoted in G. Doctor, 'Celebrating Janaki Ammal'.
57 Davis (1911); also see Harman, *The Man Who Invented the Chromosome*, pp. 73–74.
58 Campos, 'Mutant Sexuality', p. 51.
59 Shackman, 'The botanical garden on Iroquois'. I am most thankful to Dorothy Blanchard for bringing this article to my attention.
60 Pycior, Slack and Abir-Am (eds.), *Creative Couples in the Sciences*, pp. 156–169.
61 One of Kulkarni's early papers was on maize: 'Inheritance Studies of White-Capping in Yellow Dent Maize', 1927. It is of interest that besides working as a research assistant at the University Botanic Gardens after a BS (Mount Morris College) and MS (Michigan State University), Kulkarni was employed as a Sanskrit instructor. He published at least one paper on the subject: 'The date of the Bhagvadgita', 1927. In her doctoral dissertation, Janaki referred to Kulkarni's paper titled, 'Meiosis in Pollen Mother Cells of Strains of *Oenothera pratincola* Bartlett', 1929.
62 Anon, 'The botanical garden: A fascinating place'.
63 *TMD*, 8 March 1930.
64 Personal communication from Dorothy Blanchard, Frieda Blanchard's eldest daughter.
65 E. K. Janaki Ammal to E. K. Parvathi (1930), extracts from the letter quoted in Doctor, 'Celebrating Janaki Ammal'.
66 *TMD*, 27 April 1930.
67 BHL, UoM: Barbour Scholarships, Minutes dated 18 December 1930.
68 Born in Chicago, B. M. Davis obtained a PhD from Harvard and taught plant morphology at the University of Chicago from 1902 to 1906, later holding positions at the Marine Biological Station, Woods Hole, Mass., the Bureau of Fisheries and the universities of Pennsylvania and Michigan. Besides publishing on the morphology and cytology of algae, fungi, liverworts and *Oenothera*, he also co-authored textbooks (with J. Y. Bergen), *Principles of Botany* (1906) and *Laboratory and Field Manual of Botany* (1907).
69 This book narrates a spectacular journey undertaken by William Bartram between 1773 and 1777 through Southeastern parts of North America, with detailed descriptions of plants, animals, American Indians and landscape.
70 BHL, UoM: Barbour Scholarships, Minutes dated 14 May 1929.
71 Vol. 19, no. 2 (August 1931): 389–391.
72 BHL, UoM: M. U. Botanical Gardens (Correspondence series), Box 5, E. K. Janaki Ammal to Frieda Blanchard, letter dated 8 May 1931.

4

MICHIGAN II

The Private Life of Plants

I can commend Miss Janaki as a quiet individual who would give no trouble and would add a bit of color to your group. She is a reasonably good worker who takes suggestions readily and with gratitude.

—B. M. Davis (Janaki's doctoral supervisor, 1931)

It was only after her arrival at the University of Michigan in 1925 that Janaki had truly fallen in love with cytology. From now on, the microscope in the laboratory would become an extension of her physical self; this was more so during her second stint at Ann Arbor, to complete her doctoral dissertation as Barbour Fellow (1929–31). Several hours of the day would be spent peering down the eye-piece at elegantly prepared slides, in an effort to count chromosomes and unravel the drama that was unfolding in the cell nucleus. To translate and communicate this scientific intimacy in a language understood by all was a huge challenge to cytologists during these times.[1] In fact, the early successes of microscopy and chromosome study led to disputes, chiefly arising from the slips 'between languages, professions and techniques.'[2]

Back in December 1925, Bradley Davis had presented Janaki with a copy of the recently published third edition of Edmund Beecher Wilson's *The Cell in Development and Heredity* (1896). The book would become central to her life. Dedicated to his friend Theodor Heinrich Boveri, Wilson's book was a tribute to the great progress cytology had made in the last three decades of the nineteenth century. Two other books were added to her collection: she purchased de Vries' *Intracellular Pangenesis* (1889), a 1910 edition, and Babcock and Clausen's *Genetics in Relation to Agriculture* (1918), both of which she studied with much interest as a postgraduate student in Michigan. Based on a modified version of Charles Darwin's 1868 theory of pangenesis (which was later replaced by Mendel's laws of inheritance), de Vries had postulated that different characters had different hereditary carriers; that specific traits in organisms were represented by 'particles' called

DOI: 10.4324/9781003267089-4

pangenes. Two decades later, the term was abbreviated to 'genes' by the Danish plant physiologist Wilhelm Johannsen (1857–1927). As for Babcock and Clausen, they had pioneered the understanding of plant evolution in terms of genetics and their book served as a basic reference manual for the foremost geneticists and plant breeders of the time. Babcock contributed in particular to the study of wild plants and their evolution, a subject that would hugely interest Janaki in her post-Michigan years.[3]

The Cell and Its Nucleus

A quick overview of the discipline, as Janaki found it when she arrived on the scene in the 1920s, is helpful. Chromosomes were first observed in plant cells by Karl Wilhelm von Nägeli in 1842. The next breakthrough came some three decades later, when in 1875, the German zoologist Oscar Hertwig observed that only one sperm entered the ovum at the moment of fertilisation; this discovery debunked the old Darwinian belief that several sperms were needed to fertilise a single egg. The certain knowledge that the basis of sexual reproduction was the cell nucleus led biologists to explore the idea further. In 1883, the Belgian cytologist Edouard van Beneden published studies of fertilisation in the horse thread-worm, demonstrating that during a cellular process (later called meiosis), the gametes (male and female reproductive cells of the organism) received only one half the chromosomes of the normal cell. A year later, Nägeli argued that the physical basis of heredity was contained within the cell nucleus, but what form this particular hereditary substance assumed, he was unable to speculate upon. It then fell on biologists Rudolph Albert von Koelliker, Eduard Strasburger (an author familiar to Janaki through his book, *A Textbook of Botany*, prescribed by the Madras University), August Weismann and Hertwig himself to conclude that the 'hereditary' material in the cell nucleus was the thread-like chromatin (so called because the threads easily absorbed coloured stain). van Beneden's observations proved useful to Wiesmann in refining his views on the process of fertilisation; he termed the physical structures responsible for the transmission of hereditary material as the 'ids,' strung along the length of the chromatin. The slender microscopic inhabitants of the cell nucleus, which soaked up coloured stain effortlessly, were referred to as the chromosomes in 1888, by the German biologist Wilhelm Waldeyer.[4]

There were researchers before them, in particular Friedrich Anton Schneider, who observed a complex movement of microscopic bodies within the nucleus during cell division. Though these abstruse nuclear happenings attracted the attention of biologists as being very significant, the possibility that these events could be mere 'artefacts' (i.e. arising from fixing and staining dead organic material) cast doubts in their minds. The function of fixing agents was after all to freeze the cell contents of the tissues as rapidly as possible, in order to retain the structure that they had when alive. It was therefore very important to observe living cells during reproduction, which was what Walther Flemming accomplished around 1882. A skilled histologist, he employed techniques that minimised the formation of 'artefacts' in different types of tissues and helped to differentiate them from normal cell structures. Flemming chose for his study the epidermal cells of salamander embryos as they contained large chromosomes, enabling live observation during cell division. He would become the first person ever to describe rigorously the nuclear event called mitosis, which although continuous in reality, happened in distinct stages, namely prophase, metaphase, anaphase and telophase. Alterations and movement in the nucleus signified mitosis, which was a nuclear division by ordinary means.

In the prophase stage, for the first time, minute threads become visible in the nucleus. These bodies become constricted and thick in the metaphase, accompanied by the vanishing nuclear membrane and the mass migration of chromosomes towards the centre of the cell; all of which happens within an elongated fibrous structure called the spindle. At the ends of the spindle are tiny granule-like bodies called centrioles. In the anaphase, the chromosomes form two groups and migrate to the opposite ends of the spindle. They then gradually retreat from the observer's view, as a nuclear membrane covers each group. The cell is now in the resting stage once again, although there are now two cells instead of the original one. Flemming's research made it possible to understand how cells *divided*, but exactly how sex cells *united* during biological reproduction to produce a daughter cell with the same number of chromosomes as the parent rather than its double remained a mystery.

Given that the number of chromosomes in any one species remained constant across generations, it was only logical to assume that a division other than the mitotic kind was occurring in special sex cells or gametes, but what puzzled biologists was the lack of microscopic evidence to establish this. It was clear that a mere sharpening of the eye, with the help of improved microscopy and fixing and staining techniques, was inadequate to do this. Weismann suggested that if there was a doubling, there had to be an equivalent process of reduction at some point in the life of the cell—one which would produce haploid gametes (containing half the number of chromosomes of a regular cell, designated by the small letter 'n'). These haploid gametes would fertilise to produce a diploid daughter cell with the normal number of chromosomes (that is containing two sets of chromosomes, one from each parent, denoted as '2n'). This cellular process was given the name meiosis by John B. Farmer and John Moore in 1905. Meiosis was a form of nuclear division distinct from mitosis and occurred at only one time in the life-cycle of every species that reproduced by sexual means. How exactly this reduction division took place puzzled biologists and became a major subject of contention at the turn of the twentieth century.[5]

Uniting Cytology with Genetics

The chromosomal theory of heredity and its role in evolution and development was the result of the collaborative efforts of several researchers over a period of five decades, beginning in the 1860s. Two remarkable breakthroughs were made in the year 1865: an Augustinian friar, the German-speaking Gregor Mendel, published his experiments in plant hybridisation, speculating that cells contained something in them that carried traits from one generation to the next; and Darwin put forward his provisional hypothesis of pangenesis, wherein he suggested that traits could be passed down via units called 'gemmules,' which travelled from every body part to the sexual organs, where they were stored. Neither of the men provided scientific evidence to prove their theories of heredity, as a result, their work, especially of Mendel, went completely unrecognised by the scientific world, even lost, for several decades. It would have to wait until 1900 (thanks to de Vries, Erich von Tschermak, Carl Correns and the English biologist William Bateson who rediscovered Mendel) before a reliable understanding of hereditary transmission could be had, and against which cytological facts could be corroborated.

Darwin had outlined the evolution of species through natural selection, Mendel had postulated his laws of inheritance, and Weismann, de Vries and other embryologists were beginning to crack the question of how an organism developed from a single cell, but what troubled cytologists of the turn of the century was the question of how evolution, heredity and development were related to each other: how latent sex characters were contained in the germ cells and how they were mobilised as the individual developed still remained unknown. A major breakthrough occurred in 1902, when the particulate conception of heredity was united with cytological evidence, thanks to the researches of Walter S. Sutton and Theodor Boveri. Sutton, working on lubber grasshoppers in Edmund Wilson's laboratory at Columbia University, observed that chromosomes occurred in matched pairs of maternal and paternal copies (homologous, these copies had the same genes in the same location, or locus, as one another), and separated during meiosis. It dawned on biologists that the phenomena of germ-cell (ovum or sperm cell) division as observed under the microscope, and heredity as demonstrated by breeding experiments, were not different processes but essentially the same. If cytology and genetics were describing the same phenomenon, the argument was that chromosomes must be the substance of heredity. Independently of Sutton, Boveri's work on sea urchins also revealed how a complete set of chromosomes was important for proper embryonic development; Correns and de Vries would too arrive at similar conclusions.

By 1890, the cell division process called meiosis was understood as involving two successive nuclear divisions, in which chromosomes divided once instead of twice as they would in ordinary mitotic division. In meiosis, during the first division, the chromosomes came together in pairs, with each splitting longitudinally in the middle to produce two strands called chromatids, attached to each other at a point, (later termed) the centromere; this was also the case with mitotic division, except that in the mitotic metaphase, chromosomes were arranged singly, next to each other, rather than in pairs as seen in meiosis. How and when the chromosomes split into chromatids, however, remained to be explained.

By the time E. B. Wilson began working on his iconic *The Cell in Development and Heredity*, in 1896, the term chromosome had gained popularity, but there was still no concrete evidence to prove that these bodies were the real physical basis of heredity, despite powerful microscopes and effective staining techniques. In his book, Wilson was only able to *describe* how and why chromosomes behaved the way they did during cell division. He concluded that the chromosome count (usually determined at mitosis by counting the chromosomes and denoted by the diploid number 2n) was a constant throughout the individual and the species; these bodies thus existed in the resting nucleus, even if invisible, and possessed the quality required to be bearers of heredity. Wilson named the path-breaking chromosomal theory of inheritance the 'Sutton-Boveri Theory' in appreciation of his brilliant student Sutton and good friend Boveri. In fact, he would dedicate his book to Boveri.

Ascribing a central place to the chromosome in heredity and development, a school of cytology evolved in America at the turn of the twentieth century, with Wilson at the helm; it focused on a study of the mechanics of the cell, in the laboratory. Contemporaneously, across the Atlantic in England, embryologist Bateson had given rise to a genetical school (later the John Innes), which argued for the central position of Mendelism in heredity (demonstrating Mendel's two laws of inheritance: the law of segregation and law of assortment, and their exceptions), confirmed through breeding experiments in the field. The theory that chromosomes carried hereditary material (the factors of Mendelian inheritance) was at first controversial, but in 1913, an exceptional but little-known woman cytologist Estrella E. Carothers of the University of Pennsylvania provided definitive evidence of Mendel's (second) law of independent assortment, in a species of grasshopper. The law described how different genes independently separated from one another when reproductive cells developed. The combination of traits in the offspring did not always match the combinations of traits in the parental organisms; chromosomes were randomly sorted from all possible combinations of maternal and paternal chromosomes. This way, the gametes ended up with a random mix, rather than a pre-defined set from either parent, and were thus said to be assorted independently. The principle of independent assortment, along with crossing-over, increased genetic diversity by producing new genetic combinations.

In a matter of a few years, the American biologist Thomas Hunt Morgan (1866–1945), working in his 'Fly Room' at the Zoology Department at Columbia with the support of a group of talented students including A.

H. Sturtevant, C. B. Bridges and H. J. Muller, would convincingly demonstrate that genes are carried on chromosomes, as if beads on a string, and are the mechanical basis of heredity, providing incontrovertible evidence for the chromosomal theory of heredity. The fruit-fly *Drosophila melanogaster* turned out to be the ideal genetic research subject because it could be bred cheaply and reproduced quickly in the laboratory. Morgan had identified the gene as the impetus for Darwinian evolution and as the switching device, controlling development. With this discovery he had transformed biology from the descriptive science it was in the nineteenth century to an experimental one by the early twentieth. Morgan's seventeen-year research, commencing in 1910 with the discovery of the white-eyed mutation in *Drosophila* formed the basis of the modern science of genetics, for which he was awarded a Nobel (in the category Physiology or Medicine) in 1933.

Crossing-Over

Cytologists in the early decades of the twentieth century were debating on whether homologous chromosomes aligned side-by-side (called parasynapsis) or end-to-end (telesynapsis). The significance of the question lay in its bearing on 'how' crossing-over (the exchange of genes) between homologous chromosomes took place during meiosis. The success of Morgan and his team stood on an acceptance of a parasynaptic alignment of chromosomes. However, the chromosome rings observed in the evening primrose *Oenothera lamarckiana*, in particular, seemed to support a telesynaptic explanation and denied the possibility of a universal crossing-over at meiosis. Working on the plant *Datura stramonium* with Albert F. Blakeslee at Cold Spring Harbour, cytologist John Belling, who developed the iron-acetocarmine staining technique, put forth a theory of segmental interchange, wherein non-homologous chromosomes (chromosomes that are not members of the same pair) exchanged segments and created chromosome rings that appeared once again to be telesynaptic in origin. As for the great E. B. Wilson, he was cautious about taking sides on the issue. On the other side of the Atlantic however, a young English cytologist Cyril D. Darlington working on the *Oenothera lamarckiana* at the John Innes in Merton (a southwest London borough) was able to provide incontrovertible evidence for the parasynaptic interpretation, settling the vexing issue once and for all. Incidentally, in his letter to the Royal Society on the eve of being elected its Fellow, Darlington confessed that *The Physical Basis of Heredity* by Morgan (1919) was the one book that had inspired his research trajectory and choice of career.

Janaki's Research Problem

Janaki had chosen to work on the plant *Nicandra physalodes* Linn. Gaertn, of the family Solanaceae comparing it with those of related genera from the standpoint of meiosis and chromosomal behaviour, in the hope that 'some light might be thrown on the problem of the affinities of this genus.'[6] An annual, reputed to be of Peruvian origin, *Nicandra* was a monotypic genus (*physalodes* being the solitary species of the genus), widely distributed both in the New (the Americas) and the Old World (Europe, Africa and Asia). In 1929, at Ann Arbor, Janaki raised her *Nicandra* plants from seeds collected in Wayanad a couple of years earlier. Over the succeeding year, she raised a second generation of plants of the Indian race (*immaculata*), besides a few plants of the American race (*typica*) from seeds extracted from an herbarium specimen in the University collection (this material had been collected in 1926 by Carl Erlanson from Tennessee).[7] The two varieties were distinguished by the presence of fine blue spots at the base of the corolla in *typica*, and their absence in *immaculata*. Janaki's plants were grown both in the greenhouse and in the open in the University Botanic Gardens. They were highly susceptible to environmental changes and easily succumbed to drought or frost; in poor dry soil, the plants hardly grew to a height of a few inches, but given a good amount of shade and water, she found, they attained a height of six feet and were broad-leaved.

Plant material was chosen from both cultures. Although considerable attention had been given to its related genera such as *Nicotiana*, the only cytological investigation on the *Nicandra* until Janaki's research was a report by Vilmorin and Simonet in 1927, which disclosed the chromosome number (through a study of mitosis). By a detailed study of the behaviour of chromosomes during meiosis, Janaki aimed to explain how the reduction division was accomplished in the genus, and thereby shed some light on the problem of the affinities of the genus.

Mastering Technique

During Janaki's first stint at Michigan (1925–26), her good friend Doris Holdrup, whose home she visited in Torquay, presented her with Charles J. Chamberlain's *Methods in Plant Histology* (1900), which discussed several reagents, stains and staining methods used in preparing plant tissues, the making of temporary mounts, the micro-chemical tests and various methods used in preparing plant material for thorough systematic observation and study. Chamberlain's book was an indispensable reference guide for postgraduate students of botany in the West. In cytology, mastering technique was crucial for the successful interpretation of chromosome morphology and complex structures during cell division; almost everything depended on how well the material was fixed and the power of the microscope. By the 1870s, greatly improved microscopes had become available, as did new techniques for staining cells, which revealed clearly their internal structure.[8] A perfect fixation of the chromosomes was a prerequisite for a correct description. It was ideal to have a fixation that penetrated the cell quickly without shrinking the cytoplasm, one that preserved the natural distribution of the chromosomes for purposes of counting and one that provided a clear definition of constrictions. It was also necessary that the material be sufficiently hardened to preserve it against further changes by the action of reagents, with which it was usually treated.[9]

Smear preparation was considered by the early twentieth century as the most effective method for the study of cell structures, even more so than the sectioning technique, which post-dated it. In the smear method of preparation, material is spread in a thin film on a cover-glass, fixed, stained and mounted in situ. The method enables one to observe entire cells rather than parts as with sectioned tissues. John Belling (1926), Barbara McClintock (1929), Karl Sax, William Campbell Steere (1931) and Darlington (1932) would employ the smear technique to great benefit in their cytological researches, especially in the study of chiasmata (the points of contact between paired chromatids during the prophase stage of meiosis), which resulted in a cross-shaped formation, representing the cytological manifestation of the crossing-over of genetic material. McClintock and Steere, both working with plant material, discovered methods by which smears could be rendered permanent.[10] McClintock adapted Belling's iron aceto-carmine smear method for the study of maize, which permitted a clear observation of individual maize chromosomes as they went about dividing and replicating. Some others however found the aceto-carmine method not very satisfactory, especially in the early prophase stage, on account of the weak affinity of the chromatic substance to carmine during this phase.[11]

Like a potter discovering wonderful glaze recipes through experimentation, the ingenious cytologist mastered her technique through trial and error. It was all about crafting a science, using the right tools for the job. *Nicandra* offered an unusual advantage for the study of the progressive stages of meiosis because of the slight difference in age among the anthers (part of the stamen that contains the pollen) of the same flower. Janaki would discover that a finer gradation in development existed in the pollen mother cells of the same anther, a phenomenon observed in other solanaceous plants, in particular by J. W. Lesley and M. M. Lesley (1929) in tomatoes, by R. E. Clausen and T. H. Goodspeed (1923) in diploid *Nicotiana tabacum* and by herself in several species of *Solanum*. Janaki fixed anthers in a number of fluids, at all times of the day, for studying meiosis; she used a modified version of the Bouin's fixative solution and found that fixations done from 11.30 am to 1.30 pm yielded the best results.[12] Fixation of cytology specimens is critical to the preservation of the cellular components; material fixed in strong Flemming's solution (a mixture of osmic, chromic and acetic acids) and in acetic alcohol (1 part acetic acid and 3 parts absolute alcohol), she found, gave good results. Janaki also experimented with Karpechenko's fluid and the combination of Carnoy's acetic-alcohol chloroform and Nawashin's chromo-acetic solution as used by T. Maeda (1928), which proved very useful to her even though the technique created fibrils in the cytoplasm. In her preliminary studies, such as when working with flower buds, she found it beneficial to use McClintock's method (1929) of staining with aceto-carmine, after fixation in acetic alcohol. In addition, she employed the sectioning method for an accurate understanding of nuclear structure and chromosomal configuration: several sections were cut out, embedded in paraffin wax and then stained with Heidenhain's iron alum haematoxylin; iodine gentian violet as used by Newton and Darlington (1929) and Flemming's triple stain also gave her good results. In 1930, Janaki experimented with Taylor's (W. R. Taylor) method of smear fixation but with limited success, although it helped her in reconfirming interpretations made using the paraffin wax method.[13]

Results

Janaki's study reconfirmed Vilmorin and Simonet's observation that the haploid number of *Nicandra* was 10 (2n = 20) and showed that there existed no cytological difference between the Indian *Nicandra physalodes immaculata* and the American *Nicandra physalodes typica*. Further, she was able to demonstrate the parasynaptic pairing of chromosomes (chromosomes conjugating side by side) during meiosis in the genus, thereby debunking the telesynaptic one (end-to-end union) propagated by cytologists studying solanaceous plants (this view would be later discarded as a mere observational artefact) and reaffirming Darlington's thesis on crossing-over. Further, Janaki's study suggested that there was a close phylogenetic affinity between the genera of *Datura* and *Nicandra* given the resemblance observed between their chromosomal organisations.

It was only natural that the budding cytogeneticist's next step was a period of study at John Innes, under the brilliant young cytologist Darlington, who was in the midst of writing his path-breaking even if controversial book, *Recent Advances in Cytology* (1932). The world of cytogenetics was akin to a small village at this time, with everyone knowing everyone else. In April 1931, Janaki's doctoral supervisor, B. M. Davis, who had himself spent the summer of 1925 at the John Innes introduced Janaki to Darlington, by way of a short letter (of which his final commendation forms the epigraph to this chapter):

> Miss E. K. Janaki who has recently passed her examination with us for the Doctorate is planning to be in England for part of the summer before returning to India. She has material of a polyploid Egg plant and would very much like to do some work on it at John Innes. The material will, I think, interest you.[14]

Darlington accepted Janaki's application, and her life would never be the same again.

Figure 4.1 *Nicandra physalodes*. Collected (F. J. Hermann) from 'Waste ground', Ann Arbor on 5 October 1934.

Source: Courtesy of the University of Michigan Herbarium.

81

Notes

1 See, for example, Keller, *A Feeling for the Organism* and Harman, *The Man Who Invented the Chromosome*.

2 Darlington, 'Genetics and Plant Breeding, 1910–80', p. 401.

3 *Genetics in Relation to Agriculture* published in 1918 soon became popular, and during the 1920s and 1930s it was one of the most widely used of genetics textbooks. A greatly expanded second edition was published in 1927. For a biographical memoir of Babcock see Stebbins, *Ernest Brown Babcock 1877–1954*.

4 For histories of cytology and genetics, see Hughes, *A History of Cytology*, Dunn, *A Short History of Genetics* and Carlson, *The Gene*.

5 Harman, *The Man Who Invented the Chromosome*, pp. 37–38; Bowler, *The Mendelian Revolution*, pp. 85–88.

6 University of Michigan, Ann Arbor (Buhr Building), Dissertations: E. K. Janaki Ammal, 'Chromosome Studies in *Nicandra physalodes* (L.) Gaertn', 1931.

7 Carl Erlanson (1901–75) received his BS and MS degrees from the University of Michigan, which also awarded him an honorary PhD. When Janaki arrived in Michigan in 1928, Carl had just returned from a University-funded scientific expedition to Greenland.

8 For a history of the developments in the field, on the 'instrumental' front, see Robinson, *A Prelude to Genetics*; also see Mayr, *The Growth of Biological Thought*.

9 La Cour, 'New fixatives for plant cytology'; 'Improvements in everyday technique in plant cytology'.

10 Belling, 'The iron-aceto-carmine method of fixing and staining tissues'; McClintock, 'A method for making aceto-carmine smear permanent'; Steere, 'A new and rapid method for making permanent acetocarmine smears'; Sax, 'The smear technique in plant cytology'; Darlington, 'The origin and behaviour of chiasmata'.

11 See for example, Yasui, 'Ethyl Alcohol as a Fixation for Smear Methods'.

12 Janaki prepared two solutions, A and B: solution A contained a saturated solution of picric acid/50cc[ml], glacial acetic acid/5cc and chromic acid/1g; solution B comprised a saturated solution of picric acid/25 cc, commercial formalin/25cc and urea crystals/1g. She then mixed equal parts of the two solutions at the time of fixation. Buds were allowed to remain in the fluid for 4 to 6 hours and were then run rapidly through alcohols, beginning with 35% and increasing to 85%, where they were left until all trace of picric acid were removed by constant change of alcohol.

13 Taylor, 'The Smear method of plant cytology'.

14 Bodleian Library, Oxford, Darlington Papers (DP): C 107 (J 33), letter dated 15 April 1931.

5

ENGLAND

Love, Tulips and Chiasmata

The study of life and of chiasmata at John Innes are both extremely interesting. C. D. Darlington is a most extraordinary individual.

—E. K. Janaki Ammal (1931)[1]

'My last day in USA!' Janaki announced to Frieda Blanchard on 8 May 1931.[2] She was at Boston, Massachusetts, from where she would board a ship to England with the intention of spending a few months at the John Innes studying cytology under Cyril Dean Darlington (1903–81), before finally returning to India.

Born in Chorley, England, Darlington had studied agriculture at Wye College, receiving a BS degree in 1923. He began his career unsalaried at the John Innes that very year, under its first Director, the legendary English geneticist William Bateson, who gave the new biological discipline of genetics its name. Four years later, when just twenty-four, Darlington was awarded a doctorate by the University of London (to which the John Innes was affiliated) for the thesis 'Genetical Studies in *Prunus:* The Cytology of Domestic Cherries.' In 1937, in a matter of only fourteen years, Darlington, the iconoclast, would head the Cytology Department at the John Innes, and make the institution a premier centre for cytogenetic research, not only of Britain but the world.[3]

Cytology and Genetics at the John Innes

The John Innes was established in 1910, on a piece of land in Merton bequeathed by John Innes, a wealthy merchant of the City of London, for the promotion of horticultural instruction, experiment and research. It was administered by a Council appointed by the Ministry of Agriculture and Fisheries, the Royal Horticultural Society, the Fruiterer's Company, the Universities of Oxford, Cambridge and London, and the Imperial College of Science and Technology. David Prain, the Scottish physician and botanist,

DOI: 10.4324/9781003267089-5

and Director of the Royal Botanic Gardens, Kew (incidentally, between 1898 and 1905, he had been Director of the Calcutta Botanic Gardens and the Botanical Survey of India) was Chairman of the Council since its inception; among the original members was cytologist John Farmer, who edited the *Annals of Botany* (who we may recall, coined the term meiosis in 1905).[4] The Institution aimed to 'carry out investigation and research, whether of a scientific or practical nature, into any matters having reference to the growth of trees and plants generally.'[5] In the early twentieth century, the John Innes was one of the only institutions in Britain devoted to research on heredity and genetics.[6]

By 1910, Bateson was pioneering the new science of genetics in Britain. His research had taken him in quite the opposite direction of Darwin, who argued that evolutionary transformation by natural selection always proceeded gradually and never in jumps. In contrast, Bateson argued that biological evolution occurred in sudden jumps (saltation) and was the result of inherited factors alone; it was entirely uninfluenced by the environment, he argued. In 1900, Bateson had rediscovered the landmark paper on pea hybridisation experiments which the Austrian monk Gregor Mendel had presented before the Natural History Society of Brünn, Moravia, in 1865.[7] Bateson had Mendel's paper translated and published in the journal of the Royal Society of London. His Cambridge friend, the evolutionary biologist W. F. R. Weldon of the Biometrical School (with Francis Galton and Karl Pearson) attacked the Mendelian notion of heredity based on 'unseen' inherited 'factors' in the first volume of *Biometrika*; they were willing to accept only a quantifiable or Galtonian take on inheritance (based on Francis Galton's statistical law of ancestral heredity, 1889). Weldon's critique of Mendelism provoked Bateson to write *Mendel's Principles of Heredity; a Defence* (1902), a clear elucidation of the principles of genetics. He also began experimental work on a number of plant and animal species, supported by Reginald C. Punnett, Leonard Doncaster, Charles C. Hurst and women researchers Edith R. Saunders and Florence M. Durham, to demonstrate the validity of Mendelian laws of inheritance, namely, the law of segregation and the law of independent assortment.[8]

The Cambridge University acknowledged Bateson's position as world leader in genetics by establishing in 1909 a special chair of biology for him, but he found it unattractive from the standpoint of research. He was keen on establishing a centre for genetic research, but the University was unwilling to allocate any money towards this.[9] Thus when a year later the John Innes offered him a large salary and splendid facilities at Merton for experimental work, he accepted the offer without the slightest hesitation. Bateson would become the John Innes' first director and his group of researchers would work on the many problems of inheritance in plants—particularly on a humble little primrose, the *Primula sinensis* (brought with him from

Cambridge) and the *Prunus*, besides Mendel's *Pisum*. By this time, he had come to accept that sex chromosomes were determiners of the sex of off-spring but was still far from accepting the role of chromosomes as carriers of hereditary material, an outlook that regressed the discipline of genetics in Britain.[10] Despite Bateson's larger-than-life reputation, his influence over British universities was negligible; even in the 1920s, genetics as a discipline did not find a place in the country's biology departments. Several of his plant species and genera, particularly flowering plants, exhibited polyploidy (the occurrence of multiple homologous sets of chromosomes in an organism) and yet he was unprepared to treat this phenomenon as relevant to the understanding of heredity and variation. Bateson obviously had a very limited view of Mendelism, which prevented him from thinking that the microscope might provide a much larger framework, something deserving of the name genetics.[11] However, there were a few British biologists, Bateson's own students among them, who acknowledged chromosomes and their role in heredity and were involved in research that explored the connection between genetics (Mendelism) and cytology. Unfortunately, the war intervened, bringing their work to a halt.[12]

After the war, only Dorothy Cayley and Caroline Pellew remained of Bateson's earliest staff. They were however soon joined by women researchers, Alice Gairdner, Irma Anderson and Dorothea de Winton, while the 'men's labs' housed W. C. F. Newton, Reginald J. Chittenden, Ernest J. Collins and Morley B. Crane. Genetical investigations were commenced with renewed vigour, in particular crossing *Pisum* and *Primula* to reaffirm the Mendelian laws of inheritance and a Genetics Society was founded by Bateson in 1919, but neither of this helped the discipline make much progress. An aesthete and a passionate collector of Japanese prints, Bateson spent the rest of his life keeping Mendelism separate from the 'chromosome cult' as he called it, only to exclaim in despair, after a visit to Woods Hole in early 1922 (subsequent to addressing a meeting of the American Association for the Advancement of Science at Toronto), that his whole life had been doomed. At the opening address of the meeting in Toronto, he professed himself 'converted' to the chromosomal theory of heredity. At last, he had acknowledged that cytology was a 'real thing.' He was now determined to find a cytologist for the John Innes; W. C. F. Newton was thus brought in, but he would not be officially appointed until June that year.

The physically frail Newton devoted his energies to the new department of cytology, producing some high-quality work, albeit interrupted by long bouts of illness. He studied the morphology of the species of *Tulipa* and discovered the existence of several polyploids. In 1923, Darlington joined Newton, oblivious of Bateson's critical attitude towards the chromosomal theory of heredity. A year later, the main garden was extended by the addition of four and a half acres and was enclosed by a high wall. Soon new laboratories and workshops were erected, the plant breeding house built and a new house of residence constructed. Newton unfortunately died in 1927, only five years after joining the John Innes, a tragic loss to Darlington, to whom he had been a true mentor. It was Newton who had shown him under the microscope, for the first time ever, the thread-like 'things' referred to as chromosomes, and brought to his attention the phenomenon called polyploidy.

Sexually reproducing organisms normally possess two sets of chromosomes and during normal sexual reproduction, each parent gives half its chromosomes (n) to the offspring. Having received one set of chromosomes from each parent, the offspring contains two sets of chromosomes and is called diploid (represented as 2n). When the offspring matures and is able to reproduce, the process of sexual reproduction continues. However, if things go askew during the process of sexual reproduction, and the offspring ends up receiving all of the chromosomes from both the parents, it will have four sets of chromosomes (4n), thus becoming tetraploid rather than diploid. In general, when an offspring receives more chromosomes than normal, it is referred to as a case of polyploidy. Polyploidy is common in plants, and even desirable, but in animals it can be cancerous and therefore lethal.

E. B. Wilson's *The Cell in Development and Heredity* would serve as the young Darlington's chief introduction to the field of cytology, as it would for Janaki in the late 1920s. Two camps, as it were, had emerged at the John Innes during Bateson's regime. The first one ascribed the role of heredity exclusively to the nucleus, at the cost of the cytoplasm, best represented by Morgan's *The Theory of the Gene* (1926). The second, having given nothing to the nucleus, also allowed nothing to the cytoplasm, as in Bateson's view of Mendelism, termed anisogeny (where the gametes, or the mature haploid male or female germ cells, fuse to form a hybrid offspring that has blended characteristics inherited from both gametes). The role of the cytoplasm, which had been emphasised by the German botanists Erwin Baur, Fritz von Wettstein and Otto Renner, went completely unacknowledged by both Morgan and Bateson, and by a generation of plant breeders.[13] On the one hand, biologists were breeding plants in the field without a bother about chromosomes, and on the other, chromosomes were being studied in isolation, disconnected from the actual breeding of plants in the experimental field. This unfortunate situation would end with the death of Bateson (1926). The opposition to the chromosomal theory at the John Innes would become a thing of the past, and plant breeding would be practised on a scientific basis, but not without its share of conflicts.

A year after Bateson's death, John Burdon Sanderson Haldane (1892–1964), Reader in Biochemistry at the Cambridge University since 1923, was appointed 'Officer in Charge of Genetical Investigations' at the John Innes. Haldane together with Ronald Aylmer Fisher (statistician at the Rothamsted

Experimental Station in Harpenden) and the American geneticist Sewell Wright founded a new field of biology called theoretical population genetics (later derisively referred to as beanbag genetics).[14] With Haldane's arrival at the John Innes, a trend towards applying mathematical theory to genetics was initiated. His research at this time centred chiefly on the problem of gene linkage (or the tendency, during meiosis, of genes proximal to each other on a chromosome to be inherited together; that is some genes are transmitted to the offspring in groups) in *Primula sinensis* and *Antirrhinum majus*, a phenomenon that had interested Bateson. Iconoclast and polymath, well known for his Royal Society speeches and Hyde Park political oratory, Haldane in the 1930s significantly contributed to the understanding of gene linkage by developing a complicated theory of polyploidy in collaboration with Dorothea de Winton, applying it to tetraploids, in comparison with the diploid *Primula sinensis*.

For Darlington, who had become a 'scientific orphan' with the deaths of Bateson and Newton, Haldane became a great source of inspiration. Haldane would become (as would Hermann Joseph Muller, the American geneticist) his staunchest supporter. Haldane's supreme confidence, nonconformity and irreverence to authority and the state deeply attracted the young Darlington. He was especially influenced by Haldane's theoretical approach to biological problems, even if the older man was no experimental biologist or cytologist. The two, irrespective of age or experience, would spend hours deliberating on a range of subjects including genetics, politics and eugenics.[15]

By the time of Newton's death, Darlington had learnt to use chromosomes as a morphological tool for taxonomic purposes, particularly of the several polyploid varieties he had come across at the John Innes, but the problem of crossing-over was too difficult for him to comprehend or pursue at this point in time, and so very cautiously but with determination he embarked on the problem of pairing of chromosomes, having before worked alongside Newton on this vexing issue. He wished to address the issue of *why* chromosomes paired by explaining *how* they paired and then offering a generalisation. Darlington chose the tetraploid *Hyacinthe* for its suitability for the observation of chromosomal pairing.[16] He would argue that in principle chromosomes always paired two by two, even if they appeared not to; if they did not in reality, the case was only an exception. The problem of trisomy, where one chromosome in the set was represented three times, would seize his interest next; the plant of interest this time was Newton's *Tulipa*.

Earlier in 1924, John Belling and Albert Blakeslee had claimed to have discovered exactly the number of trisomics one might expect from the total number of chromosomes, the plant of their study being *Datura stramonium* (American jimsonweed); this plant had twelve pairs of chromosomes and twelve clearly recognisable trisomic forms, in each of which one chromosome was represented thrice.[17] In the eight chromosome hyacinth triploid, Darlington found eight trisomics, each with one of the eight chromosomes repeated three times. It baffled him that while certain chromosomes failed to pair, certain others, which were very much like their partners, paired. In an attempt to resolve this issue, he began closely studying the manner in which chromosomes associated at meiosis. He identified two sorts of associations: Janssens' chiasmata and the terminal ones. Darlington had earlier postulated that chiasmata and points of crossing-over exhibited a constant ratio of occurrence. Taking recourse to deductive logic as he always did, he suggested 'that the terminal associations between pairing chromosomes were derived from the interstitial ones of the Janssen kind, running along the chromosomes to its ends, like a zipper opening, a process called *terminalization*.'[18] Darlington theorised that chiasmata was the condition by which chromosomes that had been paired at the meiosis prophase stage called pachytene remained paired at the metaphase stage (when the chromosomes aligned themselves in the middle of the cell); if this were true, he argued, parasynapsis was true, thus driving the last nail into the coffin of the telesynaptic theory best represented by cytologists John Farmer, R. R. Gates and E. W. MacBride. Not satisfied by the focus on individual plants or problems, Darlington turned to critically examine and review the entire literature on the cytology of polyploids, hybrids and sex chromosomes, an exercise that would result in *Recent Advances in Cytology*.

Between 1927 and 1931, Haldane began genetical and biochemical studies of plant pigments in collaboration with another woman collaborator, Rose Scott-Moncrieff. Minor alterations were made to the old laboratories at the John Innes and a chemical laboratory was fitted up on the ground floor. During this period, the room in the annex to the Manor House previously used by Bateson as a study was made into a lecture room and the greenhouse was further extended and updated with two unheated houses for the tulip collection and other plants requiring cool-house conditions; besides a fern pit and a lean-to house suitable for tomatoes were erected. X-ray treatment of seeds was introduced, aiding both genetical and cytological studies. In addition to the above, varieties of commercial value, discovered in the course of pure research, were placed in the hands of commercial breeders and market gardeners.

A Cytology School

From its very inception, the John Innes had aimed to provide facilities to volunteer workers qualified to conduct research and to a great number of investigators visiting from both abroad and Britain. It was a cytology school like none other in the country. In the late 1920s, the young Leonard Francis La Cour, an expert in the preparation of chromosomes of both plants and animals, helped visiting students 'see' chromosomes. La Cour was a pioneer in the development of techniques, which helped disclose the inner structure and the entwining of chromosomes and also contributed to cytological theory.[19]

Incidentally, the John Innes was recognised by the University of London as a place of research for students seeking to qualify for a higher degree. This was probably how C. K. Kausalya, Lecturer at the QMC, Madras, and Janaki's family friend, spent a brief time at the John Innes in late November 1922 (during Bateson's tenure).[20] The first Indian woman to visit the institution as a volunteer worker, Kausalya was at this time pursuing a second BSc degree at London's Bedford College for Women (today's Royal Holloway College) as a government scholar. Janaki would arrive at the John Innes nine years after Kausalya, and after much had changed at the institution, but certainly for the better.

Janaki Arrives at the John Innes

After a fairly comfortable Atlantic crossing in early May 1931, Janaki arrived in Galway on Ireland's west coast (where the river Corrib meets the Atlantic Ocean), where she engaged in some plant collecting, an activity that gave her immense happiness and peace; between 16 and 20 May, she collected several algae specimens. She would reach England a week later and reunite with friends Katherine Fellows and Doris Holdrup. When Janaki was told that they were about to leave on a two-week trip to Devonshire (where Holdrup's parents lived), she jumped at the idea, to the extent of leaving her bags still unpacked at 62, Lebanon Park, Twickenham, her residence for the coming months. She thoroughly enjoyed the trip to Torquay (Fellows and she resided at a hotel on Avenue Road this time) despite Fellows being an unspirited companion (Janaki would later remark to Frieda: '[Fellows] imagines she is a hopeless invalid. It is rather difficult to make her enjoy life').[21]

It all ended sadly however when they received news of the tragic car accident in the Mojave Desert in the southwest United States, involving the Dobzhanskys (the Ukranian geneticist Theodosius Dobzhansky and his wife), Karl Franz Josef Belar (of the University of Berlin, who was based at the John Innes at this time) and their good friend, Eileen Erlanson. Janaki would have more news of the incident when she reached the John Innes; George W. Beadle, the American cytogeneticist (co-author with Sturtevant of the standard text, *An Introduction to Genetics*) had provided Darlington with a detailed account of the mishap. While the Dobzhanskys escaped mostly unhurt, Belar had succumbed, and Eileen was badly injured. Belar had spent some months at the John Innes helping the young Darlington 'with his cytological technique' and was strongly of the view that cytology 'should not be the ancilla of genetics.'[22] At the time of his death, Belar had been working towards establishing the universal character of mitosis, and the chromosome, which he believed was the basis of the uniformity of development in living creatures, whether plant, animal or a uni-cellular organism. 'Belar's death is a terrible loss to cytology,' Janaki would remark.[23] Darlington's *Recent Advances in Cytology* would be dedicated to Belar, Frans A. Janssens (the first to explain the formation of chiasmata during meiosis) and W. C. F. Newton (his late mentor).

As for Eileen Erlanson, she was at the time of the accident, a visiting National Research Fellow at the California Institute of Technology, Pasadena in Los Angeles. In March 1930, she had received a third National Fellowship in botany (two was customary, but only rarely and in unusual circumstances were researchers awarded the fellowship a third time), which had taken her to the John Innes, where she continued with her work on the genetics of the *Rosa*. When Morgan heard from Bartlett of her rose investigations, his interest was so piqued that he promptly provided her with research facilities at Pasadena, which she accepted in early 1931. A woman of great vitality and daring, Eileen was on a solitary rose-collecting trip in Western United States, at the time of the accident.

Nostalgic for Ann Arbor

Janaki already missed Ann Arbor and was anxious. She had left behind at Ann Arbor her box of blotters, presses, a fur coat and typewriter; she hoped Eileen would not mind taking them to the freight office, when she was fit enough and had returned to Ann Arbor. She missed the Blanchard girls very much: 'I think of Grace & Dorothy each time I see a beautiful child—and there are so many in England.'[24] Of little Grace, she had earlier remarked, the 'image of her beauty haunts me. It is about a spiritual experience to have known the little one.'[25]

The Blanchard girls remembered Janaki with equal fondness. In her reply, Frieda recalled an amusing incident: 'One morning when Grace came downstairs before the rest of the family, I heard her talking and thought it must be to herself, but when I came down, I found that she was having a conversation with your picture! And recently when she was watering the window plants I pointed to the primrose and said "That one needs water". She answered, "That's my Janaki plant".' The girls were thankful for the 'card with the sea gulls' that Janaki had sent them as she was leaving New York. The ever-generous and affectionate Frieda reminded Janaki that she was always there for her: 'Be sure to write me asking for anything which you want done, whatever it is.'[26] Janaki wondered if Bartlett had returned from his expedition in the 'wilds of Central America': 'I think the photograph he promised would help a great deal to dissociate him from the menagerie he was supposed to carry with him. Do remind him gently about visiting his photographer in Detroit,' she had written to Frieda Blanchard.[27] Frieda had by this time given birth to a son, Frank Nelson; ever the geneticist, Janaki commented: 'I was delighted . . . with news of the baby. . . . I am so happy at least one of the three has the Y chromosome of the family. I wish I had stayed a little longer in Ann Arbor to see the little bundle of perfection. I know, you will send me a photograph by & by. It is nice to think Dorothy & Grace have not forgotten me. I do miss them—especially little Grace—to behold whom was a strange inward joy to me—my love to both of them!'[28]

Working with Darlington

Janaki's arrival at the John Innes as research worker in mid-1931 was at a momentous time. Darlington was engaged in writing his path-breaking book on cytology, which he would finish in autumn that year. His early cytological investigations were concerned with the explanation of the movements, divisions and pairing of chromosomes before and during the formation of germ cells (cells that give rise to gametes). The study involved a comparison of the behaviour of chromosomes in pure forms with those of hybrids and polyploids, where the processes were less regular. Confirming the validity of the chromosome theory of heredity, his investigations concluded that the hereditary properties carried by the chromosomes not only determined the characters displayed by the organism but also the behaviour of the chromosomes themselves. There would be extended discussions on the book with colleagues and friends during leisure time, among who was Janaki. In the shortest of time, Janaki and Darlington became inseparable. At least for her, the relationship was also a deeply emotional one. Despite his unpredictable, difficult and provocative nature, she would become one of his most loyal friends, remaining close to him till the end of his life.

An entry in Darlington's pocket diary dated 2 June 1931, recorded: 'Janaki comes to Merton.'[29] She was a great attraction on the campus, exotic-looking with her long and luxuriant tresses, and colourful saris, a glowing face and easy smile. She made friends instantly, with her confident, vivacious and generous disposition, and most of these relationships would last a lifetime. 'I am enjoying John Innes very very much,' she wrote to Frieda, 'Dr Darlington is awfully nice—all the cytologists meet during lunch and tea—and there is a great deal of spicy discussions and talk. . . . I certainly feel like stretching my stay to the utmost.'[30] As for Darlington, he was a great charmer; he was tall, good-looking, brilliant and supremely confident, even cocky. The John Innes staff included several women, who looked up to him for intellectual or moral support: Miss Pellew, Dr Irma Andersson-Kottö and Dr R. Scott-Moncrieff, Miss Cayley, Miss D. de Winton, Miss A. E. Gairdner, Miss Brenhilda Schafer (a research worker with a BSc from Birkbeck College, she had been appointed Librarian and Registrar in 1926), Miss E. Sutton and later Miss Margaret B. Upcott. Eileen Erlanson would also spend substantial amounts of time at the place, as a research worker. It was also at the John Innes that Janaki first met Haldane, besides G. H. Beale, F. G. Brieger, M. B. Crane, A. C. Faberge, W. J. C. Lawrence and D. Lewis.

To be surrounded by so much natural beauty also filled her with joy. Janaki lived very close to the Kew Gardens, and often visited the place: 'England is lovely now with roses in bloom everywhere. I think English are marvelous gardeners.' It was rather cool for June that year and particularly chilly on some days, and 'with no type of heating in buildings,' it was far from comfortable. She was also in desperate need of money and would write to brother Kittu for a loan of Rs 500, but none of this would make her want to leave England. She found the John Innes, which she visited almost every day, an ideal place for research, loving every bit of its 'atmosphere of quiet & friendliness' and found Darlington very attractive. She was completely in awe of him: 'a brilliant man—with a delightful sense of humour and an infinite capacity for talk & discussion,' she would remark to Frieda. She would even take him and Brenhilda (with whom Darlington had a romantic entanglement) out to dinner once; 'Indigestible but interesting' was how Darlington described the food at the Indian restaurant! It impressed Janaki so much that Darlington was able to write 'a textbook of Cytology and a book of travels in Persia at the same time,' about both of which she got to hear so much 'at tea time & lunch.'[31] Life was suddenly so full of excitement.

Working in close proximity to Haldane, an intellectual giant of the times, in early 1930, Darlington had begun writing what would become the *Recent Advances in Cytology*. The book aimed to provide the reader with a 'general analytical account of the chromosomes'; he had realised that 'the whole of cytology (on top of which genetics had to be built) was a jumble of sound and unsound theory and observation, inconsistent with one another and with genetics.'[32] With the publication of the book in 1932, Darlington had unexpectedly transformed the field of cytology from a 'largely descriptive, empirical field of research, to a highly theoretical, speculative science based on deductions from genetic first principles.' While Bateson denied the role of chromosomes in heredity (to him the phenomenon of polyploidy was irrelevant to speciation), and Newton denied their role in evolution, Darlington boldly proclaimed the centrality of chromosomes in evolution and in generating variation in nature.

Conference on the History of Science

The Second International Conference on the History and Philosophy of Science, with Charles Singer as President, took place in London between 29

June and 4 July 1931, an event that coincided with Janaki's arrival at the John Innes. One of the Vice-Presidents of the Conference was George Sarton of the Harvard University, who was also the founder of *Isis*, an academic journal (started in 1913) devoted to the history of science, medicine and technology. There was a strong presence of a Russian delegation at the Conference led by N. I. Bukharin. The papers delivered by Boris Hessen on the socio-economic roots of the *Principia*, N. I. Vavilov on the origin of agriculture in the world, E. Colman on the crisis in mathematical science and B. Zavadosky on physical and biological factors in the process of evolution were thought to have been particularly impressive. Among the English participants were J. D. Bernal, Haldane, Hyman Levy, Lancelot Hogben, Benjamin Farrington and Joseph Needham; Needham had prepared a wall chart illustrating the history of physiology and biochemistry. Bukharin would later edit and publish the papers presented at the Conference, as *Science at the Cross Roads* (1971).

Polyploidy in the Eggplant

In the summer of 1929, at the University Botanical Gardens, Michigan, Janaki had observed a number of unfruitful plants in a large culture of F2 (second filial generation) plants of a cross between a dark purple and an 'ivory' fruited variety of *Solanum melongena* (eggplant, brinjal or aubergine, of the nightshade family, the Solanaceae, like the *Nicandra*). The plants were conspicuous for their unusually large leaves, stout stems and erect habit of growth. Although they were taken into the greenhouse in the fall, the plants had died before she had done any cytological work on them. Exactly a year later, a single 'abnormal' plant that showed similar characteristics to those ones appeared in a culture of some 100 F1 (first filial generation) plants, of a cross between a dark purple and a green fruited eggplant. A cytological examination of the cells of the root tip of the 'abnormal' plant showed 36 chromosomes, while the somatic number in the eggplant was usually 24; the evidence for the last of course came from the root tips of the two parent plants and their normal hybrids. Janaki's immediate conclusion was that the 'abnormal' plant was a triploid. The normal F1 plants were intermediate in regard to the development of anthocyanin in the stem and leaves and the fruit; the fruits were themselves abundant, and in colour was a medium purple on green. The triploid plant on the other hand had dark green leaves and stem like those of the male parent, and a single fruit produced in the field by open pollination showed faint streaks of purple on green. Even if genetical and cytological studies remained to be done, the hybrid nature of the triploid plant was beyond doubt.[33]

At the John Innes, Janaki began studying the triploid eggplant slides she had brought with her from Michigan, prepared from the root tip sections of fifteen progenies of the plant. She discovered to her pleasant surprise that they contained 46–48 chromosomes, which meant that they were all tetraploids and not triploids as she had imagined them to be! Excited, she sent instructions to Bill Steere (bryologist William C. Steere, who worked as Instructor) at the University Botanic Gardens at Ann Arbor, to fix some buds from her triploid progenies, which she fervently hoped had not died from lack of care.[34] In an earlier letter, Frieda had assured her that the plants were being looked after well and that Lois Ehlers had taken charge of plant-ing them out.[35] Unfortunately for Janaki, her single triploid had died, much as she had predicted, or at least Frieda had no clue as to which one it really was. 'How was it labelled,' Frieda queried, 'You have told us the numbers of its parents and of its offspring, but nowhere have you mentioned its own number.'[36] Steere had by this time moved out of the Gardens to the Depart-ment, putting Janaki's plant material under the care of Anderson or Dale. Janaki clarified that the so-called triploid plant was the one in a very large pot in room B or C, marked F1, but she wondered if she had taken care to mark it as a triploid! She lost all hope for it because the plant had already been looking poorly when she left Michigan. Her next alternative was to have the anthers fixed separately for the 15 progenies of the hybrid (num-bers 113.1 to 113.15 at the Gardens), but the stages for fixing these were the diakinesis (the fifth and last stage of the prophase of meiosis, following diplotene, when the separation of homologous chromosomes was complete and crossing over had occurred) and the metaphase. She just wished some-one could take a photograph of the finest looking of the fifteen progenies of the triploid, and another, of them together, for purposes of comparison, but this was difficult to be had. Morley B. Crane at the John Innes, who had produced a tetraploid *Solanum* for Carl A. Jørgensen by decapitation of the stem and generating a callus (that is inducing polyploidy through 'injury') was keen to experiment with Janaki's *Nicandra* and her eggplants, another one of the reasons she wanted her plants sent out from Michigan.[37] Janaki was however not hopeful that they would survive the long journey. Having recovered from her injuries, Eileen Erlanson, the good friend that she was, kindly offered to fix some material for her, but Janaki hoped Bill Steere would also fix some anthers for her, in Flemming's solution. All this was not easy nor indeed the right way to research, by 'remote control' from across the Atlantic. As a result of this, Janaki would end up taking a long time, about three years, to complete her work on polyploidy in the eggplant.

Joint Research on the *Tulipa*

In early August, Janaki hinted to Frieda Blanchard that she might be able to extend her stay at the John Innes until September that year. Darling-ton wanted her to work on some of Newton's tulips, she explained. She was enjoying her work very much but was also missing Ann Arbor. 'I wish I could be there too,' Janaki would remark, yearning for the impossible. The weather in England had turned 'unsummery,' with frequent cold and wet days and she was spending the weekend with Eileen Erlanson's mother: 'She is a very cheerful person and if you hear her speak, you'd think it is Eileen speaking. Almost uncanny Family resemblances!' she would exclaim to Frieda, in 'genetically-correct' speak.[38]

Janaki had by this time begun studying Newton's tulip slides at the laboratory under Darlington's direction, with an eye on chiasma frequency, terminalization coefficient, etc., ahead of the arrival of fresh tulip bulbs from the field. This piece of research would result in their first ever joint publication.[39] By the time it was published, the idea that chromosomes paired side by side (parasynaptic) in the pachytene stage of meiosis rather than end-to-end (telesynaptic) had come to be widely accepted, but their paper was taking the argument further by stating that the original location of the points at which crossing-over or exchange of genetic material occurred (points of contact or chiasmata) were always interstitial and never terminal. The terminalization theory, they argued, relied primarily on a comparison of the metaphase observations in *Hyacinthus* and *Tulipa* with those in *Tradescantia*, which had serious limitations, while their own observations on the meiosis in *Tulipa* had revealed that it underwent incomplete terminalization of chiasmata between diplotene and metaphase; that is, the original positions of the chiasmata influenced their chances of becoming terminal.

Reluctant Departure for India

It was time for Janaki to leave England, but she was still negotiating with Frieda across the seas about how her box of blotters and sundry things left behind at Ann Arbor could be retrieved. She now requested the blotters be sent to the care of 'Dr T. Ekambaram, the Dept. of Botany, The Presidency College, Madras,' in charge of the University Botanical Laboratory in Madras, where by this time she knew she would be based. Janaki was however too enamoured by Darlington and inspired by the work at the John Innes to want to leave, as evident from the epigraph to this chapter. Janaki especially missed 'loitering' about in England because she had to stay put at the John Innes to complete her experiments and write up the *Tulipa* paper.[40]

On 7 September 1931, the Second Round Table Conference had begun at the St. James's Palace in London; Mahatma Gandhi was the sole official Congress representative at the Conference, although he was accompanied by several others.[41] In Britain, the Labour government had fallen only two weeks before, and Ramsay MacDonald headed the new government dominated by the Conservatives. The world was reeling under the impact of the Depression, even before it had recovered fully from the

World War; Gandhi's Salt Satyagraha of 1930 we know was a response against the heavy taxation imposed by the colonial government during the slump. Even if the Round Table Conference was a failure as far as Gandhi was concerned, he had successfully captured public imagination in England. On the 22nd of that month, Charlie Chaplin, who was already a worldwide icon (*City Lights*, considered his best by the British Film Institute, had been released early that year), with a large fan base, would not miss an opportunity to meet Gandhi, residing at Kingsley Hall in Bow, one of London's poorest Dockland neighbourhoods. Gandhi would address enthusiastic crowds in London's East End and also make an informal visit to the Lancashire mills. Janaki must have followed these events with much interest, without doubt, but it does not appear that she saw Gandhi in London.

Only a day before the Round Table Conference, Janaki had written to Bartlett, having heard of his 'safe return from the wilds of Central America.' She wanted to see him once again, before leaving for India; she had after all prolonged her stay in England thanks to 'the excellent facilities' that John Innes had provided her for cytological work. 'I have been working with C. D. Darlington, who wanted me to look at some of Newton's unstudied slides of *Tulipa*—to note the chiasma behaviour in the genus,' she enlightened him. She was hoping to attend the meeting of the British Association for the Advancement of Science in October, and then 'leave for India by the quickest route.' Meanwhile, she had met with the High Commissioner of India (perhaps Atul Chandra Chatterjee), 'who is the official who looks after the interests of Indian students in Great Britain.' When she had approached him 'with regards to a post in India' he suggested that Janaki turn in an application, complete with a letter of recommendation, which she hoped Bartlett would provide. She had been occasionally meeting with a certain 'Mr and Mrs Sharpe' in London, Janaki informed Bartlett. 'You will be very busy interviewing new students when this reaches you. I wish I were also one,' she would playfully remark, obviously missing him and Ann Arbor very much.[42]

Janaki finally departed for India on 28th of that month (September 1931); Darlington did not fail to record the farewell moment in his pocket diary: 'EKJA kisses me good bye—"the only man I have ever kissed," she says.'[43] The prospect of returning to India was a mixed one for Janaki, for although the thought of 'wallowing in S. Indian jungles' in the company of elder brother Kittu, the forester, cheered her somewhat, she was deeply anxious about what she might find 'in the way of a post in India.' She wrote to Frieda Blanchard: 'I would feel happier if I had something definite to do as soon as I returned.' Janaki hoped Frieda and Bartlett would send her a few lines recommending her as a botanist, 'that is if you really feel you can' and wondered wistfully if the little Blanchard girls would remember her when they grew up.[44]

It appears however that the highly erudite Bartlett nursed a certain prejudice against Indians or 'Hindus,' as he referred to them. One entry in his diary, dated 4 February 1932, stated:

> There is something innately wrong with most Hindus. They are the most incompetent and inferior race on the face of the earth, barring none. Try as I will to be broadminded, this is a deep conviction that I can't shake off. To me Gandhi's maudlin inconsequentiality is typical of India. The Hindus say that he represents them perfectly, and he certainly does. Inane, longwinded, illogical, impractical, helpless, conceited, indirect—and I've always tried to make myself think I had no race prejudice!

To Bartlett, Janaki was an exception however; and as for her, she had immense affection and respect for Bartlett, as did his doctoral student, the phycologist Francisca Swamikannu Thivy of Madras (who taught botany at the WCC), who would spend a few years at Ann Arbor as a Barbour Scholar in the 1940s. In fact, the published version of the Bartlett diaries (1926–59) ends with Thivy's glowing tribute to her teacher: 'His brilliant mind and dedicated life imparted philosophical values to all those with whom he dealt just like the effect of things splendid or sublime in Nature.'[45]

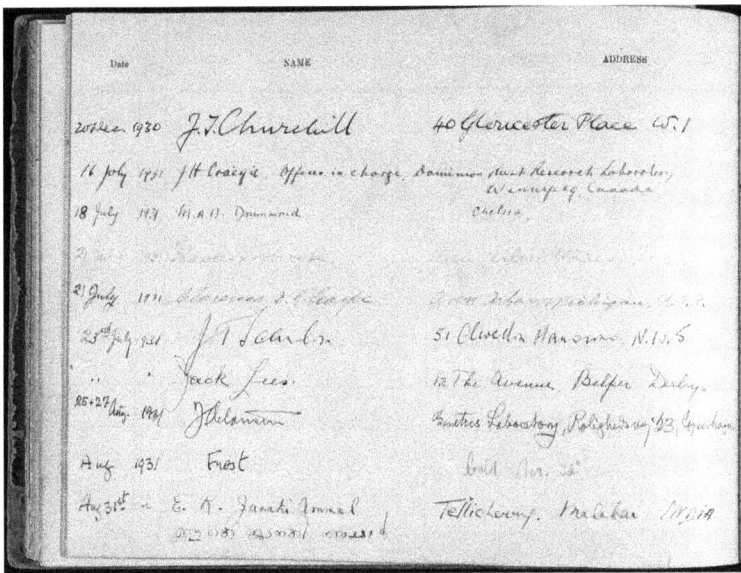

Figure 5.1 Janaki's signature in the Visitor's Register, John Innes. 31 August 1931.

Source: John Innes Archives courtesy of the John Innes Foundation.

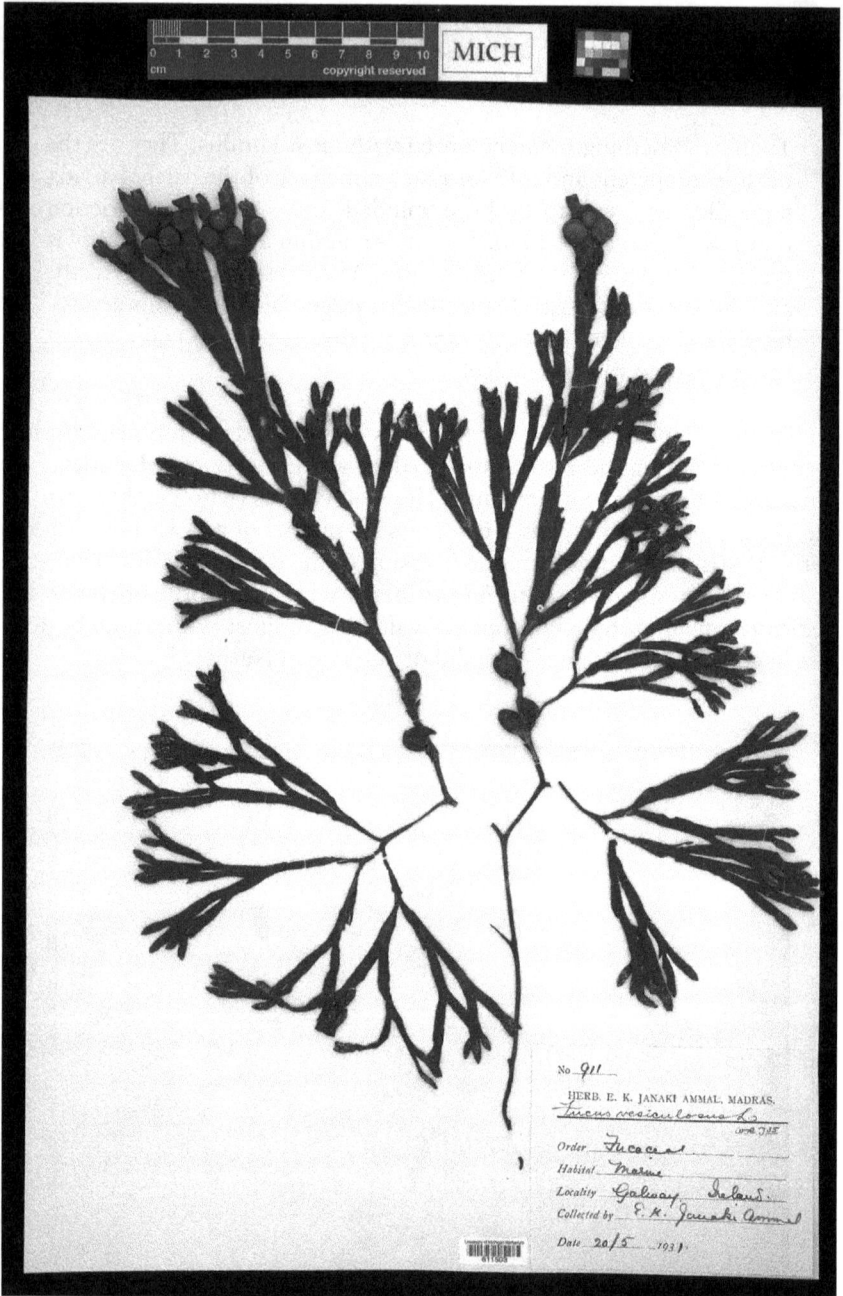

Figure 5.2 Fucus vesiculosus collected by Janaki from Galway, Ireland. 20 May 1931.
Source: Courtesy of University of Michigan Herbarium.

Figure 5.3 Janaki in the glass house (perhaps at the John Innes). 1931.
Source: Courtesy of the EK Family.

Figure 5.4 Cyril Dean Darlington. (1903–1981).

Source: John Innes Archives courtesy of the John Innes Foundation.

Notes

1 BHL, UoM: M. U. Botanical Gardens (Correspondence series), Box 5, E. K. Janaki Ammal to Frieda Blanchard, letter dated 2 August 1931.
2 Ibid., letter dated 8 May 1931.
3 Lewis, 'Cyril Dean Darlington 1903–1981'; for a recent biography see Harman, *The Man Who Invented the Chromosome*.
4 Among John Farmer's earliest doctoral students from India was cytologist P. C. Sarbadhikari of the Imperial College of Science and Technology, London, who later succeeded S. P. Agharkar as Ghosh Professor of Botany at the University College of Science, Calcutta. Sarbadhikari's original work was on the cytology of the fungi, ferns and flowering plants. Both as a student and while on leave as a teacher in Colombo, he made wide contacts, working at the Royal Botanic Gardens, Kew, at the John Innes during the time of Bateson, at the Jodrell Laboratory with Miss Digby, and in Paris under French cytologist A. Guillermond. For many years he was associated with the University of Ceylon, first as a lecturer and later as professor of botany.
5 John Innes Horticultural Institution (The John Innes), Norwich: Record of Work of the John Innes Horticultural Institution, 1910–1935, p. 2.
6 For an early history of the John Innes, see Harman, *The Man Who Invented the Chromosome*, p. 17.
7 Mendel, '*Versucheüber Pflanzen-Hybriden*'. About the same time as Bateson, Hugo de Vries (Amsterdam), Carl Correns (Tübingen) and Eric von Tschermak (Vienna) had independently rediscovered Mendel.
8 Mendel's first law, the law of segregation, states that the two alleles (copies) for a heritable character segregate (separate from each other) during gamete formation and end up in different gametes; the second law, the law of assortment (also referred to as the 'inheritance law'), states that separate genes for separate traits are passed independently of one another from parents to offspring.
9 In 1912, the Cambridge Chair of Biology was renamed the Chair of Genetics and occupied by Bateson's friend Reginald Punnett.
10 Harman, *The Man Who Invented the Chromosome*, pp. 18–19.
11 Darlington, 'Genetics and Plant Breeding, 1910–80', p. 401.
12 Harman, *The Man Who Invented the Chromosome*, p. 28.
13 Darlington, 'Genetics and Plant Breeding, 1910–80', p. 402.
14 Fisher's 1930 work, *The Genetical Theory of Natural Selection*, which put forth a mathematical model of a population of hypothetical organisms, using complex and innovative mathematical techniques, was a major contribution to the theory of evolution. The eugenicist that he was, Fisher aimed to demonstrate how favourable genes spread through a population and how unfavourable variations could survive to maintain overall genetic diversity.
15 Harman, *The Man Who Invented the Chromosome*, pp. 55–56.
16 Three years later, in 1930, Darlington would publish 'A cytological demonstration of "Genetic" Crossing-over (*Hyacinthus*)' in the *Proceedings of the Royal Society of London*.
17 Belling and Blakeslee, 'The configurations and sizes of the chromosomes in the trivalents of 25-chromosome *Daturas*.'
18 Harman, *The Man Who Invented the Chromosome*, pp. 58–60.
19 Lewis, 'Leonard Francis La Cour 1907–1984.'
20 The John Innes, Norwich: Visitor's Register.
21 BHL, UoM: M. U. Botanical Gardens (Correspondence series), Box 5, E. K. Janaki Ammal to Frieda Blanchard, letter dated 12 June 1931.

22 Harman, *The Man Who Invented the Chromosome*, p. 106.
23 BHL, UoM: M. U. Botanical Gardens (Correspondence series), Box 5, E. K. Janaki Ammal to Frieda Blanchard, letter dated 12 June 1931.
24 Ibid.
25 Ibid., letter dated 8 May 1931.
26 Ibid., Frieda Blanchard to E. K. Janaki Ammal, letter dated 8 June 1931.
27 Ibid., E. K. Janaki Ammal to Frieda Blanchard, letter dated 12 June 1931.
28 Ibid., letter dated 28 June 1931.
29 DP: Darlington's pocket diary, g. 3, 1931.
30 BHL, UoM: M. U. Botanical Gardens (Correspondence series), Box 5, E. K. Janaki Ammal to Frieda Blanchard, letter dated, letter dated 12 June 1931.
31 Ibid., letter dated 28 June 1931.
32 Harman, *The Man Who Invented the Chromosome*, p. 77.
33 Janaki Ammal, 'A polyploid eggplant, *Solanum melongena* Linn.', 1931, p. 81.
34 BHL, UoM: M. U. Botanical Gardens (Correspondence series), Box 5, E. K. Janaki Ammal to Frieda Blanchard, letter dated 28 June 1931.
35 Ibid., Frieda Blanchard to E. K. Janaki Ammal, letter dated 8 June 1931.
36 Ibid., 20 July 1931.
37 Jorgensen and Crane, 'Formation and morphology of Solanum chimaeras'.
38 BHL, UoM: M. U. Botanical Gardens (Correspondence series), Box 5, E. K. Janaki Ammal to Frieda Blanchard, letter dated 2 August 1931.
39 Darlington and Janaki Ammal, 'The origin and behaviour of Chiasmata, I. Diploid and tetraploid tulips', 1932.
40 BHL, UoM: M. U. Botanical Gardens (Correspondence series), Box 5, E. K. Janaki Ammal to Frieda Blanchard, letter dated 2 August 1931.
41 Gandhi was accompanied by Sarojini Naidu, Madan Mohan Malaviya, Ghanshyam Das Birla, Muhammad Iqbal, Mirza Ismail, S. K. Dutta and Syed Ali Imam.
42 BHL, UoM: University Herbarium Records (University of Michigan) H. H. Bartlett series, Box 11, E. K. Janaki Ammal to H. H. Bartlett, letter dated 6 September 1931.
43 DP: Darlington's pocket diary, g. 3, 1931.
44 BHL, UoM: M. U. Botanical Gardens (Correspondence series), Box 5, E. K. Janaki Ammal to Frieda Blanchard, letter dated 17 August 1931.
45 *The Harley Harris Bartlett Diaries (1926–59)*, p. 323.

6

MADRAS II
A Flora of South India

I find I could slip into old Malabar ways with the greatest
ease but it is not so easy to slip into the ways of other people's
thinking. I am trying not to be too critical when I find men and
women looking at things with hundred-year-old spectacles.
                                          —E. K. Janaki Ammal (1932)[1]

On arrival from England in late October 1931, Janaki headed straight to
Tellicherry, to reunite with her family, who she had not seen for more than
three years now. There was every reason to celebrate, most of all her success; over the next few days, she would do everything to rejuvenate herself:
indulge in medicated oil baths, dine on the sumptuous Thiya fare prepared
lovingly by her mother, and catch up with news on the family front. She
would also almost immediately immerse herself in what she enjoyed and
missed the most, plant hunting in the Edam surrounds. From the wet fields
close by in the first week of November, she collected grasses such as the
Arthraxon hispidus (the small carpet grass), *Paspalum scrobiculatum* (Kodo
millet) and *Ischaemum rugosum* (saramolla grass) and, from the mud-walls,
Arundinella pumila (dwarf reed grass).

After a week at Edam, Janaki visited Shoranur, where sister Cousalya
and brother Vasudevan lived with their families; eldest brother Raghavan
was still in Burma. Apart from enjoying her time with them, she would
spend several hours of the day wandering in the open hill sides, looking
for grasses. About the 21st of that month, she would leave for Mangalore, a few hours away by train from Shoranur, to meet youngest sister
Devayani, who was employed as a teacher; Devayani was married only
a year ago to a Thiya doctor employed in Burma.[2] No matter where she
went and the reason for it, Janaki always slipped away for a ramble or
two to hunt for plants; at this time, her mind was on grasses. Among the
plants she collected from this place were several specimens of the *Dimeria*.
Distinguishing species within this genus was never easy, sometimes even
with a lens; Chase's help was of great help in such cases. After a couple

DOI: 10.4324/9781003267089-6   103

of weeks with her sister, Janaki travelled through North Malabar, to the tablelands of Mattanur, where more plants would be collected. Once she had formed a substantial collection, she would despatch them to Chase at the Smithsonian Herbarium as an offering, and mostly in exchange for species determination.

Research Fellow at the Presidency College

By the time she left England, Janaki had heard from the Madras University about the decision to award her a year's fellowship worth Rs 150 per month, for the year 1932–33.[3] This was a saving grace, as she had arrived penniless from England and had been forced to borrow again from her brother Kittu. Janaki was determined to make the best of this opportunity but did not cease to look out for suitable government positions. 'I have all facilities for work, which is a very great thing. I am hoping to get into the agricultural department by and by,' she would inform Frieda Blanchard. In fact, the Government Sugarcane Expert, T. S. Venkatraman had requested help with the cytology of Indian sugarcane and had already sent her some wild specimens for the purpose. With Janaki in mind, he had approached the Imperial Council of Agricultural Research for a cytological assistant and was hopeful of hearing in the affirmative. Moreover, Ekambaram had put her in charge of the newly formed botanical garden at the Presidency College, which she found 'great fun' to collect for; as a matter of fact, over and above the collecting she had accomplished in North Malabar and Mangalore in the weeks after her arrival, she had collected intensively at the Coimbatore Agricultural Farm, en route to Madras (mid-December), where without doubt she met with Venkatraman at the Sugarcane Breeding Station; this would perhaps have been their first ever meeting. Janaki was quite content with this temporary arrangement with the Madras University, even though she could only barely eke out a living on the fellowship. She had also by this time received news of being elected to the Fellowship of the Linnean Society. It was in August that Janaki had turned in an application, supported by geneticist Frederick Whalley Sansome of the John Innes, mycologist J. Ramsbottom of the British Museum (Nat. Hist.) and plant geneticist D. G. Catcheside. Her request would be accepted after a ballot held on 3 December 1931 at the Society.[4]

Empowering Indian Women

Over a period of two weeks, 28 December 1931–11 January 1932, the sixth session of the All India Women's Conference (AIWC) was held at the Senate House, Madras, with Mrs P. K. Ray as President and Muthulakshmi Reddy (1886–1968), one of the earliest Indian women to graduate in medicine (1912), as one of the Vice-Presidents.[5] The brain-child of the suffragist Margaret Cousins, the AIWC had held its first meeting in Poona (Pune) in January 1927, attended by a great number of social reformers, professional educationists and women associated with the national movement, from all castes and classes. Numerous resolutions were passed outlining the best sort of education for women that would allow the most complete development of the individual while at the same time teaching the ideals of motherhood and good housekeeping. Mrs Ray's predecessor, at the fifth session of the AIWC at Lahore, was Reddy. Under the influence of Annie Besant, Mahatma Gandhi and Sarojini Naidu in her college years, Reddy had become a champion of women's rights.[6]

'Life is very interesting in Madras,' Janaki would write to Frieda from the Department of Botany, Presidency College, 'We are having our All India Women's Educational Conference [All India Women's Conference for Education and Social Reform] and most of the delegates from the North are being housed at Queen Mary's College, where I am now. Miss Agha[7] is expected here tomorrow.'[8] We have already encountered the Kashmir-born Sharkeshwari Agha, the Barbour scholar (1928–29) and Janaki's contemporary at Ann Arbor. Since her arrival in Madras, Janaki had been residing at QMC, Madras, where her close family friend C. K. Kausalya was professor of Botany; both were also members of the Board of Studies for Natural Science of the Madras University, which at this time also included, among others, T. Ekambaram and M. O. P. Aiyengar.[9] Incidentally, also teaching at the College was Maneck M. Mehta, the first Indian woman to earn a doctorate in the sciences, and a member of the Board of Studies (chemistry).

As a matter of fact, Janaki had already begun hunting for plants in the sprawling QMC campus by this time.[10] 'You and the little girls have been very much in my thoughts ever since I came back but I have done so little letter writing. I lead a very gypsy life in Malabar visiting all the members of the large EK clan, and being introduced to all the new arrivals. I am now back in Madras. The Syndicate of University of Madras has allowed me to work at the Presidency College, my old College, as a Research fellow,' she would tell Frieda.[11] Although happy workwise, she was struggling to fit into a world that was so vastly different from the one she had experienced in the past few years, at Ann Arbor and England: 'I suppose it takes some time to feel at home even in one's home after leaving it for long absences,' she would comment. Janaki found that people's attitudes had hardly changed; they were still 'looking at things with hundred-year-old spectacles. I must say we are throwing them off very fast but not quick enough for me,' she would remark. Meeting and interacting with members of the AIWC would have not only been an empowering experience for

Janaki but also very reassuring, at a time when she was coming to terms with her own inner demons.

In February 1932, Janaki received photographs of the three Blanchard children, who she missed very much; the one of a smiling Grace was promptly installed on her writing table and would remain a source of constant joy to her.[12] In her reply envelope, she enclosed seeds of *Ephedra* (the medicinal plant, *E. sinica*) for Frieda, which she had received from Lahore for the botanic garden, and promised to send seeds of perennials in exchange for those of solanaceous plants. She also hoped to make a collection of herbarium specimens for Bartlett while gathering material enough for a small handbook of South Indian plants, but wondered if she was being too ambitious. 'Do you still have tea in the Dept? I wish I could drop in,' she wrote to Frieda, as she introduced friend and colleague, Kausalya: 'One of my friends—Miss C. K. Kausalya is hoping to spend a year in USA. She may spend it at Michigan. She is a very charming person and the professor of Botany at Queen Mary's College, where I am staying. I am sure the department will enjoy having her in their midst.'[13]

Incidentally, Janaki had only recently recommended Kausalya's name to the Barbour Committee: 'One of the senior professors at Queen Mary's; one of the first women to take a science degree from the U of Madras; later received the degree of BSc (Hons) from the U of London; has helped very much to encourage the study of biology in our schools and colleges; charming personality; excellent representation of Indian womanhood, comes from a distinguished and cultured family, being the niece of the late Sir C. K. Krishnan, one of the justices of Madras; a very suitable person to be the Barbour Fellow from India this year.'[14] The petite Kausalya would spend a year in the United States, but it does not appear this was in the capacity of a Barbour Fellow. In June 1933, she would visit the University of Wisconsin-Madison and the purpose of her visit was rather than botanical, a direct influence of the proceedings of the AIWC (1931). Kausalya wished to begin 'home economics extension work' in India similar to that in America. Although all families maintained sufficiently large homesteads, 'very little gardening is done, and that chiefly by the men of the family.' In the women's colleges in India, she observed, the study of cooking, sewing, gardening, health and hygiene was still being defined simplistically, as domestic 'science.' The first step, Kausalya suggested, was to take the subject to a scholarly level by conceiving it as 'home economics.'[15]

Drawing Women to Science

Close by on Cathedral Road, at the Agri-Horticultural Society Gardens (founded by Robert Wight about a century earlier) was situated the

University Botanical Laboratory (UBL), a modest unit established by Vice-Chancellor K. Ramunni Menon in 1930. Janaki's former teacher, the plant physiologist Todla Ekambaram, a student of the Cambridge biologist Frederick Frost Blackman, was its first honorary director. Ekambaram incidentally served as president of the Indian Botanical Society (founded in 1920) in 1931–32. The UBL was moved to its own building behind the Senate House on the Marina in 1933, and another of Janaki's teachers, Mandayam Osuri Parthasarathy Aiyengar (1886–1963) was appointed its first permanent director; Aiyengar had studied under algologist Felix Eugen Fritsch in London.[16] Both Ekambaram and Aiyengar encouraged women to take up research in science, evident from the great number of papers contributed by women, singly and in collaboration, in the early decades of UBL.[17] Mercia Janet and Rachel P. John were two of the earliest among women at the UBL to complete their postgraduate theses (on the algae, in 1936). Among the men researching at the UBL at this time were T. N. Ranganathan, I. Madhusudana Rao and Rama Rao Panje; Panje would soon find employment at the Sugarcane Breeding Station at Coimbatore.

A Cytology Laboratory at the Presidency

Instead of the UBL, which was chiefly devoted to morphology and physiology, Janaki set up a small cytological laboratory in the Botany Department of the Presidency College, where Ekambaram taught, to work on Venkatraman's sugarcane material. Ekambaram and another of her former teachers, the systematist Philip F. Fyson,[18] who was at this time Principal of the College, were keen to encourage cytological studies in Madras and suggested that Janaki guide three male students working on their MSc thesis, and also deliver lectures in cytology to the Honours (final degree) students at the Presidency. She also found herself busy collecting in the Madras neighbourhood, including the suburbs; in February 1932, she collected specimens of *Sesuvium* (sea purslane) from the salty marshes at Adyar, and *Aristida* and *Chrysopogon* grasses from Tambaram, perhaps from the florally rich and rambling Madras Christian College campus.

It would be early March, by the time Janaki received her blotting papers, presses (to dry her specimens), box of labels (to affix on the herbarium sheet) and typewriter that she had left behind in Michigan. 'A great wave of homesickness for Ann Arbor, especially for the Botanical Garden, swept over me as I unpacked,' she remarked to Frieda; although the package had reached the Madras port as early as January, she had only received news of its arrival much later. Janaki was thrilled that at last she could get on with her crosses and selfings, as the solanaceous seeds she so wished for were also in the package.[19] The *Vinca* seeds she had brought back from Ann Arbor were refusing to germinate; she was however hopeful that the ones Frieda had sent would certainly do, and help her save time in producing pure lines.

She was also putting together a collection of economic plants with a view to encouraging genetical studies in Madras. 'At present there is nothing done. All the genetical work is done by the agricultural Dept. and they are interested only in crop plants.'[20] Janaki was 'perfectly happy' at the Presidency College and was excited by the interesting results she was obtaining from her sugarcane material. She was also delivering a few lectures in cytology and genetics 'as there is nobody here to do that phase of botany' but was often filled with guilt about 'not taking part in the struggle for . . . freedom.'[21]

Overall, she was pleased with life, so much so that she dreaded the prospect of anything altering it, not even the Coimbatore job. 'Madras is a very cosmopolitan city and I prefer it to a narrow Brahmanical one like Coimbatore,' she would comment.[22] All the same, Janaki felt it was time she earned a 'little money' and bought herself a 'little car to rush round the country' to collect plants, which the Coimbatore job would make possible. She was expecting to appear before the Public Service Commission for the post of Sugarcane Cytologist, and hoped Bartlett would send her a strong letter of recommendation as these 'play a great part in these assemblies.'[23]

The Botanising Family

Janaki did not attend the entire two-week session of the AIWC held in Madras. Instead, she travelled to Salem a few days after the New Year (1932) to meet elder brother Kittu, who she had not yet met after her return. Kittu was based at the Hosur Cattle Farm, not far from Bangalore, and as was usual when they got together, went on plant collecting excursions. Janaki and Kittu enjoyed camp life immensely. At Salem, he had been put in charge of forest settlement work in what was a dry, deciduous scrub jungle and was expected to stay in tents two or three weeks at a time. He always carried his 12-bore Westley Richards shotgun, 'an old trusty weapon . . . acquired from the family armoury,' when he joined the forest service in April 1910, and a .22 bore BSA rifle, besides a clasp knife and a Lauder Brunton lancet, kept in his coat-pocket, the last used in the event of a snake bite.

On the 10th of that month (January), among the grasses Janaki collected from Hosur were *Aristida*, *Urochloa* and *Echinochloa* ('jungle rice'), besides an *Imperata cylindrica* (Cogon grass), which would eventually reach Chase at the Smithsonian. Kittu would himself despatch some specimens to Ekambaram, at the Presidency College, collected from Hosur. Shortly after this visit from Janaki, Kittu published a note in the *Indian Forester* on a biological method for containing the weed *Lantana*, so prolific in the forests of the

Mysore plateau, using the plant *Dodonaea viscosa*.[24] There is no doubt he discussed such matters with his highly accomplished botanist sister. While in Bangalore, Kittu rarely ever missed visits to the Cubbon Park, the Lal Bagh and the several garden nurseries of the city, looking for plants. Incidentally, in May, niece Leela (the younger daughter of sister Cousalya), who had recently completed an MA in Botany from the Presidency College (she was however disappointed with the third class she had been awarded) would call on her uncle; only two years earlier as a graduate student of St Agnes College, Mangalore she had won a gold medal of the Madras University for excellence in the English language (the Dr T. M. Nair Gold Medal). As always, uncle and niece would go plant hunting. Other siblings would also visit him over the year. Varadan, the youngest, would arrive that summer to indulge in his favourite recreation of hunting in the company of his elder brother; a few months later, he would be married. Janaki would again not attend, as she was away in Lahore to set the university examination papers in botany.[25]

Island archipelagos were for the naturalist ideal places (as it was famously for Darwin) to discover a range of forms, in various stages of speciation, with potential to shed light on the evolutionary process. Janaki much wished to visit the Laccadives (Lakshadweep, an archipelago comprising of twelve atolls, three reefs and five submerged banks) in pursuit of her research, but this would not happen in her lifetime. However, her adventurous younger brother, E. K. Madhavan, a master boat-builder and great lover of islands, would send her collections from remote islands. In late June 1932, Janaki travelled to Rameshwaram, to visit the Krusadai Island, near Pamban, a veritable paradise for biologists and to spend some days with her brother Madhavan, who was at this time, the Island's chief care-taker; he was employed with the Department of Fisheries, Madras Government. Madhavan and Janaki were very close in age and enjoyed each other's company immensely. She would spend several hours rambling along the coast and collecting marine specimens from the sandy lagoon and the coral reefs—algae chiefly—all of which would go into the making of her 'Flora of South India.' It may be stated here that Janaki maintained an abiding interest, even if largely unrealised academically speaking, in the study of what is today referred as island biogeography—examining the factors that affect species richness and speciation in isolated natural communities such as remote islands—a branch of study that would later, in the 1960s, become associated with the work of ecologists such as E. O. Wilson and Robert H. MacArthur.[26]

Tropical Plant Material for Bartlett

Some weeks later, Janaki received 'a really good likeness' of Bartlett and in her reply to his, reminded him that she was still attached to her old college (Presidency College) as a Research Fellow and that she was 'enjoying' her work very much. 'Looking after the new botanical garden is my pet job. I am often reminded of the garden at Ann Arbor and try to follow what is being done there,' she told Bartlett. In her letter, she had enclosed 'a small parcel of seeds of some Indian shrubs & trees' collected by her brother Kittu from 'the forests of North Salem—close to Mysore province.' 'I thought you would like to try growing these in the tropical house!' She was expecting to remain in Madras until January the next year (1933). 'The Agricultural Council is very slow in sanctioning a cytologist for the Imperial Sugarcane Station at Coimbatore. The Cane expert [T. S. Venkatraman] has me in view for the post. I am helping him with the cytology of some species of Saccharum even now and he has permitted me to make a cytological study of the Sorghum x Saccharum hybrid,' she explained. Janaki hoped Bartlett would visit India, at least during his next expedition to Sumatra. 'It is a great thing to look forward to. Please come before our castes and tribes become too civilised. You may like us even less otherwise,' she joked. She had also collected some herbarium specimens, she informed him, 'including some of marine algae' from Krusadai Island in the summer. 'That is as far as I got to the Laccadives this year,' she added.[27]

Fixing Material at the Sugarcane Breeding Station

The Imperial Council of Agricultural Research had allotted a prime place to the production of sugar in their development programme, which revolved around three aspects: the selection of canes better suited to the important sugar-growing districts than the existing local varieties and the improvement of their cultivation; the improvement of the local methods of making *gur* or jaggery; and the development of an advisory and research service for the sugar factories. The existing Sugarcane Breeding Station at Coimbatore was geared towards serving this three-pronged aim. Breeding and selection of sugarcane was a 'slow business' (there was a period of seven years between the production of new seed and the availability of the fully tested new variety in sufficient quantities), making government support crucial for its improvement. The Imperial Council under its acting director B. Vishwanath thus allocated a grant of Rs 37,400 towards hiring a geneticist (spread over five years), on the understanding that cytological studies at experimental stations provided valuable information to the breeder about the genetics of crops. It was under this Government of India scheme that Janaki was to be considered for employment at the Sugarcane Breeding Station.

In November, Janaki visited the Sugarcane Breeding Station to fix some more material for research, especially of Venkatraman's intergeneric

Saccharum-Sorghum' hybrid. *Sorghum* belongs to the family Gramineae (to which belong the most important economic plants including cereal grasses such as barley, maize, rye and wheat, besides bamboos and sugarcane), the subfamily Panicoideae and tribe Andropogoneae, and is thus closely related to sugarcane. It was in 1929 that Venkatraman had successfully crossed *Saccharum officinarum* (clone POJ 2725) and *Sorghum bicolor* (the 'peri-amanjal' variety) to produce hybrid plants, with the aim of producing an early maturing cane.[28] However, successive attempts to breed early canes from derivatives of this intergeneric hybrid failed.

We may recall that one of the first books that Janaki consulted on grasses was Mary Agnes Chase's *The First Book of Grasses* (1922). Incidentally, in the 1930s, the American evolutionary biologist, Edgar Anderson (1897–1969), an exact contemporary of Janaki, who we will encounter several times in the present biography, would begin his study on the maize (a grass, like the sugarcane) by turning to a book by another woman plant scientist, Agnes Arber; Arber's book *The Gramineae* (1934), was 'so fundamental in its simplicity that it [was] the scorn of college students but the delight of scholars.' She was incidentally the first woman botanist (awarded a doctorate in botany in 1905) to be elected Fellow of the Royal Society (in 1946). Anderson tells us that with time it began to dawn on him that 'like a good sonnet the book had more than just words—it transmitted an attitude, and with the new attitude one could look down whole new vistas of experience, and old facts took on new significance.'[29] That Janaki found Chase's book extremely useful is also beyond doubt and among grasses, the sugarcane would become her first love.

Research Plans

Janaki kept Darlington informed of her research plans. Venkatraman had already selected a few early maturing types from among hundreds of his F1 plants of the intergeneric cross (*Saccharum-Sorghum*), the ones which promised 'to be very valuable from the economic point of view.' There were also a number of interspecific hybrids in *Saccharum* that were awaiting cytological study, and the reason Venkatraman had requested the Government for a 'scientific worker,' she explained to him in a letter posted only after her return to Madras. Janaki had fixed root tips and buds of wild, cultivated and hybrid sugarcanes at the Breeding Station to take back with her and was excited by the 'terrific amount of variation in the F1s.' She felt that 'the dwarfs and pale fellows' were 'aching to disclose something very interesting cytologically and genetically.'[30] To the cytogeneticist, the sugarcane was an unknown grass, at least almost. Nothing had been done with these from the 'scientific point of view' except determining chromosome counts of the parents and the hybrid, with the help of Gustav Bremer of the Wageningen Agricultural College in the Netherlands.

Darlington had by this time sent away Janaki's paper on the *Nicandra* (based on her doctoral thesis) to cytologist V. Gregoirè for publication in the *La Cellule*. She would comment: 'I hope he will accept it. I think most of the drawings are better in the waste paper basket.'[31] The drawings would indeed be discarded, but the paper published.[32]

Some Family Time

Janaki would break journey at Shoranur on her return journey to Madras from Coimbatore, where elder sister Cousalya and her litterateur husband K. Sukumaran lived in their house, Coustubham. On one occasion, Janaki would gift her brother-in-law (who was also a cousin, being their paternal aunt Neeli's son), who was a voracious reader, a copy of James Frazer's *The Golden Bough: A Study in Comparative Religion* (1890).[33] 'I am in the most delightful part of old Malabar practically untouched by modern civilisation. You will be surprised how very "modern" it is consequently,' a perceptive Janaki would write to Darlington from Shoranur. She was pleased that at least Brenhilda Schaffer of the John Innes wrote her 'nice letters.' 'I am rushing about so much these days that I can scarcely find time to sit down and write a proper letter. John Innes is always in my mind!', she told him.[34] She had indeed become a nomad, ever since her return from England, combining social visits with plant collecting for her ambitious Floras.

Science, State and Society

The year 1932 was an important one for Indian science. It saw the birth of the *Current Science*, an Indian journal of science founded by a working committee comprising pioneering Indian scientists C.V. Raman, Birbal Sahni, Meghnad Saha and S. S. Bhatnagar in consultation with the chemist and director of the Indian Institute of Science, Martin O. Forster. It was decided that the journal would be published on a monthly basis from Bangalore with the editorial cooperation of a large number of scientists. News and discussions revolving around the relationship between science, society and state were regularly reported on the pages of the *Current Science*. Very importantly, the *Current Science* offered the much-needed platform to make contributions of women scientists known more widely and speedily. For instance, a review of a scientific paper, by Janaki's former colleague at the WCC, the zoologist Eleanor D. Mason titled 'Some Aspects of Racial Anthropology,' and read before the Society of Biological Chemists (established in India in 1930) was published in the very first issue of the *Current Science* (July 1932).[35] Very importantly, the founding of the journal coincided with the beginning of the 'social relations of science movement' in Britain under the aegis of the British Association of Advancement of Science, an organisation founded in 1831. Modelled after Richard Gregory's *Nature*, Arnold Berliner's *Die Naturwissenchaften* and James McKeen Cattell's *Science*, Indian scientists looked up to the *Current Science* for scientific news, especially from abroad. Referring to scientists as 'scientific workers,' the journal strived to be India's own *Nature*, and claimed to accurately represent the progress of Indian science.[36]

Slant Towards Eugenics

It was again in 1932 that Janaki became a member of the British Eugenics Society, probably the first Indian to do so. Interestingly, this was also the year that the Society's membership peaked to an all-time high. The Eugenics Society, today called the Galton Institute, was founded in London in 1926 (in continuation of the Eugenics Education Society established in 1907, with Francis Galton as its honorary President), and from its very early days was successful in recruiting members who were highly regarded academics and scientists, especially geneticists who were in the business of making improved plants and animals. Prominent members included evolutionary biologist Julian Huxley ('no-one doubts the wisdom of managing the germ-plasm of agricultural stocks, so why not apply the same concept to human stocks?'),[37] besides economist and Bloomsbury socialite John Maynard Keynes, anthropologists C. G. Seligman and A. C. Haddon, political scientist Harold J. Laski, physician Havelock Ellis, the biometricians Ronald Fisher and Karl Pearson, psychiatrist C. P. Blacker, geneticists MacBride and R. R. Gates and biologist and geographer Patrick Geddes. Among well-known women eugenicists were birth-control campaigners Marie Stopes (wife of R. R. Gates) and Margaret Sanger.[38] While Haldane does not appear in the membership list, his mother and sister do. However, Haldane published a piece titled 'If you were alive in 2023 A D,' in which he spoke prophetically of the birth in 1951 of the world's first 'ectogenetic child,' produced from an embryo grown for nine months in the biologist's laboratory and then brought into the open; the closest to such a birth was the so-called 'test-tube baby' in 1978. Haldane certainly did not think that the idea of eugenics was anything novel nor did he think it was tremendously exciting in its present state, but did see its potential in 'crafting' the desirable individual.[39]

Interestingly, Darlington cautiously kept away from eugenics like his teacher Bateson, who believed that it was a useless distraction, 'giving a doubtful flavor to good material,' although in his later years he would join the society.[40] There is no evidence that Janaki's friend Eileen Erlanson was a member either, despite her turning to the study of human genetics in the 1930s. What motivated Janaki then to become a member of the society? Was it her mixed-race origin or the social class she belonged to? Or was it the impact of the recent AIWC conference held in Madras, which she had attended? In 1931, the AIWC despite protests had passed resolutions to open birth-control clinics influenced by Stopes and Sangers, and a year later appealed to civic institutions to educate newly married couples in sexual matters.[41] Janaki was deeply anxious about population explosion, a concern frequently voiced at the AIWC conferences; she had also imbibed this from Bartlett, her mentor at Michigan, who was a eugenicist himself. Among the books added to her personal collection in the recent times was John Berry Haycraft's *Darwinism and Race Progress* (1900), presented to her. Haycraft, a professor of physiology at the University College, Cardiff was an advocate of negative eugenics in Britain, and applied the ideology of social Darwinism to argue that the only sensible option available to man was to apply the 'same care and attention to our own race propagation that a gardener does to his roses or chrysanthemums, or a dog-fancier to his hounds or terriers.'[42] We also know that in 1933, Janaki purchased a copy of Leonard Darwin's *The Need for Eugenic Reforms* (1926), a five-hundred-odd paged heavy tome; she was certainly re-examining her position with respect to eugenics at this time.[43] At no point in her life did she explicitly discuss the subject of eugenics, but we know for sure that she was a keen student of human genetics and anthropology, and often supplied Darlington and Eileen with useful information drawn from the Indian context.

A Scientist Among the Soviets

The year 1932 also saw the publication of eugenicist Julian Huxley's *A Scientist Among the Soviets* and that of the *Brave New World* by his brother Aldous Huxley on the evils of totalitarianism and the use of technology to support it. Janaki would buy her copy of Julian Huxley's book within months of its publication, from the Higginbotham outlet in Madras, which provided her with a wonderful introduction to social and economic planning in the Soviet Union under Joseph Stalin. That same month, the Sixth International Congress of Genetics was held at the Cornell University but Janaki would not attend. Her doctoral supervisor B. M. Davis was Chairman, and cytogeneticist McClintock was Vice-Chairperson of the cytology section of the Congress. Davis delivered a paper on the genetics and cytology of triploids and tetraploids of the *Oenothera franciscana*, while Eileen Erlanson presented her latest research on the *Rosa*. Among the other participants at the Congress was the Soviet geneticist Nikolai Vavilov, who spoke on evolution in cultivated plants, a subject that would totally capture Janaki's attention in the years following.

Penniless in Madras

Late that year (1932), Janaki was appointed Reader in Botany, Chairman of the Board of Examiners in Botany, Member of the Board of Studies and Academic Council and Member of the College Council. She was still struggling to make ends meet though, because but for a University Fellowship, there was little by way of a salary. There was also no news still about the post of cytologist at Coimbatore. So when she received the offer of a Professorship at the Maharaja's College of Science in Trivandrum in Travancore State, it seemed like a blessing in disguise. This is not to say that she was over the moon with the news, because it also implied relegating her research to the back-burner. It was out of desperation that Janaki had accepted the offer, and so when brother Raghavan's letter of congratulation reached her from Burma, she could not but respond angrily about her awful predicament, and with an unmistakeable sense of urgency about her life's chief purpose, which was striving for work rather than material comfort or happiness:

> Thanks for your letter & congratulations—and the clipping—Everybody seems happy about this post—except myself. It brings me more money. My pay will be Rs 450—but it means divorce from my life's ambition—scientific research—very few know my mind in this direction. I am taking the post as a detour towards the end in view. Money I suppose we all need—I most of all because there is so much to be done but to sell our ideal for money is almost like prostitution. However I shall try my best to keep up my work and achieve the end—be a good scientist, a recognized scientist. . . . I have very little time to write letters. You must excuse my silence. I go to Trivandrum. . . . Shall write from there.[44]

Janaki's was the first instance of a woman appointed to the professorial chair in what was a men's college, and newspapers publicised her appointment. She was eager to take her nieces Ganga and Yamuna (daughters of her brothers, E. K. Raghavan and E. K. Vasudevan, respectively) with her to her newly adopted city, the capital of the princely state of Travancore. She would enquire with Raghavan: 'Do you object to my taking Ganga with me to Travancore? I am drawn to the girl in a strange way—and would like to have a hand in her training. I mean to take Yamuna also with me. They were both with me at summer time and we had jolly times together.'[45]

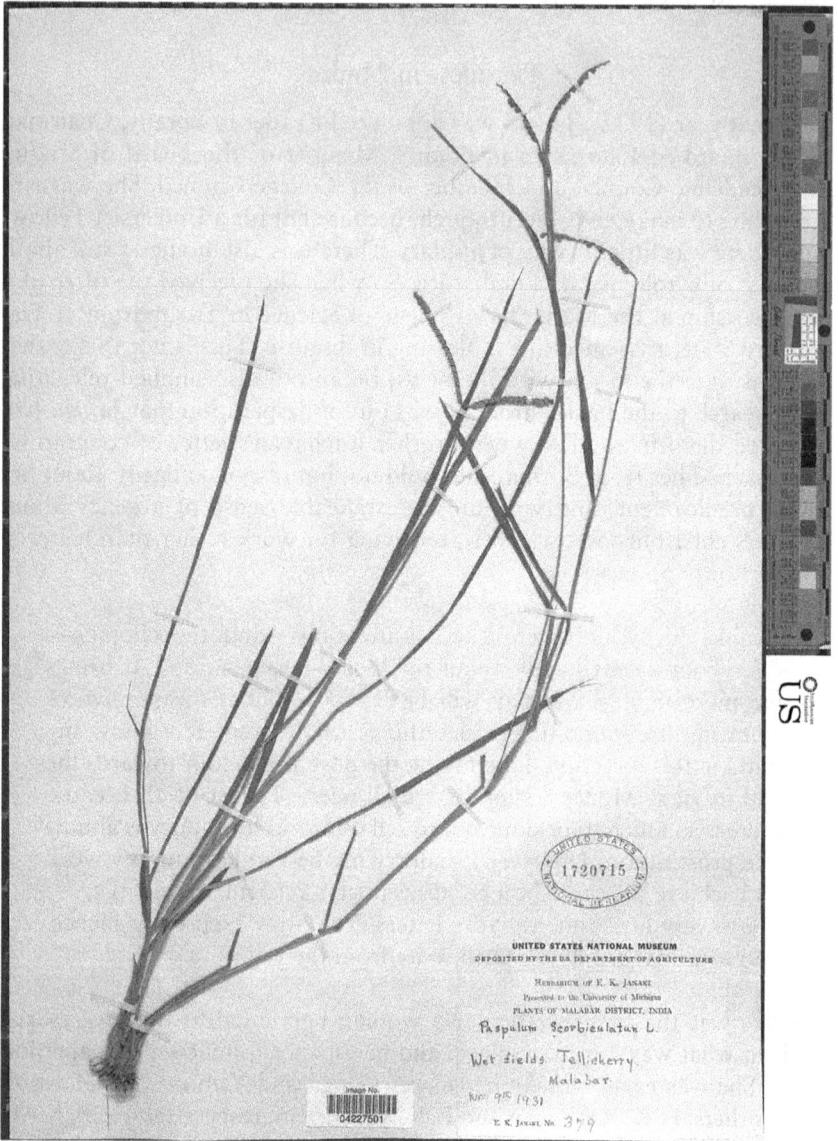

Figure 6.1 Paspalum scrobiculatum collected by Janaki from the wet fields of Tellicherry on 9 November 1931.

Source: 'courtesy of the United States National Herbarium (US)' 04227501.

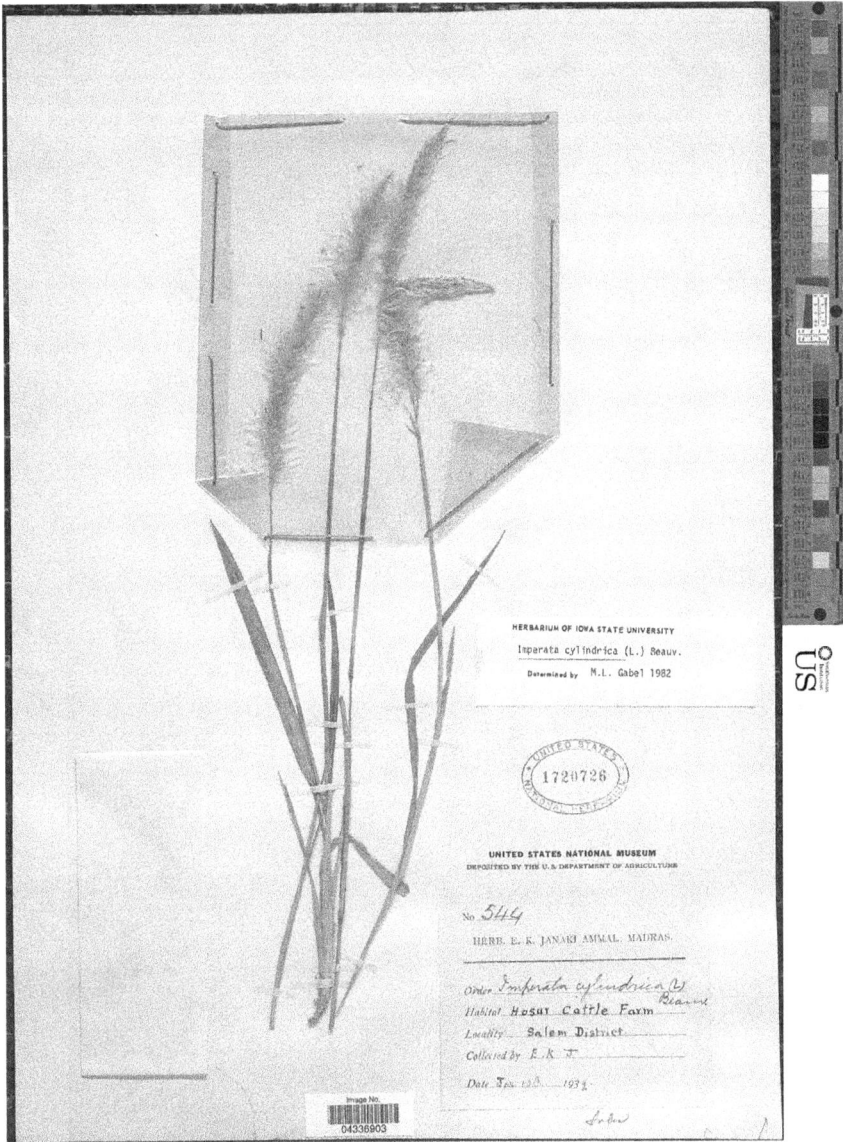

Figure 6.2 Imperata cylindrica (Cogon Grass) collected by Janaki from Hosur Cattle Farm on 10 January 1932.

Source: 'courtesy of the United States National Herbarium (US)' 04336903.

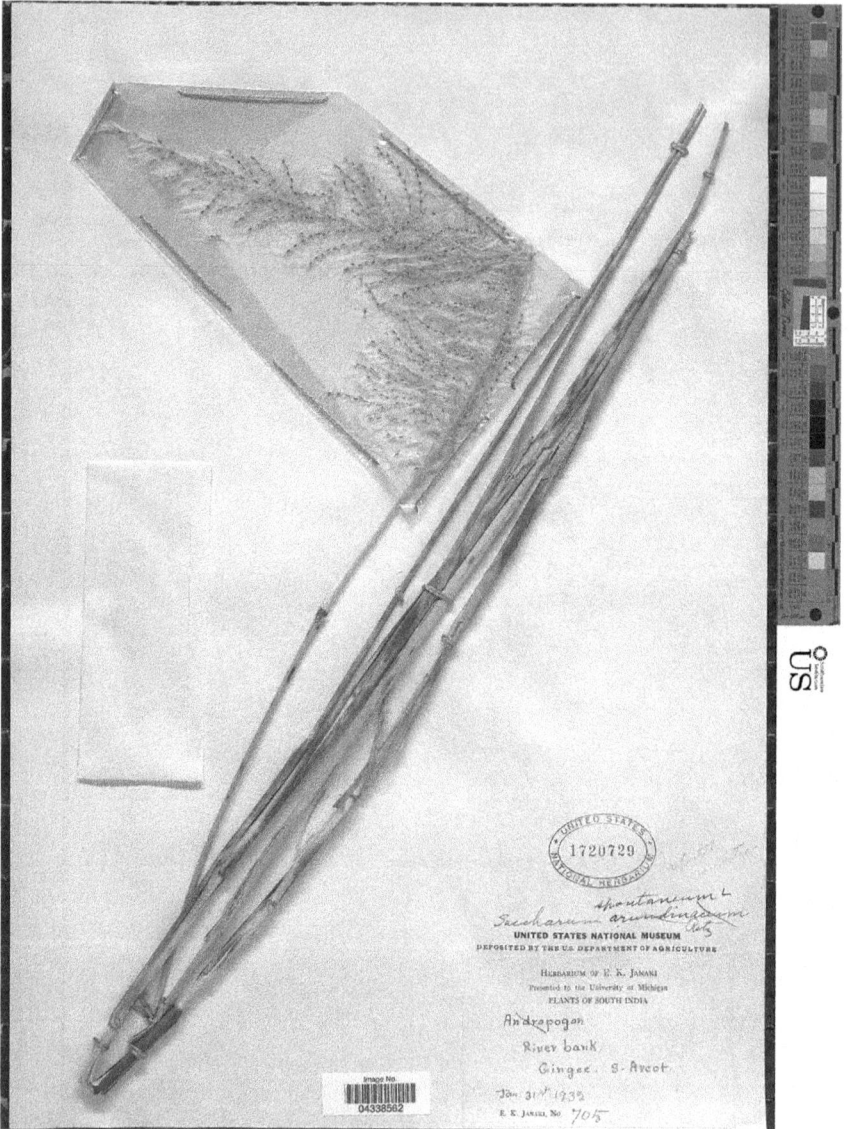

Figure 6.3 Saccharum spontaneum collected from the river bank at Gingee on 31 January 1932.

Source: 'courtesy of the United States National Herbarium (US)' 04338562.

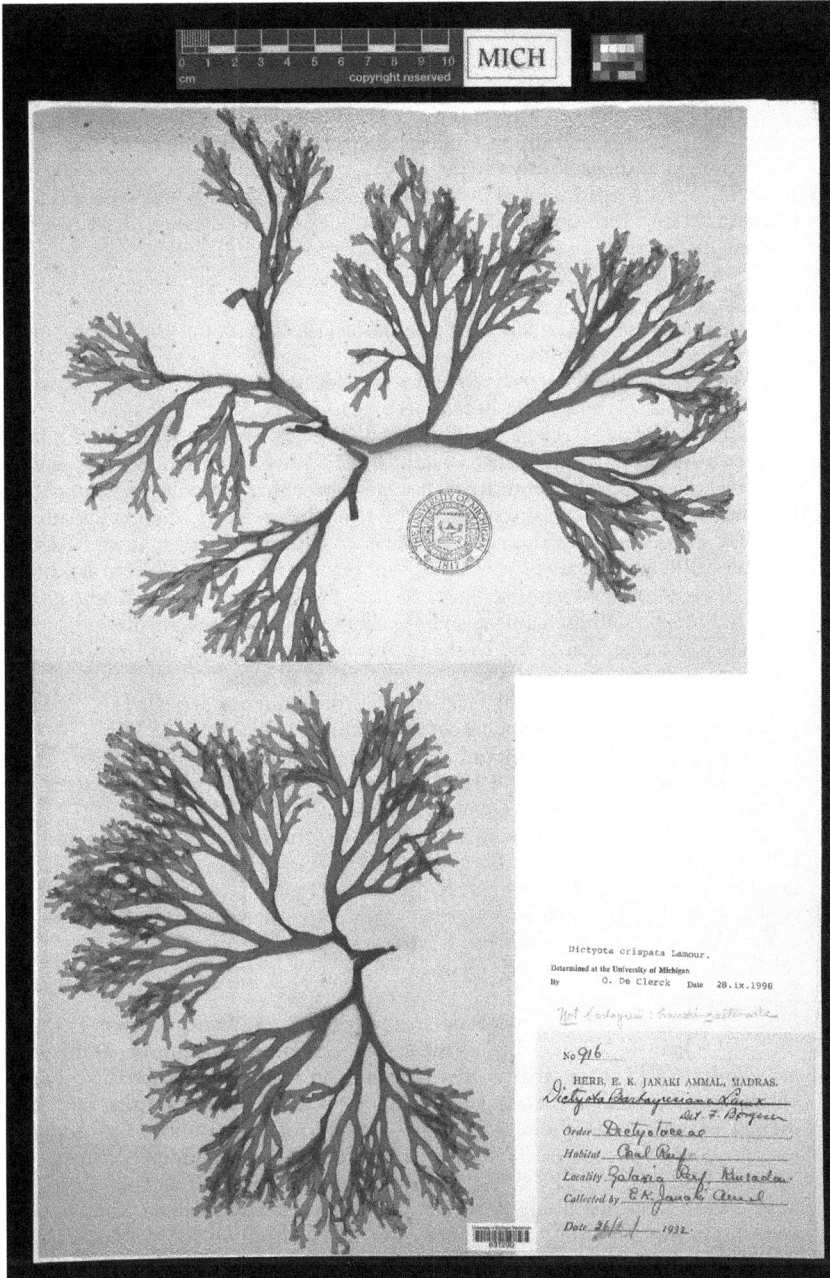

Figure 6.4 Dictyota cryspata collected by Janaki from Galaxia Reef, Krusadai. 26
  June 1932.

Source: Courtesy of University of Michigan Herbarium.

Notes

1 BHL, UoM: M. U. Botanical Gardens (Correspondence series), Box 5, E. K. Janaki Ammal to Frieda Blanchard, letter dated 10 February 1932.

2 Devayani had completed a Licentiate in Teaching course in the late 1920s from the St. Christopher's Training College, Kilpauk, a sister college of the WCC, where Janaki had previously taught.

3 In 1932, Anna K. Joshua was awarded a similar Fellowship, but worth Rs 125 per month, and for zoological research. Besides the Fellowship, the University offered studentships usually worth Rs 75 per month.

4 The Linnean Society, London: CR/144 (1931)

5 See Ray (ed.), *Women of India*, p. 195.

6 Reddy's book, *My Experience as a Legislator* (1930), recounts her efforts in the Madras Legislature to initiate social reforms.

7 *Michigan Alumnus Quarterly Review*, 19 December 1942. On her return to India from Ann Arbor, Agha became the head of the Teacher Training Department of Crosthwaite College for Women, Allahabad. Further, when the All India Women's Association planned to establish the Lady Irwin College for teacher training, domestic science and psychological research, Miss Agha was chosen to conduct a study of girls' schools in the United Provinces, Madras Presidency, Bombay Presidency and the princely states of Mysore, Bhavnagar and Baroda, which resulted in a series of very useful publications from the educational policy point of view (see for example, *Some Aspects of the Education of Women in the United Provinces*, Indian Press, 1933). Agha served on several national committees, was a member of the Board of Studies of the Allahabad University and served as secretary of the AIWC.

8 BHL, UoM: M. U. Botanical Gardens (Correspondence series), Box 5, E. K. Janaki Ammal to Frieda Blanchard, letter dated 24 December 1931.

9 M. O. P. Aiyengar and T. Ekambaram won Madras University's Pulney Andy Medal for botany in 1910 and 1913, respectively. In 1919, the medal went to a woman student, Miss M. A. Dweltz of the WCC.

10 Some of these would be later sent to Chase. See for instance, Smithsonian Institution Collection, Specimen Barcode: 04216576, the grass *Aristida* sp. (of an unknown species), collected by E. K. Janaki on 15 January 1932.

11 BHL, UoM: M. U. Botanical Gardens (Correspondence series), Box 5, E. K. Janaki Ammal to Frieda Blanchard, letter dated 24 December 1931.

12 Ibid., letter dated 10 February 1932.

13 Ibid.

14 BHL, UoM: Bimu C11, Barbour Scholarships for Oriental Women. Papers 1918–69.

15 'Indian Scientist "Home Ec" Visitor on Campus', *Wisconsin State Journal*, 12 June 1933. She also visited Redlands in California, where at a Forum Club meeting held at Beach City (Laguna Beach), to which she had been invited as guest, she spoke about the customs of her country. See *San Bernardino Sun*, Vol. 40, 12 September 1933, p. 13, https://cdnc.ucr.edu/cgi-bin/cdnc?a=d&d=SBS19330912.1.13 accessed online on 6 November 2017. By May 1934, Kausalya along with her young niece, Miss Vimala Karunakaran were homeward bound on the *Hakozaki Maru*, heading for Colombo, from where they would travel by boat and then train to reach Madras. *The Singapore Free Press and Mercantile Advertiser (1884–1942)*, 19 May 1934.

16 For a brief history of the UBL, see Anon, *University Botany Laboratory*. I am grateful to the renowned mycologist late Dr C. V. Subramanian for a copy of this publication.

17 In his honour, the University would institute the Dr Todla Ekambaram Endowment Lectures in 1953.

18 Philip Furley Fyson (1877–1947) is best known as the author of the popular textbook *A Botany for India* (1912), a series of articles titled 'Madras Flowers' published in the *Botanical Bulletin* of the Presidency College Magazine from 1912, and *The Flora of the South Indian Hill Stations* (1932), 3 Vols. Fyson was the Principal of the Presidency College from 1925 to 1932.

19 BHL, UoM: M. U. Botanical Gardens (Correspondence series), Box 5, E. K. Janaki Ammal to Frieda Blanchard, letter dated 1 March 1932.

20 Ibid., letter dated 10 Feb 1932.

21 Ibid., letter dated 6 April 1932.

22 Ibid.

23 Ibid., letter dated 1 March 1932.

24 Krishnan, 'Control of Lantana'.

25 E. K. Krishnan Jr's Diaries, private collection.

26 MacArthur and Wilson, *The Theory of Island Biogeography*.

27 BHL, UoM: University Herbarium Records (University of Michigan) H. H. Bartlett series, Box 11, E. K. Janaki Ammal to H. H. Bartlett, letter dated 25 August 1932.

28 Thomas and Venkatraman, 'Sugarcane—Sorghum Hybrids'.

29 Anderson, *Plants, Man and Life*, p. 211.

30 DP: C 107 (J112), E. K. Janaki Anmal to C. D. Darlington, letter dated 14 November 1932.

31 Ibid.

32 Janaki Ammal, 'Chromosome studies in *Nicandra physalodes*', 1932.

33 Among Sukumaran's early publications are *Arāntekuṭṭitutaninñañca cerukathakaḷ* (1932), *Cilaanyāpadeśanniạl* (1933) and *Asuyāmayaṃ: orukatha* (1936), all published at Calicut by K. R. Brothers Accukūtam.

34 DP: C 107 (J112), E. K. Janaki Ammal to C. D. Darlington, letter dated 14 November 1932.

35 It contained the results of some of the experiments conducted by Miss Mason on the relative vital capacity and basal metabolism of South Indian girls, chiefly students of the WCC. The preliminary results of her experiments (with F. G. Benedict) had been published in the *Indian Journal of Medical Research* 19, 1931, p. 75.

36 *Current Science* 3, July 1934, p. 1.

37 Allen, 'Julian Huxley and the eugenical view of human evolution', p. 221.

38 The term 'eugenics' was coined by Francis Galton in his *Inquiries into Human Faculty and Its Development* (1883). For eugenics in Britain in the 1930s see Searle, 'Eugenics and politics in Britain in the 1930s'; for more recent studies on the history of eugenics see Kevles, *In the Name of Eugenics*.

39 *The Century Magazine*, August 1923, pp. 549–566.

40 Haldane, 'Forty Years of Genetics', p. 238, cited in Harman, *The Man Who Invented the Chromosome*, p. 51.

41 Anandhi, 'Reproductive bodies and regulated sexuality', p. 150.

42 Haycraft, *Darwinism and Race Progress*; also see Kemp, *Merciful Release*, p. 47.

43 Son of Charles Darwin, Major Leonard Darwin was a soldier, politician, economist and mentor of statistician and geneticist Ronald A. Fisher. He was chairman of the Eugenics Society between 1911 and 1928.

44 E. K. Janaki Ammal to E. K. Raghavan, dated 9 October 1932 private collection.

45 Ibid.

7

TRIVANDRUM

A Teaching Interlude

Janaki Ammal comes to Trivandrum and the breath of Queen
Mary's comes with her. We connect up promptly . . . and feel
we have caught a glimpse of a truly academic world.
—Louise Carolina Maria Ouwerkerk,
Trivandrum, 5 February 1933[1]

The College of Science, Trivandrum was founded as the HH the Raja's
Free School in 1834, with the chief aim of teaching English. In 1869 it was
upgraded to an intermediate college affiliated to the Madras University, and
a few years later, began to offer graduate-level courses. In May 1924, it was
divided into a college of science and a college of arts (for teaching languages,
economics and history); the latter was moved to a new building in the Training College campus, leaving the college of science in the old building under
the name, 'HH the Maharaja's College of Science.' During Janaki's tenure
at the college, the Principal and Professor of Natural Science was the plant
physiologist John Pryde, best known for his textbook *Recent Advances in
Biochemistry* (1928); in Pryde's absence, K. S. Padmanabha Aiyyar, the Professor of Zoology, or T. K. Koshy, Professor of Botany, acted as department head. The Physics Department was headed by S. Ramakrishna Ayyar
and Chemistry by the Cambridge-educated K. L. Moudgill. The only other
woman on the science faculty during this period was Assistant Professor of
Physics, Miss A. R. John.[2]

The Very Oily *Cleome viscosa*

Janaki arrived in Trivandrum in late October (1932), to join the College of
Science as Professor of Botany. It appears that Janaki had learnt to drive
a car by this time, for she had bought herself a car to get around the city,
and to go to the beaches in the company of her girlfriends. She was determined to carry out research despite the limited facilities and time available
to her outside of teaching. Using only her microscope and some basic fixing

122   DOI: 10.4324/9781003267089-7

material, she would settle down to prepare slides at the college laboratory, of the crushed pollen mother cells (PMC) of the commonly found medicinal plant, *Cleome viscosa* (*Naikkaduku* in Tamil) using a drop of Belling's ace-tocarmine solution; the plant incidentally was one among those described by her Professor at the Madras University, Philip Fyson, in his published series titled 'Madras Flowers.' At first the chromosomes of the *Cleome* were invisible, occluded by the large number of oil globules so typical of the plant, but when the slide was left to rest for a couple of hours, the globules disappeared, much like a moving cloud, bringing the chromosomes magi-cally into view. The metaphase plate revealed the haploid number of *Cleome viscosa* to be 10. Janaki observed the reduction division (meiosis) religiously 'every morning at about 9 am,' for the first two months of joining the Col-lege. Her diligence paid off, and she was able to send a research note to the *Current Science* for publication (it would appear in the section, 'Sci-ence News,' and it may be noted, the journal had been existence only some months now) within a short time of arriving in Trivandrum.[3] At this time, Janaki was the only other woman scientist (after Eleanor Mason), whose name had appeared on its pages. More than anything else, it made her feel like a real scientist, after several months.

Darlington's American Tour

Janaki's letter to Darlington in July 1932 had found him in America; he was on a Rockefeller scholarship visiting Woods Hole, Massachusetts, where in continuation of his work at the John Innes he was engaged in cytologi-cal research and preparing to attend the International Congress of Genetics to be held in the month after. It was at this Congress that he would have the occasion to listen to T. H. Morgan lecture on the future of genetics, besides the opportunity to examine the highly refined slide preparations of McClintock, showing the pairing of maize chromosomes at the prophase stage of meiosis (revealing the existence of structural differences); the last would inspire him to do something similar with his *Paeonia* hybrids. Dur-ing this tour, Darlington also visited the botanical departments of Michigan, Wisconsin and the Berkeley universities to give seminars and demonstra-tions. At the last was where the insightful but cautious cytologist of the old school John Belling was based; Belling had severely criticised Darlington's *Recent Advances in Cytology* when he had been shown the unpublished version (also sent to Morgan, Muller, Bridges, Dobzhansky and Babcock) for its arrogant cytological deductions. He was even preparing a devastating review of Darlington's book when he suddenly passed away. Besides young researchers like G. Ledyard Stebbins, Darlington befriended during his visit all the big or rising stars in the field, including the ageing E. B. Wilson, T. H. Morgan, Alfred Strutevant, E. B. Babcock, J. Clausen, T. Dobzhan-sky, George Beadle and Calvin Bridges, all of who shared Belling's opinion and believed his book to be nothing less than dangerous for young aspiring cytogeneticists.

It was at the California Institute of Technology's Division of Biology in Pasadena, where Darlington spent almost six months as a postdoctoral visitor (between November 1932 and early May 1933 at the Kerckhoff Laboratory), working on the cytology of the maize and the *Drosophila*, that he had had the opportunity to discuss genetical issues at length with biologist Morgan. Morgan had laid the foundation for the Caltech Division of Biology in 1928 at the behest of the institute. The first faculty recruitments were to the genetics department, and Morgan was soon joined by Strutevant from his core *Drosophila* group at the Columbia University and the young and promising cytogeneticist Karl J. Belar from the Kaiser Wilhelm Institute for Biology in Berlin-Dahlem. Morgan's goal (he won the Nobel in 1933) was to collect the most brilliant minds around him, 'representing the most modern lines of biological research,' making sure to give them an atmosphere of utmost freedom to do research; the Division would soon become the world centre for research in genetics.[4]

Among the postdoctoral fellows at the time of Darlington's visit to Pasadena were two future Nobel Prize winners, George Beadle and Barbara McClintock (a 1927 PhD from Cornell), on the verge of accepting a postdoctoral position at the University of Freiburg, Germany. In 1928, Eileen Erlanson became the first woman postdoc of the Caltech Division of Biology and the one to give the first seminar (we may recall that in 1928, she received a doctorate from the University of Michigan for her work on the genetics of the native American Rose); all three, Beadle, McClintock and Erlanson were Fellows of the National Research Council in the Natural and Medical Sciences.

Keen on travelling to the East, to Japan and India in particular, Darlington applied for an extension of his fellowship, which fortuitously for him was accepted by the Rockefeller Foundation; this would be his first ever visit to the Asian continent. In late May 1933, he reached Japan, where he would spend the major part of two months with the amiable and brilliant wheat geneticist Kihara Hitoshi (1893–1986) at the University of Kyoto, besides a short stint at the Departments of Botany and Agriculture of the Imperial University of Tokyo.

Darlington Visits Trivandrum

Darlington left Japan for India in late July (1933), to arrive at Colombo on the 26th of that month. After much wrangling and a stringent customs examination, he managed to catch a late evening train to Talaimannar, on the Northwestern coast of Ceylon. At Talaimannar, he got on a boat to Dhanushkodi,[5] where he boarded a train to Madurai ('Monkeys on the train at Madura,' he recorded in his pocket diary) from which place a connecting train brought him to Trivandrum. Incidentally, also on the boat to Dhanushkodi was Mohan Dnyaneshwar Shahane of the Servants of India Society (founded in 1905 by Gopal Krishna Gokhale), returning home after completing postgraduate studies as Vincent Massey Fellow (1932–33) at the School of Politics and Economics, University of Toronto; Darlington and Shahane would discuss politics during their short journey together.[6] In India, Janaki would be his host and, without doubt, one of the chief reasons for his visit. He had informed her of his plans sufficiently early, but it does not appear that he had revealed his, now year-old, secret marriage to the academic Kate Pinsdorf, at a place not far from Woods Hole. An intelligent, charming and free-spirited woman, Pinsdorf had grown up in Brazil and studied history at Stanford.[7] Both partners wished to treat the relationship as a non-permanent one.

Darlington's train pulled into Trivandrum Central early in the morning of 3 August 1933 (at forty-five minutes past five, to be precise), where he was met with by a beaming and visibly excited Janaki. For the next ten days, they would be inseparable, of which almost a week would be spent in Trivandrum. Janaki had ensured that Darlington was treated as a State Guest, which meant he would be accommodated at the Mascot Hotel—a grand colonial building erected to house the British Army during World War I and now owned by the state—and given other conveniences such as a car for commuting in and around the city. Trivandrum was at this time quite cosmopolitan, with several Britishers and Europeans living in the city, who regularly met up at the European club over bridge, games and drinks.[8] Some of the city's elite were close to the Travancore royal family, visiting the palace to grace social occasions or, if they were fortunate enough, to play a game of tennis with the young Raja (Chithira Thirunal Rama Varma, who had been invested with full ruling powers in 1930)[9] or his family members. There was electric street lighting even in the narrow lanes of the city, which was well-connected to other municipalities by bus services.

Official and Social Meetings

Darlington's first official meeting on 4 August 1933 as State Guest, as was customary, was with the Diwan, Thomas Austin, at his official residence, the Bhakti Vilas palace at Vazhuthacaud; he was accompanied by Janaki. Darlington was amazed that the Diwan had ninety-nine servants. After tea with Austin, they visited the College of Science, where she was professor, and Darlington was formally introduced to her colleagues. Later the two went out for a 'walk in the hills' (the grounds of the Kanakakunnu palace, or perhaps it was the hill opposite, which housed the Meteorological Department, not far from the College) for a tête-à-tête. There was so much to talk about—the virulent attack on his book, his meetings with the leading figures of American genetics and cytology, the Ithaca conference, genetics and society in Japan and, not the least, his newfound conjugality, which would no doubt have shocked her but she cared far too much for him to be bothered by his impetuousness. The dinner party at Janaki's bungalow in honour of Darlington was an elegant one; 'she has three servants,' he remarked in surprise. At the 'very quiet' party (incidentally, he had been faced with the 'problem of dressing for dinner'), he had the opportunity to meet with Janaki's close friend Louise Carolina Maria Ouwerkerk, 'the Dutch Newnham girl who is Prof of Economics' at the Maharaja's College for Women in the city.[10]

Born to Dutch parents living in London, Ouwerkerk had completed an MA in Economics from Newnham College, Cambridge in 1925. Four years later, she had set sail for India to take up a Professorship at the Maharaja's College for Women (simply referred to as the Women's College).[11] Ouwerkerk was an active member of the European Club, where she had the opportunity to meet people who mattered in the Travancore administration and discuss politics with them.[12] She would also visit the Club for 'a hand of bridge and a spot of dancing' and enjoyed playing the piano herself. In fact, she bought herself a piano despite it being a major squeeze on her far from substantial finances.[13] In 1930, within a year of arriving, Ouwerkerk had founded the Trivandrum branch of the International Fellowship (established in 1922), aimed at bridging the gap between Britishers and Indians. If it was anxiety over being jobless (the Great Depression was looming large) that drove Janaki to seize the College of Science opportunity in 1933, so it was for Ouwerkerk. While Janaki's stint at Trivandrum would be short-lived, lasting no more than a year, Ouwerkerk would go on to work in the city until 1939, or until she was unfairly dismissed. By this time, she was also Professor of History and Economics at the newly formed University of Travancore (founded in 1937), the brainchild of Diwan C. P. Ramaswami Aiyer.

When they met in Trivandrum in late October 1932, Janaki and Ouwerkerk were instantly drawn to each other. An elated Ouwerkerk noted in her diary: 'Janaki Ammal comes to Trivandrum and the breath of Queen Mary's comes with her. We connect up promptly, Eunice [Eunice Gomez, Assistant Professor of English, Maharaja's College for Women] and she and I, and feel we have caught a glimpse of a truly academic world.' Janaki thought of herself as a daughter of the ocean. These women faculty members seem to have had a great time, with frequent excursions to the beach: at Shankhumukham not far from the city, Kovalam on the outskirts and the more distant Cape Comorin or Kanyakumari. The Smithsonian Institution Herbarium collections reveal that Janaki was at Cape Comorin on the 17 December (1932), within two months of joining the College; she collected marine algae from the place. On 5 February 1933, the 'jolly group' visited Kovalam—Lakshmi, Daisy [Daisy Muthunayagom, Lecturer in English], Pansy, Eunice, Ouwerkerk and Janaki—'such a colourful splash they made on the sands in their bright saris!' and Ouwerkerk contributed to this with her 'green swimming costume.' After a swim they 'had an incredibly large tea, went for a walk and drew up a Five-Year Plan for Brighter Trivandrum. The programme is, laughter, song and dance—the more mixed the better.'[14]

Darlington's second day in Trivandrum, 5 August 1933, was spent wholly on work, chiefly reviewing Janaki's paper on the *Solanum melongena* to be sent to the *Cytologia* for publication; the paper would be submitted on the last day of October.[15] This paper merely aimed at revisiting the research undertaken by her at Michigan and the John Innes in 1931 (in the light of recent developments in cytogenetics), on the diploid eggplant, wherein she had discovered a mutant, an abnormal plant among the population, which was triploid and highly sterile.[16] A shorter preliminary version of this work had appeared in the journal of the Michigan Academy of Science, Arts and Letters (established in 1894), a regional professional and interdisciplinary organisation of scholars, which aimed at disseminating new and outstanding research undertaken at institutions affiliated to the University of Michigan. She had presented a paper on the subject at the 19th Indian Science Congress held in Bangalore (1932),[17] and two years later, in July 1934, would do so at a meeting of the Association of Economic Biologists, Coimbatore. Perhaps it was at the suggestion of Darlington that Janaki had begun rewriting the paper for publication in the *Cytologia*, a new international journal of cytology published from Japan, which country he had recently visited. Her paper would cite Darlington's published papers over the period 1929–31, including his work on the meiosis of polyploids and the *Fritillaria*, their joint paper on the *Tulipa* and, importantly, his new book *Recent Advances in Cytology*; this last had not figured in the earlier versions of her paper, because it was yet to be published.

We may recall that in the course of her research at Michigan in 1929–31 on the diploid eggplant *Solanum*, Janaki had found an abnormal specimen/a mutant, whose occurrence had led her to undertake a cytological study of the plants used in its breeding. Material had been obtained from the cultures grown at the Botanical Garden of the University of Michigan. The diploid plants investigated included the dark purple variety (J22), seeds of which were sent by Professor Kakizaki of the Saitama Agricultural Farm of Japan, the oval, green fruited true-breeding variety, which occurred among a culture of 'Long White' (515B) sent by Messrs Suttons, the British seed merchants, in 1925, and the F1 hybrids between 515B and J22, which had medium purple skin colour together with the green flesh colour found in both parents; her unpublished genetical work had shown that these colours were independently inherited. That the mutant, which had appeared in a culture of the above cross, was triploid, she had already communicated in an earlier paper.[18] This was found to be highly sterile; the high sterility in the triploid was indicated by the few seeds present in the fruit.

Ploidy

A little genetics primer on 'ploidy' may be helpful now. Cells are described according to the number of sets of chromosomes present in the nucleus, or what is called the ploidy level (represented by the letter 'x'). To distinguish between the ploidy of a species as it presently breeds and that of an ancestor, the x and n symbols are employed. Reproductive (sex) cells or gametes contain one set of genetic information, while somatic or body cells contain two sets of genetic information; somatic cells or individuals can be described according to their ploidy levels, that is the number of sets of chromosomes present in the nucleus—monoploid (x), meaning one set of chromosomes; diploid (2x), two sets; triploid (3x), three sets, etc. The number of chromosomes in the ancestral set is referred to as the monoploid number (x) and is distinct from the haploid number (n) which is the total number of chromosomes found in the gametes (reproductive cells contain half the genetic material necessary to form a complete organism). During fertilisation, the male and female gametes fuse, producing a diploid. Humans for instance carry two complete sets of chromosomes (2x); that is, they have a ploidy level of 2 (one set of 23 chromosomes from the father and another set of 23 chromosomes from the mother, making a total of 46 chromosomes, referred to as the chromosome number). The haploid number (n) for humans is 23 (half the chromosome number), while the monoploid number (x) equals the chromosome number (46) divided by the ploidy level (2), which also happens to be 23 in this case. The rise in ploidy level is today considered to be an important evolutionary mechanism in both plants and animals, but in the 1930s the link was still largely unexplained.

Publishing Her Eggplant Research

Through studies of the diploid (2x=24) and triploid (3x=36) *Solanum melongena* (both the somatic and meiotic divisions), Janaki was able to throw light on the origin of the triploid eggplant and on the origin of the mutant, namely, the tetraploid progeny of a triploid. She suggested that the triploid was the result of the fertilisation of a normal female germ by a diploid male germ (which arose by the union of the two nuclei before fertilisation; she had observed the presence of exceptional binucleatic pollen grains in one of the diploid parents); it was a case of doubling where reunion took place not at the last division before meiosis but at the formation of the generative and tube nuclei in the pollen grain. In simpler terms, the triploid arose from the functioning of a diploid pollen grain. Further, when the triploid was 'selfed' (that is inbred, or crossed with another of the same), she discovered, it gave two triploid and eleven tetraploid or nearly tetraploid seedlings.

'Polyploidy in *Solanum melongena* Linn.' (1934) was her first ever single author publication (she had only been a co-author previously) in a peer-reviewed international journal exclusively devoted to cytology. Her earlier paper on the *Nicandra* (1932) had been published in *La Cellule*, a Belgium-based journal founded in 1884 (with articles chiefly in French), the first such to carry papers in cytology, but not exclusively so. The journal was by this time regarded as 'old-fashioned' when compared to *Cytologia*, located at the cutting-edge of the nascent field of cytogenetics and perhaps with a larger circulation and impact in the English-speaking world. The paper was an important milestone in her career as a cytogeneticist and an ode to Darlington's Trivandrum visit. With its publication, Janaki was announcing to the world of cytogenetics that she had arrived.

A Full Diary

On the morning (6 August 1933) after he had finished reviewing Janaki's paper, Darlington and Janaki set out for Cape Comorin with the State Archaeologist R. Vasudeva Poduval; Poduval would show them around the Padmanabhapuram palace in Thuckalay near Cape Comorin, being restored by the Archaeological Survey. They would also visit the Thanumalayan temple at Suchindram, and see its *teppakulam* (large water tank) and *theru* (festival car). Darlington thoroughly enjoyed himself, taking in the abundant architectural beauty, relishing tender coconut water and eating 'curry on plantain leaves.' On return to Trivandrum, they visited the museum and the zoological gardens; a curious Darlington quizzed the very 'communicative' K. P. Padmanabhan Tampy (in charge of the Sri Chithira Art Gallery) about the panther-lion hybrid in the zoo. For Darlington, the rest of the day, or whatever remained of it, would be devoted to work primarily, but for attending a dinner party at Janaki's bungalow. At the party, Darlington played bridge with two Travancore ministers and the acting Principal of the College of Science, the chemistry professor K. L. Moudgill who, he was glad to learn, had studied at Christ College, Cambridge.

Darlington visited the HH The Maharaja's College of Science, where Janaki taught, on the following day. Everything had been well-arranged beforehand. He was welcomed with garlands by her students. He was there to address the Botanical Society of the College, which he formally inaugurated and would be presented with a 'Malayalam Document and Oil Lamp as used by the wise virgins,' chosen thoughtfully by Janaki.[19] Incidentally, T. K. Koshy was the only other teacher (besides Janaki) at the department when Darlington visited; Koshy had published his first paper in cytology, on the *Allium*, at this point.[20] Over the next couple of days, again in Janaki's company, Darlington visited the Record Office located within the fort (where its staff presented him with a palm leaf manuscript), the 'closed palaces' with the 'crystal throne' (the *Kuthiramalika*), the Treasury, the Art School, the Armoury and the Legislative Assembly, where a debate on education was progressing; a Travancore Education Reforms Committee had just been formed that year. Darlington also made time to go on a romantic night-time canoe-ride with Janaki along the river (perhaps the Karamana River) and to the beach to see the phosphorescent breakers. In the day he viewed the state elephants and attended a recital of Indian music, on which he commented disparagingly: 'no crisis, no harmony, usual oriental.' Four days later, on 11 August 1933, at daybreak, Darlington accompanied by Janaki left for Coimbatore by train, breaking journey at Madurai to visit the Meenakshi temple, which he described as 'a little city with 5 gates & 100 shops & the usual bathing tank.' After an uneventful overnight journey, they pulled into the Coimbatore Railway Station nice and early the next day.[21]

Notes

1 BL: Ouwerkerk Papers Mss Eur F232/1 (Diary no. 3, July 1931–February 1933).
2 At the Maharaja's College for Women in the city, Natural Science was at this time being taught by one Mr A. P. Mathew, and later by Miss Sosa P. John of Aymanam, Kottayam (sister of Rachel P. John, a researcher at the UBL, Madras, who went on to complete a doctorate from the Queen Mary's College, London in the 1940s).
3 *Current Science* 1, no. 10 (April 1933), p. 28.
4 Allen, *Thomas Hunt Morgan*, p. 334.
5 DP: Darlington's pocket diary, g. 4, India, 1933.
6 Ibid.
7 Pinsdorf was at this time history instructor at the Vassar College and already had a book to her name: *Relations between Argentina and Brazil* (1929).
8 The Irish literary figure and art historian James Cousins and his wife Margaret, the women's rights activist, the German horticulturist Gustav Krumbiegel, the Dutch ethnomusicologist Arnold Bake and his wife Corrie, visited Travancore in the 1930s.
9 He would establish the University of Travancore in 1937.
10 DP: Darlington's pocket diary, g. 4, India, 1933.
11 The Maharaja's College for Women originated as a school for Christian girls. In 1864, when the Travancore state opened it out to all castes, it became the Sirkar Girls' School, and later in 1890 upgraded by the Madras University to a High School. After a series of reorganisations, the institution became HH Maharaja's College for Women in June 1921.
12 For more on Ouwerkerk, see Kooiman's introduction to Ouwerkerk, *No Elephants for the Maharaja*, pp. 5–14.
13 BL: Ouwerkerk Papers Mss Eur F232/1, Diary no. 3, July 1931–February 1933.
14 Ibid.
15 Janaki Ammal, 'Polyploidy in *Solanum melongena* Linn.', 1934.
16 Ibid., 'A Polyploid Eggplant, *Solanum melongena* Linn.', 1931, p. 81.
17 Ibid., 'Polyploidy in *Solanum melongena* Linn.', 1932.
18 Ibid., 'A Polyploid Eggplant, *Solanum melongena* Linn.', 1931, p. 81.
19 DP: Darlington's pocket diary, g. 4, India, 1933.
20 Koshy, 'Chromosome studies in allium I: The somatic chromosomes'.
21 DP: Darlington's pocket diary, g. 4, India, 1933.

8

TRIVANDRUM–COIMBATORE–KRUSADAI

Unforgettable Sojourn

> I shan't give up research & Botany ever for a mint of money.
> I am in fact prepared to be a beggar but a happy beggar.
> Trivandrum is a very difficult land for one like me whose ideas
> of right & wrong are sharply defined.
>
> —E. K. Janaki Ammal (November, 1933)

One of the major branches of the biological sciences to emerge in the early twentieth century, economic biology, dealt primarily with the application of biological knowledge to economic goals, as in agriculture and industry. This gave rise to the subdisciplines of entomology, agricultural chemistry, soil microbiology and mycology. In England, an association of economic biologists was formed in 1904, following the rediscovery of Mendelian genetics to promote and advance the science of economic biology in its agricultural, horticultural, medical and commercial aspects.[1] Several of the papers presented at the association's meetings were considered valuable not only for their economic applications but also for their clear demonstration or confirmation of the Mendelian laws of heredity.[2]

Economic biology found fertile ground for development and institutionalisation in colonial settings. In India, at the opening of the twentieth century, new agricultural departments were formed and economic biologists trained in the new science of genetics were appointed. It had by now become evident that substantial improvements could be made if scientific methods were systematically applied to agriculture. In 1903, American philanthropist Henry Phipps donated a large sum to Lord Curzon, the Viceroy of India, towards scientific research of a public nature. While part of the amount was devoted to the building of a Pasteur Institute at Coonoor in South India, the rest was utilised towards establishing a laboratory of agricultural research which it was hoped would become a centre of economic science.[3] In due course, the Government of India established an agricultural college and research institute with a laboratory at Pusa in Samastipur (Bihar), for the diffusion of knowledge of scientific and practical agriculture (together called the Phipps

DOI: 10.4324/9781003267089-8

laboratory, Naulakha or Pusa Institute), attached to which was a large farm of some 1,300 acres for the sole purpose of agricultural experiments.[4]

Of the two-year course taught at the Pusa Institute, the first was devoted to physiology (based on Darwin and Acton's 1894 textbook) and the improvement of plants, and the second to practical applications of the principles of plant improvement.[5] The lectures on plant improvement dealt firstly with the principles of plant breeding (in particular Mendel's laws and the ideas of evolution, variation, mutation, selection and hybridisation) and secondly with specific methods conducive to Indian conditions. It was hoped that the agricultural trainees would return to their provinces at the end of the course and contribute to the diffusion of modern technologies of plant improvement. In 1911, the Pusa Institute began to be referred to as the Imperial Institute of Agricultural Research, and in 1919, renamed the Imperial Agricultural Research Institute (IARI). The formation of the Royal Commission on Agriculture in 1926 ushered in 'a new era in the life of the Indian countryside'; it dawned on the government that the problem of bringing about agricultural improvement was intimately linked to the problem of how village life could be bettered and that this demanded a more holistic approach to agriculture. Thus, in addition to the IARI, an Imperial Council of Agricultural Research (ICAR) for coordinating, guiding and managing agriculture research and education in India was established in Delhi in 1929.

The Sugarcane Breeding Station

In the first half of the twentieth century, the chief sites of the practice of biological sciences in India were government-run agricultural farms and breeding stations, such as the Coimbatore Sugarcane Breeding Station, where boundaries between pure and applied sciences were constantly redrawn. The history of the institution goes back to 1907, when a report from the Board of Agriculture recommended that a sugarcane breeding and acclimatisation station be established in the Madras Presidency for the production of improved canes suitable to subtropical conditions.[6] India was at this time meeting its entire sugar requirements through imports from Java, an unfavourable scenario, which had to be urgently addressed. The Government of India was quick to accept the Board's recommendation and within a year a sugarcane breeding station was established in Coimbatore under the direction of Charles Alfred Barber (1860–1933). Barber had done something path-breaking, when faced with the sugarcane rust disease in the Godavari delta. He tackled the problem not through curative methods but by the replacement of the disease-prone sugarcanes with disease-resistant varieties, paving the way for a major economic breakthrough in sugarcane cultivation not just in the region, but the whole of the Indian subcontinent. Barber's name would forever be linked to the sugarcane, with at least one indigenous cultivated cane named after him, the *Saccharum barberi*.

Coimbatore was chosen as the site for the Imperial Sugarcane Breeding Station because its natural conditions allowed sugarcane varieties to flower profusely and set seed early. The cultivated sugarcanes at Coimbatore belonged to two groups, the first being the thin and hardy indigenous canes growing in North India such as the *Saccharum barberi* and *Saccharum sinense*, while the second included the thick *Saccharum officinarum*, the origin of which was believed by Barber and Jesweit to be the mountainous islands of the Malayan Archipelago, New Guinea and Polynesia. *S. officinarum* had been introduced to India relatively recently and was referred to as the 'introduced' or 'noble' cane on account of its stout form and abundant sweet juice.[7] Crossing the 'noble' *S. officinarum* with the North Indian *S. barberi* (a process called 'nobilisation') did not prove successful, and in a few years the experiment was entirely given up. Barber then hit upon the bright idea of crossing the 'noble' with *S. spontaneum* (Khas or Kahi ban grass), a wild sugarcane that yielded little juice, and grew along the channel bunds adjoining the Coimbatore Agricultural College, to produce the desired hardy varieties suitable for subtropical regions.

The result of this experiment was the interspecific hybrid cane (a cross between two sugarcane species), Co 205 (Co for Coimbatore) created in 1914, which soon proved to be a great success in North India, especially Punjab, where it recorded a 50% greater yield than the low-yielding indigenous varieties (which it soon replaced) in cultivation. This revolutionary hybrid was found conducive to the climate and soil conditions there, owing to its wild ancestry. The successful utilisation by Barber of the wild species of a plant for crop improvement (*S. spontaneum* contributed to improvement in sugarcane vigour, hardness, tillering, ratooning ability and resistance to diseases) and varietal evolution was considered a path-breaking episode in the history of sugarcane breeding. The credit for this epoch-making interspecific hybrid should not only go to Barber but also his able Indian assistant, T. S. Venkatraman, who began his career at the Station in 1912.[8]

Tiruvadi Sambasiva Venkatraman (1884–1961) ably continued Barber's work until his retirement in 1942, creating new and improved sugarcane varieties that led to the remarkable growth of the sugar industry in India. A field-man through and through, and master-breeder, Venkatraman evolved ingenious methods to surmount practical problems specifically connected to the breeding of sugarcane: for instance, he chose varieties that had no pollen (he raised about 200 seedlings annually), because the sugarcane flower had about 6,000 to 10,000 florets and it would be impossible to emasculate every one of them. In order to test the suitability of a variety across different soils, he would develop about six 'root-eyes' from one node by moistening the bud, from which twenty or more roots could be produced and grown in different soils. He would put four into each soil type contained in a paraffin paper bag 2–4 feet long and about 2 inches wide. Further, the problem of handling a mature cane, which could easily reach 16 feet, was tackled by evolving a technique of inducing root formation at one of the nodes and then cutting off the cane at a reasonable length, resulting in a shortened plant that was convenient to work with.

To induce better vigour and a robust root system (when crossing the 'noble' and cultivated *S. officinarum* with the wild *S. spontaneum*), he devised a gunny-bag technique, wherein pieces of wire-netting wrapped in gunny bags would be put into the soil around the plant. If a variety demonstrated quick development of roots and good growth, it would be included in the breeding programme.[9] When it came to dispatching cane to sugar factories, once again he came up with innovative ideas, such as applying paraffin to the ends of the canes and packing them up in straw or wood-charcoal in the case of overseas parcels. His focus was on developing fibre- and sucrose-rich canes, since both factories and peasants preferred these. Aided by sufficient tariff protection, and the patronage of such people as the educationist Madan Mohan Malaviya and capitalist Walchand Hirachand, his sugarcane breeding work transformed India from an importer of white sugar to a position where it began to look for export markets. A perfect marriage between science and industry, Venkatraman's work received accolades from the colonial state, including a knighthood in 1942; he would be awarded the Padma Bhushan in 1956.

An Association of Economic Biologists

In 1930, an Association of Economic Biologists had been founded in Coimbatore, with its headquarters at the Agricultural College. Paddy specialist Krishnaswamy Ramiah, the sugarcane expert T. S. Venkatraman and millet specialist G. N. Rangaswamy Ayyangar were among those closely involved with its birth. The association's chief objective was to promote the cause of applied biology in all its aspects, develop an *esprit de corps* among those engaged in the biological sciences and facilitate the diffusion of biological knowledge among the public, and the exchange of ideas among members through presentations of papers and discussions of agricultural problems. Original investigations were encouraged and distinguished biologists invited to address the association.[10] Until 1933, when the Association began a journal of its own, its proceedings were published in the *Madras Agricultural Journal*. Papers presented before the association addressed several biological aspects—physiological, entomological, mycological and very importantly, cytological and genetical—of all the important agricultural crops of South India. Public lectures delivered in the early years of its existence included those by biochemist Gilbert J. Fowler, Director of the Pusa Institute, Bernard A. Keen, and the Dutch scientist Van der Veen. It was at the behest of

Janaki and Ramiah that Darlington had been invited to speak before the association in August 1933.

Darlington in Coimbatore

On arrival in Coimbatore on 12 August 1933, Darlington, accompanied by Janaki, was shown around the sugarcane, rice and cotton breeding stations. He also had the opportunity to meet one of Janaki's nephews at the guest house of the Sugarcane Station, where they were put up; perhaps this was Hari, forester brother Kittu's eldest son who had a passion for nature, and ornithology in particular. In his pocket diary, Darlington recorded that he had 'a long tete-a-tete with Ayyangar, the Millet man';[11] the Government Millet Specialist, Rangaswamy Ayyangar was at this time Principal of the Coimbatore Agricultural College and President of the Association of Economic Biologists. On the following evening, Darlington delivered two lectures at the Agricultural College, under the aegis of the Association, one on chromosomes and plant breeding, and the other, on sterility. The Madras University had fixed a remuneration of Rs 250 for public lectures, which included a travel allowance.[12] Additionally, a special tea session was organised in Darlington's honour and he was gifted a box of Trichinopoly cigars (called 'Trichies'). The gift idea was of course Janaki's; only she among them knew of Darlington's great love for cigars.[13]

In his report to the Rockefeller Foundation sent at the culmination of his thirteen-month fellowship, Darlington would state that 'extensive cytological and genetic work was being done' at Coimbatore, but had 'escaped general notice owing to inaccessible publications.' He was particularly impressed by the excellent propagation of Venkatraman's *Saccharum-Sorghum* crosses, which promised to bring about 'a revolution in sugar growing in India & elsewhere.' This 'method of reproduction is therefore of considerable interest,' he noted. From Coimbatore, Darlington collected crop plant-material including that of rice for purposes of research and hoped to use the connection he had established with the Association to also obtain material relating to tropical fruits, for he had discovered that fruits like the mangosteen displayed sterility. He was curious to find out if the sterility was associated with the phenomenon of triploploidy, as indeed the case was with temperate fruits such as apples and pears.[14]

Off to Krusadai

Darlington and Janaki departed for the coast on the afternoon of 13 August 1933, to reach Krusadai Island the next morning, where Janaki's younger brother, the free-spirited E. K. Madhavan lived 'Crusoe style,' as its sole official caretaker. The Madras Presidency was the first to conduct marine fisheries research in India on an organised scale, for which the Marine

Biological Station at Krusadai in the Gulf of Mannar, West Hill Fishery Research Station in South Malabar (Calicut) and the Ennore Fishery Station near Madras had been established. Madhavan, with a BA (Zoology) from Presidency College, Madras, was initially appointed to the post of Research Assistant at Krusadai and later as Inspector of Inland Fisheries at the Nilgiris (Ootacamund), in place of the retired T. R. Lakshmana Ayyar. An ingenious boat-maker, Madhavan was also a talented musician and a veritable non-conformist, with a penchant for adventure. He was fond of his elder sister Janaki and, like elder brother Kittu, collected plants for her from every station he was posted.[15]

The Krusadai Island had been attracting biologists from India and abroad as early as 1898.[16] Botany and zoology graduate students accompanied by their teachers, and researchers regularly visited the island for collection and study. In September 1931, Miss Ethel Prem Singh and Mrs Francisca Thivy, Professor and Assistant Professor of Botany respectively, of the WCC, had officially visited the island for this purpose.[17] We may recall that Janaki had visited the island herself, in the summer of 1932, and made a collection of marine algae for the Herbarium at the University of Michigan. In 1933, A. R. Gopal Ayyar of the Department of Botany, Central College (Bangalore University) arrived in the company of several students to collect specimens. Official visitors to the island in 1933–34 included John E. Chelladurai of the College of Science, Trivandrum; Dr H. K. Mukherjee, Head of the Department of Zoology, the University College of Science and Technology, Calcutta; Col. Harold Charles Winckworth (a senior army officer, he was an authority on marine shells and molluscs); and Professor C. P. Gnanamuthu of the American College, Madura.[18]

Janaki and Darlington were not however visiting Krusadai in any official capacity; they were on the island primarily to meet Madhavan and to spend some time taking in the beauty and biological richness of the place. In the company of Madhavan they explored the island's coral reefs, fringing reefs and mangroves, and bathed in the shark-free sea, 'a Revolution for J,' Darlington would remark.[19] To keep Madhavan entertained on what was a remote island (an outlying island, Krusadai was separated from Rameshwaram Island by half a mile of sea), his eldest brother Raghavan periodically sent him from Burma his compilations on curious subjects, which he titled, 'Tit Bits' and 'My Magazine.' 'You cannot imagine what a great treat they were to me in my lonely isolation,' Madhavan would later comment. He had been having a hard time, being unwell and having 'to dive two fathoms among the coral reefs and grope for an anchor' attached to the pearl oyster research cages, which had run adrift. At this time, the Krusadai Station was engaged in research on pearl oysters and sacred chanks. In late 1932, following the discovery of a small bed of oysters in the pearl banks, a beginning had been made in establishing a pearl oyster park in close proximity to the island; five hundred and fifty-eight oysterlings were successfully transported from the banks to the island and placed in specially designed oyster cages for purposes of observation. That same year, a farm for the culture of an edible fish *Chanos chanos* (Milk Fish) with the help of the Philippine Bureau of Fisheries (a lucrative industry established in the Philippines) was also initiated; Milk Fish had been discovered in a swamp on the Krusadai Island, in all probability at Watchman's Bay.[20] Darlington would obtain a quick overview of all the fisheries related activities on the Island.

On days of rough and squally weather, the island could be cut off from the rest of the world and on several such occasions staff would be forced to subsist on the bare minimum. The zoology graduates among the Research Assistants and Inspectors, like Madhavan, were junior men and any reduction in staff affected them instantly.[21] In late November 1932, Madhavan had gone on sick leave for more than four months, during which time another zoology graduate had acted for him. He had returned to work only months before Janaki and Darlington arrived on the island.[22] These difficulties were compounded by the frequent threat of malaria. Sometimes Madhavan's wife Kanthi, son Mahadevan and little daughter Shashi joined him. The family enjoyed themselves singing and playing music on the Hawaiian guitar, gardening (Madhavan introduced the bougainvillea among other plants to the island) or tending to their pets.[23] A wild and musical child, Shashi was most happy among the poultry and the goats and when in the company of her pet Stone Plover named 'Billy Boy,' inspired by the American nursery rhyme (published around this time) which went, 'Oh, where have you been, Billy Boy, Billy Boy?' 'Janthi has been on a 3 day visit with one Dr Darlington one of her colleagues during her work in London,' Madhavan wrote to Raghavan on 18 August 1933, enclosing a description of the island and a rough pen-sketch of the place (Madhavan was a natural when it came to plotting maps and rendering technical drawings).[24]

The strong winds and the Biological Station's 'unsatisfactory sailing canoe' however had made Janaki and Darlington's return journey an ordeal; the three-mile travel to the Mandapam camp situated on the Indian mainland 'nearly caused Dr Darlington to miss his train [to Dhanushkodi] and consequently the steamer from Colombo.' To add to his distress, he discovered his medical chest was missing but was forced to let go of it so as not to miss the South Indian Railway's Ferry Steamer leaving Dhanushkodi for Talaimannar; he was to clear the Ceylon immigration regulations at the last point. 'Such is life in Krusadai,' Madhavan would comment wistfully; he would promptly despatch a petition to the Director of Fisheries, with a request for a proper canoe for the station, but the ongoing economic depression would delay its sanction.[25] 'All is well that ends well,' Janaki wired Darlington, on her return journey to Madras; a voracious reader, Darlington was already by this time, immersed in Dostoevsky's *The Brothers Karamazov* on the steamer from Colombo.[26]

The Aftermath

Darlington's India trip left Janaki emotionally drained, confused and distracted. Addressing him 'Dearest Cyril' (Darlington would playfully call her Janthi, in the manner of her younger brother Madhavan) in her letters, she unabashedly expressed her yearning for him. It would soon dawn on her that he was too free-spirited a person to expect commitments out of, a painful realisation that would transform the relationship, at least for her, into something deeply platonic. Janaki would henceforth imagine herself as a renunciate of sorts and sometimes even dressing like one. While she criticised his sometimes highhanded and insensitive ways, she would never cease to admire Darlington's intellectual brilliance, moral courage and irreverence to authority.

On the Social Front

By late 1933, Janaki's brother, Kittu had been transferred with promotion from the Hosur Cattle Farm to Salem as Conservator of Forests and, like his father before him, had become a Freemason.[27] Janaki was in Madras on a short break (after seeing Darlington off), residing at QMC with friends. A visit to the QMC on the Marina was customary for Kittu; there was always somebody or the other from the family or among friends studying there. Niece E. U. Shantha was doing a BA in Botany at the college and then there was family friend C. K. Kausalya who was teaching, but away in America at this time. When Kittu and family visited the college, they ran into a busy Janaki and her friend Miss K. Shanti Ranga Rao, who had recently arrived from England. The daughter of Kudmul Ranga Rao of Mangalore, who had devoted his life to social reform, and the sister of Mrs Subbaroyan (Radha Bai K. Ranga Rao), the Zamindarini of Kumaramangalam, near Erode, Miss Ranga Rao was a postgraduate of the Madras University and an educator and social worker. A member of the staff of the QMC, teaching geography, she had spent almost seven years pursuing higher studies abroad but for a short break, when she returned to Madras to organise the Geography course at the College. In 1931, she had been made a Fellow of the Royal Geographical Society, perhaps the first Indian woman to be so elected.[28] Eileen Erlanson was also in the city and staying at the University Guest House but does not appear to have met Darlington during his recent visit. At the request of Janaki, who was to leave soon for Trivandrum, Kittu and family played host to Eileen and drove her around the city, taking her shopping and showing her sights.

Grasses for Chase

Grasses (family Gramineae, also called Poaceae) were what Janaki had the greatest passion for. Throughout the 1930s, she maintained a correspondence with the American agrostologist Mary Agnes Chase, whom we met with earlier. From Madras, where she was still in charge of the University Botanical Garden, Janaki sent Chase (working at the Smithsonian, in the Herbarium of the U.S. Department of Agriculture), a package of seventy grasses. 'We are very glad to have them. Seeing *Spinifex*, I looked up your locality on the map and find you are almost at the tip of the peninsula. It must be an interesting region,' Chase remarked. Sending Janaki, a 'general paper on grasses,' in return, she promised a collection of grasses of the United States. She also gave Janaki news of Carl Erlanson ('Mr Erlanson has just returned in good health from his South American hunt for potatoes') and of Bartlett ('Professor Bartlett is now interested in Guatemala and Yucatan'), and conveyed colleague Hitchcock's warm wishes to her.

Janaki had in fact collected 'hundreds of grass specimens' for Chase, which she hoped to despatch in batches.[29] She had informed Chase about her new job and relocation to Trivandrum, which would 'give [her] opportunities of getting acquainted with the rich flora of Travancore.' Interesting experiments were being conducted at Coimbatore with sugarcane and Indian cereals, and Venkatraman had allowed her to work on the cytology of his intergeneric *Saccharum-Sorghum* hybrid, but the government being slow in making her Coimbatore post official, she had been forced to take up the teaching job, she explained. 'Things move very slowly in this country,' Janaki added wistfully. 'I carry with me pleasant recollections of the morning I spent with you and Dr Hitchcock at the Smithsonian Institute. Kindly give my salaams to the Doctor and to Prof. Bartlett,' she wrote to Chase. Janaki had one more request: to goad Bartlett to undertake a comparative study of the flora of the Malay regions (which he was after all familiar with) and Travancore. There was bound to be a strong floral affinity between the two regions, she hinted.[30] In fact, her future researches would consider this important question of distribution of flora across the humid tropics. That she was already thinking along these lines, and precociously so, is to be carefully noted.

On Marriage

On her thirty-sixth birthday (4 November 1933), in Trivandrum, Janaki received a bag and a pair of slippers from her brother Raghavan in Rangoon. 'It is very sweet of you to have remembered my birthday. The bag is very welcome. I still have one you sent me while I was in U.S.A. but the slippers, I am afraid will not fit a *sanyasin* (renunciate) like me. So with your leave, I shall present them to someone whose feet it might adorn better,' she remarked. She was the only one still single among the Edathil sisters. The independent-minded Devayani, who worked as a teacher at Mangalore, was however unhappy, 'finding it difficult to get along with her husband's people.' Some members of her family ascribed this to her stubbornness, but Janaki believed Devayani had brought this upon herself, for despite being so well educated, had chosen to marry a man she had hardly known. Even if she sounded somewhat excessively harsh, her letter to brother Raghavan reveals her incredibly progressive stand, and not just for her times, on marriage:

> rather than blame . . . his people I am most ready to blame my own sister. Well the worst I can say is she has made her bed & must lie on it. Her consent to marrying a man she had never known is to me a most bewildering thing on the part of one so educated & old to do. I suppose one does get adventurous sometimes . . . in Malabar we not only marry a man but we [also] marry his family. . . . There is an innate culture and dignity one looks for . . . the girl has been independent and expects to be independent to some extent. However, my point of view about these things is very different to the ordinary Malayali concepts—where women occupy very back position as wife . . . She better find her remedies herself.[31]

Devayani's husband Anandan, a Licentiate in Medicine, was the brother of N. K. Narayani, Janaki's classmate from her FA days at the Madras College for Women. Janaki had suggested the name of Narayani as a match for her brother Raghavan, we may recall, although she had reservations about the girl's family, which was not 'very modern in their outlook' or as highly placed socially and culturally as theirs. The Nattiyala Keloth family however could claim an untainted (by 'white blood') social reputation, a matter of envy for the EK family, Janaki included. Besides they were all in agreement that Anandan was a generous person and, moreover, a selfless doctor. With time however, Devayani would learn to cope with the situation, by pursuing a teaching career in Malabar, and leaving Anandan to carry on with his in Rangoon, for the major part of their lives.

At Trivandrum, Janaki was doing very well but her contract was not on a permanent footing; neither did she wish it to be. In fact, she had been asked by the Travancore government if she would accept the post of Chief Inspectoress, in charge of women's education in the state, but had rejected it without hesitation. 'I shan't give up research & Botany ever for a mint of money. I am in fact prepared to be a beggar but a happy beggar. Trivandrum is a very difficult land for one like me whose ideas of right & wrong are sharply defined. I am not as bad as Sumthi [sister Sumithra] in this respect but there are certain things in which I shall never compromise—for instance honesty,' she would remark to Raghavan. Janaki was glad that he was coming home soon from Burma. 'I hope we shall have some good times together during the summer vac[ation]—must think seriously about our plantation.'

A most affectionate and caring sibling, she was glad that Sumithra was going to Burma to be with sister Devayani, who had just had a baby. 'Do bring Devayani also with you. Has Devayani registered her marriage,' she queried with much concern.[32]

Eileen Erlanson Visits Trivandrum

Eileen, who had been teaching at the Kent State College, Ohio, and was also part of the summer biology staff at Michigan University since 1930, arrived in Trivandrum by the end of that year (1933), on Janaki's invitation. She would spend 'a year of research and travel in Travancore and Malabar' in the capacity of honorary professor of Botany at the College. Eileen remarked in an interview given at Ann Arbor, just prior to her departure for India:

> Dr Janaki expects to obtain some honorary appointment from the Maharani for me, to make my position official. It may be in connection with the herbarium of the College or as the head of a botanical survey of Travancore, which I shall have to organise. During the minority of the Maharaja of Travancore, his mother, the Maharani, is ruling. She is eager to advance science and scientific research and appointed my friend, Dr Janaki, to a high position in what had formerly been strictly a men's College.[33]

Janaki was of the opinion that biology was 'stagnant in South India' and that Eileen could help make a difference, by demonstrating that research was more than just teaching.[34]

No sooner than she had arrived in Trivandrum, the outgoing and spirited Eileen, sometimes to the point of annoyance to Janaki, began exploring the place mostly on her own (Janaki, it appears, had lent her her car), and at other times with botanists Mercia Janet and Sosa P. John (teaching at the Women's College in the city). Algal collections, in the joint names of Janaki and Eileen, would be despatched to the herbaria of the Harvard University, the Field Museum of Natural History (Chicago), the University of California, Berkeley and the University of Michigan, Ann Arbor; among the ones sent to Harvard was a *Syzygium caryophyllatum* from the Pulayanar Kotta hill in the suburbs. On one occasion, Eileen drove thirteen miles to collect in the area around Aruvikkara, where a dam of granite was in the process of being built; she had taken a 'small boy' along to carry her plant-press. That same evening, Janaki drove down to Kovalam beach, with Eileen and Ouwerkerk. While the last two enjoyed a 'lovely swim, and rode on a log boat (Katamaran) paddled by Mahomedan fishermen,' Janaki 'got wet collecting sea-weeds.' The three then had 'tea on rocks & drove back at dusk.'[35]

Eileen's honorary professorship would take her to Madras and Malabar by the end of the year, and surely to Edam, where she appears to have spent the Christmas holidays with Janaki. For both Janaki and Eileen, holidays were invariably associated with collecting excursions, and there was nothing more joyous than wandering in each other's company hunting for plants. Eileen appears to have also visited Janaki's siblings at Shoranur; from the Shoranur neighbourhood, she would collect a specimen of *Tragia hispida* (a kind of climbing nettle) on 18 December 1933. In early January 1934, after their return to Trivandrum from Madras via Malabar, Eileen and Janaki were engaged in extensive collecting on the coast, of marine algae.[36]

Gandhi Visits Trivandrum

On 20 January 1934, Gandhi visited Trivandrum, part of his fund-raising campaign for the Harijan Sahaya Samithi. Eileen, out of curiosity, joined the procession of young women, who were to receive Gandhi, when he arrived at the Maidan. Gandhi was on this occasion accompanied by Miss Madeleine Slade (Mirabehn) and two other women. On the following morning, Janaki visited Miss Slade, and through her, met and conversed with Gandhi for a few minutes. After wishing him success with his campaign, she respectfully touched his feet. This was without doubt a momentous occasion for her, one that she would never ever forget.[37]

News from Coimbatore

In April 1934, at Trivandrum, a few months after Darlington's departure, Janaki received news from the Imperial Council of Agricultural Research that she had been appointed cytologist to the Sugarcane Breeding Station, Coimbatore. In the time she had before departing for Coimbatore, she was joined (once again) by Eileen, who had arrived in the capacity of honorary Visiting Professor of Genetics, and 'guest of E. K Janaki Ammal of the College of Science, Trivandrum.' This time, Eileen's focus was the diets of vegetarian and non-vegetarian school-going children of Trivandrum, among the fisher community of Vettukadu, the 'low-caste' Hindu girls from schools in the Trivandrum Fort and Chalai areas, and the Brahman and Nair children from the Fort Government School, Christian Mission Zenana School and the Mahila Mandiram Orphanage. Three years later, she would publish the results of her study (on their differential growth), as 'collaborator in Asiatic Research for the University of Michigan.'[38] During her time in Travancore, and Malabar, Eileen not only collected plants but also information on castes and tribes; her love for the anthropology of the East was perhaps imbibed from her doctoral supervisor, Bartlett, much as Janaki herself had. Eileen would deliver an illustrated lecture on 'the botany and anthropology of Indians' on her return to Ann Arbor, where she was a research fellow at the University Botanic Gardens.[39]

Figure 8.1 Janaki (standing middle) with friends; Eileen Erlanson draped in a sari (right), and possibly, N. K. Narayani (left) and seated, Miss G. C. McCormick. c. 1933.

Source: Courtesy of the EK family.

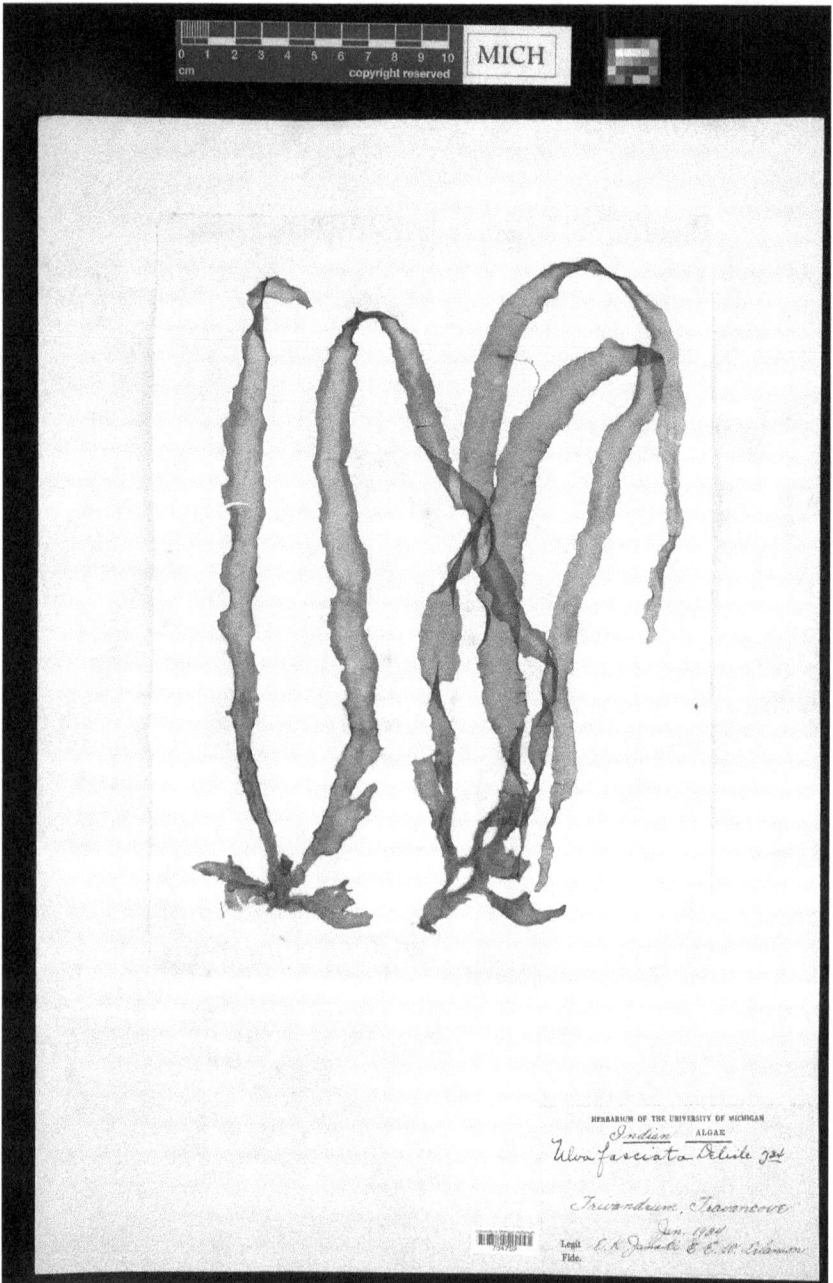

Figure 8.2 Ulva fasciata collected by Janaki and Eileen Erlanson. Trivandrum, Travancore. January 1934.

Source: Courtesy of University of Michigan Herbarium.

Notes

1 Minutes of the Annual Meeting of the Association of Economic Biologists held at Cambridge, 1907. It was at the inaugural address of the third international conference on hybridisation and plant breeding in 1907 that Bateson outlined the progress of genetic research in the aftermath of the rediscovery of Mendelian genetics. See Wilks (ed.), *Report of the 3rd International Conference on Genetics, Hybridisation and General Plant-Breeding*, pp. 90–97. Also see, Kraft, 'Pragmatism, patronage and politics in English biology'.

2 See for instance, R. H. Biffen's papers on wheat breeding: 'Experiments with wheat and barley hybrids illustration illustrating Mendel's laws of heredity'; 'Mendel's laws of inheritance and wheat breeding'; 'The application of Mendel's laws of inheritance to breeding problems'.

3 *Prospectus of the Agricultural Research Institute and College, Pusa*, Calcutta, 1909, p. 1.

4 Ibid.; the staff at the Pusa institution included an agricultural chemist, mycologist, agriculturist, agricultural bacteriologist, two entomologists and an economic botanist. For a history of agricultural departments and colleges in colonial India see Borthakur and Singh, 'History of Agricultural Research in India'; the institute was shifted to Delhi in 1936 following the devastating earthquake in Bihar in January 1934.

5 Darwin and Acton, *Practical Physiology of Plants*, based on a course of instruction given at the Cambridge University.

6 In 1924, the Government of India vested technical control of the Coimbatore institution with the Pusa Institute.

7 Parthasarathy, 'Origin of noble sugar-canes (*Saccharum officinarum*)'.

8 For a biography see Thuljarama Rao, *Biographical Memoirs*, Indian National Science Academy; and for a more recent account, partly based on Rao's account, see Maheshwari and Raman, 'The knight of sugar industry'.

9 Venkatraman, 'Simple contrivances for studying root development in agricultural crops'.

10 Rules of the Association, *Proceedings of the Association of Economic Biologists, Coimbatore*, Vol. 1, 1930–33, published by the Association, 1934.

11 DP: Darlington's pocket diary, g 4, India, 1933.

12 *Madras Agricultural Journal* 21 (1933): 355–356; DP: C. 98 (H. 50).

13 The Trichinopoly cigar was manufactured at a place called Woraiyur, from tobacco grown near the town of Dindigul, near Tiruchirappalli. Incidentally, a reference to Trichinopoly cigars appears in Jorge Luis Borges' *Ficciones*, in the short story 'The Approach to Al-Mu'tasim' (1935) and in *A Study in Scarlet* by Arthur Conan Doyle. Besides, Dorothy L. Sayers (a writer Janaki was familiar with) described Lord Peter Wimsy, one of her chief characters, as smoking a 'Trichinopoly' in *The Unpleasantness at the Bellona Club* (1928), a mystery novel. G. K. Chesterton and Rudyard Kipling also mention the cigar in some of their writings.

14 DP: C. 98 (H. 47).

15 In 1945, a special unit was organized in the Laccadive Islands (Lakshadweep) under the leadership of E. K. Madhavan for designing and building small-motorized boats for fishermen; the first such vessel designed by him was named *Ajith* after his youngest son. Madhavan had three children: two sons and a daughter.

16 In 1914, impressed by the biological diversity of the island, James Hornell, the Director of Fisheries in Madras, had recommended the establishment of a

Biological Station at Krusadai, an island owned by the Raja of Ramnad. See Jeyabaskaran and Lyla, 'Krusadai Island, the biologist's paradise', pp. 63–72.

17 Madras Fisheries Department, *Administration Report for the Year 1931–32* (Madras: Government Press, 1933), p. 12.

18 Madras Fisheries Department, *Administration Report for the Year 1933–34*.

19 DP: Darlington's pocket diary, g. 4, India, 1933.

20 Madras Fisheries Department, *Administration Report for the Year 1932–33* (Madras: Government Press, 1934).

21 Ibid., *Administration Report for the Year 1933–34*.

22 Ibid., *Administration Report for the Year 1932–33*.

23 Ukuleles were being sold at Misquith & Co., Madras for Rs 20 in the 1920s; sales-advertisements regularly appeared in the *Madras Mail*.

24 E. K. Madhavan to E. K. Raghavan, letter dated 18 August 1933, private collection.

25 Ibid.

26 DP: Darlington's pocket diary, g. 4, 1933, India. The first English translation of the book, by Constance Garnett, appeared in 1912.

27 Janaki's father, Sub-Judge E. K. Krishnan's social milieu was closely linked to the world of freemasonry; he was perhaps one of the first Malayali gentlemen to become a freemason. Krishnan was initiated to the Lodge Western Star, Cannanore on 21 November 1874, passed on 23 December and raised on the 20 February 1875. When Krishnan moved to Calicut in 1884, he appears to have become a member of Lodge Southern Cross, and a Past Master.

28 In 1935, Miss Ranga Rao went on to become the Principal (she was also one of its founders) of Nagpur's College of Arts for Women (later called Central College for Women) and the first woman deputy chief of the Indian Women's Auxiliary Corps with the rank of full colonel. She was also Principal of the Indraprastha College for Women, Delhi University. In 1956, Miss Ranga Rao published a report on legal disabilities of married women in the Commonwealth based on a conference held at Newnham College, Cambridge.

29 Smithsonian Institution Archives, Washington: Record Unit 229, United States National Museum, Division of Grasses, Folder 42, Box 4, E. K. Janaki Ammal to Mary Agnes Chase, letter dated 25 February 1933.

30 Ibid., letter dated 24 November 1933.

31 E. K. Janaki to E. K. Raghavan, letter dated 15 November 1933, private collection.

32 Ibid.

33 *TMD*, 9 August 1933.

34 BHL, UoM: University Herbarium Records (University of Michigan) H. H. Bartlett series, Box 12, Eileen W. Erlanson to H. H. Bartlett, letter dated 27 January 1933.

35 BHL, UoM: University Herbarium Records (University of Michigan) H. H. Bartlett series, Box 12, Eileen Erlanson's journal of her stay in Trivandrum, sent to Bartlett, dated 14 January 1934.

36 Taylor, 'Records of Asian and Western Pacific Marine Algae, particularly algae from Indonesia and the Philippines'.

37 Ibid, dated 20 and 21 January 1934.

38 *Current Science* 6, no. 4 (October 1937): 148–151.

39 *TMD*, 9 August 1934.

9

COIMBATORE I
Dreaming of Russia

I wish I had someone to discipline me in my work. I am carried away by every new problem I encounter with the result I accomplish very little.
—E. K. Janaki Ammal (October, 1934)[1]

A month after her appointment letter reached her, Janaki arrived at the Coimbatore Sugarcane Breeding Station all excited to begin research in sugarcane cytology, but she was still officially attached to the Maharaja's College of Science, Trivandrum. Invariably devoted to crop plants, breeding stations were male bastions the world over in the first half of the twentieth century. Janaki for instance was the only woman employed on the staff of the Sugarcane Breeding Station, apart from Miss Rajul Shah, a plant physiologist, who worked there briefly. Incidentally, Shah was the first Indian woman to graduate in the Agricultural Sciences (from the Poona College of Agriculture, Bombay University); she would also earn an MSc degree from the University of Michigan as a Barbour Scholar (1931). Shah would be appointed Horticulturist to the Central Provinces in the late 1930s, and become the first woman ever to head a Breeding Station (this one dedicated to Citrus) in India, in 1938.[2]

Janaki's entry into the world of applied science coincided with what was in a way the end of an era, as far as Indian botany was concerned. On 10 May 1934, the botanist-ethnographer and photographer Diwan Bahadur Kadambi Rangachari (1868–1934), who had collaborated with Madras Museum Superintendent Edgar Thurston on the monumental *The Castes and Tribes of Southern India* (1907) was no more. K. Rangachari had been appointed Herbarium Keeper at the Madras Museum, in the year Janaki was born (he would remain attached to this post until 1907). In 1909, he was appointed Assistant Economic Botanist at the Coimbatore Agricultural College, where he remained until 1923.[3] As a graduate student, Janaki had studied Rangachari's *A Manual of Elementary Botany for India* (1921), a standard textbook for university students; his *A Handbook of Some South*

DOI: 10.4324/9781003267089-9

Indian Grasses (assisted by C. Tadulinga Mudaliar, 1921) played a major role in kindling her love for the study of Indian grasses. Hailing from starkly different social and educational backgrounds, and belonging to two different generations, Rangachari and Janaki would however share the same academic platform, as members of the Board of Studies (Natural Science) of Madras University.

Fellow of the Indian Academy of Sciences

On 31 July 1934, the Indian Academy of Sciences was inaugurated by the Diwan of Mysore, Mirza Muhammad Ismail; physicist C. V. Raman (who had won the Nobel in 1930) based at the Indian Institute of Science, Bangalore, was its president. Janaki was the sole woman among the sixty-five founding fellows of the Academy. *The Madras Agricultural Journal* expressed hope that the Academy would 'prove a force of major magnitude' in scientific advancement.[4] The *Proceedings of the Indian Academy of Sciences Section A* (covering the physical and mathematical sciences) and *Section B* (the biological sciences) published monthly by the Academy from its first year of inception would provide yet another platform for women scientists to publish their work. Two chemists, Miss Kamakshi of the University of Rangoon and Miss Kamala Bhagvat (later Sohonie) (1911–98), a graduate of the Bombay University, would be the earliest among women to publish in the *Proceedings*. Four years prior to the founding of the Indian Academy of Sciences, a National Academy of Sciences had been established at Allahabad by physicist Meghnad Saha with the aim of providing a national forum for the publication of research carried out by Indian scientists.

Hopes and Wishes

Janaki's letter to Darlington in early July 1934 found him in Russia. The Soviet State's active involvement in promoting scientific research and education, in particular the use of Soviet radio for the purpose, had impressed him very much. Darlington's Russian experience, and his discussions with the Marxist Haldane, led him to discard the divide between 'pure' and 'applied' science.[5] Janaki had enclosed for him a piece of *Saccharum munja* (a wild grass growing in arid regions and along river beds), the first sugarcane to 'arrow' (produce inflorescence) after her arrival in Coimbatore. During his visit to Trivandrum, Janaki had presented him with a book of illustrated psalms. She referred to it in her letter: 'May be in my life the microscope has to take the place of what impressed you most in the book. Might not one be the symbol of the other,' she asked in a mysterious vein. Convinced of his unconventional take on love and marriage, her decision to remain unmarried was now firmer than ever, but that did not stop her from longing for him. In fact, Darlington had by this time begun a relationship with his young doctoral student, Margaret Upcott (and whom he would later marry). Janaki was sometimes deeply pained by his brash and insensitive behaviour, but remained charitable, with only words of solace when he was distraught or unhappy.[6] Darlington would always be her intellectual lodestar.

Janaki ardently hoped Darlington would visit India again, 'perhaps,' she wrote, 'the Imperial Council of Agriculture could be persuaded to write to you to spend a year in India. Wouldn't that be jolly.' She had only recently met the Vice-Chairman of the ICAR, Diwan Bahadur T. Vijayaraghavacharya, 'a very nice old man' and had had 'a long talk about sugarcane.'[7] 'I am fairly happy in Coimbatore,' she wrote, 'both amongst Indians and Europeans but I shall be truly happy only if I do some good work cytologically. . . . Our Agricultural Director, the Madras man is just back from Russia. I would like to visit Russia too. I would even like to leave India for good and work in Russia if conditions are not too bad there. You must tell me all about it.'[8] Janaki was referring to S. V. Ramamurti, the Director of Agriculture of the Madras Presidency, who had just returned from a tour to Italy and Russia and had delivered a lecture under the auspices of the Madras Agricultural Students' Union at the Coimbatore Agricultural College on 10 July 1934. In his lecture, Ramamurti had spoken 'about the many things' he had seen in Europe, particularly in Italy under the fascist regime; he found that the quality of research work in Europe was comparable to that of India and the work done in Coimbatore was second to none, except perhaps Russia. However, the big problem in India, he explained, as opposed to Russia, was the huge gap between research and its application; in a similar vein, the paddy specialist K. Ramiah would stress on the need to ensure that results of scientific research were entirely transferred into practice.[9] On the other hand, the cotton geneticist J. B. Hutchinson would remark that 'in the 1930s, Indian agricultural research was a standing object lesson of the dangers of concentrating on immediate problems. The criterion adopted by the early I. C. S. heads of the service, that a research project must offer something of immediate applicability had resulted in the service running out of steam in 30 years.'[10]

Learning About Vavilov

Janaki was acquainted with the three-volume *History of the Russian Revolution* by Leon Trotsky (1930), translated into English by Max Eastman in 1932, but one of the chief reasons, Janaki yearned to go to Russia was Julian Huxley's *A Scientist among the Soviets*; she had read the book in the very year of its publication (1932). In 1931, Huxley had visited Russia at the invitation of Intourist, a travel agency founded in 1929 by Stalin, staffed by the NKVD (The People's Commissariat for Internal Affairs) and responsible for managing and guiding foreign visitors through the Soviet Union. It was thus no surprise that he came away with a favourable view of large-scale social and economic planning; in fact, immediately on return he became a founding member of the British policy think tank called Political and Economic Planning. Incidentally, in August 1933, the Methodist missionary Charles W. Ranson of Madras had delivered a talk titled 'Will the Nations follow Russia,' the fourth of a lecture series at Kellett Hall (named after the missionary F. W. Kellett) in Triplicane.[11] *The Madras Mail* published the text of the lecture, which provided a summary of the history of the Russian Revolution and stated that Russian Communism was the first attempt to put into practice on an extended scale the political philosophy of Karl Marx. Later that month, 'objectionable' communist books were banned by the custom authorities from entering Madras State and impounded until the government had decided on what to do with them.

There was however a more compelling reason for Janaki to seek Russia: Vavilov and his Institute of Plant Breeding in Leningrad. P. S. Hudson of the Imperial Bureau of Plant Genetics, Cambridge, had called on Coimbatore, shortly after Darlington had visited, also at the behest of the Association of Economic Biologists.[12] Hudson (with R. O. Whyte) had by this time published *Vernalization or Lyssenko's method for the Pre-treatment of Seed* (Aberystwyth, 1933). He however chose to speak chiefly on the Soviet geneticist Nikolai Ivanovich Vavilov's theory of the origin and evolution of cultivated plants (based on a translation of a paper published in a Russian agricultural bulletin) at Coimbatore, rather than on the anti-Mendelian and Lamarckist Lysenko. Bateson's student at the John Innes in the formative years of the institution, and later better known as Lysenko's staunch opponent, Vavilov was the director of the Lenin All-Union Academy of Agricultural Academy of Agricultural Sciences at Leningrad between 1921 and 1940. In the 1920s, he had proposed that the world's cultivated plants originated in five main geographical zones—Southwestern Asia, Southeastern Asia, the Mediterranean region, Northeast Africa and Central and Southern America (and later in the 1930s, upgraded this to seven)[13]—from where plants diffused naturally and through artificial agencies to fill the whole world. While developing his theory on the centres of origin of cultivated plants, Vavilov had embarked on a series of botanical-agronomic expeditions (accounts of which were published posthumously as *Five Continents*), by the end of which, he had built up (preserved in Leningrad, today's St. Petersburg) a formidable repository of edible seeds; the institution was renamed the N. I. Vavilov Institute of Plant Industry in 1968 and remains one of the world's largest collections of plant genetic material.

Thanks to Darlington's and Hudson's visits to Coimbatore, Janaki had come to learn of the rise of Lysenkoism in Russia and the unjust criticisms levelled against Vavilov. The Soviet agronomist Trofim Lysenko, who had rejected Mendelian genetics in favour of the hybridisation theories of the Russian breeder Ivan Vladimirovich Michurin, drew wide attention of the Soviet leaders in the late 1920s for advocating a radical agricultural method called 'vernalization,' which promised to increase crop-yield manifold by exposing wheat seed to high humidity and low temperature. Deceiving nature to ensure productivity was hardly a new idea, but in reality, it failed to produce high yields, although some increase in production was initially noticed on the collective farms. This was a time of severe shortage of food in the Soviet Union, owing to a violent conversion to an industrial economy and the disastrous consequences of collectivisation; the *kulaks* had been liquidated, there was widespread mismanagement of the collective farms and peasants were in general unhappy. In the eyes of the Soviet State therefore, Lysenko, a mere peasant, had provided simple, tangible and quick solutions to agricultural problems, where even academic geneticists had failed. The leadership of Stalin was thus quick to hail Lysenko, a member of the proletariat, as the hero of Soviet agriculture, while severely attacking its critics. Academic biologists who practiced a science that was time-consuming were unable to demonstrate the uselessness or harmful nature of some of Lysenko's quick-fixes, owing to the rapidity at which they were being developed and adopted. Lysenko had labelled Mendelian geneticists as enemies of Marxism, akin to peasants who still opposed the Soviet government's collectivisation strategy. Also, like the French biologist Jean-Baptiste Lamarck, he believed acquired characteristics were inherited, something Mendelian geneticists rejected outright.

India was one of the major biodiversity countries in the world, with 45,000 species of flora, and a Vavilovian centre of origin and variation of cultivated plants (including the sugarcane, mango, citrus, banana, jack-fruit, egg-plant, edible *Dioscorea*, several medicinal and aromatic plants, besides rice, some millets, jute, *Alocasia*, *Colocasia*, *Amorphophallus*, cardamom, black pepper, ginger, and turmeric). Vavilov had planned an expedition to India in the very year Hudson and Darlington had visited the country. In one of his letters to his colleagues, Vavilov wrote:

> All my concern is now [1933] with India and South-East Asia; I am summarizing the worldwide plant philosophy and distribution for we have logically approached this very point, while all other issues are more or less clear . . . On the whole it is necessary to begin serious studies of India, Indo-China and China . . . Our interests in India are growing from year to year. These days I shall send you a book with a map where our work is generalized, and it will help you to understand why our eyes are fixed on India.[14]

Vavilov's expedition to India would not be realised because he was denied permission by the Indian government to enter the country.

Restless Again

At the Breeding Station, after an initial high, Janaki began to feel out of sorts and discontented with her work environment; she was missing Darlington. Her letters to him grew more frequent and her longing desperate and even irrational. She became distressed when he did not reciprocate with the same intensity. All the same, she was busy making crosses with the wild species of the *Saccharum* (*S. spontaneum*) and hoped her *Saccharum munja* would cross with *Sorghum*. In early August that year, she sent Darlington a letter enclosing three photographs of herself (one of which is Fig. 9.1):

> Dearest Cyril . . . Rather bad taste to bombard you like this but I want you to have the one you like. You can give one to Brenhilda from me and throw the third into your waste paper basket. I hope you are doing well and happy. I am feeling very restless and I am almost at the end of my capacity to endure this hunger you have created in me much longer. You hoped 'life would mean a little more' to me. It has been just the other way. Wish you the best of holidays in Russia. You were in Coimbatore this day last year! . . . My love to you, Yours passionately, Janthi.[15]

In another letter, sounding somewhat mystical, Janaki wrote:

> I have a strange feeling about you that something not good is about to happen to you. I am not given to such feelings generally and this may sound very stupid to you but there is an urge that I must warn you. Please forgive me if it is just my imagination but take care of yourself in every way. I hope I am not getting morbid.

Something in his last letter had bothered her. Perhaps he had narrated a nightmare that he had had recently, or something similar, for she responded: 'It is strange to think of you in sack cloth and ashes, Cyril,' and very tellingly, quoted lines from Shakespeare's *Romeo and Juliet* (Act 1 Scene 5, spoken by Romeo):

> 'Give me my sin again,' is more like you. Even I feel that way sometimes- and please don't ask me to 'find someone'. Is that a hint a very broad one! Anyhow I love my loneliness more than anyone just now . . . and find there is a lot of Beauty around us to meditate and be happy about.

Undeniably Nietzschean in spirit, or even Schopenhauerian (suffering is the meaning of life), she observed that 'Unrest can be subdued—and pain is essential for growth.' She was in search of new experiences and was 'doing hard work with Russia as the goal,' but there is no doubt had she succeeded, she would have found it suffocating, even more so than Coimbatore, for the Soviet state had begun to exercise a strangle-hold over science, and not just genetics. She wanted Darlington to throw her 'a cytological thought' as he sat by his 'fireside smoking one of those cheroots.' As a matter of fact, she had just placed an order with Spencers, Madras to despatch an assorted box of cigars (his birthday which overlapped with Christmas was fast approaching), because his 'tastes must be very cosmopolitan by now.' As a return gift, Janaki demanded a picture of his taken in Russia. She would end her letter this time with, 'to you much love, of the very sisterly kind! Janthi.'[16]

Janaki was leaving for Agra (it is not clear what took her there but from her field notebook, we know that she collected *Sorghum* specimens at Fatehpur Sikhri in October 1934),[17] when she received Darlington's postcard from Russia. This cheered her up somewhat, but when she found two more letters from him on her return to Coimbatore, she was overjoyed. Darlington felt Russia would suit her. She was thrilled by the prospect of this and responded that she was 'tired of living in an environment whose values [were] so different to [hers]' and was determined 'to accomplish something worthwhile in Saccharum cytology so as to get an "open sesame" to USSR.' Janaki was obviously overwhelmed by some deep epistemic anxiety at this time: 'My new microscope arrived last week (a Leitz binocular). I have fixed enough material to keep me engaged for 3 years. I wish I had someone to discipline me in my work. I am carried away by every new problem I encounter with the result I accomplish very little.'[18] It was the sugarcane flowering season and a resolute Janaki was busy making crosses. She had succeeded in getting seeds from a cross between *Sorghum* and *Erianthes*, but to her utter dismay, a herd of stray goats had eaten up the precious plant, the chief reason, she explained, why her letters sounded so full of desperation; she had lost so much time. 'August and September were rotten months in every way. If I cannot produce good work there is no use of my living on, is there?' she asked rhetorically, sounding Nietzschean again.[19] Her complaints in life would never be about physical or material discomforts, but only about not being able to accomplish her research goals. She never strived for happiness; only for work. About the same time as Janaki, a young American agronomist Jack Harlan (1917–98) was getting impatient to go to Russia, to complete his undergraduate degree under Vavilov (his father was a close friend of the Russian geneticist). Neither Janaki nor Harlan would make it, and fortunately so, for things would change for the worse in Soviet land.

Scientific Meetings at Calcutta

At the annual meeting of the Indian Botanical Society at Calcutta in January 1935, Janaki was elected Honorary Secretary for a period of three years,[20] with J. H. Mitter of Allahabad as President, P. Parija (Professor, Ravenshaw College, Cuttack) and S. R. Bose (Professor of Botany, Carmichael Medical College, Calcutta) as Vice-Presidents and M.O. P. Aiyengar as Treasurer. As was usual for this time of the year, a large number of scientists had descended, upon Calcutta, to attend the Indian Science Congress. Janaki would attend too. This Congress was special for the number of women participating in its proceedings; the Botany Section included, besides Janaki (the only one with a doctorate), Mrs Craker (with H. P. Naskar and K. P. Biswas of the Calcutta Botanic Gardens), Smt. Usha Chatterjee of Allahabad and Mrs Sarojini Datta of the Bethune College, Calcutta. Miss Olive Joseph (with Prof. S. M. Mehta) and Miss K. D. Gavankar (with Prof. N. W. Hirwe) of the Royal Institute of Science, Bombay, presented papers in the Chemistry Section of the Congress.[21] Several women were elected members of the Indian Science Congress that year.[22]

The inaugural meeting of the National Institute of Sciences of India (later called the Indian National Science Academy or INSA) was also held that same week, on 7 January 1935, the venue being the Senate Hall of the University of Calcutta. J. H. Hutton, President of the Indian Science Congress, was in the chair, supported by L. L. Fermor, President of the National Institute of Sciences of India. It must be noted that none of the elected Fellows of the NISI were women; not even Janaki figured in the list. In the same year, the monthly journal *Science and Culture* began to be published from Calcutta under the aegis of the Indian Science News Association (ISNA), founded by physicist Meghnad Saha and chemist Profulla Chandra Ray, the chief focus of which was the public understanding of science and its social implications.

Cytology of the *S. spontaneum*

In late June 1935, Janaki presented the results of her cytological study on the wild cane *S. spontaneum* before the Association of Economic Biologists, Coimbatore; by the middle of August, she had also sent two papers to the *Indian Journal of Agricultural Science* for publication.[23] The results obtained in Java and India had fully justified the use of *S. spontaneum* (widely distributed in the Old World) for purposes of breeding. Starting with Co 205, an interspecific hybrid between the thick cane 'Vellai' (a 'true' local *S. officinarum*) and the local form of *S. spontaneum*, most canes produced in Coimbatore had in them 'traces of spontaneum "blood".' In India, the local species took on a number of distinct forms. In 1929, nine forms of *S. spontaneum* were obtainable in Coimbatore, and in 1933, R. Thomas added eight additional forms from Bihar. While hitherto the method of approaching taxonomic problems was purely on the basis of morphological studies, now the additional evidence of cytology and genetics had widened the concept of species. Janaki aimed to bring together observations on the

154

chromosome behaviour of some of the Indian forms of *S. spontaneum* in the hope that the cytological facts would help throw light on the phylogeny (the evolutionary development of a taxonomic group or the history of genetic relationship) of this widely distributed Linnaean species. Fundamental to phylogeny was the proposition that plants or animals of different species descended from common ancestors.

For her microscopic study, Janaki had focused on the cells of the root tips and the pollen mother cells; the root tips had been fixed in Allen's modification of Bouin's solution (invented by French biologist Pol Bouin) and in La Cour's fixative (Len La Cour of the John Innes), a combination she had developed through trial and error. Janaki had found a modification of Bouin's solution most satisfactory for the study of the metaphase plates. Further, she discovered that freezing the root tips for two to three minutes by surrounding them with crushed ice, before fixing, had the effect of spacing the chromosomes evenly, making counting much easier. This was particularly important in a genus like *Saccharum* in which chromosomes were many. She prepared aceto-carmine smears and sections and made drawings 'at bench level' with a Spencer Abbé camera lucida, a Leitz objective, and a Zeiss x 25 compensating eye-piece, to give a magnification of 4,000.

The different sugarcane types that Janaki studied (all grown at the Coimbatore Station) were assigned names after the places they were collected from such as the Lahore, Dehra Dun, Coimbatore, Godavari, Dacca, Bihar and Cochin canes. Only a couple of years previously, in 1933, Rama Rao Panje (trained at the UBL, Madras) had provided morphological descriptions of a few forms growing at the Station[24] and further back in 1925, Bremer had counted the chromosome numbers of the Java (*Glagah*, n = 56) and the Celebes forms (*Glagah tabongo*, n = 40). N. L. Dutt and K. S. Subba Rao were the first (in 1932) to point out that the Coimbatore form of *S. spontaneum* differed from the Java and the Celebes forms in having 64 chromosomes in its somatic cells[25]; two years after their note was published, T. S. N. Singh reported 32 bivalents (paired homologous chromosomes of the type found in diploids) in the Godavari form, 27 in the Dehra Dun and 39 in the Dacca form.[26] Chromosomes previously diverged, typically segregated as bivalents when brought together by hybridisation.

Janaki's contribution was the determination of chromosome counts of hitherto unexamined forms, with an eye on polyploidy, to shed light on speciation in the *Saccharum* family. *Saccharum* presented high ploidy levels, like the others in the '*Saccharum* complex' (by which was meant the subset of the Saccharinae including *Erianthus, Miscanthus, Narenga* and *Sclerostachys*), and exhibited a marked capacity for intergeneric hybridisation.

It was Bremer who had first turned his attention to the study of the transmission of chromosomes in the interspecific hybrids of *Saccharum*. The cross between the 80-chromosome *S. officinarum* and the 112-chromosome *S. spontaneum* revealed a count of 136 chromosomes, which was certainly not the sum total of the parental haploid numbers (n+n = 40 + 56 = 96). Instead, it was the sum of the maternal diploid and paternal haploid numbers (2n+ n = 80 + 56 = 136).[27] Stimulated by Bremer's study, several papers emerged on the ploidy level in the interspecific and intergeneric hybrids of *Saccharum*. That polyploidy was indeed the most important force by which the organism adapted itself to the environment and maintained a stable chromosome balance would become a widely accepted theoretical idea only much later, but the practical plant breeder on the field began exploiting its benefits very early. Polyploidy's importance in the evolution of economic crops is best exemplified by food crops such as wheat, potatoes and oats, and cash crops like tobacco, cotton and sugarcane, which are all naturally occurring polyploids. They were selected for being successful as invaders of new habitats and tolerant of extreme environmental conditions, thereby shedding light on the evolutionary possibilities that chromosomes opened out.

Earliest *Saccharum* Publications

Janaki's first ever publication on the sugarcane dealt with her investigations of the different types of *Saccharum*; she was able to generate a polyploid series, with 8 as the basic number (2n = 48, 2n = 56, 2n = 64 and 2n = 80), suggesting that it was highly probable that the more primitive types were those with a lower number of chromosomes. Her paper also indicated that the primitive forms occurred chiefly in India; she speculated that a comprehensive study of the species would lead to the discovery of more 'eco types' (i.e. adapted to specific environmental conditions), with chromosome numbers 48 and 56.[28] Further, on examining the division of the pollen mother cells, she observed that the pairing of chromosomes during meiotic division occurred by means of chiasmata, but the number of chiasmata (or chiasmata frequency) was not always proportional to the length of chromosomes, as would normally be expected.

Janaki's second paper (co-authored with T. S. N. Singh) was concerned with the cytogenetic analysis of a *S. spontaneum* form from Burma. The plant rarely ever flowered in Coimbatore, with only a single inflorescence observed in 1933. As it did not arrow in 1934–35, she was only able to examine its root tips and not the pollen mother cells; she found it to be a triploid with a basic chromosome number x=32. As soon as her papers were published, Janaki would send copies to Chase, now associate agrostologist with the Bureau of Plant Industry at the Smithsonian Institution, Washington. Chase was glad to have them for the library of the Grass Herbarium.[29]

Cambridge and Amsterdam

About mid-August 1935, immediately after she had sent away her papers for publication, Janaki left for England to attend the Imperial Botanical

Congress at Cambridge and the Sixth International Botanical Congress in Amsterdam; the *Times of India* reported that she was 'the only woman scientist in the Agricultural Dept of India' to be deputed by the Government of India to attend the Cambridge conference.[30] At this time, T. S. Venkatraman was in Brisbane to attend the Sugar Technologists Conference (he had combined this trip with a visit to the Sugarcane Station at Pasoeroean, Java). Janaki's renewed passport issued at Ootacamund by order of the Viceroy and Governor-General of India, dated 15 July 1935, stated that she be allowed to 'pass freely without let or hindrance,' and to be afforded 'every assistance and protection of which she may stand in need' in all countries in Europe and the United States. The passport indicated that she was a 'Sugarcane Geneticist (Research),' had dark brown eyes and black hair and her visible distinguishing mark being a 'mole on tip of nose.'[31]

At the Cambridge Congress, whose chief focus was economic biology, Janaki and the biologist Joseph Burtt Hutchinson (1902–88), known as 'Hutch' in official circles, represented the Imperial Council of Agricultural Research. Hutchinson's research in India, from his arrival in 1933 until his departure four years later, was devoted to the genetics of Asiatic cottons (*Gossypium*). Other members of the Indian delegation included W. J. Jenkins (chief agricultural officer in Sind), D. G. Munro (deputy director of agriculture, Madras), D. Rhind (economic botanist, Burma), M. Bashir (assistant registrar, Punjab University), Birbal Sahni of the Lucknow University and R. K. Saksena and S. P. Naithani of the Allahabad University.

As for the Amsterdam Congress, it was chiefly devoted to academic botany, and no matters of policy were discussed. Janaki would run into the American grass expert Albert Hitchcock at the Congress, almost four years after she had first met him in Washington; his wife had accompanied him to Amsterdam. The Indian participants (which included the director of agriculture of Bombay, W. Burns; Burma's deputy director of agriculture, A. McLean; S. P. Agharkar representing the National Institute of Sciences, besides Janaki, Hutchinson, Saksena, Naithani and Bashir) were all researchers in the fields of cytology and genetics. While leading cytologists Victor Gregoirè, Emil Heitz, G. A. Lewitzky and the British-American woman biochemist Dorothy M. Wrinch presented papers on the structure of the chromosome and the nucleus, Darlington and Karl Sax discussed the mechanism of crossing-over in relation to the internal mechanisms of the chromosomes. W. J. C. Lawrence and Olavi Meurman spoke on the pairing of chromosomes in hybrids and heteroploids and Thomas Goodspeed of the University of California on the nature and significance of chromosomal transformations induced by high-frequency radiation.

The genetics section of the Congress focused on the genetics and breeding of immense variations, on inbreeding and incompatibility, on lethal factors and of the species concept in the light of modern research.[32] Incidentally, Janaki had carried to the Congress, as per the suggestion of Darlington, her

Saccharum-Sorghum hybrid (cross between *S. spontaneum* and *Sorghum*) for purposes of demonstration. She could be proud of herself, for not only had she made conspicuous advances in *Saccharum* cytology but also demonstrated that she was a competent breeder of intergeneric crosses, much as she had hoped.

At the John Innes Again

In the period between August and December 1935 Janaki resided at Crosby Hall, Chelsea. She had decided to spend three months as visiting researcher at the John Innes, making the best of her trip abroad as a government scientist.[33] The John Innes now was no longer the isolated institution of Bateson's days, nor even the place Janaki had known it to be in 1931. It was a world-ranking institution with Darlington, M. B. Crane, Kenneth Mather, Dan Lewis, Geoffrey Beale and Irma Andersson-Kottö among its staff.[34]

Lewis vividly described the ambience of the cytology laboratory at the time:

> The not unpleasant smell of clove oil. Alcohol and xylol, the noise at first of the microtome and later the woodpecker-like tapping of the glass rods on slides for squash preparations and the clink of slides going through their alcoholic series were the constant sensory and musical background; each worker had enough room for a microscope, slide jars and a notebook. . . . Darlington, in his separate room with the door open, had his ancient brass microscope with its long tube, originally bought second-hand by Bateson. The two, simple home-made lensless cameras were constantly passing from one microscope to another to record the latest chromosome. On the walls of the laboratory hung a few portrait photographs, one of which was of the Russian geneticist Nikolai Vavilov, who according to Darlington was distinguished for his theoretical basis of plant breeding, his collection of genetic material and as the leader of genetics in the U. S. S. R. [There were frequent visits] to the laboratory from the dark room of the presiding and charming genius of illustration and photography, the congenitally deaf H. C. Osterstock, with whom everybody soon learned to converse in his own self-taught language. Darlington's room was lined with reprint boxes, the spines of which were wittily illustrated with the most telling caricatures; one member of the early Drosophila school received H. J. M. and a crown, a sincere and not cynical comment; the several boxes of another showed a profile expanding from box to box with age, which matched the ever increasing size and number of his publications. The first was H. J. Muller and the second Th. Dobzhansky.[35]

Besides engaging in path-breaking science, staff members enjoyed staging Shakespeare, playing ping-pong and generally having an exciting social life. When Janaki arrived at the John Innes in August, she found Darlington in the company of Margaret Upcott, his young doctoral student; together with the Hungarian cytogeneticist Pio Koller (Pius Charles Koller, 1904–79), Darlington and Upcott had become an inseparable threesome.[36] Photographs of the John Innes staff from the period, taken by the plant physiologist Sansome in November that year, include a beautiful one of Janaki standing in thick snow; each photograph in the album was accompanied by a quotation from Shakespeare, and hers carried the line, 'She came adorned hither like sweet May' (*Richard II*, Act V. Scene 1).[37] Darlington visited her at Crosby Hall; on 5 October 1935, his pocket diary records that they lunched on 'crumpets and curry,' at Chelsea (perhaps at an Indian restaurant).[38]

Unlike in 1931, this time her departure from England would be marked by a deep sense of sadness and loss. This was also when Janaki completely renounced her colourful saris for the saffron, off-white or pale-yellow silk ones, sourced chiefly from Bangalore. It was as if she was ready to free herself of this 'burden' of youth (akin to McClintock's to be 'free of the body').[39] Her personality would now begin to exhibit a strange mystical streak, something that would become more pronounced in her later years. It is not as if she discussed such matters with anyone; she was too private a person for that, but we do get a sense of this in her letters to Darlington and her family members.

Janaki would also visit the Kew Gardens during this trip. There, she would run into Hitchcock again (she had met him only recently, at the Amsterdam Congress), who sadly, would be no more in just a matter of weeks (he died on 16 December 1935). Incidentally, it would be months before she would learn of his death, and when she did, would convey her deepest shock and regret 'at the death of the Great botanist' in a letter to Chase.[40]

Renews Eugenics Membership

During her visit to England in late 1935 Janaki renewed her membership of the British Eugenics Society. Among other Indian members of the Society at this time were the anthropologist Diwan Bahadur L. K. Ananthakrishna Iyer (1934, 1937, 1939) of Malabar, engineer theosophist Rao Bahadur K. V. Natesa Aiyar of Travancore (1937, 1938–39), mathematician Raghunath Dhondo Karve (1938–39),[41] who edited *Samaj Swasthya* and was the Indian counterpart to Sanger and Stopes, medical practitioner-sexologist Dr A. P. Pillay of Bombay who founded *The Journal of Marriage Hygiene*[42] and zoologist-cytologist J. J. Asana of the Gujarat College, Ahmedabad (1936–39). It may be noted that not a single botanist or agricultural scientist from India, except Janaki, was a member of the Eugenics Society.

Back to India

On 15 December that year, Janaki was well on her way home aboard the *S. S. Kaiser-i-Hind*. She announced to brother Raghavan: 'I shall be with you all soon after this arrives. The boat is at last in the Arabian Sea!—and have been looking forward to my return.'[43] Janaki had flown the Imperial Airways from London to Marseilles to board the ship to India. Of her maiden flight she remarked: 'It was a very interesting experience—the only trouble of air journey is that we can't bring much luggage—only 33 lbs were allowed.' Janaki's friends, the Cockrams of Twickenham (Benjamin Cockram and his wife Doris nee Holdrup, from her University of Michigan days), 'had a lovely time weighing everything' in her suitcase (the heavier bits of her baggage had been shipped out earlier), but she grew despondent when several things had to be bailed out at the last minute; she always took presents for her family and friends, but now she felt she was going back to India empty-handed.

The Aden port where the ship had called was 'full of battleships all lying in readiness to pump on the Italians—in case of any hostilities.'[44] A year before, a border dispute had arisen between Abyssinia (Ethiopia) and the Italian Somaliland, which had led to a brutal invasion of the country by Italy. Britain and France refused to intervene and in December 1935, a secret plan had been drawn out by the foreign secretary of Britain, Samuel Hoare and the Prime Minister of France, Pierre Laval for bringing to a halt, what has been referred to as, the second Italo-Abyssinian war. The Hoare-Laval pact offered to partition Abyssinia, but the proposal stirred up so much hostility in Britain and France that it was never executed. Eventually, in May 1936, Italy would conquer Abyssinia. At the League of Nations, Abyssinian emperor Haile Selassie condemned the use of chemical weapons by Italy against his people. 'I wonder if Aby[ssinia] will accept the terms of Hoare & Laval. It will be unfair to them I think,' Janaki commented from aboard the ship. At Aden, she saw things of great interest to her, such as sea shells, like those in her father's collection: 'I saw shells exactly like the one father collected & we have on the table,' an excited Janaki commented. 'They must have been given to him by his Moplah friends. The Arabs there look exactly like our Moplahs—in fact I think most of them are . . . same dress and cap!' she added.[45]

The Nedumpoyil Model Farm

Janaki was much looking forward to seeing Raghavan's newly constructed house, which stood on a hilltop at Chuduvalathoor in Shoranur, not far from the railway station. 'It will be such a happy sight,' she exclaimed.[46] As early as 1932, Raghavan had been on the lookout for a suitable piece of land to build a house and to cultivate, when he ended up a 'Burma pensioner.' During one vacation, he chanced upon a hillside property, close to sister Cousalya's house in Shoranur. Raghavan had mixed feelings about it initially and Janaki was hardly in its favour, because she thought it was a bad investment from the point of view of productivity. She had instead drawn his attention to a highland place in Cannanore district called Nedumpoyil (a small border village surrounded by forests, between the towns of Tellicherry and Manantoddy or Mananthavadi in Wayanad), which she came to learn about thanks to a casual meeting with one Kanatty Govindan on a bus journey to Tellicherry. Janaki had visited Nedumpoyil in the company of younger brother Varadan, a few days after meeting Govindan. She was completely taken in by the hundred-acre model farm she saw at the place, growing a range of crops including pineapples, papayas and vegetables, besides paddy, cultivated by the *poonam* method, a traditional form of shifting cultivation, involving slash and burn, and practised in north Malabar. This kind of agricultural practice (called *kumari* in South Kanara and *jhum* in the Assam Hills) involved growing rice as a mixed crop, by first cutting and burning bushes in the forest, ploughing with the pre-monsoon showers, and then sowing. The land was abandoned after the harvest of rice and allowed to recover its fertility, and a fresh area of scrubland was broken up for cultivation each year. Janaki was particularly impressed by the farm's successful application of the ideas of sustainability and polycropping, typical of local agricultural practice. On returning to Madras, an excited Janaki had described her visit to Raghavan, insisting that he emulate the Nedumpoyil example: 'You should do something of the sort. Government will easily give you forest land. Rent will be Rs 1.20 per acre per year. I found out all details from Mr Govindan with you in mind. Perhaps we shall both be partners in this work. Shall we—if so I shall go ahead and apply for some land. It's very profitable . . . Mr Kanatty has promised me all help . . . now I have a little money. I shall be able to do something constructive.'[47]

Janaki's plan would not materialise, and Raghavan would choose the Shoranur property instead, a six-acre parcel of land on a hill, where he would develop a fine garden, including a section devoted to vegetables, under the direction of Alfred Soans of the well-known Mangalore Soans Nursery (established with the help of the Basel Mission). He would also build an extravagantly grand house for his family, variously called Kunnumal House (from its location on the top of the hill), Mingaladon (a fond reference to his Burma days), Sugvir (an acronym formed by stringing the first letters of the names of the family members: *S*reedevi, *U*ma, *G*anga, *V*asumathi, *I*ra and *R*aghavan) or Udayagiri (literally, the hill of the rising sun), the last being Janaki's contribution. There were two houses on the hill already, Sri Nivas and Gokulam: the former a traditional wooden house that came with the land (belonging to a Namboothiri family), and the latter a house built by brother Vasudevan for his family. At the foot of the hill were the twin houses, Coustubham (a reference to Cousalya) and Sukumaram (after Sukumaran), where sister Cousalya lived with her husband (also cousin), the humourist and short-story writer Malabar K. Sukumaran.

Figure 9.1 Janaki. Studio portrait (three quarters' length), Coimbatore. August 1934.
Source: Courtesy of Clare Passingham.

Figure 9.2 Janaki at the John Innes. 1935.
Source: Courtesy of Clare Passingham.

Figure 9.3 Janaki in the snow at the John Innes.
Source: From the Sansome Album, 1935.

Notes

1 DP: C 109 (J 112), E. K. Janaki Ammal to C. D. Darlington, letter dated 31 October 1934.
2 Prashad, *The Progress of Science in India*, p. 177; Prashad however makes no mention of Janaki in his account.
3 For a biography, see *The Madras Agricultural Journal* 22, no. 6: 219–221.
4 *The Madras Agricultural Journal* 22, no. 8 (August 1934): 265.
5 Harman, *The Man Who Invented the Chromosome*, p. 173.
6 For his sake, Janaki even undertook a discrete investigation into whether Eileen was a lesbian; Darlington appears to have been deeply anxious for some reason. At the end of her probe, Janaki would tell him: 'No—I made no discoveries about EE [Eileen Erlanson] being Lesbian. She denied it once when I asked her about Beatrice, her Negro friend. I wonder if you are fretting about EE. You can think of her as having a different personality—more like her mother or sister. . . . I often wonder "who" it is with you now. Think of me sometimes Cyril D: Yours always, Janthi'. DP: C 109 (J 112), letter dated 8 July 1934.
7 For more see, *Current Science* 4, no. 5 (November 1935): 307–308. Vijayaraghavacharya first gained prominence as Commissioner for India at the British Empire Exhibition, Wembley (1924). He was an active member of the Indian Science Congress, presiding over its agricultural section in 1930, and a foundation Fellow of the National Institute of Sciences of India in 1935.
8 DP: C 109 (J 112), letter dated 8 July 1934.
9 *The Madras Agricultural Journal* 22, no. 8 (August 1934): 297–83.
10 Cited in Arnold, 'Joseph Burtt Hutchinson 21 March 1902–16 January 1988', p. 282.
11 Incidentally, it was Ranson, a keen follower of Gandhi's thoughts and nationalist ideals, who founded the Triplicane Sociological Brotherhood, an organisation for social change, which included among its activities building sewers, working towards a minimum wage law and social-housing, and opposing child labour. It was also Ranson who helped establish the first radio station in India, later to become part of the All India Radio. After a period of study at Oxford, Ranson published a book in 1938 based on his Madras experience and on the social outreach programme of Kellett titled, *A City in Transition*. Mathews, 'The legacy of Charles W. Ranson'.
12 Filed with the MS Darlington C. 98, Bodleain Library, Oxford, is a booklet, *Proceedings of the Association of Economic Biologists, Coimbatore*, Vol. 1, 1930–33, Coimbatore: The Association, 1934. This was perhaps a copy presented to Darlington by the Association during his visit to Coimbatore in late August 1933.
13 Vavilov identified seven centres of origin of cultivated plants in the 1930s: 1. The Tropical Centre, including India, Indo-China, Southern China and the islands of Southeastern Asia: of rice, sugarcane and the majority of tropical vegetable and fruit crops; one-third of the plants cultivated in the world emerged from this Centre; 2. The East Asiatic Centre, including the central and western parts of China, Korea, Japan and the major portion of the island of Taiwan: plants such as soyabeans, millets, many vegetable crops and several fruits; 20% of all cultivated flora came from this Centre; 3. The Southwest Asiatic Centre, including the interior mountains of Asia Minor (Anatolia), Iran, Afghanistan, Inner Asia and Northwestern India, and joined by the Caucasus, whose cultivated flora is genetically linked to the Near East and Middle East, is subdivided into: 3 a. the Caucasian Centre with its species of wheat, rye and fruit trees; 3 b. The Near East Centre, comprising Asia Minor, interior Syria and Palestine, TransJordania,

Iran, Northern Afghanistan and Inner Asia together with Chinese Turkestan; 3 c. The Northwestern Indian Centre including Peshawar, the adjacent provinces of Northern India and Kashmir, besides Beluchistan and Southern Afghanistan; 14–15% of all cultivated plants came from this region, with a high concentration of the wild species of wheat, rye and several European fruit trees; 4. The Mediterranean Centre, include the countries spread along the coast of the Mediterranean; originally supplied 10–11% of all cultivated plants such as the olives, the Aarob tree and a range of fruits and forage crops; 5. The Centre of Abyssinia, with a number of endemic genera and species of cultivated plants such as coffee, a grain called teff, an oil plant named Ramtil and a special kind of banana; contains 3–4% of cultivated plants of the world; 6. The Central American Centre, including a large portion of North America and Southern Mexico, and divided into three sub-Centres: 6 a. the mountains of Southern Mexico; 6 b. the Central American Centre; and 6 c. the West Indian Islands; 8% of cultivated plants came originally from this Centre such as maize, long-staple cotton and other American cottons, several species of beans, pumpkins, the cocoa tree and several fruit crops such as guava, cherimoya and avocados; 7. the Andean centre within South America: 7 a. the Andean focus proper, including the mountainous regions of Peru, Bolivia and Ecuador; 7 b. the Chilean focus, including Southern Chile and the neighbouring islands, which produced the common potato, *Solanum tuberosum*; 7 c. the Bogota focus in Eastern Columbia. These seven large centres corresponded to 'the sites of ancient agricultural civilizations'. See Loskutov *Vavilov and his Institute*, pp. 87–88.

14 Ibid., p. 78.
15 DP: C 109 (J 112), E. K. Janaki Ammal to C. D. Darlington, letter dated 10 August 1934.
16 Ibid, letter dated 22 November 1934.
17 RBG Kew, Library & Archives: Printed Notebooks, Herbarium E. K. Janaki Ammal, Madras.
18 DP: C 109 (J 112), E. K. Janaki Ammal to C. D. Darlington, letter dated 31 October 1934.
19 Ibid.
20 In August 1938, Janaki resigned from her post as secretary of the Indian Botanical Society; she was replaced by botanist Yajnavalkya Bharadwaja of the Banaras Hindu University.
21 The Royal Institute of Science was founded in 1920, supported by private donations. In 1926, it was affiliated to the Bombay University for the BSc degree course in mathematics, physics, chemistry, botany and zoology. In 1933, the east wing of the Institute was handed over to the Bombay University to establish the new Department of Chemical Technology.
22 Women members of the Congress included Mrs Sarojini Datta who had an MSc from Manchester, Miss Swarnalatha Ghosh, Miss Suniti Bala Gupta, Miss Rachel P. John of Madras, Miss P. M. Kanga of Bombay, Mrs S. R. Kashyap, Miss Maneck M. Mehta, Miss Nirupama Sen, Dr E. K. Janaki Ammal, Mrs S. L. Hora, Mrs R. B. Lal, Mrs H. Krall, Mrs B. Sahni, Mrs V. Sethi, Miss K. D. Gavankar, Miss Sally Meyer and Miss Olive Joseph. See Datta, 'Indian Women in Science'. A few of them, such as Mrs S. L. Hora, Mrs B. Sahni and Mrs S. R. Kashyap, were wives of scientists well-established in their respective fields.
23 *The Indian Journal of Agricultural Sciences* 6 (February 1936): 1–8 and 9–10.
24 Panje, '*Saccharum spontaneum* Linn., A Comparative Study of the Forms Grown at the Imperial Sugarcane Breeding Station, Coimbatore.'
25 *Indian Journal of Agricultural Sciences* 3, no. 6 (1933): 41–42.

26 During the prophase I of meiosis the process of synapsis or the pairing of two non-sister homologous chromosomes occurs in which bivalents are formed. In other words, bivalents or tetrads are formed as two homologous chromosomes (containing four non-sister chromatids) undergo recombination or 'crossing-over'.

27 Bremer, 'Cytology of the sugarcane'.

28 Janaki Ammal, 'Chromosome Studies in *Saccharum arundinaceum* L.', 1937. Later in 1942, T. S. Venkatraman and N. Parthasarathy came up with corrections to Janaki's chromosome counts: the Lahore clone of the *S. spontaneum*, which Janaki had shown to contain 48 chromosomes (2n = 48), they claimed was 2n = 54, and as for the Dehra Dun form, which she reported to contain n = 28, according to them, was 27 ('Chromosome counts in the sugarcane and its hybrids,' *Current Science* 11, no. 5 (May 1942): 194–195).

29 Smithsonian Institution Archives, Washington (SIAW): Record Unit 229, United States National Museum, Division of Grasses, Folder 42, Box 4, Mary Agnes Chase to E. K. Janaki Ammal, letter dated 16 June 1936.

30 'Woman delegate to botany conference', *The Times of India*, 25 July 1935, p. 9.

31 See BL, IOR: L/P & J/11/1/1523 (?) for a copy of her blue passport (no. 12187).

32 *Annual Report of the Imperial Council of Agricultural Research, 1935–36*, New Delhi: Government of India, 1937, Appendix V.

33 Incidentally, in April 1935, Walter Gawen King, the Surgeon General of Burma and a one-time partner of Janaki's maternal grandmother Kunhi Kurumbi, died at the age of 83.

34 A month after Janaki left the John Innes in late 1931, Mather arrived to study genetics and cytology under Darlington, funded by a three-year scholarship from the Ministry of Agriculture. In less than four months, he was ready with his first research paper: 'The origin and behaviour of chiasmata.' His third scholarship year was spent at Svalöf in Sweden, with geneticist Hermann Nilsson-Ehle. Mather received a PhD in 1933 for his work on the cytology of the *Lilium* and genetics of the *Antirrhinum*, after which he joined R. A. Fisher at the Galton Laboratory, University College London. Mather continued to collaborate with researchers at the John Innes, and five years later, joined the institution as Head of its newly formed Genetics Department. Janaki would meet him the first time in 1939.

35 Lewis, 'Cyril Dean Darlington, 9 Dec 1903–26 March 1981': 113–157, p. 144.

36 Harman, *The Man Who Invented the Chromosome*, p. 129.

37 John Innes, Norwich: The F. W. Sansome Album of Merton Staff, 1935; Sansome left the John Innes in November 1935 to join the University of Manchester as Senior Lecturer in Horticulture.

38 Detail of the lunch, drawn from Damodaran (2017).

39 Keller, *A Feeling for the Organism*, p. 36.

40 SIAW: Record Unit 229, United States National Museum, Division of Grasses, Folder 42, Box 4, E. K. Janaki Ammal to Mary Agnes Chase, letter dated 14 April 1937.

41 Raghunath was the elder brother of Dinkar Dhondo Karve, who was married to the trailblazing Indian woman anthropologist, Irawati Karve (1905–1970).

42 The pioneering sexologist A. P. Pillay of Bombay was the author of *Welfare Problems in Rural India* (1931).

43 E. K. Janaki Ammal to E. K. Raghavan, 15 December 1935, private collection.

44 Ibid.

45 Ibid.

46 Ibid.

47 Janaki to Raghavan, letter dated Queen Mary's College, Mylapore, 9 October 1932, private collection.

10

COIMBATORE II
Making Order Out of Chaos

I am a born wanderer and shall soon have nothing to keep me in India! . . . there is a Great restlessness within me . . . because of its 'unscientific atmosphere- India is not a good place for scientific research- too much Red Tape in Govt departments & Institutions.

—E. K. Janaki Ammal (1939)[1]

On 22 December 1935, Janaki was back in Coimbatore. Also, unlike her previous stint at the Breeding Station, she was now officially relieved of her post at the Department of Botany of the HH Maharaja's College of Science, Trivandrum. Janaki would leave for Shoranur, as soon as she completed a few official formalities, to reunite with brothers Raghavan and Vasudevan and sister Cousalya, as she always did, before her onward journey to Tellicherry to spend the Christmas vacation with her mother. She would return to Coimbatore in the new year, after a much-deserved holiday at Edam, soaking in the love and affection her ageing mother showered on her.

Janaki was now financially independent; she had bought herself a car and moved into a lovely bungalow in Coimbatore (in R. S. Puram), an exclusive locality, that had just been electrified. Here, she would have several occasions to host friends and family. Brother Kittu was the first sibling to visit her: 'Janaki's house is very spacious,' he would remark. With Janaki behind the wheel, the two would drive to Palghat to visit niece Tara's new house, Raghukula. Tara (the elder of the two daughters of Cousalya and Sukumaran) was married to Dr A. K. Raghavan of Palghat, later bestowed with the title Rao Bahadur.

During the summer vacations, Janaki's more members of the family would visit or stay a few days with her. Kittu's eldest son Hari Krishnan recalled one such summer vacation:

I spent a long summer holiday, Damtan [E. U. Damodaran, Lekshmi's son], my senior cousin also stayed, with Nachee [as Janaki

DOI: 10.4324/9781003267089-10

was addressed, by the younger generations of the EK family]. D was young impetuous, brilliant, witty and held a newly appointed post of a Labour Relations Officer. In many ways, D & N had many Edam characteristics, both were outstanding in their professions & both were exciting conversationalists. This set the ideal situation for controversial & challenging discussions. Sometimes this left N in a foul temper like an exploding volcano, but it often fizzled out as quickly as it erupted. D could goad & tease her, all in excellent humour which at times resembled a wayward mongoose after a cobra! N kept a beautiful house & demanded high standards of behaviour & etiquette for us all, which at one stage included three or four other cousins . . . on holiday as well.

On most days, Janaki would be back home for lunch. The meal would be 'laid out in the cleanest of kitchens' and they all ate 'Malayali fashion sitting on a "paaya" [straw or pandanus mat] with the food served on banana leaves. . . . At dinner it was a Western affair with proper crockery & cutlery with lessons on their proper manipulation.' 'While we enjoyed this very much, we were somewhat overawed in her presence. By then she had started to be clad in saffron sarees. We often spent our evenings in her research station imbibing a love for nature that remains with me to this day,' Hari reflected. Janaki had by this time also turned a veritable vegetarian.

Report on Agricultural Research

In late 1936, E. John Russell, Director of the Rothamsted Experimental Station, Harpenden, was invited to review the status of 'applying science to crop production in India': at the experimental stations under the Imperial Council of Agricultural Research, that had been established on the recommendation of the Royal Commission on Agriculture to lead and coordinate agricultural research in the country, and which involved large sums of money and was time-consuming, with results not immediately forthcoming. The ICAR's activities were to be periodically reviewed by disinterested experts, and the first set of people chosen for this onerous task were Russell (focusing on agricultural experiment stations) and his colleague N. C. Wright (on animal husbandry).

In his assessment of the state of affairs at the various scientific institutions in India, Russell concluded that the conditions were extremely favourable for the application of science to the problems of Indian agriculture. However, there were issues with the mode of running experimental stations, whose chief purpose was to apply the methods and results of science to the problems confronted by cultivators. According to Russell, the work being conducted at these stations was 'too diffuse,' a problem which he believed was not limited to these institutions alone. His chief criticism was that the work was confined too much to the laboratories, at the expense of experiments in the field. Further, trial experiments were to be carried out in a more rounded manner—that is on other soils and at other centres not far removed from the station—to ensure that all variables were taken into account. Very importantly, the two experts felt that although experimental stations in India had been functioning for several decades, the results of their trials were still to find their way into the general body of agricultural knowledge 'as expounded in standard treatises.' The chief reason why references to Indian researches were rarely found in publications outside India was that researchers published their findings in largely insignificant places 'having the character of journalism than of serious scientific literature. . . . Good memoirs compel attention and give recognition wherever, and by whomsoever they are published,' they jointly observed.[2]

Venkatraman was working on two sets of problems at this time: finding early canes that promised high tonnage, and finding late canes that could withstand the Indian summer (the sign of degeneration being the shooting of the buds and the inversion of sucrose). But the more thrilling development at the Station that year (1936) was Venkatraman's intergeneric *Saccharum-Bambusa* cross. 'What the economic benefit will be from such a cross is hard to say- but it certainly is a very revolutionary step in cane breeding!' Janaki admitted to Chase.[3] Within a year, on the basis of somatic chromosome counts, she would establish that it was a genuine cross.[4] Four years later, in 1940, through a critical study of the anatomy and morphology of the hybrid's seedlings, Ekambaram and his assistant P. R. Bhagavathi Kutty Amma at the UBL[5] would reconfirm Janaki's finding.[6] The Sugarcane x Bamboo cross would continue to interest cytologists and geneticists for many more years to come.[7] This was the first time ever such an intergeneric hybrid had been created anywhere in the world; 'I shall send you some herbarium specimens if you care to have them,' Janaki would write to the American grass-expert Chase.[8]

Walter Norman Koelz

At the same time as Russell and Wright, the American zoologist and museum collector Walter Norman Koelz of the University of Michigan, a close associate of Bartlett, was travelling through India. In 1932, the Regents of the University of Michigan had appointed him a research fellow supported by the Charles L. Freer Fund to travel to India to collect biological specimens and material culture for the University's museums of zoology and anthropology.[9] The plant-collecting expedition would last seven years, taking him to India, Nepal and Iran, and was undertaken on behalf of the United States Department of Agriculture, Washington, where agrostologist Chase was based.[10] He was in Tellicherry in late February 1937, and most certainly resided at Edam during this visit and knowing Janaki, she would have accompanied him during his excursions in North Malabar. Among the collections Koelz sent to Michigan from Tellicherry were a *Nymphaea stellata* and a *Loranthus longiflorus*.[11] Considered one of the last in the tradition of Victorian explorers, Koelz was often accompanied by Thakur Rup Chand (1902–94) from the Lahaul district of Punjab Province (today in the state of Himachal Pradesh), who would later be employed by the University of Michigan as collector and labeller of plants and birds from South Asia.[12]

Janaki met with Koelz twice during this visit; it was from him that she learnt that Bartlett was unwell. 'I wonder if you know anything about it,' she enquired with Chase; in her letter, Janaki did not forget to convey her 'salaams' to Carl Erlanson, if he was 'anywhere around.'[13] 'Sugarcane-bamboo is a most amazing cross. What sort of plant results?' Chase queried with obvious curiosity. She promised to send Janaki the bamboo seeds she required for cytological examination, but nevertheless directed her attention to the American botanist and bamboo specialist Floyd Alonzo McClure (1897–1970) of Lingnan University, Canton, who had a bamboo garden and was more than willing to send seeds to anyone interested. Chase also shared with Janaki news about Bartlett; that he had undergone a 'severe operation' and that he was slowly recovering.[14] Typical of Bartlett, he would clinically record the details of the surgery in his diary: '21 December 1936, Ann Arbor: Operation bright and early. Dr Waldron removed right testicle and everything thereunto appertaining, consisting for the greater part of a 4 lobed teratoma or a string of 4 teratoma, for practically nothing was left of the testicular tissue.'[15]

Russell on Janaki

On return to England in May 1937, after the all-India tour of inspection, John Russell sent a confidential letter to Darlington:

> I was recently in India visiting the agricultural experimental stations there, with a view to reporting to the Government their work, and in the sugarcane section I met a Miss Janaki-Ammal, who has been in touch with you in regard to her work. In case you do not recollect her I am sending you her two papers. Could you kindly tell me whether her work is sound, and whether also it is simply in the nature of a student's exercise or can be identified with the title of research? I should be grateful for your views. She is working alone and may be getting on to some sterile track.
>
> The only other Indian at an experiment station, who is doing anything in the nature of genetical work is K. Ramiah, paddy specialist, also at Coimbatore, who sums up his work in the enclosed address. Would you kindly tell me what you think of that also? He has now, as I think you know, gone to Indore to succeed Hutchinson, who was, of course, a first rate man.[16]

Russell had discovered that the selection of canes at Coimbatore was being done largely in a mechanical manner, with valuable new varieties being overlooked and perhaps lost in the process. This situation could be remedied somewhat, he had suggested in his *Report*, if trials were made independently at a number of stations, but the real need of the hour, he would note, was 'a really first class geneticist, who could substitute proper scientific tests for the present mechanical ones.'[17] A thorough knowledge of the nature of inheritance and segregation of plant characters required a continuous study of more than one generation from seed, and in the case of sugarcane this was especially difficult because of the generally non-flowering nature of hybrids, and if they flowered at all, the issue of infertility had to be tackled. Clearly, Russell had not been impressed by what Janaki was doing or had achieved at this point in time, and the reason he felt 'a really first class geneticist' was a serious desideratum at the Station. However, to Janaki's credit she had successfully crossed *S. spontaneum* with *Sorghum* by the time of Russell's visit, an extremely challenging task; her earlier attempts to cross *S. munja* with *Sorghum*, for instance, had not met with success.

Darlington of course had been in regular touch with Janaki and knew exactly what she was trying to achieve. Trying to be as objective as possible on the matter, he wrote to Russell that he had been in India a few years ago (1933) and that he had had the opportunity to acquaint himself with the work of both Janaki and Ramiah [the rice expert] at Coimbatore: 'The case of Ramiah is the simpler. From what I saw then, and from his paper, it seems to me that he has a clear idea of his job and a very enterprising way of settling about it. I think he might well prove to be as useful as Hutchinson [the cotton specialist].' However, Janaki's case, Darlington explained, was part of 'a larger problem.' He believed Indian cytologists were in general better technicians than thinkers: 'Practitioners of cytology in India are very numerous, but cytological work of outstanding interest is unknown. The reason for this seems to be that Indians go in for cytology because they think it is a matter of technique and needs no thought otherwise. But cytology has begun to require a good deal of thought during the last few years. The job hasn't turned out to be what they thought it was. Therefore when I say Janaki Ammal understands her work better than anyone else in India, I do not mean to pay her a vast compliment.' Darlington was however confident that she was 'doing sound work' and that she 'would continue to do so for some time, just because a great deal of elementary exploration in this field is necessary and she can't fail to be of value to the geneticist working with her.'[18]

This indeed was an objective reading of her work, however imperious it sounded, for Janaki was at this time chiefly focused on rigorous exploratory research (generating data and addressing specific concrete/cytological problems), essential to make local order out of the chaos that was *Saccharum* cytology. It was not as if mastering cytological techniques was a small matter either and Janaki was already far superior to most cytologists working in India in this department but what the laboratory-bound analytical Darlington had also chosen to turn a blind eye to was Janaki's amazing skills on the field (something he had never demonstrated himself)—her ability to breed new plants, including useful intergeneric crosses, on the strength of her cytological investigations, unlike Venkatraman who was doing it through trial and error, backed by years of field experience rather than science. The true significance of her work would only be unfolded in the years to come, and contrary to what Darlington believed, she would have a life-long career, and not as an ancillary to some male (white or coloured) geneticist, but as a cytogeneticist in her own right. The crux of the matter was really that their ways and goals of doing science were fundamentally different—if to Darlington, it was all about using deductive logic to impatiently reduce large amounts of cytological data to construct theories or elicit universals, and forcing a model of interpretation from above, for Janaki, it was about generating data by solving problems one at a time, slowly and progressively (of plant relationships across space and time), through a mixed border practice involving the lab and the field, and a creative navigation of diverse knowledge streams, including the humanities—this perspective will become more evident in the subsequent chapters of her life.

A few months after writing this letter to Russell, Darlington would start for India to attend the joint meeting of the Indian Science Congress and the British Association for the Advancement of Science (henceforth, British Association) to be held at Calcutta.

Seasonal Field Excursions

An intrepid plant hunter, Janaki dedicated at least a month each, during two seasons of the year, to collecting; this was an extremely important component of her scientific practice. Roadsides, railroad yards and dump heaps or waste grounds were favourite collecting sites, where the possibility of finding migrant weeds and mutants was very high. The area around the Tellicherry railway station was one of her favourite hunting grounds for example. As a constant plant collector she carried with her at all times, a field notebook; it contained a number list, which corresponded to specimens collected, against which ecological and other data, such as name, order and habitat to which the specimen belonged, locality from which collected and date of collection, were recorded.[19] Field notebooks and herbarium sheets over and above being scientific documents also provide information to the historian, on the scientist's physical coordinates at a given point in time. In the months of January–February 1937, for example, we know from Janaki's field notebook that she was busy collecting grasses, chiefly *S. spontaneum*, from various locations in Eastern India: the banks of the Mahanadi

173

(Cuttack), at Malatipur (near Puri), at Lutuma (from this place, on 23 January she also collected wild rice), Kamakhya and a farm at Khanapara in Guwahati, and other parts of Assam like Kamarkuchi (from near a stream on 24 January, and along the Shillong Road (until 30 January), and then down south, from farms at Anakapalli (2 February), Dummugudem (5 February), Bhadrachalam (from rock crevices on the Ramdoss Island) and Parnasala (Andropogoneae and other grasses) on the river Godavari (from the 7 February until 11 February) in the Telugu-speaking region of the Madras Presidency, and finally concluding the collecting excursion at Vedapathi (on 14 February), the environs of the Breeding Station (Kamakoti Hill of the Botanic Garden), the Forestry College, Coimbatore (15 to 17 February) and the South Central Railway Station yard.[20]

Some Preliminary Findings

At the meeting of the Association of Economic Biologists, Coimbatore in September 1937, Janaki presented her preliminary findings on the origin of indigenous canes of India, the so-called *S. barberi* and *S. sinense*. Among her singular discoveries were the occasional giant triploids found among 'selfed' progenies (progenies obtained by self-fertilisation) of *S. spontaneum*, with strong resemblance to *S. barberi*; of the occurrence of giant interspecific hybrids caused by the fertilisation of unreduced gametes in one of the parents; the phenomenon of heterosis (the improved function of a biological quality in a hybrid offspring, also called hybrid vigour) met with in crosses between widely separated chromosomal types of *S. spontaneum*; and lastly, the variation in sucrose content in populations of *S. spontaneum* seedlings. On the basis of these, she argued that, cytotaxonomically, *S. barberi* and *S. sinense* occupied a position intermediate between the noble cane *S. officinarum* and the wild species *S. spontaneum*. In other words, she had arrived at a new order for arranging the *Saccharum* species and their hybrids. Cytologically, these hybrids represented a polyploid series ranging from $2n = 48$ to $2n = 80$ in India, and $2n = 80$ to $2n = 124$ in Greater India and the East Indies (or the Indian subcontinent and Southeast Asia together); Janaki was clearly suggesting that interspecific and intergeneric hybridisation played important roles in the origin of some Indian canes.

Incidentally, some of the *Saccharum-Sorghum* hybrids evolved by Venkatraman would by this time be raised at the Colonial Sugar Refining Company's quarantine home in Sydney. He would also be hired as a consultant by Indian entrepreneurs involved in the manufacture of sugar, such as industrialist Walchand Hirachand.[21] Venkatraman who had amazed the world with his *Saccharum*-Bamboo hybrid in December 1936, after several failed attempts, was elected the President of the Indian Science Congress at Hyderabad (1937) and an honorary member of the South African

174

Sugar Technologist's Association—the only person to be so elected outside of South Africa.

24th Indian Science Congress

Janaki attended the 24th Indian Science Congress held in Hyderabad in January 1937, at which T. S. Venkatraman delivered the Presidential lecture titled 'The Indian Village—Its Past, Present and Future,' a tribute to the Royal Commission on Agriculture, with its focus on the Indian countryside. The section-head of botany that year was H. G. Champion, Sylviculturist at the Forest Research Institute, Dehra Dun; among the botanists who presented papers at the Congress besides Janaki included M. O. P. Aiyengar, T. S. Mahabale, Miss R. Shah (plant physiologist at the Coimbatore Sugarcane Breeding Station), P. N. Bhaduri, N. K. Tiwary and A. C. Joshi. Janaki presented her work on the *S. arundinaceum* (hardy sugarcane) collected from near Coimbatore: on the nature and frequency of chiasma at meiosis (the point at which, during metaphase, paired chromosomes remained in contact and where crossing-over and exchange of genetic material occurred), and on its chromosome count.[22]

The Joint Session, 1938

In December 1937, Darlington, who had succeeded Daniel Hall as director of the John Innes, arrived in India as a delegate of the joint session of the Indian Science Congress (founded in 1913, the year marked its Silver Jubilee) and the British Association, to be held in Calcutta between 3 and 9 January 1938; this was the first time such an event was being organised—a gathering of some of the most important minds in the world of science from across the globe. The sugarcane expert Venkatraman was the Indian Science Congress Association's sole elected representative at the meeting. The official authorities, the Calcutta University, the Corporation of Calcutta, several scientific institutions and private hosts extended generous hospitality to the delegates. A tour through Central India before the session (16–29 December) and one through Southern India, post-session (11–13 January) had been arranged (this was how some got to visit the Sugarcane Breeding Station, Coimbatore). In addition, shorter excursions to sites of scientific interest in Bengal and Bihar had been organised.[23] The New Zealand physicist Lord Ernest Rutherford, known as the father of nuclear physics, was to be President, but having passed away in October, was replaced by James Jeans. Some of the foreign delegates were invited to deliver popular evening lectures: Arthur Stanley Eddington spoke on 'The Milky Way and Beyond,' F. A. E. Crew on 'The Biology of Death' and H. J. Fleure on 'Stages in the Growth of Civilization.' Incidentally, Eddington's visit to India came three years after his public (even racist) ridiculing of astrophysicist

S. Chandrasekhar's views on the white dwarf or degenerate dwarf (first communicated to the world in his 'The Maximum Mass of Ideal White Dwarfs,' 1931), at a talk given at the Royal Astronomical Society in January 1935.

Darlington, one of the star participants, arrived a few weeks earlier with the intention of travelling around a bit, in the company of Janaki. She would receive him at the Delhi airport, and together, they would go on excursions in and around Delhi and then travel to Allahabad to collect *Saccharum* and *Sorghum* for Janaki (this was her second collecting season of the year, and she had already done some collecting in Coimbatore in November, including of the grass *Cenchrus ciliaris*) and certain grass specimens for Darlington; they collected Munja grass (*Saccharum munja*) from the muddy banks of the Jamuna (Yamuna) and from the wayside in Delhi (24 December), *Sorghum halapense* (specifically) at Joshi and the banks of the Macpherson Lake, and sugarcane bud sports[24] from the bed of the Ganges (Ganga) in Allahabad (26–27 Dec).[25] They also visited Cuttack and Banaras, before reaching Calcutta in time for the Jubilee session. Incidentally, during neither of his visits to India (1933, 1937), would Darlington get to meet Janaki's family at Tellicherry or Shoranur. At Cuttack, they were guests of Prankrushna Parija, a former research student of Frederick Blackman (as was T. Ekambaram of Presidency College, Madras) and a close friend of M. O. P. Aiyengar; and at Calcutta, of botanist Boshi Sen of the Vivekananda Laboratory (Almora), K. P. Biswas of the Sibpur Botanic Gardens (recently returned from Kew). Eileen Erlanson now Macfarlane, who had made the city her home to pursue her research in human genetics (on the blood group distribution among the different castes of Bengal) would also play host to them in Calcutta; she had only recently married her third husband, James Macfarlane, an engineer with Burmah Shell. Incidentally, the founder of analytical psychology, Carl Jung, was in India at this time (December 1937–February 1938), invited by the British Government to participate in the 25th anniversary celebrations of the University of Calcutta. Jung used the opportunity to travel extensively (with Fowler McCormick); he was awarded honorary doctorates by the Universities of Banaras, Allahabad and Calcutta, the last one in absentia, as he had taken seriously ill and was admitted to the English hospital in Calcutta.[26]

Women Science Researchers

By the late 1930s, papers by at least a few Indian women scientists, mostly postgraduates (none yet with a doctorate, except for Janaki (1931) and Kamala Bhagvat (1939)) began to appear in Indian science journals like the *Current Science*: biochemists R. Karnad and Kamala Bhagvat (later Sohonie); zoologists Ciriaca Vales, T. P. Vanajakshi and K. P. Nalini; botanists P. R. Parukutty (later Baruah) and Janaki; chemists Kamakshi, Ione Nitravati Dharam Dass, K. D. Gavankar; mathematician S. Pankajam; and geologist Chinna Virkki, had scientific papers to their credit. In fact, a report on the joint meeting published in the *Current Science* stated: 'In a world that has for long been dominated by the idea of the intellectual superiority of man over the opposite sex, it is easy to fail to appreciate the part that Indian women have played in the advancement of scientific knowledge. The proceedings of the Indian Science Congress, however bear ample evidence to the fact that Indian women are not far behind their Western sisters in regard to active participation in scientific research. It will perhaps come as an agreeable surprise to many when told that Indian women scientists have contributed original papers to the Jubilee session [1938] on a variety of subjects ranging from astrophysics, physical, organic and industrial chemistry and geography to the biological sciences such as morphology, ecology, physiology, anthropology, genetics and psychology.'[27] It may be noted that the Palaeobotany subsection of the Jubilee session, chaired by Janaki's good friend, Birbal Sahni, had the young Miss Chinna Virkki present a paper, alongside geologist K. Jacob (Virkki and Jacob would become a couple soon) and K. N. Kaul (brother of Kamala Nehru). Also, the section devoted to Zoology, headed by the Cambridge-trained George Matthai, was conspicuous for the number of women presenters, among who were C. K. Ratnavathi, K. P. Nalini and Miss G. Mahadevan from Madras, and Miss B. K. Dhillon from the University of Lahore, where Matthai headed the Zoology Department.

Plant Cytology and Genetics at the Congress

The President of the botany section of the joint meeting was the Cambridge-trained palaeobotanist Birbal Sahni. Janaki was Chairperson of the subsection devoted to genetics and cytology; among the chief papers presented under this category were those by Darlington and his arch-rival, R. R Gates, besides Janaki, of course. In his paper, 'The Biology of Crossing-Over,' Darlington stated that crossing-over was not 'merely a genetical incident' in the segregation of chromosomes at meiosis, but 'the mechanical condition sine qua non of meiosis and hence of sexual reproduction.' It was at this session that Gates had presented his controversial paper, explaining some finer cytological points: he claimed that a large sample of direct observational evidence went against the idea that chromosomes were single. Chromosomes are, Gates argued, double structures consisting of two chromonemata variously intertwined in all stages of mitosis, the new split occurring at or about prometaphase. Only a few weeks before the Congress, he had published a paper on the subject in *Nature*. Darlington would demolish this claim mercilessly, laying out incontrovertible proof, to support his argument, an act that would lead to greater animosity between the two men and unpleasantness for all those around them.

As for Janaki's paper, it presented the results of a comparative study of meiosis and an analysis of chromosome pairing, in two types of intergeneric hybrids ($2n = 38$ and $2n = 66$) produced by crossing *S. spontaneum*

$(2n = 56)$ and *Sorghum durra* $(2n = 20)$.[28] The genus *Saccharum* was of great importance economically because it included all the noble sugarcanes, the *S. officinarum*. The wild Saccharums comprised two species—*S. spontaneum* and *S. robustum*—confined to Asia and New Guinea respectively. In Venkatraman's breeding work, he had observed that of the two wild canes, *S. robustum* had a closer affinity to the thick class of canes. The economic results derived from raising seed canes had been so remarkable and the demands of the sugar industry so acute that the Breeding Station, Venkatraman (as the President of the Agriculture session) confessed, had focused on economic gains alone 'unmindful of the scientific explanations behind the production of these types.'[29] In his presentation, to his credit, he did not hesitate to acknowledge the importance of Janaki's scientific work at the Coimbatore station:

> With funds from the Imperial Council of Agricultural Research, somewhat intensive work on the subject has been in progress at Coimbatore under a special duty officer, Dr E. K. Janaki Ammal, for the last three years. Properly controlled crosses have been made and are being studied and various interesting features associated with meiosis in the Saccharum are revealing themselves. This work is throwing light on the origin and distribution of S. spontaneum types, the probable origin of certain Indian canes and similar problems. The large numbers associated with Saccharum chromosomes—particularly in the later seedlings—render the work both difficult and time-consuming. A fund of knowledge is waiting to be revealed when, besides the number, the study of chromosome morphology advances further than at present.[30]

As far as he was concerned, however, such academic work was better done in a university ambience; at a breeding station it could only be treated as a minor activity, best reserved for the sidelines, he observed. While the choice of the hybrids, for Venkatraman, depended on their use in the sugar industry (early maturing canes with high-sucrose levels), for Janaki, they were all 'epistemic things,' with the potential to throw light on the origin and distribution of the cultivated cane.[31] Venkatraman would continue to hold on to this opinion. In a lecture delivered in Baroda in January 1942, under the title, 'The Message of the Sugarcane,' his congenital dislike for academic plant breeders would manifest again. Perhaps referring to Janaki and the rising number of young cytologists in India, he would observe, that 'foreign-trained laboratory technicians have a tendency to do some amount of copying work forgetting the very vital differences that exist between the agricultural conditions served by foreign laboratories and those in India. This results more in publications than in types badly needed by the country.' He remained convinced that academic research should be confined to universities and not breeding stations: 'it is within the province of a university teacher to study the effect on plants of all substances irrespective of their presence or absence in the soils of a tract. Agricultural research, on the other hand, should always confine itself to the effect on plant of such elements as are found in the soils, as otherwise the investigation would cease to have practical value,' he would conclude.[32]

Evolution and the Species Concept

Common to the sections on botany, zoology and agriculture was a section devoted entirely to 'The Species Concept in the Light of Cytology and Genetics,' which saw the triumvirate Janaki, Darlington and Eileen Macfarlane come together. While Sahni and later S. R. Bose presided over the panel, Janaki set the tone of the discussion: 'It has been said that the test of the creed of a biologist is his [she used masculine pronouns habitually, a common practice during these times] definition of species,' she began. Although modern genetics had shed light on the nature of genetical difference between related species, for more than two decades the concept (of species) itself had eluded a clear-cut definition, despite knowledge of the underlying causes of variation or discontinuity between two forms; and of late it had also become known that the causes themselves varied across different species and genera, she qualified. The cytogenetic analyses of plants and animals revealed that although the chief differences between some species were genetic, in certain others they were cytological, linked to changes in the number and structure of chromosomes. The discovery of flowering plants with multiple sets of chromosomes (polyploidy) across genera was of vital importance to evolution and to botanists tracing the phylogeny of species. Polyploidy was the agent behind species formation, but this was still a contentious viewpoint at this point. Today we know that a large percentage of modern plants are the result of polyploidy.

Janaki opened the discussion with a useful review of literature on the subject. She stated that while Darlington was the first to interpret the history of *Pomoidae* (the subfamily of pome fruits like apple) by studying the 'secondary pairing' of chromosomes at meiosis, it was the Swedish cytologist Arne Müntzing who had first argued that the Linnaean species had arisen through hybridisation and doubling of chromosomes (polyploidy). She drew attention to the ease with which it was possible to hybridise widely different genera in the family Gramineae, evidence being the several hybrids she had produced at Coimbatore between species of *Saccharum* and those of *Sorghum, Erianthus, Narenga, Bambusa* and *Imperata*. By studying the occasional giant triploids (2n = 84), which had arisen among 'selfed' progenies of *S. spontaneum* (2n = 56) in Coimbatore, she had been able to establish that *S. barberi* originated from *S. spontaneum*. These were found to be thicker and had more sugar than the type from which they arose, with the resemblance to *S. barberi* stark; the physiological effect of polyploidy was an increase in cell size, reduced growth rate and increased adaptability.

She observed that the occurrence of giant triplo-polyploids (a term suggested to her by Darlington at Calcutta) among intraspecific hybrids (between the different forms of *S. spontaneum*), and the phenomenon of heterosis or hybrid vigour in crosses between widely separated forms of *S. spontaneum*, seemed to suggest that hybridisation played an important part in the evolution of the cultivated cane. Although the existence of a polyploid series with $2n = 48, 56$ and so on suggested 8 as the basic number in the species, certain other features (the absence of univalent and multivalents in the meiotic figures of forms with 28 and 36 bivalents) proved that these were dibasic, that is, having resulted from the hybridisation of a form with $x = 10$ and one with $x = 6$. The triplo-polyploid was the 84-chromosome plant, which Janaki discovered among the 'selfed' offsprings of the 56-chromosome Dehra Dun form. In vegetative characters, the triplo-polyploidal plants stood intermediate between *S. spontaneum* of India and the indigenous cultivated sugarcanes which they resembled, leading her to suggest that it was unlikely that some of the sugarcanes of India had arisen from *S. spontaneum* as triplo-polyploids. And if this were true, she went on to speculate, there was a striking parallel between the chromosomal history of a cultivated plant like sugarcane and the triploid mutants, whose propagation as clones had likewise brought forth the best varieties of apples, pears, tulips and hyacinths.[33]

Darlington offered that they were all in general agreement that species arose, as Darwin had said, by the selection of hereditary variations, but the more challenging question was how selection itself occurred. Resorting to deductive logic, and ever the theoretician, thinking in terms of a (genetic) "system" or an overarching explanatory model, he stated that it obviously depended on the nature of variations available for selection, which were of three sorts: changes in the internal properties, in linear positions and the quantitative proportions of the genes making up the chromosomes. These three would be selected to provide the means of isolation and adaptation. A comparative study of species and their hybrids revealed that chromosomes were continually undergoing structural rearrangements, which in different ways prevented crossing-over between the genes in the rearranged parts, thus enabling the preservation of good combinations. This to him was the means by which integration of species was first achieved. The second step involved changes in the proportion of the genes in the complement (caused by structural changes), and from adapting to different environments, but structural changes could also inhibit the pairing of the relatively changed chromosome, leading to intersterility (sterility of individuals when interbred), the third step in integration.

In response, Eileen Macfarlane (at this time based at the Bose Research Institute, Calcutta), stated that genetics had aided the explanation of variation within species and parallelism in related species. This improved understanding had reduced the number of the American species of *Rosa*, her speciality, from 115 (she would refer to this as a 'taxonomic inflation') to 20. Cytology, she said, had not only shed light on phylogeny but also helped solve some problems of phytogeography, that is, the spatial relationships of plants both in the present and the past. In her words: 'We have finally broken away from the concept of a morphologically delimited species with a definite range and have adopted a three-dimensional fundamental unit the "line of evolution" in both space and time for each species.'[34]

Blood Group Genetics, Racial Controversies

Eileen Macfarlane had made Calcutta her home between 1935 and the fall of 1941. She held the position of honorary collaborator in Asiatic Research of the University of Michigan and Research Assistant at the Botanic Gardens in Sibpur, while also being attached to the Bose Research Institute in Calcutta. Eileen and her husband initially resided at Budge Budge in the 24 Parganas district of Bengal, about 25 miles south of the city on the Hooghly, where Burmah-Shell, Standard Oil and Indo-Burma Petroleum companies had installations with storage and distribution facilities. A few months later they would move into 'a fine airy furnished house, with marble floors, 3 bedrooms and 3 bathrooms, & a tennis court, within walking distance of shops and the museum,' at Middleton Row, near Park Street, Calcutta. In the company of her new husband and 'twin white and grey kittens,' Tweedledee and Tweedledum, Eileen was immensely happy in 'the city of joy.'[35] It was in this house that she entertained Janaki and Darlington, who were in the city to attend the joint meeting. Eileen lived in close proximity to the Royal Asiatic Society of Bengal, where she would frequently present results of her ongoing research in sero anthropology (blood group genetics). At the Society's ordinary monthly meeting held on 5 April 1937, she read her first paper on human genetics, focusing on the Jewish families of Cochin.[36]

Blood group genetics had paved way for a modern, guiltless and objective study of race and it is not surprising that the joint meeting organised a special section, 'Blood Groupings and Racial Classification.' A brief history of seroanthropology as a sub-discipline of genetics is worth charting here. The Polish microbiologist Ludwik Hirszfeld of the Institute for Experimental Cancer Research, Heidelberg, together with his wife Hanka, was the first to study blood groups in a large number of World War I soldiers on the Macedonian front. They observed significant differences in the distribution of the human ABO blood groups discovered by Karl Landsteiner in 1901; type A was more common in soldiers from North Central Europe, while type B was prevalent among those from Eastern Europe. They also drew attention to the high percentage of blood group B in Indians and to the descending numbers of B as one passed westwards from India. In 1910, with Emil von Dungern, they published their fundamental work on the inheritance of ABO blood groups, which is recognised as a major milestone in human genetics and the onset of forensic genetics, drawing scholars towards the new field of seroanthropology, such as Arthur Ernest Mourant, Robert Russell Race and Ruth Ann Sanger. The Hirszfeld data was however misused later by the German nationalists advocating Aryan supremacy.

To shed light on heredity and human variation, blood groups had become objects of investigation in the hands of statisticians R. A. Fisher and Haldane in the early 1930s; they projected blood groups with the ability to provide credible knowledge of human heredity (blood groups were known to be inherited in keeping with Mendelian laws). Serological discoveries began to find anthropological applications in most countries, including India, to define scientifically the concept of race based on the distribution of blood groups, but by the early 1940s, it was clear that they had failed in achieving their goal.[37] For some cytogeneticists, like Eileen Macfarlane and Darlington, the study of heredity in the organism and heredity in society were intertwined. Eileen was among the first geneticists to turn to the study of human inheritance of blood groups in India.[38]

Through her researches, Eileen reconfirmed that group O was the oldest and original ancestral group; while group A was also very old, group B, she suggested, was only of recent origin. Although it was long known that the greatest concentration of group B was in India, it was for the first time that someone had collected data from each caste and tribe, and from the point of view of human genetics; it was hoped that the serological data would throw light on the origin of the B gene and also upon the rate of mutation. Eileen however warned that blood grouping was only an aid to anthropology and that it could not be used by itself to 'solve a relationship.' It was to be considered alongside other racial characters. Meanwhile, Eileen had failed to win an American Association of University Women (AAUW) scholarship; but not one to give up research for want of funds, she turned to investigating the blood groups of mothers and babies, including the M and N antigens, at a Calcutta maternity hospital, and 'studying serology, Hindustani & first aid.'[39]

In his discussion of Eileen's presentation at the Jubilee session and the subject in general, the controversial geneticist R. R. Gates noted that blood groups were definite units, whose method of inheritance was known, which therefore enjoyed a greater advantage over other indices of racial relationships. For the kind of scientific views Gates held, it was not surprising that he was often entangled in controversies on racial origins, racial differences, race-crossing and so on, although he was opposed to the nineteenth-century obsession with craniometry and believed that the study of other physical characteristics was more important. The criticism however persisted because he attempted to correlate physical characteristics and moral character; he believed in the intellectual superiority of some races and the undesirability of racial intermarriage.

Gates' academic career incidentally began in 1919 as reader in botany at King's College, London; in a matter of only two years, he was appointed Professor and over the next few years became a Fellow of the Royal Society (1931), the President of the Microscopical Society, the Vice-President of the Linnean Society and the Royal Anthropological Institute, and secretary of the Society for Experimental Biology. Unsurprisingly, he was also a long-standing member of the British Eugenics Society; while Gates spoke of the new science of genetics and what it could do to shed light on the evolution of man, in his book *Heredity and Eugenics* (1923) he also harked back to the nineteenth-century assumption that racial groups could be classified.[40] By the time Gates arrived in India for the joint meeting, he was at the height of his professional career despite the controversies surrounding his work. As a result, he had gathered a large number of students at King's College, London, including a good number of Indians like P. N. Bhaduri, S. M. Sikha, S. Ramanujam, H. K. Nandi, T. S. Raghavan, K.V. Srinath, G. N. Pathak and S. P. Naithani. It was while they were at King's that Ramanujam and Raghavan attended the summer course at the John Innes (in 1936), along with P. K. Sen and P. Maheshwari of the Botanisches Institut, Kiel (Germany); H. K. Nandi and S. P. Agharkar of Calcutta had attended the summer course a year before.

The Canadian-born Gates often faced a cold welcome and even direct expressions of displeasure at scientific meetings, for his 'racist' outlook. Darlington for one completely ignored him at the Calcutta meeting. In fact, Gates had alerted Birbal Sahni as early as February 1937 about an alleged conspiracy brewing among members of the British Association to exclude him from the delegation to India in December. He reminded Sahni that he had 'students all over India,' and that he could do much for them but only if 'cytologists, geneticists and anthropologists in India' would insist his name be included as a delegate.[41] Boshi Sen (Basiswar Sen), however, was among those few in India critical of Gates. Sen was a student of Jagdish Chandra Bose, and married to the American geographer Gertrude Emerson, who counted among his friends such luminaries as Julian Huxley, D. H. Lawrence, Rabindranath Tagore, Carl Jung and Jawaharlal Nehru.[42] At the conclusion of the meeting in Calcutta, Sen wrote to Darlington: 'I am looking forward to getting reports of Gates. Nandi [Gates's student, H. K. Nandi] didn't even invite me to the Bose Institute lecture of Gates. The general impression in Calcutta is that you completely *ignored* Gates. And the little dig I could administer nonplussed Kosi [T. K. Koshy of the Maharaja's College of Science, Trivandrum, where Janaki had taught]! The result was that the Research Director of Trivandrum, who made a previous engagement to visit this Laboratory, cancelled it!'[43] It must be admitted however that Gates had a far greater influence on the scientific fraternity in India than Darlington ever did. Janaki and Boshi Sen were perhaps Gates' fiercest opponents and the only loyal supporters of Darlington in the country, and in the former's case, this would work much against her.

Gates Attacks Again

At the conclusion of the joint meeting in January 1938, most of the foreign delegates travelled southwards, visiting research institutions at Madras, Coimbatore, Bangalore and Mysore. Darlington did not join the group however, perhaps because Gates was part of the group. When they called on the Sugarcane Breeding Station, Janaki presented before the gathering the cytological evidence that helped her trace the origin of several forms of *S. spontaneum*, growing wild in various parts of India and other countries like New Guinea, the hybrids between *Saccharum* and *Sorghum*; among her listeners were Gates and Venkatraman.[44] Janaki had only recently created a new intergeneric *Saccharum-Zea* cross (between sugarcane and maize); wanting an 'expert' opinion on its authenticity, he approached Gates, clearly taken in by the latter's 'keen interest in the work done at Coimbatore,' his 'fund of information' and his 'gracious manners.' After a quick look at the printed image of the chromosomes, Gates expressed serious doubts on its validity.[45] The result was that Venkatraman refused to recommend Janaki's note announcing the cross to the Director of Agriculture at Delhi for the necessary permission to publish in *Nature*. In distress, Janaki very nearly left the station but, to her elation, after seven long months of deliberation, Venkatraman conceded that the cross was genuine and her note was processed for publication.[46]

Gates had failed to notice the maize chromosomes in the nucleolus of the hybrid, but to be fair to him, the image accompanying her note was not a well-defined one; instead of demanding a clearer image, before he passed a verdict, he had declared the hybrid invalid, thereby giving away a measure of the man that he sometimes was—a geneticist who arrived at hasty conclusions, backed more by prejudice than evidence. Once the issue was resolved, a hurt and peeved Janaki would remark to Darlington: 'I have known in the course of these six months what Oriental persecution is (and when it is directed against a woman it has a flavour all its own!).'[47] She was thankful to him for referring to her *Saccharum-Zea* cross (she would aptly refer to it as 'the Cinderella of the Sugarcane Station') in his forthcoming book, *The Evolution of Genetic Systems* (1939).[48] Both *Saccharum* and *Zea mays* (maize) were distinguished for the readiness with which they crossed with related genera. We may recall that as early as 1913, Barber had produced an intergeneric hybrid of the male sterile clone of *S. officinarum* called 'Vellai' (or the Otaheite cane) with the grass *Narenga narenga*. Several years later, in 1929, Venkatraman (and Thomas) crossed *S. officinarum* with a species of *Sorghum* and in 1936, even with the remotely related *Bambusa*; in 1938, he would obtain a cross with *Erianthus arundinaceus*, a wild relative of the genus *Saccharum*.[49]

In 1936, Janaki had crossed several inflorescences of the *S. officinarum* clone, 'Vellai' (after pre-treating the root tips with ice) with the pollen of

Zea mays ('Golden Beauty,' sold in Poona as a type suitable for cultivation in India) but managed to obtain only a single seedling.[50] A true hybrid, this plant had received 40 chromosomes as expected from the *Saccharum* parent and 12 from *Zea* and resembled *Saccharum* more closely, as one would expect from the chromosomal contributions, but with the characteristic epidermal hair found on the upper side of the leaf in *Zea* and related genera. The hybrid did not produce flowering canes even after twenty-two months and it lacked the vigour and early maturity found in *Saccharum-Sorghum* hybrids. A second attempt in 1938 again produced a single seedling, and this did not survive. Although *S. officinarum* and *Zea* were in different sections of the Gramineae (Andropogoneae and Maydeae respectively), Janaki had thought it was worthwhile crossing them. The biggest obstacle in crossing *Saccharum* with *Zea* was encountered in the first stage of the operation owing to the widely different sugar concentrations required by the germinating pollen; *Zea* pollen germinated in concentrations, much below the concentration required for sugarcane pollen. The seedling in its early stages being very weak, nutrient solutions were necessary to keep it alive. The vegetative abnormality of the hybrid, Janaki attributed to the remoteness of the parents.

Janaki collected several clones of the tall perennial grass *Narenga porphyrocoma* (previously known as *Saccharum narenga*) flowering copiously on the banks of the Brahmaputra in Assam a year later. Fortunately for her, the male parent (raised from seeds collected in Northern Bihar) that Barber had used in 1913 to cross with *S. officinarum* had been propagated from cuttings at the Coimbatore Station, which allowed her to examine the identical clone used by Barber as well as those collected from Assam, and from the herbarium sheets at Kew; six of the clones revealed a count of 30 chromosomes. The *S. officinarum* clone 'Vellai' (2n = 80) that she examined was again the same clone used by Barber in his cross and the same that she had used in obtaining a cross with *Zea*, thus bringing her hybrids in line with Barber's. All the hybrids between *Saccharum* and *Narenga* were very cane-like and, unlike the *Saccharum-Zea* hybrid, extremely vigorous. In the root tips of sixteen of the hybrids, Janaki found 55 chromosomes, the sum of the haploid numbers of 'Vellai' and *Narenga*. She had previously crossed 'Vellai' with *Sorghum durra*, which produced both diploid and triploid hybrids (2n = 50, 90), leading her to infer that only the haploid egg cells of 'Vellai' were fertilised by *Narenga*.

Family Matters

Mid-April 1938, Janaki spent a few days with her family at Edam. Devi Amma was now seventy-four and increasingly frail, but a matriarch nevertheless presiding over the large family that had gathered at this time of year, cooking and feeding them sumptuous meals, all despite the severe strain on Edam's finances. Together the family would visit picturesque Koduvalli and enjoy some pleasurable hours boating—the group included Janaki, Sumithra, Raghavan's daughter Sridevi, Kittu and his children, Vasudevan's daughter Yamuna, Cousalya's daughters Tara and Leela, and Varadan and family. Kittu's eldest son Hari Krishnan recalled his aunt Janaki as 'always full of life and laughter' at such family gatherings. The lot would walk towards the seaside through the narrow by lanes (*edavazhi*), a ritual which 'happened almost every evening with Dad [Kittu] and the aunts Chumai [Sumithra] and Nachee [Janaki] . . . they conversed mostly in English.'[51] If on one day it was a grand picnic on the beach at Dharmadom, it would be car rides (the two cars, belonging to Kittu and Janaki) around town on another. Listening to the gramophone sitting in the *mittam* on hot summer nights was something they invariably did every time the family united. Kittu was doing particularly well; he had recently purchased a *paramba* of 3 acres at Cannanore, containing 123 coconut trees and a wetland equivalent to 550 measures of paddy. He would also write occasional pieces for the *Indian Forester*; a recent one dealt with the patches of evergreen forest that he had discovered in an otherwise arid Cudappah district,[52] another on the adhesive properties of the prickly pear, useful in mending a punctured bicycle tube for instance,[53] a third on forest fires, a subject on which he regularly lectured to the forest-dwellers within his jurisdiction.[54]

After a couple of days of bonding with her family, Janaki drove back to Coimbatore, taking younger brother Varadan and his children along; they were to be dropped off en route, at Shoranur. At Shoranur, unfortunately for Raghavan, his many attempts of living off his property were proving unsuccessful, much as Janaki had anticipated. He had, rather unwisely, spent much of his money on building a grand house. The short-lived baking business drained his finances even more. As for his vegetable garden, it was virtually destroyed when the tomato plants were attacked by a peculiar disease, surprising even the famed Albert Soans (under whose direction, he had laid out the kitchen garden), whose many attempts at remedying the situation yielded no results. In quite a similar fashion, a poultry farm of about seventy birds 'of country and good breeds,' which had begun to thrive under wife Vasumathi's care, fell prey to a contagious illness and the birds began to die, '5 to 10 a day,' despite all the necessary precautions taken; the dead 'fowls were thrown on the waste land in the sundry hills.'[55]

Experimenting with Colchicine

In the 1930s, a poisonous alkaloid drug called colchicine, extracted from the seed and corm of the Autumn Crocus (*Colchicum autumnale*), had become an indispensable new tool of the cytogeneticists, revolutionising the world of plant breeding. From 1938 until about 1942, a flood of publications on colchicine-induced mutations poured in from around the world.

The chemical's first known use was in medicine, in the treatment of gout, but later in the 1930s was discovered to be a mutagen, at the University of Brussels, impacting upon mitosis in animal and plant tissues. The year 1937 marked the dawn of a new era in polyploidy, thanks to the geneticist Albert F. Blakeslee of the Carnegie Institution at Cold Spring Harbour, New York, who discovered its use in inducing polyploidy in several species.[56] Polyploidal plants were known for their large size, robustness and vigour. Colchicine was easy to use; it could be conveniently applied to young growing plants, without damaging them. The chemical acted by inhibiting the spindle fibres such that the sets of divided chromosomes failed to separate and became encased in a common nuclear membrane. It was highly soluble in water, was not toxic to plant cells even in strong doses, was effective even when concentrations varied from 1 to 0.01%, was soluble in lipoids and very importantly, the resulting effect was totally reversible, making colchicine 'almost "made to order" for changing diploids into polyploids.' Several researchers in the United States demonstrated the uses of colchicine in plant breeding about the same time as Blakeslee, such as his colleagues Amos Avery and O. J. Eigsti, also of Cold Spring Harbour, and Bernard Nebel and Mabel Ruttle of the New York State Agricultural Experiment Station.[57] Colchicine did not produce mutations, unless mutation was understood in a larger sense to include chromosomal doubling, Eigsti warned. Therefore, it was wrong, he stated, to include colchicine among mutagens.[58] Having made sense of what polyploidy was, and the ways and means by which polyploids could be created at will, it became easier to understand the history of cultivated plants; one only needed to count the chromosomes in the root tips to establish polyploidy in a plant.

From as early as 1926, new tetraploids had begun to be synthesised, which demonstrated that evolution could be catapulted; the examples included Arne Müntzing's synthetic *Galeopsis tetrahit* (common hemp-nettle), the *Primula kewensis* cultured at the Kew Gardens, Georgii D. Karpechenko's *Raphano brassica* (an intergeneric hybrid between *Brassica* and *Raphanus*) and a doubled intergeneric hybrid between radish and cabbage.[59] By the early 1950s, it would become evident that polyploids were plentiful in nature, with about 50% of flowering plants being polyploids. Several important economic crops were also polyploid and of ancient origin (such as bread wheat, oats, sugarcane, tobacco, grapes and so on), even without the interference of man (evolution in nature). Through a process of selection (domestication by man), evolution could be directed towards useful ends and speedily at that. Planning a new programme of hybridisation depended much on what one knew about the origin of polyploids in nature.

Colchicine also began to be used in crossing hitherto uncrossable species and in treating sterility in hybrids. In early 1938, the Cytology Department of the John Innes began to experiment with the chemical in their researches on cell division. Soon, colchicine replaced the 'heat-shock' method that was being used to induce chromosomal mutation. Geneticists in Sweden and in other locations also took quickly to colchicine; 'it appeared that the colchicine "fad" in research had arrived.'[60] A Bulgarian cytogeneticist Dontcho Kostoff (1897–1949) of the Academy of Sciences, Leningrad, and a prolific contributor to the Indian journal *Current Science* in the late 1930s had begun to experiment with colchicine in 1938, to induce polyploidy in *Nicotiana* (he worked with wild tobaccos); the number of polyploids produced in this genus would increase substantially over the years. An exact contemporary of Janaki, Kostoff shared her interest in cytogenetics, phylogenesis, evolution and interspecific (also intergeneric in the case of Janaki) hybridisation. This brilliant and prolific scientist (he published more than 200 papers, of which 50 were in English), who had worked with Vavilov, would be persecuted by the Bulgarian Communist Party and die (like Vavilov himself) tragically and prematurely in his early 50s.

Janaki began using a crude extract of colchicine by mid-1938, for the induction of polyploidy in some of her *Saccharum* hybrids, the only kind available to her in Coimbatore.[61] It is not known how or where she procured the drug from, but we know that K. Ramiah at the Paddy Station in Coimbatore, who previously used X-rays to induce chromosomal mutation, was also experimenting with the chemical. The colchicine solution had to be stored away from direct sunlight and contamination by bacteria or insects. Wartime would reduce the production of colchicine drastically, especially as a large tract in Southeastern Europe was the chief source of the raw material, and there was an unforeseen demand for it from geneticists across the world, bent on creating polyploids. Colchicine worked like a wonder drug for Janaki, enabling her to contribute significantly to experimental genetics. Incidentally, the plant she had so correctly identified for the Inspector at her school in Tellicherry, the *Gloriosa superba*, was also discovered to contain colchicine, which when applied produced positive results in maize.[62] Observing colchicine's effect on germinating seeds, Janaki was hopeful of its use in breeding sugarcane, and in August that year attempted to induce tetraploidy using colchicine in her *Saccharum-Imperata* hybrid (cross between sugarcane and the Cogon grass, *Imperata cylindrica*).

More Controversies

In the intervening period, Darlington had recommended Janaki's note on the cytology of Venkatraman's intergeneric *Saccharum-Bambusa* hybrids for publication.[63] Not only was the distance in their taxonomic positions far greater than in any hybrid recorded hitherto, they were also excellent canes from the economic point of view despite the remoteness of the cross. She had examined the pollen mother cells of some of the F1 population and found meiosis to be regular; interestingly they were also fertile. A large F2 population was in the field, all of which resembled sugarcane rather than bamboo; this was when she received news that her contract had been extended for a further period of three years, which meant that her chances of attending the International Genetics Congress at Edinburgh, due in 1939, were now remote. Rather than remain at Coimbatore, Janaki wished to leave it for the Imperial Agricultural Institute, New Delhi, to carry on with her researches: 'the atmosphere here . . . is very Gatesian [and] is getting on my nerves,' she wrote to Darlington. What irked her utmost was that Venkatraman was keeping the 'Great professor [Gates]' regularly updated on work at the Station.[64]

It was about this time that Darlington had published a counter to a piece written by Gates (in collaboration with S. V. Mensinkai), entitled 'Double Structure of Chromosomes,'[65] as a response to Darlington's attack on his paper presented at the joint meeting in Calcutta in 1937/38. Darlington's argument was that the appearance of the two threads (the 'double' structure of Gates) was only an artefact, induced by a fixative containing acetic acid, and due to the 'bubbles' arising in the chromatid. To make this more evident, Darlington published a photomicrograph of the cell in telophase,[66] which resembled Gates' own figures. Gates' response to this was a weak one, but he got his Indian student T. S. Raghavan (based at the Annamalai University) to publish a piece in support of him in the *Current Science*.[67] Janaki was livid at how Raghavan (and his colleague K. R. Venkatasubban) could claim that Darlington's criticism (of the duality as artefact) was untenable 'in the face of overarching evidence' deduced from direct observation. Their note was accompanied by a photomicrograph of *Naravelia zeylanica*, which she ridiculed as looking more like 'an astral nebulae than anything.'[68] She was not allowed to respond, and when Ramanujam of Coimbatore (though he was also one of Gates' students) sent 'a mild note of protest' to the editor, he was unceremoniously asked to send it directly to the author (Raghavan). Janaki decided to seek Birbal Sahni's support in the matter; Sahni had just returned from Vienna.

'Will I attend the genetical Congress at Edinburgh? I hardly think so,' she wrote to Darlington. 'I hope your book will soon see the light of day. I shall think of it as a souvenir of your passage to India. I hope it will make up for much that was not good about that trip.'[69] Janaki was obviously referring to Gates' annoying presence at the joint meeting and the controversy around his paper, but also on how her work (her *Saccharum-Zea* hybrid) was being treated by the likes of him. In a mood of utter dejection, compounded by the state of affairs at the Breeding Station, she wrote to Margaret Upcott: 'Cyril must have told you of the mischief Gates did during his visit to India and the difficult time I had with my *Saccharum-Zea* Cross. I am very tempted to walk out of this Station. I was happy to hear from you after Cyril's arrival in England—that was years ago. I am glad you liked the things he brought you from India. You must both come to India again sometime. Perhaps you will be seeing me as a lonely hermit on the Himalayas when you come. Wishing you both every happiness.'[70]

Yet another Gates-related controversy raised its head about this time: the use of an allegedly 'doctored' photomicrograph in support of Gates by his student, cytogeneticist Naithani, published in the *Annals of Botany*.[71] When Venkatraman was shown the image by Janaki, he said he could not comment on it unless he had seen the negative—a fair enough response—but Janaki would have none of it; she was determined to expose Gates at the earliest. Meanwhile, palaeobotanist Sahni had posted her a copy of Gates' letter to him, accompanied by a print of Naithani's image; he wanted her opinion on it. The print was compared with the one published in the *Annals* and, just as Janaki had expected, found it different! She lost no time in informing Darlington that 'the work of the knife [was] not found in the former,' but cautiously awaited further confirmation. Placing the two images side by side, Janaki invited volunteers, including a student of Gates recently returned from London, to detect any difference, if at all, between them. Everyone, without exception, pointed to 'the scrapings' in the latter confirming that the image in the *Annals* was the result of a procrustean processing in support of Gates. 'Kindly let me know what the verdict of the impartial committee of scientific men is . . . I have informed Sahni about it. On this side of the Ocean we can do very little. Nearly every "mail boat" brings a student or two from Gates' Laboratory. His interest in his Indian students only increases—and he does his best to provide them with good posts. You have seen some of the specimens!' Janaki would remark acerbically to Darlington. She added in typical style: 'The subject of Naithani's drawing will I think make an excellent theme for a detective novel of the kind in which Lord Peter figures. Ask Brenhilda [Schafer] to emulate Dorothy Sayers!'[72] Lord Peter Death Bredon Wimsey, the archetypal British gentleman detective, appeared in a series of detective novels and short stories by the poet and fiction writer, Dorothy L. Sayers, an exact contemporary of Janaki. It appears that Janaki was introduced to the Wimsey stories by someone at the John Innes during her first stint there, in 1931.

More Collecting

At the time of this correspondence (mid-September 1938), Janaki was back again at Edam, spending time with her mother and the rest of the family, while also devoting a major part of the day collecting grasses from the nearby mud banks and paddy fields close to the Tellicherry railway station. In November that year she would visit Ootacamund, again with the intention of collecting grasses,[73] even while worrying about the fate of her note on the *Saccharum* x *Zea* cross submitted for publication; her only prayer was that it should not go to Gates for a review.[74] At Ooty, Janaki would have in all probability resided with younger brother Madhavan, who was at this time, Inspector of Inland Fisheries at the station. On 12 November 1938, Janaki is found in the Dhoni Hills in the Western Ghats, 15 km from

Palghat town, collecting more grasses. A few weeks later, she was collecting plants (including Sorghum) from around the Seedling House of the Sugar-cane Breeding Station. The Christmas break saw her back at Tellicherry, where she indulged in more collecting, especially from the mud banks near Edam. She would return to Coimbatore at the end of her vacation, break-ing journey at Shoranur as always to spend a couple of days with her elder siblings, Raghavan, Vasudevan and Cousalya. On New Year's Eve, she was busy collecting grasses from Vasudevan's *paramba*.[75] A pacifist by convic-tion, Janaki was deeply distressed by the growing prospect of war. 'I wish it were given to some of us to offer our lives to stop it. I am prepared to do so,' she wrote to Darlington.[76] A death wish lurked within her, manifesting itself every now and then in letters to Darlington and brother Raghavan.

The Breeder Versus the Cytogeneticist

Breeding work in sugarcane differed from most crops. While the parents of most other crops bred more or less true when self-fertilised, in the case of cane even a single batch of 'selfed' seedlings showed wide variations in botanical, agricultural and biological characters, with none of the progeny resembling their parents. This surprising and unpredictable nature, or "dis-order," was what attracted Janaki to *Saccharum* cytology. While the seed-produced crop was of little value from the cultivator's point of view, the wide variation allowed (to a certain extent) a speedy production of seedling canes. However, by using the vegetative method (through cuttings), the orig-inal characters could be maintained over long periods, without necessitating the elaborate fixing of characters or taking precautions against undesirable cross-pollination, as with crops grown entirely from seed.[77] The work of the breeder, according to Venkatraman, was very complicated, because the envi-ronment played a major role in determining the economic character of the plant, such as early or late flowering.[78] As a breeder, his work mainly con-sisted of crossing cultivated cane with wild forms, and then comparing the wide range of hybrids with standard canes and choosing what he wanted. Venkatraman argued that to evaluate yield characters, the progenies of crosses might be grown under different soil conditions and the best chosen out of them. However, work in cytology proved that it did not always pay to reject sterile hybrid progenies, and that through polyploidy (multiplication of chromosome sets through controlled mutation) great improvements could be made to the parents. We might recall that Russell had rightly observed, during his visit to the Sugarcane Station, that the selection of canes at Coim-batore was being done largely in a mechanical manner, with valuable new varieties being overlooked and perhaps lost in the process.

In late 1938, Janaki had been investigating the so-called 'mothers' of the male sterile canes that Venkatraman was using for his crosses, namely POJ 213 (a triploid hybrid between *S. officinarum* and *S. barberi*), POJ 2725 (a complex hybrid of *S. officinarum* and Java *spontaneum*)[79] and 'Vellai' (a 'true' local *S. officinarum*). Janaki discovered that while all three crossed with widely segregated genera, 'Vellai' had not yielded an economically good cane. Venkatraman had relied wholly on the POJ canes to obtain his economic crosses. POJ 2725, although wholly pollen sterile, had produced more than 20 seedlings on bagging, which meant it was parthenogenetic in origin, and being more like sugarcane had a high sucrose content; these were automatically selected as 'economic' types, with the 'true' hybrids of 'Vellai' rejected as unsuitable. Further, Janaki found that all the crossings at the station, except hers, were being done in an unscientific manner, without bagging or emasculation (the process of removing the stamens to avoid self-pollination), thus exposing them to foreign pollen; she had found this to be the case in the majority of the F2 of the *Saccharum-Bambusa* hybrid. Venkatraman on the other hand had discovered painstakingly, by trial and error, that bagging of flowers against unintended pollen caused an adverse impact on seed setting of the enclosed arrows.[80] Janaki's findings and claims annoyed Venkatraman, leaving the breeder and cytogeneticist, at loggerheads with one another. She was distressed by Venkatraman's practice, which in her lab-trained eyes met none of the rigorous standards required for doing science. While he was a field-man by choice, Janaki was a border-worker; much like a weed, the solitary occupant of a border zone of mixed practices, and at this point in her career, closer to the lab side than the field. She would often speak about 'her professional disagreements' with Venkatraman, 'adding credits and criticisms in equal proportion,' recalled nephew Hari Krishnan. Janaki would share her distress with Darlington:

> I am weary of the pseudo-scientific atmosphere I am working in. I am given very little chance to publish my findings!! . . . the utter loneliness of working in an unfriendly atmosphere is almost unnerving me Cyril. . . . I feel I must get away from this place. I shall try to go to Delhi next year.[81]

In December, it was exactly a year since she had met Darlington, and spent several days travelling with him, visiting old friends and collecting plants. She missed him very much but wished him well in his newfound love and domesticity; she hoped Margaret and he had 'settled down very happily to the almost enviable and peaceful married life one generally finds amongst you English.' This time the annual conference of the Indian Science Congress was to be held in Bangalore. Attending it uplifted her spirits somewhat, for the President of the Biological Section was none other than her friend, Birbal Sahni. 'It was jolly to get back amongst the few real scientists of India,' she remarked to Darlington, 'especially after the pseudo-scientific atmosphere of Coimbatore. The contrast between Raman and Venkataraman is very very great . . . I marvelled how the Indian Science

Congress chose the latter to be its President at the Calcutta meeting last year.'[82] In Janaki's view, a lab scientist like C. V. Raman was a true scientist in contrast to a field-man like Venkatraman (without a doctorate or commensurate academic qualification), even if the latter had the skills required to breed highly economic canes. Incidentally, Raman had been President of the Congress in 1929, held in Madras, which Janaki did not attend, being in Michigan at the time.

In her letter, Janaki was drawing a clear distinction between practitioners of academic science like herself—inspired both by Vavilov, the intrepid plant collector, who travelled widely in search of the wild relatives of domesticated plants, and Darlington, located in the research laboratory, devoting all his time and energy to studying chromosomes and making breakthrough theoretical discoveries—and those (like Venkatraman) employed as breeders (invariably all were male) at government stations, working solely towards the economic interests of the state. Venkatraman to her was a mere plant breeder, albeit with the wizardry of a Luther Burbank (in the matter of the sugarcane), but a servant of the state nevertheless and not a scientist. By her definition, a true scientist had the freedom to choose a research problem, irrespective of economic gain. Janaki referred to Venkatraman's practice as pseudo-science, implying that it hardly met with the stringent standards of laboratory science, in methodology and interpretation, and merely catered to the state's economic interests rather than of science.[83] With only government institutions to depend on for employment and research in the field of cytogenetics in India, in particular breeding stations with their state-led agenda, Janaki was left with not much to choose from. As for teaching institutions, which provided relatively more freedom for research, they were poorly equipped, and their grounds unsuited to field experiments, an important component of a cytogeneticist's practice. Janaki had already explored the last option, as we know, taking up a teaching position at the Maharaja's College of Science, Trivandrum in the early 1930s, and had found it uninspiring as far as research was concerned.

Call of the Nomad

Janaki had lost all hopes of attending the International Congress of Genetics at Edinburgh in 1939, expecting too much opposition from Venkatraman. Early that year, extremely anxious about her future, and in one of her 'mad moods,' she wrote to Arthur William Hill of the Royal Botanic Gardens, Kew, enquiring whether there was a post available to work on the Indian grasses in the Kew collection; she had met Hill at the joint meeting of the Indian Science Congress and the British Association, in Calcutta in January 1938. In fact, precisely with the view of expanding Kew's imperial networks, Hill had presented a paper at the meeting on the Indian collections at Kew, and the relations between Kew and Sibpur. Since the time of botanist

Joseph Dalton Hooker (Director, 1865–85), an arrangement had been in place for an Indian botanist to work on plants of the subcontinent in the Kew herbarium (founded in 1853). Incidentally, just ahead of the meeting, the *Current Science* had published a letter penned by Birbal Sahni and S. P. Agharkar (Ghosh Professor of Botany at Calcutta University) in favour of retrieving precious Indian plant material held by the herbarium of the Kew Gardens, and placing an Indian botanist there 'as part of a programme to have Indian botanists trained for work in India.' Janaki (at this time Secretary of the Indian Botanical Society, founded by the trio Sahni, P. F. Fyson and M. O. P. Aiyengar) and P. Parija were sent copies of the letter by Sahni with the comment: 'Full publicity to the facts should be given without much comment in *Current Science*. . . . This will prepare the ground before the Congress.' The letter had also been sent to Richard Gregory for publication in *Nature*; Gregory instead forwarded it to Hill and, much as expected, the latter rejected outright the idea. It would take extended negotiations before anything positive would emerge from this campaign.[84]

Hill was however happy to support Janaki's proposal and recommended that she be given a subsistence allowance to work at the Gardens for a year, provided there be sufficient funds for the purpose.[85] He also hoped Darlington would help her with a grant from the John Innes and/or that she should explore the option of a fellowship at the Carnegie Institution, Cold Spring Harbour, where incidentally a coeval of Janaki, the maize geneticist McClintock, was based at this time.[86] Janaki remained anxious and confused not knowing what the future held for her. She was sure of one thing, that she wanted to leave Coimbatore for good, even leave India for ever. She wrote to Darlington:

> I am very anxious to get on with the study of grasses instead of sitting under Venkataraman's thumb all my life . . . if I come to England I hope you will allow me to spend some time at Merton. . . . This is building castles in the air—but why not? I am a born wanderer and shall soon have nothing to keep me in India! When I am too exhausted there is always the sea to drown myself in.[87]

Around this time, she heard from Pio Koller, of the Institute of Animal Genetics, Edinburgh; a friend from her John Innes days, Koller had sent her a formal invitation to attend the genetics congress. She learnt from him that F. A. E. Crew, Director of the Institute and President of the Genetics Congress (1939), who had visited Calcutta for the joint meeting, had written to the 'G. S. Expert' (a term she often used to refer to Venkatraman, the Government Sugarcane Expert) to nominate a delegate. 'He [Crew] hopes I will be selected but I have no such hopes—because Venkatraman is anxious to attend the Congress,' she replied to Koller. She knew for certain that if she wished to go, it would have to be on her own accord, which might even mean resigning from her present position. 'I am 95% prepared to do both,' she informed Darlington. 'All the same, there is a Great restlessness within me.'[88]

Resolves to Leave Coimbatore

Janaki never wished to remain at any one place too long; this was really the heart of the matter. A scientist among nomads, or a nomad among scientists, Janaki coveted freedom far too much to be controlled by the state. Her nomadic subjectivity subverted conventions and categories such as gender, debunking the traditional view of the nomad as embodied in the male persona. Janaki's existential life was one of radical rootlessness, pervaded by a strong sense of the *unheimlich* (never fully at home),[89] combined with a deep-seated death wish and a prodigious curiosity for the world.[90] In early 1939, she decided to refuse the extension offered by the Government of India, choosing academic freedom over financial stability: 'Rather a daring thing to do at this time of my life but I refuse to be bullied by the G. S. expert. I'd rather collect grasses in the forests of India than have my papers put away in files and submit to his ruling in the matter of my work here,' she remarked to Darlington.[91] She was already looking forward to meeting friends at the John Innes but hoped the international situation would clear up soon, even wishing someone would kidnap Hitler 'and take him off to a St. Helena.'[92] Her mentor Bartlett expressed a similar sentiment when he wrote:

> The world is certainly in a grand mess, getting less civilized every day. It would be a grand deed for some patriotic Austrian to take a few pot shots at Hitler during his "triumph" in Vienna, and to bag him with a choice selection of gangsters. Alas nothing so good can be hoped for.[93]

Janaki would soon receive news of Darlington's appointment as Director of the John Innes and of Margaret expecting their first child. She was thrilled but only a week earlier, she had had a bad fall at the Station and fractured her elbow, and was in deep pain. With her left arm in a sling, she wrote to Darlington: 'I hope you will revolutionize research departments in England & India too and put Gates in his place! . . . To Margaret, I feel like saying—Hail full of Grace—Blessed art thou amongst women and blessed be the fruit etc etc . . . The pain is horrible and I have not been able to think clearly.' Janaki was hoping to send her abstract to the Genetics Congress committee, as soon as she felt 'a bit normal.' She had also not done anything yet with regard to the Carnegie Institution application suggested by Hill, but her spirits were high. 'I hope I will be successful. I want you to send me a testimonial. I am getting my grass material together,' she wrote to Darlington.[94] Eileen Macfarlane was also planning to leave Calcutta for England

soon. Janaki however often found Eileen's loud and out-going nature somewhat 'disturbing' (she would confess to Darlington, their common node of
communication) and at this time was frankly not looking forward to her
company, but the latter was not one to be put off. Her response was in fact
simple and practical: 'We are all cracked in places so may as well overlook
each other's deficiencies.' It however made her 'unhappy to be alienated
from . . . old friends because they [were] irreplaceable.' In fact, it was Eileen
who had informed Darlington of Devi Amma's failing health—'still hanging
on in a feeble condition,' she would tell him.[95]

Slowly, and painfully, Janaki resumed work; she began preparing a paper
for the upcoming Congress, a comparative study of her two intergeneric
hybrids: the *Saccharum* x *Zea* (a dwarf, a cross between the noble *S. officinarum* and *Zea mays*) and *Saccharum* x *Imperata* (a good cane, involving
the wild *S. spontaneum*); over the past three years (1936–39), she had collected various forms of *S. spontaneum* from the sub-Himalayan range (the
hills between Punjab and Assam).[96] *S. spontaneum* and *S. officinarum* and
the hybrids between them were invariably used in her experiments; incidentally, it was Barber who in 1914 had made the first successful interspecific
cross, involving *S. officinarum* and *S. spontaneum*.[97] *S. spontaneum* was a
polymorphic species, of which she had collected clones of 48, 56, 64, 72 and
80 chromosomes within India; the ones collected from Assam and Burma
however had 96 and those from Southeast Asia had 112, while Bremer in
1929 had found forms with 80 in Celebes and the Philippines. *S. officinarum* or the 'noble' sugarcane was commonly an octoploid (2n = 80) and
in common with several important cultivated plants, it was not known in
the wild, with its nearest wild relative being *S. robustum* discovered by E.
W. Brandes in New Guinea (chromosome numbers, 2n = 84 for Brandes and
2n = 80 for Janaki). Unfortunately for Janaki, her *Saccharum* x *Zea* cross
had not yet flowered despite the buds being treated with colchicine, and she
was running out of time; the Edinburgh Congress was less than a couple of
months away. 'It would be rather nice to have a fertile *Saccharum* x *Zea*
hybrid!' she would comment to Darlington.[98]

A Bouquet of Grasses for Chase

Janaki had kept up communication with the American grass expert Chase
throughout her tenure at the Coimbatore station, often sending the latter
grasses for identification. In early August 1938, she was putting together a
collection of fodder grass to send to Chase, after working out their chromosome numbers. In a letter accompanying the parcel, Janaki wrote about her
success in crossing one of the Java hybrid canes, POJ 2725 (from the seed
supplied by the Department of Agriculture, Kuala Lumpur) with *Imperata cylindrica*, a bothersome weed (this cross, unbeknownst to her would
turn out to be a veritable "stud bull" of the Station.), and another crossing

involving one of the local canes and *Zea mays*; her *Saccharum* x *Zea* hybrid was a dwarf and useless from the point of view of the sugar industry, while the *Imperata* hybrid was a very good cane with high sucrose content; a few of the latter were even fertile, throwing out *Imperata* like F2s. The true octoploid species *S. officinarum*, when used as the female parent in intergeneric crosses, had given disappointing results from the economic point of view, but when in 1930 Thomas and Venkatraman (and Bourne in 1935) crossed POJ 2725 (2n = 106) with *Sorghum*, they obtained some economically valuable seedlings, besides several others, generally considered useless from the breeder's perspective.[99] Ever since this success, POJ 2725 and another Java cane POJ 213 (2n = 24) came to be widely used in breeding, especially in obtaining intergeneric crosses with widely different genera, such as *Bambusa*.

Janaki used the occasion to convey to Chase her deep yearning to go to the United States to do some work on the phylogeny of grasses in relation to chromosome studies. She would be free to leave India in May 1939, she told her, when her five-year contract would end: 'If there is any scope for work in U.S.A. for a couple of years, I would be delighted to come. One is very isolated here and I would like to return to Ann Arbor to see old friends & publish some of my findings on Grass Cytology. I shall be very thankful if you could help me to get to U.S.A.'[100] Chase had identified certain Indian grasses for her earlier, and Janaki was very thankful for this; she told Chase how her interest in grasses was 'increasing daily' and that she had a 'living herbarium of Coimbatore grasses' in her garden and that it was her aim to 'work out the cytology of some of them.' Apart from her interest in the Andropogoneae as likely parents for intergeneric crosses with *Saccharum*, the Bamboo and the Maydeae (maize for example) had become her 'special favourites.' The Forest Department of India had been helping her form a representative collection of *Bambusa*.[101]

Janaki also shared with Chase her excitement at having produced a cross between *S. spontaneum* (from Java) and the *Erianthus ravennae* (also called *Saccharum ravennae*, or elephant grass, this was the variety 'purpurascens' collected from the Punjab and designated '*S. munja*, spiny' at Coimbatore, until correctly identified by Kew's C. E. Hubbard, a world authority on grasses) in 1935. The second generation had produced amphiploids (an interspecific hybrid with a complete diploid chromosome set from each parent, AABB, where A and B represent the parents), which resembled the sugarcane in thickness but contained no sugar! She found it to be quite fertile and 'breeding true' (that is a pure line, wherein parents produced offspring only of one variety, a trait for which the parents were homozygous—that is both were dominant or both were recessive for that particular trait—and was passed down to all subsequent generations).[102]

Janaki had been attempting a series of crosses between *S. spontaneum* and related grasses since 1934, of which the most successful hybrids were those between two types of this species (with 56 and 112 chromosomes) and *Erianthus ravennae*, 2n = 20, as the pollen or male parent; the first of these had not flowered. Her present research was on the second hybrid, in which the Java clone of *S. spontaneum* 'Glagah' (acquired from the Pasoeroean Experimental Station in 1919 and propagated vegetatively at the Sugarcane Station, Coimbatore) was used as the female parent. The resultant cross (between 'Glagah' and *Erianthus*) was fertile with 66 chromosomes; the F1 hybrids resembled the two parents in proportion to their chromosome contributions (56 and 10, that is more like sugarcane than elephant grass), while the F2 seedlings consisted of three groups: diploids (68–76), triploids (104–08) and a tetraploid (136).

These new creations were valuable objects of exchange, and they demanded appropriate names. Janaki sought Chase's help with naming her new hybrid: 'I wish to give it a name. Will *Erio-Saccharum* be correct or should it be *Eriantho-Saccharum*. In appearance it is more *Saccharum* than *Erianthus* (only a small awn of Erianthus was present).'[103] Chase thought *Eriantho-Saccharum* suited the cross better because there were already several generic grass names beginning with Erio, and also that it was pronounceable and was a combination of the two names. As always, Janaki hoped to send Chase a pressed specimen of the cross, besides one of her *Saccharum-Zea* hybrid, which was yet to flower. Although Chase was pleased to hear that Janaki wished to come to America, she warned that at the Smithsonian, there were no laboratory facilities suitable for chromosomal work; all the same, she was aware that the University of Michigan could facilitate that if they so wished. Chase was now seventy and had retired, but continued to be the Custodian of Grasses at the Smithsonian and had just embarked on a new project on the grasses of Brazil.[104]

Janaki replied to Chase, just as she was heading 'Westward' on the T. M. S. Cilicia, to attend the Genetics Congress in Edinburgh and later 'look round in the British Isles for some time working at Kew and John Innes Hort.' She had been asked by the government of India to represent the country at the Congress, along with a few others; her secret plan however was to stay on in Britain or America. Janaki had applied to the Carnegie Corporation for a research grant, but the terms of their charter restricted their activities to the United States. 'I do not see why they cannot give me the grant if I work in England or U.S.A. I am very anxious to get on with my grass hybrids,' she complained to Chase.[105] She had left Coimbatore for good, because of its 'unscientific atmosphere—India is not a good place for scientific research—too much Red Tape in Govt departments & Institutions.' Janaki had collected all the money she could lay her hands on—she had sold her car and used up her provident fund to buy her passage and to cover her living expenses in England—and was even prepared, she wrote to Chase, 'to go to the South Sea Island [where sugarcane abounds] if it will carry me there which I doubt very much.'[106]

Figure 10.1 Family group with Janaki (standing with her brothers). December 1935.
Source: Courtesy of the EK family.

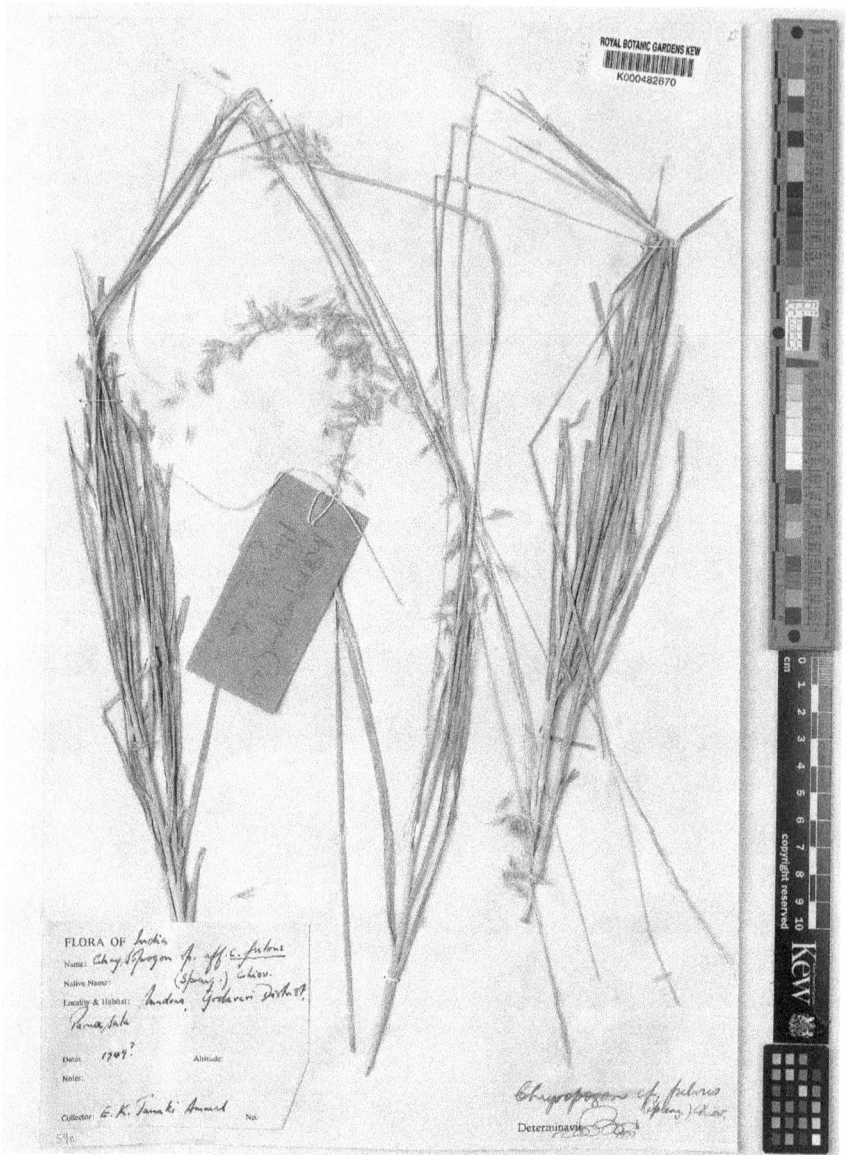

Figure 10.2 A species of *Chrysopogon* collected by Janaki from Parnasala, Godavari, in 1937 (wrongly dated by library staff as '1949?').

Source: © copyright of the Board of Trustees of the Royal Botanic Gardens. http://specimens.
kew.org/herbarium/K000482670

Notes

1 DP: C. 98 (H. 60), E. K. Janaki Ammal to C. D. Darlington, letter dated 14 March 1939.
2 Russell, *Report on the Work of the Imperial Council of Agricultural Research*, p. 6.
3 Smithsonian Institution Archives, Washington (SIAW): Record Unit 229, United States National Museum, Division of Grasses, Folder 42, Box 4, E. K. Janaki Ammal to Mary Agnes Chase, letter dated 14 April 1937.
4 Janaki Ammal, 'Chromosome numbers in sugarcane x bamboo hybrids', 1938.
5 Incidentally, as part of his itinerary, Russell visited the UBL on 13–14 December 1936, then under the directorship of M. O. P. Aiyengar, where much of the work done was concerned with the study of algae, and of soil. He would also visit the Botany Department of the Presidency College; the chief focus of this department, where Janaki had set up a small cytology laboratory on her return from America (and the John Innes), was plant morphology and physiology. A government grant of Rs 8,600, spread over three years, had been sanctioned to the department, for investigating the developmental morphology and anatomy of the intergeneric *Saccharum-Sorghum* hybrids produced at the Coimbatore Sugarcane Breeding Institute by Venkatraman and Janaki individually. The research was to be carried out at the laboratory, under the supervision of T. Ekambaram, with Venkatraman and Janaki acting as advisers; the results were to be judged not from the point of view of practical value but their methodological 'soundness'.
6 Bhagavathi Kutty Amma and Ekambaram, 'Sugarcane x Bamboo Hybrids.'
7 Incidentally, in 1952, T. S. Raghavan of the Sugarcane Breeding Institute, Coimbatore reconfirmed (once again) the genuineness of Venkatraman's intergeneric *Saccharum* x *Bambusa* cross in a note published in *Nature*. See Raghavan, 'Sugarcane x bamboo hybrids'.
8 Smithsonian Institution Archives, Washington (SIAW): Record Unit 229, United States National Museum, Division of Grasses, Folder 42, Box 4, E. K. Janaki Ammal to Mary Agnes Chase, letter dated 14 April 1937.
9 Koelz had conducted botanical explorations in 1930 in the area around Kulu, while holding a post with the Himalayan Research Institute of the Roerich Museum.
10 For seven years, from 1939, Koelz explored Persia, Nepal and parts of India including Assam and amassed a large ornithological collection. He had collected nearly 30,000 bird specimens for the University of Michigan's zoology museum and some 30,000 plants for the university herbarium.
11 University of Michigan, Herbarium: Catalogues nos. 1498518, and 1497193 respectively.
12 Barndt and Sinopoli (eds.), *Object Lessons and the Formation of Knowledge*.
13 SIAW: Record Unit 229, United States National Museum, Division of Grasses, Folder 42, Box 4, E. K. Janaki Ammal to Mary Agnes Chase, letter dated 14 April 1937.
14 Ibid., Mary Agnes Chase to E. K. Janaki Ammal, letter dated 17 May 1937.
15 *The Harley Harris Bartlett Diaries (1926–1959)*, p. 106.
16 DP: C. 109 (J. 112), John Russell to C. D. Darlington, letter dated 10 May 1937.
17 Ibid., p. 12.
18 Ibid., C. D. Darlington to John Russell, letter dated 13 May 1937.
19 Fifteen of Janaki's printed field notebooks are held by the Kew Gardens Herbarium, and cover the period 09/06/1934 to 11/7/1951, with some gaps between. The rest of her field notebooks remain untraceable.

20 Serial numbers 1001–1100 of her herbarium records/field notebook pertain to a collection of grasses made between January and February 1937. RBG Kew, Library & Archives: Printed Notebooks, Herbarium E. K. Janaki Ammal, Madras.
21 See Piramal, *Business Legends*, p. 272.
22 Janaki Ammal, 'Chromosome Studies in *Saccharum arundinaceum* L.', 1937. Janaki also participated in the proceedings of the Agricultural Section of the Congress, with papers, 'The Inheritance of Habit in *Saccharum spontaneum* L.' and 'Tetrasomic Inheritance in Two *Saccharum officinarum* and *Saccharum spontaneum* Hybrids'.
23 For more details see Gates, 'The Jubilee Meeting of the Indian Science Congress'.
24 Bud sports refer to an inflorescence that differs genetically from the rest of the plant, with the differences persisting when the plant is vegetatively propagated from the inflorescence.
25 Serial numbers 1101–1200 of her herbarium records/field notebook pertaining to a collection of grasses made between February and December 1937 in Delhi, Allahabad, Coimbatore again and Tellicherry. RBG Kew, Library & Archives: Printed Notebooks, Herbarium E. K. Janaki Ammal, Madras.
26 Shamdasani (ed.), *The Psychology of Kundalini Yoga*, 'Introduction'.
27 *Current Science* 6, no. 8 (February 1938): 368.
28 Janaki-Ammal, 'Chromosome Behaviour in *S. spontaneum* x *Sorghum durra* hybrids', (Abstract), 1938.
29 Venkatraman, 'Presidential address', p. 275.
30 Ibid., p. 283.
31 A reference to Hans Rheinberger's concept, explored in his *Towards a History of Epistemic Things*.
32 Venkatraman, 'Message of the sugarcane', p. 8.
33 In late 1938, Janaki had sent a short note to *Current Science* for publication, but before seeking the approval of Darlington and Margaret. DP: C. 109 (J. 112), E. K. Janaki Ammal to Margaret Upcott, letter dated 30 August 1938; Janaki Ammal, 'Triplo-polyploidy in *Saccharum spontaneum* L.', 1939.
34 Eileen Macfarlane, discussing the species concept, *Proceedings of the Indian Science Congress, 1938*, pp. 207–208.
35 DP: C. 9 (B. 8), Eileen Macfarlane to C. D. Darlington and Margaret Upcott, letter dated 28 April 1939, 9/2 Middleton Row, Calcutta.
36 Macfarlane's review of Holmes' *Human Genetics and Its Social Import* appeared in *Current Science*, March 1937. For her early published articles on the subject of blood groups, see Bibliography.
37 See Schneider, 'Blood Group Research'.
38 For an early history of blood group genetics, see Schneider, 'The History of Research on Blood Group Genetics.' Contemporaneously, A. Aiyappan of the Madras Museum was working on the blood groups of South India, but from the physical anthropology perspective; he had received a PhD from the London School of Economics in 1937 for a thesis prepared under the supervision of anthropologist Raymond Firth. See Aiyappan, 'Blood Groups of the Paniyans of the Wynaad Plateau'.
39 DP: C. 9 (B. 8), Eileen Macfarlane to C.D. Darlington and Margaret Upcott, letter dated 28 April 1939, 9/2 Middleton Row, Calcutta.
40 For a biography and an exhaustive bibliography of Gates' work, see Fraser Roberts, 'Reginald Ruggles Gates, 1882–1962'.
41 DP: C. 108 (J. 64), R. R. Gates to Birbal Sahni, letter dated February 1937.
42 Also see DP: C. 98 (H. 53), Boshi Sen to C. D. Darlington, letter dated 3 February 1938, 8 Bosepara Lane. For a biography see, Mehra.

43 Ibid.
44 Janaki Ammal, 'Sugarcane-Sorghum Hybrids in Wild State' (1937). An abstract of the paper was published in *The Madras Agricultural Journal* XXVI (June 1938): 288.
45 DP: C. 109 (J. 112), E. K. Janaki Ammal to C. D. Darlington, letter dated 8 August 1938.
46 Janaki Ammal, 'A *Saccharum-Zea* cross', 1938.
47 DP: C. 109 (J. 112), E. K. Janaki Ammal to C. D. Darlington, letter dated 8 August 1938.
48 Ibid., E. K. Janaki Ammal to C. D. Darlington, letter dated 28 August 1938; see Darlington, *The Evolution of Genetic Systems*, p. 173: 'Hybrids between *Saccharum* and *Bambusa* are fertile. Only that between *Saccharum* and *Zea* shows vegetative abnormality (Janaki Ammal, 1938)'.
49 Venkatraman and Thomas, 'Brief note on Sugarcane-Sorghum hybrids'.
50 This was grown for several generations without the risk of cross-pollination, being the only one cultivated in the neighbourhood.
51 Hari Krishnan, 'Nachee', July 1993, private collection.
52 Krishnan, 'Evergreens in Cudappah'.
53 Krishnan, 'Prickly pear for gum'.
54 Krishnan, 'Forest fires'.
55 E. K. Raghavan, 'Random notes', 24 June 1971, private collection.
56 If the basic number was 7, then the polyploid series would be 21 (triploid), 28 (tetraploid), 42 (hexaploid) and 56 (octoploid). The term autoploidy is used to denote those polyploids formed by multiplication of sets of chromosomes within the species.
57 For sources of the drug and its various uses, see the classic by Eigsti and Dustin, *Colchicine in Agriculture, Medicine, Biology and Chemistry*.
58 Ibid., p. 275.
59 Ibid, p. 279.
60 Ibid., p. 274; In the 1940s–50s, the chemical began to be widely used by horticulturists and amateur gardeners in America, to produce large-sized flowers of ornamental plants such as the marigold. See Curry, 'Making Marigolds'.
61 From 1946 Indian pharmacopoeia allowed the use of *Crocus luteum*, the spring-flowering species, in place of *Crocus autumnale*, owing to its availability. However, British pharmacopoeia standards would not allow its use because the alkaloid content was not high enough. Eigsti and Dustin, *Colchicine in Agriculture, Medicine, Biology and Chemistry*, p. 141.
62 Parthasarathy, 'An Indian Source for Colchicine'.
63 Janaki-Ammal, 'Chromosome numbers in sugarcane x bamboo hybrids', 1938.
64 DP: C. 109 (J. 112), E. K. Janaki Ammal to C. D. Darlington, letter dated 28 August 1938.
65 Published in *Nature* 140 (11 December 1937): 1013–1014; Darlington's response, *Nature* 141 (26 February 1938): 371–372; Gates' counter-response was simply that Darlington's photomicrograph of the cell in telophase hardly displayed any resemblance to the figures: *Nature* 141 (2 April 1938): 607.
66 A photograph of a microscopic object, taken with the help of a microscope.
67 *Current Science* 6, no. 12 (June 1938): 613–614.
68 DP: C. 109 (J. 112), E. K. Janaki Ammal to C. D. Darlington, letter dated 28 August 1938.
69 Ibid.
70 Ibid., Janaki Ammal to Margaret Upcott, letter dated 30 August 1938.
71 Ibid., E. K. Janaki Ammal to C. D. Darlington, letter dated 28 August 1938.

72 DP: C. 108 (J. 64), E. K. Janaki Ammal to C. D. Darlington, letter dated 20 September 1938.

73 Serial numbers, 1101–1200 of her herbarium records/field notebook pertaining to a collection of grasses from September to November 1938. See RBG Kew, Library & Archives: Printed Notebooks, Herbarium E. K. Janaki Ammal, Madras.

74 DP: C. 108 (J. 64), E. K. Janaki Ammal to C. D. Darlington, letter dated 20 September 1938.

75 Serial numbers, 1201–1300 of her herbarium records/field notebook pertaining to a collection of grasses made in November–December 1938. See RBG Kew, Library & Archives: Printed Notebooks, Herbarium E. K. Janaki Ammal, Madras.

76 DP: C. 108 (J. 64), E. K. Janaki Ammal to C. D. Darlington, letter dated 20 September 1938.

77 Venkatraman, 'Sugarcane Breeding in India'.

78 Venkatraman expressed this view as part of a discussion on a paper presented by V. Ramanathan (Cotton Specialist to the Government of Madras), before the Association of Economic Biologists in 1930. See *Proceedings of the Association of Economic Biologists, Coimbatore*, Vol. 1, 1930–33, published by the Association, 1934.

79 POJ was the abbreviation for Profestation Oost Java, just as Co was for Coimbatore.

80 Venkatraman, 'Sugarcane Breeding, Indications of Inheritance'.

81 DP: C. 109 (J. 112), E. K. Janaki Ammal to C. D. Darlington, letter dated 23 December 1938.

82 Ibid.

83 Luther Burbank (1849–1926) was an American botanist, horticulturist and a pioneer in agricultural science, who developed more than 800 strains and varieties of plants, including fruits (the Santa Rosa plum and the exceptional blackberry, for instance), vegetables such as the Burbank potato, grains, and ornamentals like the Shasta daisy) over his fifty-five-year career. For more, see Crow, 'Plant breeding giants: Burbank, the artist; Vavilov, the Scientist'.

84 Subramanian, 'Professor Birbal Sahni'.

85 DP: C. 98 (H. 60), E. K. Janaki Ammal to C. D. Darlington, letter dated 14 March 1939.

86 DP: C. 109 (J. 113), A. V. Hill to C. D. Darlington, letter dated 28 March 1940; Hill would unfortunately pass away in November 1941, following a tragic riding accident.

87 DP: C. 98 (H. 60), E. K. Janaki Ammal to C. D. Darlington, letter dated 14 March 1939.

88 Ibid.

89 In his *Being and Time*, Martin Heidegger used the term *unheimlich* (see Freud's *Das Unheimliche*, 1919) to mean 'uncanny', 'being unsettled', 'never fully at home', or the strange mix of familiarity and estrangement, and the fundamental groundlessness of our existence.

90 The notion of 'nomadic subjectivity' is central to the feminist philosopher Rosi Braidotti's work, reflected in her trilogy of books: *Nomadic Subjects: Embodiment and Difference in Contemporary Feminist Theory* (1994), *Metamorphoses: Towards a Materialist Theory of Becoming (*2002) and *Transpositions: On Nomadic Ethics* (2006).

91 DP: C. 98 (H. 60), E. K. Janaki Ammal to C. D. Darlington, letter dated 17 March 1939.

92 DP: C9 (B. 8), E. K. Janaki Ammal to C. D. Darlington, letter dated 29 March 1939.

93 *The Harley Harris Bartlett Diaries (1926–59)*, pp. 114–115.

94 DP: C. 9 (B. 8), E. K. Janaki Ammal to C. D. Darlington, letter dated 29 March 1939.

95 Ibid, Eileen Macfarlane to C. D. Darlington and Margaret Upcott, letter dated 28 April 1939, 9/2, Middleton Row, Calcutta.

96 Later, between 1947 and 1956, under an exploration and collection programme called the '*Spontaneum* Expedition Scheme', inaugurated by S. K. Mukherjee, and funded by the Indian Sugar-cane Committee, over 500 clones representing *S. spontaneum*, *Sclerostachya* spp., *Narenga* spp. and *Erianthus* spp. were collected from India and the adjoining countries.

97 In 1927, C. L. Rümke crossed another clone 'EK28' with *Erianthus sara* (Roxb.). See Rümke, '*Saccharum-Erianthus Bastaarden*'; since then, a number of intergeneric hybrids of *S. officinarum* were produced, such as the ones by Venkatraman and Janaki (1938). *Erianthus* is considered a close relative of *Saccharum*, and many species have been assigned to either of these genera, depending on the criteria used; several however believed that the two genera were distinct.

98 DP: C. 9 (B. 8), E. K. Janaki Ammal to C. D. Darlington, letter dated 29 March 1939.

99 The clone POJ 2725 was derived from the *cross S. officinarum* x *S. spontaneum* backcrossed twice with *S. officinarum*.

100 SIAW: Record Unit 229, United States National Museum, Folder 42, Division of Grasses, Box 4, E. K. Janaki Ammal to Mary Agnes Chase, letter dated 8 August 1938.

101 Ibid., letter dated 25 May 1939.

102 Ibid.

103 Ibid.

104 Ibid., Mary Agnes Chase to E. K. Janaki Ammal, letter dated 15 June 1939.

105 Ibid., E. K. Janaki Ammal to Mary Agnes Chase, letter dated 16 August 1939.

106 Ibid.

11

GREAT BRITAIN I

Doing Science in the War Years

It is a strange feeling to be homeless in this wide world but it is as it were the home has extended to fill the whole world and I meet with nothing but kindness wherever I go—more kindness than from the people of my own land.
—E. K. Janaki Ammal (1939)[1]

The Seventh International Genetics Congress held at the Institute of Animal Genetics, Edinburgh (23–30 August 1939), on the eve of World War II, was perhaps the perfect illustration of the deplorable intrusion of the Communist Party in matters that were purely scientific. Already, with the 'red terror' inaugurated by Lenin following the Russian Revolution in 1917, thousands of leading figures of the arts and sciences, including geneticists, had emigrated from Soviet Russia. However, by 1921, when the turmoil ended, genetics had evolved into a full-fledged discipline in the new Soviet Russia, with laboratories, and departments and journals devoted to research in genetics, so much so that when the first national conference on genetics was held in Leningrad in 1929, almost 1,500 delegates took part in its proceedings. The initial thrust for the meteoric institutional growth of Soviet genetics under its three founding fathers, Nikolai Vavilov, Nikolai Kol'stov and Iurii Filipchenko, came from the fields of plant and animal breeding, experimental biology and eugenics, as it was indeed the case the world over. The three men spared no effort in establishing close ties with their Western peers.

While Kol'stov developed strong links with German geneticists such as Richard Goldschmidt and Max Hartmann, and Filipchenko with Richard Hertwig in 1911–12, Vavilov studied in England with William Bateson and R. C. Punnett at the John Innes and Cambridge respectively, and with Phillipe de Vilmorin in France and Ernst Haeckel on the continent, over the period, 1913–14. In 1921, Vavilov went to the United States to take part in an International Congress on Phytopathology and used the opportunity to visit agricultural stations and university departments all over the States. On his return to Russia, he stopped by in Germany and Sweden and also

DOI: 10.4324/9781003267089-11

paid a visit to Bateson in England. Western geneticists reciprocated this visit and several including Erwin Bauer, Calvin Bridges, Leslie C. Dunn, Sidney Harland, Cyril Darlington, Julian Huxley, Goldschmidt, Dontcho Kostoff and Hermann Muller visited Russia in the 1920s and 1930s, by which time Soviet genetics had received considerable acclaim in the international arena.

Even as Soviet geneticists were developing international ties, the Bolsheviks under Stalin initiated an ambitious scheme of speedy industrialisation to construct the 'material-economic basis of socialism.' The peasantry was collectivised and the state given absolute monopoly over resources and production. The 'Great Break' of 1929 also had a major impact on Soviet science; the Bolsheviks set up an enormous, centralised hierarchical network of institutions and an administrative system to oversee and control it. They mobilised science for socialist reconstruction, which invigorated agricultural institutions. Despite the dismissal of powerful supporters of Soviet genetics from their posts and the dissolution of the Russian Eugenics Society, the institutional base of the discipline expanded during this period. Genetics thrived within the Academy of Sciences. With Filipchenko's death in 1930, Vavilov inherited his genetics laboratory and transformed the place over three years into an institute of genetics, where the former's students continued to work on their research projects. He also founded the colossal All-Union Institute of Plant Breeding in Leningrad, where seeds of cultivated and wild plants collected worldwide were used in the breeding of new varieties for Soviet agriculture. Till date, the institution remains the world's largest collection of plant seeds. Vavilov spared no effort in introducing genetics courses in agricultural schools, as a result of which the institutions became a bastion of Soviet genetics in the 1930s, and the target of the Lysenkoist debate.[2]

The Ithaca Congress

In 1932, as the worldwide economic crisis unfolded, Vavilov was the sole geneticist from Soviet Russia to attend the Sixth International Genetics Congress in Ithaca, New York, despite the organising committee inviting about twenty Soviet geneticists. It was claimed that certain unsurmountable bureaucratic barriers had prevented the Soviet delegation from attending; Vavilov, the Vice-President of the Congress, had himself been granted permission only a couple of weeks before the event. At the Ithaca Congress session on 'genetics and evolution' (he shared the platform with Haldane, R. A. Fisher and Sewall Wright, the three founding pillars of evolutionary genetics), he outlined his theory of the centres of origin of cultivated plants, an encapsulation of the results of his plant expeditions and the research done by his co-workers at his Institute of Plant Breeding. He also used this occasion to announce to the world the 'big' discovery made by a younger

colleague from Odessa, Trofim Lysenko, of a technique called vernalization which made it possible to switch the vegetation period of a range of
cultivated plants (subtropical wheat for instance could be grown in cold
climates), triggering tremendous excitement in the American media. Reject-
ing the concept of the gene as a material unit of heredity, Lysenko and his
disciples had debunked Mendelian laws of inheritance and claimed that the
environment could directly impact upon heredity.

Vernalization provided the experimental basis for Lysenko's views; a
plant's development could be altered by changing its external conditions.
Lysenko believed that plants such as wheat had a specific ontogenetic stage,
which he referred to as 'the stage of vernalization,' when seeds required low
temperature and high levels of moisture to develop into regular plants and
produce seeds. The spring varieties of wheat according to him had a very
short stage of vernalization and required high temperatures relatively speaking, while winter varieties had a longer vernalization stage and required
much lower temperatures. Normally, it was not possible to sow winter varieties in spring or vice versa, but Lysenko proposed a way of 'skipping' the
vernalization stage of the winter variety by wetting seeds and exposing them
to low temperatures in storage. These vernalized seeds were then used for
sowing in spring.[3] Lysenko would also draw theoretical conclusions that
smacked of neo-Lamarckism, that is a belief in evolution based on use and
disuse. He argued that the alteration of a winter form into a spring variety
by vernalization also altered its hereditary traits: 'the acquired characteristics of the spring form became hereditary in the progeny of the "vernalized"
winter form.' While Lysenko saw vernalization as an agricultural technique
helpful in increasing the yield of agricultural crops, to Vavilov it was a laboratory method useful in hybridising varieties with different vegetation periods. The Soviet authorities were quick in adopting Lysenko's technique to
remedy the disastrous situation in agriculture caused by the forced collectivisation of peasantry. Lysenko was lauded by the state, and appointed chair
of a special department of vernalization at the Odessa Institute of Genetics and Plant Breeding, and later made the Director of the Institute.[4] Riding high on his success, Lysenko traced his intellectual trajectory to that of
the Russian amateur breeder Ivan Michurin, besides the American botanist
Luther Burbank; the Lysenkoists called themselves Michurinists, and their
opponents, who represented bourgeois or 'formal' genetics, Mendelians or
Morganists. In 1936, when Michurin died, the Soviet state hailed him as one
of the founding fathers of Soviet science.

Wishing to give as much coverage as possible to the exciting new developments in Soviet genetics, Vavilov had carried with him to Ithaca, Karpechenko's
new polyploid hybrids; a former Rockefeller Fellow, Karpechenko was
Vavilov's colleague at Leningrad. Vavilov also used the opportunity to renew
his friendship with some of the world's leading geneticists (Congresses might
be read as epistemic communities) including Morgan, Edward Murray East,

Darlington, Muller and Haldane, besides his former fellow-countrymen Theodosius Dobzhansky, who had become a major figure in Morgan's laboratory at Caltech, and Nikolai Timofeeff-Ressovsky (Kol'tsov's student), who was at this time a leading figure in *Drosophila* genetics in Germany. At this meeting, Vavilov was elected the Soviet representative of the International Organizing Committee for Genetics Congresses (IOC), in the place of Kol'tsov who had held the position since the Berlin Congress in 1927.[5] Vavilov had arrived in Ithaca with permission from the Soviet authorities to invite the next Congress to meet at Moscow. The Ithaca Congress having been a major disappointment for the Soviet delegation, they spared no effort in strengthening international links. In 1932, the USSR Academy of Sciences was persuaded to elect geneticists Morgan, N. H. Nilsson-Elle, E. von Tschermak and de Vries as honorary members and a year later, Muller as a corresponding member.

That same year, on Vavilov's invitation, Muller arrived in the Soviet Union to work at the former's genetics laboratory at the Academy of Sciences (after a few months it would be moved to Moscow to become the Institute of Genetics), so did the Bulgarian geneticist Dontcho Kostoff, who would publish regularly in the *Current Science* in the 1930s–40s. In 1934, we know, Darlington had visited Vavilov's Institute and delivered a set of lectures. The party apparatus was however slowly strengthening its grip over the world of science, including restricting foreign travel, the communication and publication of scientific works abroad and the exchange of books and specimens. The Politburo also prohibited Soviet scientists from accepting Rockefeller Foundation fellowships. When in May 1935, the Norwegian geneticist Otto Mohr, President of the upcoming Congress, revived the possibility of Moscow being the next venue of the Genetics Congress to be held in 1937, Vavilov set out right away to secure the government's permission, but this was a long-winded process steeped in bureaucracy. By August that year, Vavilov's lobbying had paid off and he informed Mohr that the government had agreed, and that it was now the turn of the IOC to accept the official invitation sent out by the Academy of Sciences; the acceptance letter was received from Mohr in mid-November and the presidium began to discuss a preliminary programme and membership of the Soviet organising committee.

At the end of a series of meetings of the committee, which had appointed Vavilov as the Congress President and Morgan as Honorary President, the preliminary programme was amended in April 1936 at the request of more than thirty American geneticists who wished to include a discussion of 'questions relating to racial and eugenic problems.'[6] This posed a serious problem to the organising committee as eugenics was a bad word in the era of the 'Great Break' (the 1930s, when the Russian peasant became the collective farmer); it would eventually resolve the issue by agreeing to devote a session to 'human genetics and race theory,' without any mention of the term eugenics. Over the next few months, preparations would gain momentum, but all of a sudden, on 14 November, the Politburo announced the cancellation of the Moscow Genetics Congress of 1937. The decision came as a rude blow, and nobody, not even members of the organising committee knew the reason for this. On the instructions of Stalin, the Academy of Sciences would send an official letter to Mohr informing the IOC Chairman, that 'due to a number of unforeseen circumstances . . . it is impossible to convene the Seventh International Genetics Congress in the USSR in 1937.'[7]

Soviet geneticists hoped that the announcement could be interpreted to mean that the Congress had merely been postponed and not entirely abandoned, and began a combined effort to remedy the situation. But a month later, the *New York Times* published a 'wireless' communiqué from its Moscow correspondent, which not only mentioned the cancellation of the Congress, but also announced that Vavilov and Isaak Agol (a former Rockefeller fellow, like Karpechenko) had been arrested and that the party had strongly criticised the Congress' general secretary Solomon Levit. It also provided a clue to the reason behind the Politburo's decision: 'A schism among Soviet geneticists, some of the most prominent among whom are accused by Communist party authorities of holding German Fascist views on genetics and even being shielders of "Trotskyists", lay behind the cancellation. The fact that so many of the Soviet Union's most distinguished geneticists are under fire is believed to be [the] motive for the government action.'[8] The said 'schism' had to do with the growing controversy on 'issues in modern genetics' between two members of the organising committee of the Congress: Vavilov and Lysenko.

The report immediately stirred American and British scientists into action, with several of them including Morgan, Darlington and the anthropologist Franz Boas persistently bombarding the Soviet Ambassadors to the United States, the United Kingdom and Sweden with letters and cables, for details on the fate of Vavilov and Agol, and of the Congress. Darlington contacted the Foreign Office and lobbied to gather the support of the British Association in the matter; while the American and British diplomats in Russia chose to stay clear of the whole affair (they did not think it was proper or useful to intervene), the Soviet diplomats forwarded all the correspondence they had received to authorities in Moscow; the Western scientists had also tapped into their own scientific networks round the world.

In their reply, the Soviet authorities however denied all accusations of curtailing academic freedom: 'real intellectual freedom exists only in the USSR where science works not for the benefit . . . of capitalists, but for the good . . . of the peoples of all mankind,' they shouted.[9] Stalin claimed that the *New York Times* had lied about Vavilov's arrest and that although Agol had been arrested it had nothing to do with the genetics of the Congress. The Soviet state claimed that the Congress had merely been postponed and not

cancelled, because its scientists needed more time for preparation. Unsatis-
fied with these answers, and after much debate, Western scientists under the
leadership of the IOC chairman, Otto Mohr, began a search for an alternative venue for the Congress. Mohr put before the members two alternatives:
to accept the new Soviet invitation to hold the Congress in Moscow in 1938
or to hold it in Britain in 1939. The majority chose the second alternative.
By this time, incidentally, Muller had accepted Francis A. Crew's invitation
to work at the Institute of Animal Genetics in Edinburgh, and left Russia
for good.[10]

The Edinburgh Congress

In November 1937, the Genetical Society of Great Britain announced that
the seventh International Genetics Congress would be held at Edinburgh in
August 1939. In face of the very real possibility that Soviet delegates would
be barred from attending the meeting, they unanimously elected Vavilov
as Congress President. Vavilov accepted the privilege and began earnestly
lobbying support to send a large Soviet delegation to Edinburgh; about fifty
Soviet delegates began preparing abstracts and exhibits for the Congress,
with several also asked to deliver keynote addresses. Vavilov had earlier
requested Crew to organise a large session on the subject of 'plant and
animal breeding in the light of genetics,' as many felt a major gap existed
between academic genetics and practical breeding work (a situation not different from the Indian scenario). This was also accepted and preparations
gathered momentum. Crew was expecting about seven hundred geneticists
from fifty different countries to attend, and almost four hundred papers presented, of which those by Soviet geneticists formed the most prominent part.
However, as it turned out, despite Vavilov's efforts and the campaign on the
part of British geneticists, including Darlington and Daniel Hall (Vavilov's
friend), permission was denied to the Soviet delegation and the entire schedule had to be redrawn and reprinted.

Crew was now elected General Secretary of the Congress, which was to
last a week, 23–30 August 1939; the membership fee was fixed at two guineas. The University of Edinburgh made available the five University Departments of the King's Buildings group, and the Committee of the Student's
King's Buildings' Common Room agreed to accept all Congress members
as honorary members for the duration of the meeting. Six student hostels,
located with easy access to the Congress venue, were set aside to accommodate four hundred participants, and Pickford's Travel Service was appointed
official travel agent to the Congress. As a large number of the members, especially those from America and Europe, were expected to travel to Edinburgh
via London, a pre-Congress tour was arranged and reception committees
formed both in London and Cambridge. On 15 August 1939, a week before
the Congress was scheduled to begin, the Congress' London office opened to

receive its first guests; the Convener of the reception committee was UCL's R. A. Fisher, with Brenhilda Schafer of the John Innes, as Secretary.

To kickstart the pre-Congress activities, the Royal Horticultural Society, the host and initiator of the international genetics conference, had arranged a reception for the delegates, followed by a visit to important centres of genetics research in Britain, on 16 and 17 August: the group visited the Whipsnade Zoological Park, East Malling Research Station, Rothamsted Experimental Station, the RHS Garden Wisley, John Innes, Courtauld Genetical Laboratory, Natural History Museum, South Kensington, Royal Botanical Gardens, Kew, Zoological Society Gardens, Regent's Park, Galton Laboratory and Department of Biochemistry, University College, Bureau of Human Heredity and the Chelsea Physic Garden. On 18 August, the office of the Congress moved to Edinburgh along with a large group of about a hundred in a motor coach. En route, they stopped at Cambridge, where they visited the School of Agriculture, Horticultural Research Station and Animal Research Station, and attended a reception at St John's College. This was followed a day later by visits to the Potato Virus Station and the Botany School Field Station. The group eventually reached Edinburgh on 22 August, where an informal reception hosted by the President and Committee awaited them at the Student's Common Room. As a reflection of the times, women participants were treated to a separate entertainment by the Ladies' Committee of the Congress, which included visits to the Zoological Park and the Royal Botanic Garden, besides the relevant departments of the University of Edinburgh.

On 23 August, at 10.30 am, Otto Mohr opened the Congress plenary session at the University of Edinburgh's McEwan Hall with a formal address, narrating the sequence of events leading to the pull-out from Moscow and relocation to Edinburgh. Every country had an elected Vice-President, and for India it was T. S. Venkatraman; he had retired by this time but had been given a three-year extension. Among the other official delegates from India, nominated by their universities, institutes or societies were B. N. Singh and G. N. Pathak (Banaras Hindu University), T. S. Venkatraman and Wynne Sayer (IARI), Captain S. C. A. Dutta (Indian Veterinary Research Institute), Lt. Col. S. L. Bhatia, E. K. Janaki Ammal, A. C. Joshi, V. R. Khanolkar, K. Ramiah (he also represented the Institute of Plant Industry), Shri Ranjan (and Higginbotham, also representing the University of Allahabad) and G. S. Thapar (Indian Academy of Sciences), Col. Arthur Oliver (National Institute of Sciences of India), Thomas Holland, Col. Greig and Col. A. D. Stewart (Royal Asiatic Society of Bengal), G. P. Majumdar (University of Calcutta) and Lt. Col. A. N. Bose (University of Patna).

Incidentally, by the time the Congress opened, Darlington's wife, Margaret Upcott, gave birth to a baby boy, Oliver Franklin Darlington. Upcott was an exhibitor at the Congress, but remained at home with the newborn, while her husband attended the Congress.

Hybridisation, Polyploidy and Fertility

Darlington and Pio Koller were respectively appointed the Recorder and Secretary of the session 'Cytology' (Section B), and Mather and Ellerton of 'Plant Cytology in the light of Genetics' (Section E). Theme 4 of Section E, held at the Chemistry Lecture Theatre No. 2 on 29 August, was devoted to 'Reproduction and Species Hybrids,' which included Janaki's paper titled 'Triplopolyploidy and the Production of Fertile Intergeneric Hybrids of *Saccharum*.' Her focus was the intergeneric hybrids (*Saccharum-Erianthus*, *Saccharum-Imperata*, *Saccharum-Zea* and *Saccharum-Narenga*) that she had produced at the Sugarcane Breeding Station, details of which we already know. While the first two involved the wild *S. spontaneum* or its cultivated derivatives as parent, the third and fourth had the noble sugarcane *S. officinarum* as parent. Janaki's aim was to describe what happened when a high polyploid species of *Saccharum* (viz. *S. spontaneum* 'Glagah,' 2n = 112 or its derivatives like POJ 2725, 2n = 106, and *S. officinarum*, an octopolyploid, 2n = 8x = 80), was crossed with the diploid species of *Erianthus* (2n = 20), *Imperata* (2n = 20), *Zea* (2n = 20+2B) and *Narenga* (2n = 30).

She argued that *Saccharum*'s high ploidy level was the reason why all obstacles to hybridisation with other groups of Gramineae had been cleared. This was well demonstrated by the versatility of reproduction of the 'nobilised' hybrids of *S. officinarum* and *S. spontaneum* (they were capable of producing true diploid crosses, true triploid crosses and diploids and triploids which were not crosses at all, from apparent hybridisation with diploid species of other genera). Importantly, her work also showed that the fertility of the progeny depended not so much on the nearness or remoteness of the cross but on their capacity for autosyndesis, by which was meant the pairing of homologous chromosomes from the same parental gamete during meiosis, in polyploids.[11]

Darlington in his introduction to the section devoted to cytology, identified three basic trends in recent research, which pointed the unity of cytology and genetics (cytogenetics), 'enabling them to be used as a joint tool of evolutionary research': first, a growing dependence on direct study of meiosis as a basis of genetic prediction (such as work by Pio Koller on sex chromosomes, and C. L. Huskins and H. B. Newcombe on the relations of chromatids at chiasmata) and as a means of testing the reproductive methods of particular species (Janaki's *Saccharum* research for instance); second, the increased application of experimental methods to cytology (the various kinds of 'upsets' induced in cell division such as by heat treatment or colchicine to give rise to polyploidy), both to solve theoretical problems and to produce results of practical relevance to plant breeding (which was again what Janaki's research was all about); and finally, the increasing linkage of *Drosophila* genetics with the micro-study of salivary chromosomes, which had its widest bearings on the genetic structure of populations described by geneticists like Dobzhansky. Darlington also drew attention to the technical side of cytology, the great progress made in the recent years by extensive microscopic demonstration of chromosomes of *Drosophila*, *Zea*, *Osmunda*, *Trillium*, and species of Orthoptera and Mammalia, including man. The section on cytology, he observed, also revealed a growing interest in demonstrating the 'connection between genes and geography,' an aspect which especially interested Janaki. Further, the extensive manifestation of the phenomena of polyploidy and apomixes in plants (the asexual formation of seeds, avoiding the process of meiosis and fertilisation, useful in maintaining hybrid vigour, lost otherwise in successive generations) demonstrated why it was difficult to accept 'species' as a clear-cut taxonomic unit.[12]

Delegates Make a Hasty Retreat

This was all well and exciting, but Hitler was knocking at the door, or almost. Britain was expected to join the war very soon. For a day and a half after the inauguration, the Congress participants were totally absorbed in the various sessions and post-session activities, which included some merry dancing, but on the evening of the 24th everything changed: 'War, that outmoded utility of irrational immaturity, the antithesis of everything [they] represented, was about to overwhelm [them].'[13] British citizens in Germany were advised to leave the country by the next evening, the same was issued to Germans in Britain. Needless to say, the rapid downturn of events brought immense anxiety to the Congress participants. Now, there was an exodus of delegates leaving for home; however, the difficulty of arranging immediate transportation for the large contingent of nearly two hundred American geneticists, and others from Canada, Australia, New Zealand, South Africa and India, led to the decision to continue sessions, but to shorten the proceedings by a day.

'At the farewell party . . . the rebellious rump was most unwilling to depart, and it was not until the early hours of the 30th that it was forced to accept the view that all good things must come to an end. Glasses were filled and [they] drank to absent friends, those who had shared [their] bread and wine and who were in danger and distress.'[14]

The delegates received news that certain sailings to America had been cancelled and many more were expected to follow suit; a committee was formed at an urgent meeting convened by the official American delegates of the Congress to make arrangements for their travel back home. The Poles were not that fortunate; their return was almost impossible. Some among them became 'refugee' scientists, finding temporary positions in Britain such as at the Kew Gardens or institutions in Edinburgh, and remaining there until the end of the war. Writing about the Congress, Punnett's account concluded: 'The external circumstances which attended the Congress, gave it a special quality. [They] had met as geneticists sharing the same interest and enthusiasms. Suddenly [they] were required to behave as nationals with fiercely conflicting views. [They] found this demand difficult, irritating, saddening. [Their] memories of this Congress [would make them] even more fervent servants of peace and of scientific humanism.'[15]

Refuge at the Institute of Animal Genetics

At Edam, Devi Amma was on her deathbed; the aging matriarch had been heartbroken to see Janaki leave and knew for certain she would never see her daughter again. On the other side of the world, at Edinburgh, Janaki was on the verge of an emotional breakdown thinking about her mother, and praying there would be no war. She had been provided refuge at the Institute for Animal Genetics by Crew (who had been drafted to command a military hospital at the Edinburgh Castle and also function as Director of Medical Research at the War Office), but she remained anxious. She was also suffering from a painful left elbow, the result of a badly mended fracture in India. Immediately after her fall at the Coimbatore Breeding Station, Janaki had left for Tellicherry, which had several highly accomplished vydians or practitioners of indigenous medicine (her own father, Sub-Judge E. K. Krishnan, we know hailed from one such illustrious family). In a personal note, Janaki's nephew recounted how 'N [Nachee] with tears in her eyes,' had instructed the man 'with an elbow skeleton in her hands'; the vydian in response 'stretched a reluctant joint up and down,' but which worsened the condition.[16] She had moreover burnt her bridges with many at the Coimbatore Station and was also running out of time; determined to attend the Congress, she had stoically borne the pain and boarded the ship to Britain.

Missing her mother badly, and in great physical pain, and distress caused by the uncertain political situation, Janaki's death wish would resurface again, this time in a letter from Edinburgh to brother Raghavan:[17]

My dear Ragton

Here I am in Edinburgh and this is the last day of the Congress. It went off well even though the German delegates had to leave. Everybody is anxiously waiting for Hitler's answer—England is quite prepared for War—I am not a bit perturbed and await for anything that might turn up but I feel there will not be a War—If there is I do not mind being blown up for I have ceased to want to hold my life.

I hope you stayed with mother and cheered her up. I wonder how she is taking my departure—I know she must leave us soon and my one wish is to join her and father as soon as possible—until then I live for science—you can all think of me as already dead.

I am doing very well—Scotland is cold—but beautiful—I am very comfortably put up at the Institute of Animal Genetics. They work on dogs, cats & mice and you all will be interested in the breeding work done here. I shall however be working on my grasses. The Director

[Francis A. Crew] is going to make all arrangements for my elbow operation—everybody is concerned about my arm—I am waiting for the Congress to end to go to hospital.

I hope to be in Edinburgh till the end of the year, and will then work in London. The German professor [Tischler] who wants me in Germany came though he left the very next day. He thinks it will be possible for me to go to Germany in spring—but America also calls me and I think I shall be going that side after I finish my work in London. I go wherever . . . by the Grace of mother in whose hands I am now having left my Earthly mother forever. It is a strange feeling to be homeless in this wide world but it is as it were the home has extended to fill the whole world and I meet with nothing but kindness wherever I go—more kindness than from the people of my own land. My love to you all and God keep you all well & happy,

Affectionately yours
Janaki

Lines from the *Three Guineas* (1938) couldn't ring truer in Janaki's case: 'As a woman I have no country. As a woman I want no country. As a woman my country is the whole world.'[18] Janaki, we know, had begun to look out for a research position either in America or Britain, preferably the former, even before she had left India for Edinburgh; she had been writing to Bartlett, Chase and Darlington about this. It was in response to her request that Bartlett had influenced the university authorities to permit her as a 'holder of the degree of Doctor of Science of the University of Michigan (1931)' to enjoy guest privileges at the Department of Botany for the academic year 1939–40, which included the use of libraries, attending classes and working in laboratories without the payment of fees.[19] Her appointment as 'Honorary Research Fellow at the University of Michigan' was timely, reaching her as soon as the fateful Congress ended.

Although pleased to escape the misery that was the war and the dreaded return to the Coimbatore Station, in early October (she had not yet resigned, but was prepared to), Janaki wrote to Bartlett from the Institute of Animal Genetics, Edinburgh, that all was not well at her end: 'Things have been so unsettled in this part of the world that it has been very difficult to decide when it will be possible for me to leave Britain. I find now that it is impossible to leave in time to be there for the Fall semester. So I hope I can come there in Spring— that is if all goes well here.' She was also planning to write to Carl Rufus, of the Barbour Fund and the Dean of the Graduate School, explaining her situation.[20] *The Michigan Daily* reported that Janaki had postponed her visit to Ann Arbor on account of the war; it reminded readers that she 'is the first Indian woman ever to have held a chair in an Indian University attended by men.'[21]

Moreover, her left arm, Janaki explained to Bartlett, had to be 'straightened out through an operation on her fractured elbow.' A qualified medical

216

doctor himself, Crew had taken particular care to ensure that her elbow was fixed properly; an operation was imperative 'to remove a skeletal growth at the broken joint.' Unable to wait for the hospitals to resume surgeries, Janaki prevailed upon Dr William Alexander Cochrane, the chief bone surgeon in Edinburgh (perhaps through Crew), to operate upon her. This would mean being confined to a 'nursing home' and being 'helpless for some weeks,' she told Bartlett.[22] Meanwhile, she had begun doing some animal cytology with Pio Koller and writing up some of her work on the intergeneric hybrids of *Saccharum*. She was hoping to continue to work on genera related to *Saccharum* with the 'co-operation of Dr Brandes' (E. W. Brandes 1861–1964, Principal Pathologist at the US Bureau of Plant Industry) who she had heard had a 'good collection of wild Saccharum' collected by him from New Guinea and other places. 'If you can help me to get in touch with some of the sugarcane workers in USA, I would be very grateful,' she begged Bartlett,[23] who would soon be deputed to the Office of Rubber Investigations of the Department of Agriculture (1940–44), as principal botanist responsible for exploring rubber sources in Central and South America.

Tormented by her mother's impending death and her own uncertain future, Janaki resolved not to return to India; she was just about to turn forty-two years old. She saw herself as a scientist-refugee like many others at this time of war, her exile however was a self-imposed one, a retreat from an imperial breeding station where patriarchy was rife and which she believed was run on unscientific lines, and its location in Coimbatore, too 'Brahmanical' for her taste (in contrast to cosmopolitan Madras). Janaki felt India had nothing to offer her—an independent woman and a border-crossing plant cytogeneticist—by way of facilities, a stimulating atmosphere or a research career. Her exile to the West would last nine long years, during which time she suffered the ravages of war and extreme loneliness, but it was also the period that saw her evolve into a world-ranking cytogeneticist and a veritable citizen of the world.

At the John Innes

Janaki continued to stay in touch with the American grass-expert Mary Agnes Chase during the war. Darlington had arranged for her to work temporarily at the John Innes, and she was writing up her study of the intergeneric *Saccharum* hybrids, presented at the Congress. The war had altered her plans, she wrote to Chase, of going to the continent and working on the cytotaxonomy of grasses with Georg F. L. Tischler at the Botanical Institute, University of Kiel, a Baltic port city in Germany; only a few years before, she had received from Paul J. Brühl, the German botanist attached to the Calcutta University, Tischler's *Handbuch der Pflanzen* (1922) as a present.[24] She was also unable to accept the offer from the University of Michigan, because it did not provide her with a stipend or salary. She wrote once again to Chase appealing to her to find her a research position in America, revealing her utter desperation:

> I shall be able to stand on my own legs for some time only and wish I could enlist the interest of some Foundation or Department for my work. I am prepared to come to U.S.A. to continue my work. Prof. Bartlett has already sent me the papers admitting me as Hon[orary] Research Fellow at the University of Michigan. Nothing would please me than to carry out my work at the University of Michigan and come to you for my systematic work. Do you think the United States Dept. of Agri[culture] esp[ecially] the Division of Plant Exploration and Intr[Introduction] would be interested to employ me on work in grasses, relative to Sugarcane. I have as you know—my DSc from Michigan and am a Sigma Xi member—as well as a member of the American Botanical Society. They should not consider me altogether as an outsider![25]

'Edinburgh is a safer place than Kiel these days,' Chase would reply. Unfortunately for Janaki, the rate of unemployment was soaring in America, and with Chase no longer attached to the Department of Agriculture, her prospects on the other side of the Atlantic were dim. Chase however promised to forward Janaki's letter to her successor, Jason Richard Swallen (1903–1991), to see if he could do anything for her.[26]

Janaki was weighing all her options very carefully; from a letter to Bartlett written in February 1940, we gather that she was still 'stranded in England with no hopes' of going to America. 'Though I have some money, I am afraid it would not be enough to carry me very far in America, so if I cannot replenish my purse there I do not think it will be advisable to come. I wrote to the Bureau of Plant Industry, Washington [Chase] but have heard nothing [from Mr Swallen]. I shall try the Carnegie people once more and make my final decision after I hear from them. I wonder if you wrote to Dr Brandes about me. Apart from my great longing to see you all at Michigan, I do want to get on with my work on *Saccharum* and related grasses,' she explained. She was at the John Innes, she told him, doing some work on the Indian Maydeae, chiefly *Coix*, of which she had collected a tetraploid form 2n = 40 from India. She was also 'putting together data on [her] intergeneric Saccharum hybrids for publication. I am hoping to be here until I am able to decide about my next move—back to India or to USA. I wish it could be the latter,' she wrote.[27]

Months passed before Bartlett forwarded Janaki's letter to Brandes, accompanied by lines strongly recommending her employment at the Canal Point Sugarcane Field Station, Florida: 'We have the highest regard for Dr Janaki and her work. I wish she could first be here for a time to become familiar with Dr J. T. Baldwin's cytological methods using leaf tissue, and

I have invited her to be our guest while here in Ann Arbor if she can get to this country. She holds our current appointment as Hon. Barbour Fellow, to serve as credential for getting a visitor's visa, but we have no stipend for her. She might be able to do something well worthwhile for you, perhaps not only at Canal Point Station but at Summit. She merely needs to break even financially and her application for employment on a temporary basis is therefore a very meritorious one.'[28]

Anxious, Janaki decided to next try her luck at Kew. In fact, it had been her plan all along to study certain grasses in the Kew Gardens collection, once the Congress ended, in continuation of her work at Coimbatore. She had also brought with her for this purpose seeds, chiefly of the 'Indian Maydeae—Coix ... to grow in Kew ... There is a diploid and a tetraploid *Coix lachryma-jobi* L. [commonly named Job's Tear].'[29] Moreover, she had cytologically examined a number of plants grown from seeds (received from Kew and originally collected from Sudan) of the drought-resistant fodder grass, *Sorghum purpureo-sericeum* (related names being *Andropogon pappii* and *Sorghum versicolor*), native to Africa and India. She had discovered that the chromosome number in this species varied from 2n = 10 to 2n = 14, indicating the stages in the evolutionary process whereby chromosomes were gradually eliminated.

In March 1940, Kew's Arthur Hill had written to Darlington, who was now Director and Head of the Cytology Department of the John Innes, that he was 'quite prepared to support the proposal that Dr E. K. Janaki Ammal should be given subsistence allowance to work at the Institution for a year,' provided there were sufficient funds for the purpose. When Janaki had met him earlier that year to explore 'possible ways and means' of working in England, Hill had suggested that it would be worth her while contacting Darlington regarding funding opportunities.[30] In any case, by the time Janaki had sent the final draft of her paper on the intergeneric hybrids of *Saccharum* to the *Journal of Genetics* in December,[31] we gather that she had already spent several months at the John Innes (perhaps from April 1940), as a Visiting Research Fellow (at £120 per annum, it was to last until April 1941) alongside Horace Newton Barber and Harold Garnet Callan; La Cour and J. Rutland were the technical assistants at the department and Kenneth Mather was the head of the Genetics Department. Incidentally, Janaki had smuggled into the country a palm squirrel, later named 'Kapok'; a John Innes scientist Nobby Clarke kept it for several years, while another, after it was dead, operated upon it for chromosome counts.[32]

Discovering Supernumerary Chromosomes

At the John Innes, Janaki was studying the wild species of *Sorghum*, *Sorghum purpureo-sericeum*, seeds of which she had obtained from Kew's C. E. Hubbard; she found none to six 'extra' chromosomes. The genus *Sorghum* included species with haploid chromosomes of 5, 10 and 20. In 1936, the genus was divided on a morphological basis into two subsections, Para-sorghum and Eu-Sorghum; all the species in the former group exhibited a haploid chromosome number of 5, and in the latter of 10, except for S. hala-pense (n = 20). In fact, just before leaving for Edinburgh, Janaki had sent a note to the *Current Science* on the supernumerary chromosomes observed by her in Para-Sorghum.[33] That these chromosomes were entirely lost in the development of tissues other than the germ line (for instance, they were never found in the roots) was truly remarkable.[34] The phenomenon was well-known in animals, but this was the first time ever that a regular differentiation of chromosome content had been observed in a plant. Such 'plus' plants, with extra chromosomes, tended to undergo extra divisions of the vegetative nucleus in the pollen, sometimes leading to malignancy and the eventual death of the pollen grain, as P. T. Thomas of the Pomology Department at the John Innes had discovered.

Another Saccharum Publication

Earlier that year, Janaki's note on the cytology of the *Sclerostachya fusca* [*Saccharum fuscum*] had appeared in *Nature*.[35] She had collected several clones of the grass (used locally for roofing, making the armature for mud houses and for fencing) from Orissa and Assam in early 1937, which grew in association with S. *spontaneum*. Janaki discovered the Orissa clone to contain 48 chromosomes, while the larger Assam forms revealed double the number, 96 chromosomes; the doubling within the species was akin to the condition earlier witnessed by her in S. *spontaneum*, with its west to east geographical transition, from 48 to 112 chromosomes.[36] Most of the Andropogoneae, including S. *officinarum*, had 10 as their basic number, exceptions being *Miscanthus* (with 36 chromosomes) and the dibasic S. *spontaneum* (in which forms with x = 6 and x = 10 had been found). Now, with the discovery of yet another genus, more closely related to *Saccharum* than *Miscanthus*, with the basic number of 6, it had become easier to trace the origin of the dibasic sugarcane.

N. L. Bor of the Indian Forest Service

Janaki shared her findings on the cytology of the *Saccharum* and its closely allied genera with her friend N. L. Bor, based at this time at the Dehra Dun Forest Research Institute, and an expert on grasses himself. She listed them out for him, providing their chromosome numbers, thus: *Saccharum arundi-naceum* (2n = 40), S. *officinarum* (2n = 80), *Sclerostachya fusca* (2n = 48 or 96), *Imperata cylindrica* (2n = 20) and *Narenga porphyrocoma* (2n = 30). Bor found this useful in his own research; he was able to confirm the discovery of three new genera of Indian grasses, named after him: the *Narenga*

Bor, *Eragrostiella* Bor and *Pseudodichanthium* Bor.[37] The Irish botanist and tropical ecologist, Norman Loftus Bor (1893–1972), had a long and successful career in India and Britain. Hubbard of Kew, described him as a 'tall well-built man of powerful physique, with a great strength of character.' He was known for his generosity, friendliness, daring and forthrightness. Good humoured and also a great raconteur, Bor was always most willing to help when it came to solving botanical and even personal problems. In 1931, he had married Eleanor C. Rundall in Assam, and with whom he travelled widely in India and Europe, also visiting America, Malaya and Hong Kong. He had joined the Indian Forest Service in late 1921, and during the 25 years he had until he retired in 1946, Bor held several important posts; his facility with languages proved to be most useful in relating with several indigenous tribes in the North-east of India. He is believed to have nursed a stray baby rhinoceros to health, which he later gifted to the Paris Zoo. After adorning the post of Political Officer of the Balipora Frontier Tract from 1931 to 1934, he was appointed Deputy Commissioner of Forests in the Naga Hills, and later Forest Botanist and Silviculturist at Shillong. His five-year tenure as Forest Botanist at the Forest Research Institute, Dehra Dun between 1937 and 1942, was devoted to compiling lists of grasses and other plants, besides editing the *Indian Forester* for a year.[38] In 1938, Bor had published a paper on the flora of the Nilgiris.[39] Perhaps, it was their mutual love for agrostology, Assam and adventure, that brought Janaki and Bor together.

Janaki's Merton Grass

In the summer that same year (1940), Janaki produced a singular tetraploid cross, between *Euchlaena* and maize. She had begun growing *Euchlaena perennis* (the perennial Teosinte) at the John Innes from seeds sent to her by the United States Department of Agriculture, to attempt crosses between *Euchlaena* and *Coix* (both belonging to the Maydeae). This had however proved unsuccessful. She then crossed *Euchlaena perennis* (2n = 40) with *Zea mays* (2n = 20), using *Euchlaena* as the female parent; she got twenty seeds. These when sown produced twelve plants, of which one was distinct for its shortness and maize-like characteristics, while the remaining eleven were very tall (12–14 ft); these had thirty chromosomes, the sum of the haploid numbers of their parents. The lone plant was found to be a tetraploid (2n = 40), having been fertilised by an unreduced germ cell of the maize. It had viable pollen but had not produced seeds, which Janaki attributed to the climate; she was hopeful a tropical clime would help it seed. All the hybrids displayed great vigour and had inherited the perennial habit of *Euchlaena perennis* and were grown both at the John Innes and Kew. It may be noted that although several hybrids had been produced by crossing *Euchlaena* and *Zea* in the context of maize studies in America,

221

only one was found to be tetraploid (produced by Rollins A. Emerson, one of the men who rediscovered Mendel, and considered a parent of maize genetics), and even this had failed to survive, making Janaki's tetraploid hybrid unique, and therefore precious. She named it *Euchlaezea mertonensis* (Merton grass) in honour of Merton, where the John Innes was located. Her plant would flourish in pots with minimal care until 1946, impervious to the war and all the bombing.[40]

War Work at the John Innes

Meanwhile, the Council of the John Innes approved of Darlington's plans in the event of war-related emergency to move temporarily to a site (the Waterperry House), seven miles east of Oxford, which could also serve as the institution's permanent home.[41] However, the fall in staff numbers owing to war work led to the postponement of the move, which ultimately never took place. The biochemist J. R. Price and cytologists H. N. Barber and H. G. Callan were engaged in temporary technical work for the government, while G. H. Beale, several student gardeners and three of the permanent garden staff enrolled in the army. The garden staff had been reduced to twenty-two from thirty-three before the war. The severely diminished staff was also forced to take up Air Raid Protection and other duties, leaving very little time for scientific work; the time-consuming genetic experiments had now been replaced by research on food crops such as onions, tomatoes, leeks, carrots, beetroots and cabbages.

In collaboration with the Therapeutic Requirements Committee of the Medical Research Council, the John Innes was also involved in cultivating drug plants such as the *Digitalis purpurea* (the common foxglove) for the production of digitalin, besides publishing leaflets and presenting radio broadcasts by M. B. Crane and W. J. C. Lawrence on improved methods of cultivation. Between 1939 and 1942, the Pomology Department had to deal with more than five thousand public enquiries and a year later, an instructional film aimed at fruit farmers and teachers was made to illustrate the John Innes Leaflet no. 4, 'The Fertility Rules in Fruit Planting.'

Blitz and After

On 7 September (1940) precisely, a period of sustained strategic bombing of the United Kingdom by Nazi Germany, referred to as the 'Blitz,'[42] began, lasting until 11 May 1941. Food rationing had been introduced in Britain, and this would continue well after the war had ended. There were heavy aerial raids over sixteen British cities; more than 40,000 civilians killed, almost half of which occurred in the city of London, which had been bombarded for fifty-seven consecutive nights. Only a day before the Blitz began, Janaki wrote to Miss Elva M. Forncrook, a former Director of the Martha

Cook dormitory (where Janaki had resided in 1928–31) that she was in the thick of the battle over Great Britain, having returned to London from Edinburgh in February 1940. She narrated to Forncrook on how she had burnt her bridges with India and was anxious to go to America to continue with her researches on the *Saccharum*:

> I left India in August and told you all about it in my letter. I completed my five years agreement with the Imperial Department of Agriculture, India in May 1938. They requested me to sign a further contract for 3 years with them. I refused because I found the Breeding Station I was at was being run very unscientifically. I decided to accept the invitation of the International Genetics Conference in September last—sold my car, got my passage for the money and came away. I also had my provident fund to fall back upon. That was being reckless was it not? Everyone in India thought so, but the spirit of adventure is not quite dead in me.[43]

She had worked in isolation, she told Forncrook, at the Institute of Animal Genetics at Edinburgh for more than five months, and had her broken elbow 'repaired' during this time. Having joined the John Innes as a Visiting Research Fellow on £120 per year, Janaki continued, 'This is just enough to keep me and I am very happy,' but what she really wanted was 'to come to USA to work on sugarcane at the Canal Point Sugarcane Station in Florida. The University of Michigan has appointed me Honorary Research Fellow but with the war on I shouldn't be able to bring more than £10 out of England so unless I get a Fellowship I can't come.'

The letter also vividly captured for Forncrook the threat against Britain, London in particular, at this point in time, the composure of the British public in the face of adversity, besides her own disregard of danger and the ever-present death wish:

> Meanwhile I live in great danger Air raids day & night—There goes the siren. I must seek shelter. You can't imagine what a time London is having, but we are all cheerful and getting used to bombs of all description. There is no panic—the British are a fine race—and their high sense of duty is wonderful. You must be getting all news from the radio. I am not a bit sorry I am in the thick of all this. Life isn't worth it without a sense of danger. At home they think I am mad, and I get cables asking me to return, but I say very soon the war will spread all over the world and it is best to remain at headquarters. Besides I value my life very lightly—in fact I would be delighted if a bomb could end my life. After a certain stage one does not mind popping off—I don't want to live too long. But I would like to live to see you and all my friends in Michigan.[44]

Janaki reiterated that she was desperate to go to America but was sadly unable to find 'a fairy godmother or father' to help her. She revealed that Brandes was doing his best to get the Rockefellers interested in her sugar-cane work. 'He is himself very interested in it and wants me there,' Janaki added. 'Cannot you find a millionaire in Detroit [where Forncrook was now stationed] to be kind to a mad scientific worker. I would come like a shot if I could be assured money on the other side.' She wanted to return to India, Janaki told her, only 'after visiting all the cane growing stations in USA and Hawaii.' She had written to Bartlett as well, but he was in Panama at this time, she explained. 'I hope USA will help Britain to win the war. It is very severe now!' she would remark. She also sent Forncrook news of her Martha Cook mate, Doris Holdrup, who was in South Africa at this time, and her husband, Ben Cockram, private secretary to the Agent General for the Union of S. Africa. 'I miss her very much in London,' she wrote, speaking of Holdrup. 'How are you. I have not forgotten the days at Martha Cook with you as our Director. I want those days to come back. O for a chat with— smoking a cigar at your fireside,' she remarked missing Ann Arbor terribly, before concluding her long letter.[45]

At the University of Michigan, Levi Barbour had always hoped that Barbour Scholars would work towards peace in their countries and play a major role in preventing international conflicts, but how far this dream was realised remains to be assessed. Janaki, one 'of the most illustrious Barbour Scholars' (as Barbour would remark) would periodically update the Barbour Scholarships Office at Ann Arbor about her whereabouts. In early 1941, she would write: 'I am still alive in London and getting along with my work as well as one could. I have just come from Edinburgh where I went to rest after months of broken sleep. This part of London has had a lot of bombing.'[46] This was also when she received, perhaps, a terribly delayed piece of news, of her mother Devi Amma's sad demise; she had been laid up for at least a year. Like her son, Vasudevan, Devi Amma would be buried in the grounds of Varada Mandiram in Dharmadom, a beautiful house by the sea in Tellicherry.

Just prior to leaving on government service, Bartlett forwarded to Brandes, Janaki's letter to Forncrook (Forncrook having left for Detroit to take up a new position, Janaki's letter was handed over to Bartlett), because it set forth 'the circumstances underlying [her] desire to work in Florida at the Canal Point Sugar Cane Experiment Station more graphically and more at length' than her previous correspondence with him. He would also send Janaki a copy of his letter to Brandes, 'which may not reach him at once' for when he met the latter at Washington a week ago, he was 'contemplating an important business trip,' and also because he was 'one of the busiest men in the Department of Agriculture during this period of high-pressure preparation for whatever may happen.' However, Bartlett was hopeful that Brandes 'may have time for renewed attention to [her] application.' He also very kindly made arrangements with his niece Rachel in whose care he was to leave his house, to welcome Janaki if she arrived, and to allow her to occupy his room as long as she wanted.[47]

Janaki was by now at the end of her tether; she was losing resilience and her sense of humour. The war had exhausted her, and her mother's loss left her deeply distraught; her future seemed bleak. Despondent, she wrote to Bartlett in late June 1941: 'I feel I have had enough of this nerve-wracking life in England—and I want to come over to USA as soon as it is possible to get a passage across the Atlantic. I hope if I do arrive there you will let me stay with your niece—and perhaps weed in the Botanic Garden. I am prepared to do anything. I lost my mother early this year—so the pull towards India is not so intense—besides I must get to the Sugarcanes before I return to my country. I am sorry to say that Dr Brandes has not given any encouraging news about a fellowship so far. Perhaps everybody in USA is busy with work of national importance or is it because I am Asiatic. London is having a lull from bombs but a month ago we had one on the tennis court of our house. No one was hurt—but it was a great shock.' She wondered if Bartlett had returned from his expedition in Central America and on a positive note, let him know that she had 'just finished a list of chromosome numbers of economic plants chiefly tropical—most of [her] counts were from plants grown at Kew.' She added, 'From the point of work—this [John Innes] is an ideal place but all the same I would like to come over!'[48]

Compiling Chromosome Counts

Even though war time and its hostile conditions of living seemingly exhausted Janaki, truth could not be farther from that; in actual fact, they served to heighten her productivity and help her evolve into a cytogeneticist of consequence. It was during the Blitz that Janaki had discovered a new kind of chromosome organisation called the iso-chromosome, in *Nicandra*,[49] the monotypic genus she had examined as a doctoral student. Already during the war, she had successfully bred the tetraploid *Euchlaena-Zea* cross, which she dedicated to Merton, and the colchicine-induced tetraploid *Solanum melongena*. Janaki would go on to produce eight more colchicine-induced polyploids—of the *Nicotiana*, *Cucumis*, *Cucurbita* and *Nicandra*. It was also during these terrible times that she forwarded to the Indian Academy of Sciences a note by her friend, the eugenicist and German refugee geneticist, Ursula Philip of the School of Biometry, University College London (headed by Haldane), on the genetical analysis of three small populations of the *Dermestes vulpinus* (a species of beetle).[50]

Importantly, throughout the year 1940, Janaki had been busy working up a list of chromosome numbers of cultivated species of tropical plants, a hundred of which she had determined herself from the material collected at the Kew Gardens. Isolated attempts at establishing chromosome numbers of tropical plants were being attempted in some parts of the world, but nothing substantial had been accomplished, least of all a painstaking compilation such as the one Janaki had embarked upon during the war.[51] The annual report of the John Innes for the year (1941) would state that Janaki's list would be an 'indispensable guide for tropical plant breeders in the future.' To her great joy, the Imperial Bureau of Plant Breeding and Genetics (IPBG), Cambridge, offered to publish her list of chromosome numbers, which by far was the most complete one of economic plants ever compiled. This work would go into the making of her (with Darlington), *Chromosome Atlas of Cultivated Plants*. In 1942, her list expanded to include not only more economic plants but also ornamental and instructional ones, and those of their wild relatives.

Janaki would often travel to Cambridge, with her updated list; like all places in England, Cambridge was experiencing rounds of sirens, black-outs, air-raid precautions and drastic food shortages. From the School of Agriculture there, amidst the escalating war situation, Janaki sent her New Year wishes for 1943 to brother Raghavan, scribbled on a wartime postcard:

My dear Brother

This is to wish the E. K. Colony in Shoranur a happy and prosperous New Year. Good crops. Good Larder (no rationing) and good humour! Good everything. I am doing quite well in both mind and body with plenty of interesting work, time flies and I try not to think. So long as I know all are well and happy, I am happy too. So here's good luck to all as I sit by the fire on a cold wintry night. There is a thick fall of snow outside and the wind is howling. My New Year resolution is to write oftener to India and I hope it will be reciprocated. I am not dead yet! So here is the best of everything to all!

Love from Janaki[52]

During these times, Janaki was visiting Kew regularly to collect plant material, and where, *en passant*, hardly anything experimental was going on; it would for instance fall on her to introduce Kew to colchicine and its use in producing polyploids. When she visited the Gardens on 22 January 1943, to collect some Russian dandelion rubber (to determine its chromosome count), she ran into the Acting Director, Geoffrey Evans, from whom she requisitioned assorted soybeans, for cytological examination. This proved timely, for when on the following day, Evans received a letter from E. G. Hopper, the American agricultural scientist, asking for chromosome counts of certain soybeans, he promptly dashed off a letter to Darlington, with Janaki in mind for the task.[53] If there was one person during these bleak times, working on the cytology of such diverse genera and species, including temperate plants and tropicals, ornamentals and economic ones, it was Janaki.

Impact of War on the John Innes

Until now, the John Innes had suffered little damage from enemy attacks; only one bomb, which had fallen in the Old Garden in May 1941, had caused any direct damage to the premises. Darlington and La Cour were working towards a manual on cytological techniques, which they dedicated to John Belling 'whose ingenious invention brought the chromosomes within reach of every enquirer.'[54] By the early 1940s, experiments in producing new polyploids, the so-called 'synthetic plants,' had gained momentum, especially after Darlington announced in 1943 'the invention of new methods of making polyploid plants' (referring in particular to the work by Thomas of the Pomology Department) as an important branch of plant breeding work at the John Innes. The newly synthesised polyploids not only helped in producing new hybrids but also served to preserve existing ones by rendering them fertile.

The long-standing tradition at the John Innes of providing genetics training to serious aspirants was resumed at the institution in 1943. As for Janaki, she resigned from the John Innes to devote herself to war duty on the last day of the year, contrary to her earlier plan of resigning in the following April.[55] She deserved a break, having just completed work (with Darlington) on the manuscript of the *Chromosome Atlas of Cultivated Plants*, one that would become a bible for cytogeneticists and horticulturists alike. She soon moved into Darlington's house, helping Margaret take care of the children (by this time they had had a second child, a son named Andrew Jeremy), and preparing thrifty meals with the available rations. She would often be visited by her nephew Hari Krishnan, now an army pilot, who had arrived from Japan for training in Britain. Looking back on that period, he would remark: 'Dr D treated her as an inner member of his family & found much companionship at the intellectual level. Mrs D's . . . slovenly habits [were] aggravated with war time shortages, [and] Nachee found herself in the position of an ayah helping to clear up & bring up the lads, a situation she often recounted looking at its funny sides.'[56] It is believed that Janaki also contributed to war duty, but in what capacity is unknown.

The significant offensive of February 1944 also fortunately caused no damage to the John Innes but the flying bomb offensive from June that year was an altogether different matter. This was a severe and extended one, killing the Assistant Secretary of the John Innes Council. Incidentally, the John Innes visitor's register reveals that the daring and adventurous Eileen Macfarlane visited the institution during this very month.[57] Incidentally, we learn from the John Innes records that at the height of war, Janaki's elder brothers E. K. Raghavan and E. K. Padmanabhan were sending collections to the institution of plants and seeds from South India.

Over the months of July and August, at least eight flying bombs pounded the region, damaging buildings and private houses of the institution, but no fatalities were reported. The final shower of bombs, in August, caused extensive destruction to the glasshouses (ruining experimental plants) and the main water pipes and also the windows, roofs and ceilings of the main buildings, but no books or apparatus were damaged and no lives lost. It had become all the more urgent to relocate the John Innes. Already in 1943, the Charity Commissioners had sanctioned the sale of the land and buildings at Merton and the purchase of a new property. Eventually, in March 1945, Bayfordbury Park in Hertfordshire, overlooking the River Lea, was found most suitable. The Park contained a grand 80-room mansion built in 1759 by William Baker, nestled amidst beautiful cedars in a garden designed by John Claudius Loudon, besides several farm buildings, stables and cottages. The purchase was completed in December that year, but the John Innes would not move out from Merton until late 1949, eventually opening in early 1950.

Figure 11.1 Norman Loftus Bor (1893–1972).

Source: © copyright of the Board of Trustees of the Royal Botanic Gardens, Kew.

Notes

1 E. K. Janaki to E. K. Raghavan, letter dated Edinburgh, 30 August 1939, private collection.

2 Krementsov, *International Science between the World Wars*, pp. 34–36.

3 Ibid., p. 83.

4 Ibid.

5 Ibid., pp. 41–42.

6 Ibid., p. 45.

7 Ibid., p. 47.

8 Quoted in ibid., p. 49.

9 Ibid., p. 51.

10 Incidentally, in 1937, S. P. Ray-Chaudhuri from the University of Calcutta would enrol himself as Muller's doctoral student.

11 Her paper on the subject, 'Triplo-Polyploidy in *Saccharum spontaneum* L.', had been published in February 1939.

12 Darlington, 'Cytology'.

13 Punnett, *Proceedings of the Seventh International Congress of Genetics*, p. 6.

14 For a detailed history of the events leading up to the Congress, besides Krementsov, *International Science between the World Wars*, see Soyfer, 'Tragic History of the VII International Congress of Genetics'. Also, Punnett, *Proceedings of the Seventh International Congress of Genetics*, pp. 1–7.

15 Ibid., p. 7.

16 Hari Krishnan, 'Nachee', private collection, July 1993.

17 E. K. Janaki Ammal to E. K. Raghavan, letter dated Edinburgh, 30 August 1939, private collection.

18 Woolf, *Three Guineas*, p. 197.

19 BHL, UoM: University Herbarium Records (University of Michigan) H. H. Bartlett series, Box 11, Dean, University of Michigan to E. K. Janaki Ammal, letter dated 31 August 1939.

20 Ibid., E. K. Janaki Ammal to H. H. Bartlett, letter dated 5 October 1939.

21 *TMD*, 27 October 1939.

22 BHL, UoM: University Herbarium Records (University of Michigan) H. H. Bartlett series, Box 11, E. K. Janaki Ammal to H. H. Bartlett, letter dated 5 October 1939. It is said that she had a paranormal experience while at the nursing home, a close encounter with the 'ghost of a young girl' (perhaps a 'crippled' child), who had died a few days earlier on the very bed she was occupying. Janaki recalled this incident in the early 1960s, in a conversation with her doctoral student Usha Zutshi in Jammu; personal interview with Zutshi in Jammu in 2010.

23 BHL, UoM: University Herbarium Records (University of Michigan) H. H. Bartlett series, Box 11, Dean, University of Michigan to E. K. Janaki Ammal, letter dated 5 October 1939.

24 Incidentally, the Indian botanist Panchanan Maheshwari had worked with Tischler in 1935–36.

25 Smithsonian Institution Archives, Washington: Record Unit 229, United States National Museum, Division of Grasses, Folder 42, Box 4, E. K. Janaki Ammal to Mary Agnes Chase, letter dated 10 November 1939.

26 Ibid., Mary Agnes Chase to E. K. Janaki Ammal, letter dated 24 November 1939.

27 BHL, UoM: University Herbarium Records (University of Michigan) H. H. Bartlett series, Box 11, E. K. Janaki Ammal to H. H. Bartlett, letter dated 21 February 1940.

28 Ibid., H. H. Bartlett to E. W. Brandes, letter dated 11 October 1940.

29 Job's Tear is a distant relative of the maize in the Maydeae tribe of the family Gramineae.

30 DP: C 109 (J 113), letter dated 28 March 1940.
31 Janaki Ammal, 'Intergeneric hybrids of Saccharum' (published in three parts), 1941.
32 JIHI: Staff File for Dr Janaki Ammal.
33 Janaki Ammal, 'Supernumerary chromosomes in Para-Sorghum', 1939.
34 Ibid., 'Chromosome diminution in a plant', 1940.
35 Janaki Ammal, 'Chromosome numbers in *Sclerostachya fusca*', 1940.
36 See Janaki Ammal, 'Triplo-polyploidy in *Saccharum spontanaeum* L.', 1939.
37 Bor, 'Three new genera of Indian grasses'.
38 Hubbard, 'Norman Loftus Bor (1893–1972)'.
39 Bor, 'The vegetation of the Nilgiris', 1938.
40 RBG Kew, Library & Archives: 1/RHS/10A, Edward Salisbury to Prof. Shephard, Imperial College of Tropical Agriculture, Trinidad, letter dated 24 May 1946, enclosure.
41 The Waterperry House was a stately home and estate located in a loop of the River Thames, seven miles east of Oxford. The house was the property of Magdalen College, Oxford, and used as a Horticultural School for training girls in practical gardening.
42 Short for Blitzkrieg or lightning thunder.
43 BHL, UoM: University Herbarium Records (University of Michigan) H. H. Bartlett series, Box 11, E. K. Janaki Ammal to Miss E. M. Forncrook, letter dated 6 September 1940.
44 Ibid.
45 Ibid.
46 Dashini Jeyathurai, The Barbour Scholarship, 14 February 2012. www.saadigitalarchive.org/tides/article/20120214-619.
47 BHL, UoM: University Herbarium Records (University of Michigan) H. H. Bartlett series, Box 11, H. H. Bartlett to E. K. Janaki Ammal, letter dated 11 October 1940.
48 Ibid., E. K. Janaki Ammal to H. H. Bartlett, letter dated 25 June 1941.
49 Darlington and Janaki Ammal, 'Adaptive Iso-chromosomes in Nicandra'. An iso-chromosome is a mirror-image abnormal chromosome consisting of two copies of either a short arm or a long one, whereby both arms are genetically identical.
50 *Proceedings of the Indian Academy of Sciences, Section B* (Biological Sciences), October 1940, pp. 133–171.
51 For instance, see Krishnaswami and Rangaswami Ayyangar, 'Chromosome Numbers in *Sesbania grandiflora* PERS. The agathi plant'; 'Chromosome numbers in *Cajanus indicus* SPRENC'; also, Jacob, 'Cytological studies in the Genus *Sesbania*'.
52 E. K. Janaki Ammal to E. K. Raghavan, letter dated 1 January 1943, private collection.
53 RBG Kew, Library & Archives: 4/I/1B, Correspondence from John Innes Trust, Acting Director, Kew Gardens to C. D. Darlington, letter dated 23 January 1943.
54 Darlington and La Cour, *The Handling of Chromosomes*.
55 DP: C 109 (J 113), letter dated 24 January 1944.
56 Hari Krishnan, 'Nachee', private collection, July 1993.
57 Visitors from India to the John Innes during the war years included Diwan Bahadur S. E. Ranganadhan, Adviser, and Secretary of State for India; he visited in November 1940; previously to this, he was the Vice-Chancellor of the Madras University (1937–1940).

12

MERTON–KEW

The Chromosome Atlas of Flowering Plants

> First, what are the wild plants that will provide him with the food or fibre or drug he wants? Secondly, what variation of form, habitat and distribution do these plants show in nature? Thirdly, to what extent will related groups breed together?
> —*The Chromosome Atlas of Cultivated Plants* (1945)[1]

'I write this in what I consider the twilight of my stay in England,' announced Janaki to Bartlett in early February 1945, in a letter written from the Jodrell Laboratory of the Kew Gardens; she summed up her life until then for him, as usual laced with a wry sense of humour. She had the ability to see the funny side of things, even in the darkest of times: 'Except for the fireworks over London- first the blitz and then the V1 and V2 [the German retaliatory V-weapons][2] my days have not been "bright" by any means. After mending my broken elbow in Edinburgh in that fateful year the war began I moved on to John Innes—where I was given as assistant-ship in the cytology department under Dr Darlington. I managed to publish some of my work on intergeneric hybrids of sugarcane made in India—but even that was poor because I did not have my slides or all my data. I was glad to find some new data on the cytology of *Nicandra physalodes*—and lastly I was able to complete a book on chromosome numbers in cultivated plants—with Dr Darlington' (it would appear in a month's time). 'While preparing it I was closely connected with Kew gardens where I am doing little work now. Very little experimental work is being done at Kew. I am initiating work with colchicine . . . When I see the vast collection of tropical plants, I feel a great opportunity for working on them has been lying idle,' she remarked. Janaki had been meeting fellow Michigan geneticist Eileen (Macfarlane) 'several times' since she came to Britain; she has been doing 'very useful work with blood in Slough,' Janaki informed Bartlett. In her characteristic style, also reflective of her close relationship with him, she asked: 'Must I turn my steps eastwards without once more seeing you—my master? Answer.'[3]

DOI: 10.4324/9781003267089-12

On 18 May 1945, Britain and the Western allies celebrated victory over Hitler's army. The war was regarded as over in Europe but the nation's food problem remained a major issue. In Britain, Clement Attlee of the Labour had replaced Winston Churchill, as Prime Minister, in America, Harry S. Truman had taken the place of the deceased Franklin D. Roosevelt and in August, Hiroshima and Nagasaki had been ruthlessly bombed; this was the first time ever that the nuclear weapon had been used in mass destruction. Designed by the theoretical physicist Julius Robert Oppenheimer at the Los Alamos National Laboratory, the atomic bomb was the result of the Manhattan Project, a World War II research and development project based at Oak Ridge, Tennessee, led by the United States and supported by Britain and Canada.

For Breeders and Evolutionary Biologists

In a world marred by the atomic bombing of the cities of Hiroshima and Nagasaki, the *Chromosome Atlas* of *Cultivated Plants*, co-authored by Darlington and Janaki, was published, a benign product of the war.[4] By the mid-1940s, it had become universally accepted that chromosomal analysis could unravel the evolutionary stages of the cultivated plant. The range of processes responsible for species formation and plant improvement and the methods most effective in the breeding of an improved or new crop or the acclimatisation of an old one to new climes and its uses had already become widely known. Darlington had played a crucial role in linking the disciplines of cytology, genetics and evolutionary theory at an important time in the history of biology, when Darwinian gradual evolution (evolution as a slow, steady and continuous process; that selection and variation happened gradually rather than in jumps) and Mendelian genetics (variation as discrete or discontinuous, and unaffected by the environment) were finally being reconciled by means of natural selection, referred to as the modern evolutionary synthesis.[5] The synthesis was a coming together of ideas from across several biological disciplines including genetics, systematics, morphology, ecology and palaeontology; it was truly a product of a mixed border culture. The term first appeared in the title of Julian Huxley's 1942 publication, *Evolution: The Modern Synthesis*, dedicated to T. H. Morgan. More than anybody, it was Huxley who had done most to bring about unity in the biological sciences. As far as he was concerned, the future of modern man depended upon the construction of an evolutionary humanism, and his book was a substantial contribution to the growing literature on evolution, which included Dobzhansky's *Genetics and the Origin of Species* (1937) and the so-called Columbia classics: Ernst Mayr's *Systematics and the Origin of Species* (1942), G. G. Simpson's *Tempo and Mode in Evolution* (1944) and lastly, G. L. Stebbins' *Variation and Evolution in Plants* (1950).

232

By the mid-1930s, time was ripe to put forth evolution 'as mechanistic and materialistic a science as possible; the mechanical basis (selection) for evolutionary change was unified with the material basis—namely genes, arranged like a string of beads on the chromosome—thanks to the work of men like Sewell Wright and Dobzhansky, evolution would be defined as the 'change in gene frequencies.' The same genetic mechanisms that brought about microevolutionary change (below the level of species) also caused macroevolutionary change, or the entire gamut that extended 'from the gene to the human and to the human culture.' For eugenicists like R. A. Fisher, selection was a powerful agent of evolution, as it was for Huxley, only that for the latter, it was a more 'progressive' view on evolution. Mayr, a prac-tising ornithologist, suggested that successive populational samples rather than a single type specimen should be the working basis of the taxonomist; his book was a response to Dobzhansky's whose framework did not con-sider the origin of organic discontinuities, a crucial feature of evolution. Simpson's, on the other hand, was a model which employed palaeontologi-cal evidence to demonstrate how evolution was slow and gradual and yet accommodated those evolutionary quantum leaps. Stebbins, influenced by Darlington's notion of genetic systems (or overarching explanatory matri-ces, which argued that apomixis, hybridisation and polyploidy were also subject to the mechanism of selection) and Dobzhansky's framework, aimed to make sense of the copious amount of random data on the behaviour of chromosomes (the chaos) and the atypical reproductive habits of plants being produced over the decades. Huxley had been in dialogue with fellow American evolutionists Dobzhansky and Mayr and in Britain with Haldane, Darlington and C. H. Waddington, besides Fisher and the experimental nat-uralist, E. B. Ford. He wanted his readers to attack squarely with empiricism this 'many-sided topic of evolution.'[6] During the war years, these evolution-ists communicated with each other through a series of mimeographed bul-letins edited by Ernst Mayr and it was through these that a consensus was reached: that there was a common field (involving the disciplines of genet-ics, palaeontology and systematics) that connected them and that it needed to be institutionalised. Representing the heterogeneous practices of biology, the Columbia classics were the answer to that.[7]

As for *The Chromosome Atlas*, it was a project of assimilation, compila-tion and synthesis as well as addition, of new data and ideas, and one of immense implications for evolutionary biology. Highlighting the general principles that could be employed in improving plants, *the Atlas* worked as a reference manual for plant breeders and cytogeneticists equally, with its arrangement of species according to chromosome numbers, that shined new light on systematics, both large and small scale, between families and within species, and on the classification of genera, within and between. What was until then merely cursory and disjointed, had been arranged by the authors in the form of an atlas (systematic and geographical) based on 'the most precise and comprehensive knowledge of the conditions of breeding behaviour in the plants that matter.'[8] Reaffirming the bearing of chromosomal studies on the theory of origin of cultivated plants (that is putting classification back on a genetic basis), Darlington and Janaki's *Chromosome Atlas* was unmistakably inspired by Vavilov's plant geography approach.

Dedication to Vavilov

Undertaken in the difficult war years, it was Janaki's stint at the John Innes (April 1940–December 1943) which had culminated in the *Chromosome Atlas*. 'To the tripod of systematics: form, distribution and breeding, we have merely added a fourth foot [namely, chromosome count) and we must rest with all four feet on the solid earth,' the authors warned. It was a dictionary or an encyclopaedia of all known chromosome numbers of flowering plants, including some 8000 species, an index of their uses in agriculture, horticulture and industry and another of their natural distribution. As joint authors, Darlington and Janaki were indebted to the Department of Economic Botany at Kew for assistance and advice, but dedicated the book to their hero, the geneticist and plant explorer Vavilov, who had died in the Gulag camp of Saratov in January 1943, tragically, and ironically, of dystrophy caused by malnutrition.[9] Janaki had never met Vavilov, unlike Darlington, and the closest she ever came near him, physically speaking, was when he visited London in late June 1931, to participate in the Second International Conference on the History and Philosophy of Science, an event that coincided with her own arrival at the John Innes (from Ann Arbor), for the first ever time.

The *Chromosome Atlas* began with a discussion on Vavilov's 'Centres of Origin of Crop Plants' published in 1926.[10] Their list included: Abyssinia (Ethiopia), Mediterranean, Persia, Afghanistan, Indo-Burma, Siam-Malaya-Java, China, Mexico, Peru, Chile, Brazil-Paraguay and the United States; the book carried a reproduction of Vavilov's 1935 map (the plate at the tail end of the book) delineating the various centres of origin of the world. Genetic plant diversity was the basis of domestication and breeding in crops of economic importance. At the 'centres of origin,' diversity was highest; the American agronomist Jack Harlan would later refer to these as 'centres of diversity.' In his introductory chapter to *Five Continents* (put together in 1939),[11] Vavilov elegantly defined his method, involving extensive plant collecting expeditions, to the study of the flora of a region thus:

> When penetrating into each country, we wanted to achieve as much as possible: to understand the 'agricultural soul' of that country and its conditions; to master its specific and varietal composition, and to gain the most use for this information while integrating it and the evolution of worldwide agriculture and plant breeding into a single unit. The geographical literature is extensive,

but everybody observes different things depending on which filter the facts are strained through or how the investigator approaches them . . . the author has tried to join subjects otherwise difficult to unite, such as geography, botany, agronomy and the history of civilizations into a complete understanding of the fact that it is necessary to do even better than already done.[12]

The basic building blocks of diversity and heredity, genes became global objects of knowledge (epistemic things) and of geopolitics, after World War I and after international relations had been reinstated. A new category, genetic resources, was born, which could be utilised to improve food crops. Vavilov's expeditions were undertaken for the purpose of collecting genetic resources from specific geographic regions. He saw nature as 'a universal store of genes.' In 1926, however, he did not use the word 'gene' and his method was taxonomical rather than genetical. It would be another year before his theory was modified in genetic terms. A part of the seeds Vavilov brought back from his journeys was stored in safe places to be used in breeding programmes, and the rest conserved for the future of mankind. With his colleagues, he controlled a vast network of breeding stations and collection sites across the world. Vavilov had placed strict orders with his staff that the seeds were not to be tampered with to meet short-term demands but were to be saved for long-term humanitarian use; they followed his instructions so totally that it is said they starved through the siege of Leningrad, but dared not touch the precious germplasm, which was at hand's reach.[13]

Through their work, Vavilov with A. Serebrovsky, the first Russian geneticist to invent fruit flies as model organisms had thus added a third dimension to the gene, that of geography; the other two dimensions being Mendelian 'analytical genetics' (analysis of the inheritance of genes) and the Morgansian 'topographical genetics' (the mapping of genes).[14] Based chiefly on morphological, hybridisation and cytological studies, the Vavilovian method aimed to determine the genotypical composition of a species, the geographical localisation of hereditary forms of a species and centres of their diversity. These investigations were as much theoretical as practical in their import and shed light on the nature and role of a species as a system—he 'considered a species to be a flexible, isolated, complex, morphological system linked to a particular environment and area.' The Vavilovian approach was concerned specifically with the ecogeographical differentiation of a species—divided into such categories as ecotypes, genotypes and concultivars (breeding varieties). Scientists at the Institute of Plant Industry in Russia as a routine studied the anatomical, cytological, palaeobotanical, ontogenetic, biochemical, physiological, geographical and genetic characteristics besides the usual traits for purposes of classification. His theories were 'all interconnected and represented[ed] a complex doctrine about global genetic diversity of cultivated plants.' He was particularly proficient in translating this knowledge into practical or economic use.[15] Janaki would fully assimilate Vavilov's plant geography, or even better, his genegeography approach in her practice, but the results of this would become more clearly manifest only in the latter part of her scientific career.

Investigations on the origin and evolution of cultivated plants sought answers to such questions as: the origin of modern crop plants, where they were domesticated, when, and how these crops developed and travelled since their cultivation began. Crops were after all artefacts, moulded/ crafted/ brought into existence by human hands. Alphonse de Candolle in the late nineteenth century listed sources that could shine light on these questions, graded according to their usefulness, beginning with the archaeological, followed by the botanical, historical and lastly, philological sources. By the mid-twentieth century, however, Candolle's botany had become taxonomy, of a kind that was more than simply morphology and geographical distribution: cytotaxonomy. To their credit, Darwin and de Candolle (Augustin Pyramus de Candolle, 1778–1841, who introduced the term 'taxonomy') had noted conspicuous discontinuities in the intraspecific variation in cultivated plants, but taxonomists of the time classified these variants, as species.[16] With the rediscovery of Mendelian genetics, most were found to be only slightly different from each other, as in a gene or two, but with the development of genecology in the 1920s, more light was shed on the process of speciation, especially the role of reproductive isolation in bringing about variation, and now even fewer of these came to be recognised as species. The taxonomist's job was to tell apart discontinuities which appeared noticeable but were only important gene variants within a cultivated species, from those which, appeared insignificant but actually demarcated frontiers between species. Also, of consequence was the question of speciation and domestication in each crop—whether speciation occurred before domestication or the other way around, in which case, domestication was a one-time event.[17] These problems—of taxonomy, and of the origin and evolution of cultivated plants—were what the *Chromosome Atlas* helped cytosystematists resolve.

Janaki's and Darlington's reasons for focusing on cultivated plants were two: the first and obvious reason being that human beings depended on them for survival, and the second because they demonstrated most clearly the principles of heredity and variation controlling evolution in the botanical and zoological worlds (which Darwin and Mendel had boldly outlined). In the preface to the *Chromosome Atlas*, written at a time when 'flying bombs were very abundant,' the authors outlined their chief concerns:

> Now the foundation of botanical study is classification. We want to know what group of common descent our plant belongs to, and how and where that group exists in nature, before we go any further with it. The answers are at once important in dealing with economic

plants, for they tell us, or should tell us, what the range of variation and distribution in that group is and hence what opportunities there are likely to be for selection, for hybridisation, and for acclimatisation. The cultivator of plants has thus three demands to make of the systematic botanist. *First, what are the wild plants that will provide him with the food or fibre or drug he wants? Secondly, what variation of form, habitat and distribution do these plants show in nature? Thirdly, to what extent will related groups breed together?*[18]

These three questions, inspired by Vavilov, would form the cornerstone of Janaki's researches; she had already accomplished something along these lines for the sugarcane, and certain other grasses. While she acknowledged the significant contribution of the systematist in giving generic, specific and varietal names to plants, of given morphological types and of specific geographical and ecological distributions, she warned that plant variation laid 'many traps for the morphologist,' who unaided by analysis and experiment would flounder in the dark, when it came to the issue of speciation. In other words, she was pointing to the importance of locating oneself in the border zone of mixed laboratory and field cultures to gain a better understanding of what a species was.

Hereditary transformations were continually forming new breeding groups, which could go unnoticed by those focusing merely on variation in shape and size. Critiquing the splitters (those who split a taxon into multiple, often new taxa without taking into consideration their chromosome numbers), the authors stated: 'Morphology, particularly of the dead plant, thus often misses the functional meaning it was formerly supposed to possess. The decay of classical systematics is due to this loss in functional meaning of the over-named, and often over-divided, museum species.'[19] By putting classification back on a genetic basis, they argued (harking back to John Ray's definition of species that 'no matter what variations occur in the individuals or the species, if they spring from the seed of one and the same plant, they are accidental variations and not such as to distinguish a species . . . one species never springs from the seed of another nor vice versa,' *Historia plantarum generalis*, 1686, Chap. XXI) that one could 'restore to the species of systematics its former vitality and usefulness.'[20] A study of chromosomes, their number at mitosis and their behaviour at meiosis, and in the pollen grain could help make sense of the puzzling changes in the genetic make-up of a plant species. Thus, the first step in this direction was a chromosomal survey of cultivated plants; the *Atlas* also functioned as a chromosome log table of sorts.

Ascertaining the chromosome number alone was far from sufficient; however, it was also important to record the occurrence of genetically inert supernumeraries,[21] the forms of polyploidy within the species, including the type of polyploidy exhibited (even, odd, autopolyploid or allopolyploid/

amphiploid, associated with interspecific hybridisation), its foremost relationship with vigour, fitness and fertility and peripheral relationship with
colonisation (polyploidy is associated with evolutionary success manifested
in an ability to colonise new habitats), the degree of vegetative propagation
and other kinds of apomixes.

Although notably influenced by Vavilov's 'centres of origin' approach,
Janaki and Darlington took objection to its somewhat static and simplistic deductive rationale. The black mulberry (believed to have originated
in Persia), for instance, challenged the Vavilovian hypothesis by failing to
demonstrate variation. Another case in point was the Date Palm, with its
putative origin in the Persian Gulf, but not assigned to any of the Vavilovian
'centres,' even while the Coconut Palm was allotted to the Malayan region,
when it should by virtue of being sea-borne, belong to a 'centre' comprising the East Indian Seas.[22] On the basis of their (in particular of Janaki)
historical study of the diversifications in modern crop plants, the authors
observed that the 'centres' of many had shifted: 'There has not been one
region of hybridisation and selection but a series, sometimes simultaneous,
sometimes successive.' The connection between man and plants was indeed
very strong; all ancient empires were closely linked to their own staple crops
such as maize, rice, the soyabean and the palm nuts. The Bronze Age had
brought about cultural divisions of mankind centred around silk, cotton and
linen. This gave rise to the Persian and Roman Empires, which were able
to acclimatise, hybridise and establish temperate cultivated plants. It was in
the substitution of one by another that one located the important shifts in
diversity. The change in the 'centres' was linked to the 'shift in the centre
of greatest cultivation—as a rule followed by further hybridisation, further
variation and further improvement by selection.'[23] The authors introduced
the term 'centres of development' to inject dynamics and complexity to the
Vavilovian framework.

In the Mediterranean, cultivated plants tended to have larger flowers and
seeds compared to the Persian region. The authors suggested that this may
have happened thanks to the 'convergence, hybridisation and selection' in
early Egypt of the stocks of Persian and Abyssinian origin; they argued that
the secondary Mediterranean improvements were the reason, Dravidian
languages of Southern India used the term *sheema* meaning Western, to
refer to the larger-fruited crop plants—this input might be ascribed chiefly
to Janaki.[24] The authors also alluded to the 'world-travellers' among plants
such as the 'many-named' peanut, which went from Brazil to Peru long
before the Spanish conquest of the sixteenth century, and then travelled
to Africa and India thanks to the Portuguese and to the Philippines, with
the help of the Spaniards. In every one of these locations, it 'readapted and
specialised' only to return again with the slaves, to Tropical America and
the United States where its new strains or variants were recombined and
progeny reselected. Migrational modes of development were historically

undisputed as was their genetic significance. The authors explained that 'Movement [had] a snowball effect [intensifying or compounding effect] on variation, the snowball being the variants that are collected by hybridisation and selection on the track of the migrant . . . in all these enquiries our species names are true as representing not the fixed types known to systematics but rather the changing conglomerations of systems of diversity known to genetics.'[25]

Vavilov's demonstration of the principle of natural and unconscious selection in cultivation—such as tillage conditions (resulting in the selection of larger forms, such as polyploids promising a high nutritive value), sowing conditions (leading to the selection of forms showing uniform and speedy germination and those in which special agencies of natural seed dispersion and fertilisation are absent), and harvesting conditions (leading to a selection of those with non-dehiscent fruits, for instance, in cereals such as wheat, usually involving a toughening of the rachis and a loosening of the grain from the glumes or hull, so as to be easily released during threshing)— was as significant to Darlington and Janaki as was his theory of the centres of origin of cultivated plants. According to Vavilov, these processes were dominant during the 'neolithic, mountain stage of early cultivation.'[26] The significance of the natural processes of selection was also reflected in how crops were replaced by their own weeds as cultivation spread into regions less favourable to the primary crop. The evolutionary biologist, Edgar Anderson, would state so perceptively that the study of weeds was the study of man.[27] Tomatoes, potatoes and hemp for example naturally followed settled human habitation. Further, just as secondary crops followed primary crops, so also secondary uses followed primary uses. Hemp, for instance, was firstly grown for its seeds, secondly for its fibre and only lastly for oil and as a drug, an evolution aided by migration northwards. A further strand in Vavilov's investigation suggested, more in agreement with Darwin than de Candolle that the cultivated forms of a species or complete domestication sometimes rendered their wild ancestors extinct. The origin of variation in cultivated plants is also influenced by a fourth aspect (tillage, sowing and harvesting conditions being the first three), namely, the conditions of fertilisation, referred to by Mendel; changes in habitat could change breeding habits, such as the substitution of outbreeding (crossing between different breeds without common ancestors, thus increasing diversity) by inbreeding (breeding between closely related individuals, often resulting in decreased vigour) or vice versa.

On the Origin of Cultivated Plants

The authors of the *Atlas* stated that the factors that have directed the development of cultivated plants were covert conditions, rather than the overt and inbuilt consequences of cultivation. The covert conditions included the method of reproduction of crop plants (whether self or cross pollinated, vegetatively or by subsexual means) and the extent to which these techniques have been adapted by acclimatisation. By the eighteenth century a new technique had come into being, which had the potential to alter the still primitive world of cultivated plants, namely, the invention of plant breeding, which involved selection. The firm of Vilmorin, founded in 1727, was a pioneer in this department of replacing accident by intention in crop plants; it replaced the mixed, random/unselected heterogeneous races by homogeneous standardised, selected and inbred varieties (closely related to each other) through ambitious selection and hybridisation experiments and put them on the market. The phenomenon called polyploidy (multiplication of chromosome sets) was chiefly responsible for the rising number of species being brought forth by firms like Vilmorin through vegetative propagation or apomixis.

A century later, almost all the temperate crop plants had been altered using these revolutionary techniques. However, while the heterogeneous population had the ability to resist disease to some extent, homogeneous selected varieties succumbed to it, almost entirely, because the 'rigour of one selection precludes the possibility of another . . . Its stability if not its uniformity, condemns it.' Even the most perfectly bred variety, when subjected to an attack by a highly adaptable and quickly evolving insect, fungus or pest, had no escape because it was unequipped to resist such an onslaught. Selection not only led to this dangerous consequence, but it altered the very breeding system, as a result of which the world ended up with closely related and inter-incompatible varieties, leading to sterility. Sterility could be remedied by new and genetically appropriate varieties, which only the professional plant breeder could make available. Isolated acts of selection were thus replaced by a sustainable process of plant breeding, using the tools of cytology and genetics. By this time, even for those within the same geographical region, development through breeding was at varying stages; while tropical crops such as rubber, palm and cocoa were still in the primitive stages, with their potential as far as selection, hybridisation and standardisation were concerned, largely unexploited, crops such as tea, coffee and cotton were already in the advanced stages of development. Breeding of forest trees in countries such as Sweden and the United States was progressing rapidly and had arrived at the mid-stage of development through positive selection (better quality of timber), unlike Britain which was a case for demonstrating negative selection—the quantity of seed being the object of selection rather than the quantity or quality of timber.[28]

Cytology had reached a point, where it could help unravel the steps, namely the release of variability and selection, by which evolution occurred in cultivated plants. The authors identified four chief routes or evolutionary pathways: by selection among existing genetic differences or mutations, within a single species; by polyploidy within such a group, leading to reduced fertility (auto-triploidy, auto-tetraploidy and successive); by selection among segregates after crossing between two such groups (with both parents, diploid, or both polyploid); and finally, by polyploidy following such crossing (two-sided doubling and one-sided doubling). In the ideal scenario, every chromosome paired with its identical one, and the new form would invariably be fertile and true-breeding, but it was never so perfect in reality; some unexpected variability (but unexpected only if the chromosome behaviour had not been studied at meiosis) was always thrown up by new polyploids and polyploid crosses.[29] The study of the origin of cultivated plants was thus 'an adventure in apparent chaos.'[30]

What's in a Name?

The authors acknowledged that various processes rather than a single one may have been responsible for the origin of the wild forms of sugarcane (which Janaki had been working on for several years now) for instance, which made description in terms of formal species names deceptive and confusing. A complex of wild forms of *Saccharum* with chromosome numbers ranging between 48 and 112 and distributed between Turkestan and Polynesia were classified under the names of *spontaneum* and *robustum*; these forms were selected for their size and sweetness, and higher chromosome numbers, and which gave rise to the cultivated species named *barberi* in India, *sinensis* in China and *officinarum* in Java and the Pacific. These cultivated forms were then crossed with each other and with the wild forms to produce canes with even higher chromosome numbers, such as the 'prodigies of Passeroëan [Pasuruan, the East Java province of Indonesia, such as the POJ 2725] and Coimbatore [like the Co 205].' All of which alerted one to the genetic diversity hidden behind a species name: 'We have to know what kind of mating method, chromosome organization and internal discontinuity may be concealed by that name,' the authors observed, referring to the evolutionary pathways that were traversed before a species came into being; single or multiple, sudden, gradual or successive, with polyploidy and hybridisation, which a cytological study reveals. Within such an evolutionary framework, the stability or invariability of *Morus nigra* (when compared to *alba*) can be explained with reference to its high allopolyploidy (rather than its newness in cultivation), and the high variability of say *Prunus laurocerasus*, to its high autopolyploidy. However, even without hybridity and polyploidy, one species could be variable and another invariable, in the sweet pea for example, which after two centuries of remaining stable had suddenly begun to vary. This the authors attributed to the extra fragments (the supernumeraries), largely heterochromatic,[31] called the 'B chromosomes,' which appeared in the cell nucleus of some plants (such as maize, rye, sorghum and ornamentals like the *Fritillaria* and *Tradescantia*); Janaki had discovered extra chromosomes in Para-Sorghum in the late 1930s, and in 1942, Darlington had confirmed that these played an important role in influencing the vigour of growth and the stability of mitosis. While the heterochromatin on the major chromosomes in the primitive stocks of maize was found to be shrinking as one moved from the centre of dispersal in Mexico towards north and east, the frequency of the B chromosomes was found to be rising; in other words, the two variables (heterochromatin and the B chromosomes) were found to be inversely proportional to each other. This was a cytological discovery, of immense implication for plant breeding, and supportive of Vavilov's theory of the centres of origin of crop plants.

On How Genetic Systems Evolve

The authors in conclusion stated that the *Chromosome Atlas* besides discussing the significance of chromosomes for taxonomy (which cytosystematists had been shining light on for over three decades) offered a bird's eye-view on how genetic systems themselves evolved in flowering plants but warned that it was just that: a bird's eye-view, a top-down interpretation, 'often indistinct and remote, yet already in some sense [showing] the plan and proportions of nature.' In their list of chromosome counts, they included as far as was possible both the wild relatives and the cultivated species themselves: 'We need to know both if we are to cross the different stocks or species,' they explained. Species with the same number crossed happily, but now it was possible to predict whether the cross would be a triploid and therefore sterile, or tetraploid and fertile, or a diploid with the potential to become a tetraploid. They were hopeful that the future would unlock more secrets on cellular processes, thanks to the newer techniques available for the manipulation of chromosomes such as colchicine, X-rays and heat and cold shocks, which when subjected to genetic analysis and Vavilov's methodology for the study of the origin and development of cultivated plants would shine much light, on evolutionary history.

The Compilation

The authors acknowledged with gratitude the exceptionally useful published lists of chromosome numbers generated by Tischler (with whom Janaki had planned to work post-war) up to 1937, besides seventy-five other unpublished counts by their John Innes colleagues, H. N. Barber, P. C. Koller, L. F. La Cour, W. J. C. Lawrence, S. H. Revell and P. T. Thomas. Besides, the volume included 150 unpublished chromosomal counts reckoned by Janaki, initialled 'E. K. J' next to them in the relevant column, chiefly of grasses and the neglected economic plants of the tropics. Janaki would refer to the work of several of her Indian contemporaries—needless to say, it was an all-male world of botanists and agriculturists in India at this time.

242

Reception of the Atlas

Congratulatory letters on the *Atlas* would pour in, as soon as the *Atlas* was published. Ronald Edgar Cooper, Curator of the Royal Botanic Garden, Edinburgh, who had received a copy was very impressed and wrote to Janaki: 'I have seen the *Atlas*—congratulate its authors upon a very fine piece of work. I've already met one man—a Mr Taylor . . . keen of lilies and roses—who is already tremendously enthusiastic about it; so I let you now for your comfort. I will go through the *Atlas* again, quietly and more thoroughly & let you know any ideas that come to me in doing so . . . please accept my personal congratulations upon the *Atlas*, It may seem queer but the Professor [William Wright Smith, Regius Professor of Botany, Edinburgh and Keeper of the Botanic Garden, who was also Cooper's uncle] never mentioned it to me but then I suppose he is too old to learn new ways . . . I have often looked for some . . . reliable way of determining plant relationships and know [the *Atlas* is] the first step in doing so.'[32] A naturalist-explorer by inclination, Janaki had a special place in her heart for veteran plant hunters like Cooper. Incidentally, when Cooper joined the RBGE in 1910, he was sent to Sikkim, Bhutan and the Western Himalaya on a plant hunting expedition funded by the cotton trader A. K. Bulley. Botanising in these parts with Smith until 1917, Cooper had discovered many species and plants, which were later introduced to Britain, the published list of which was of much interest to Janaki.[33]

The German-British geneticist, Hans Grüneberg, who had just been appointed Reader in Genetics at the UCL (in 1933–38, he worked as a Research Assistant in Haldane's biometric laboratory at the UCL in an honorary capacity, after fleeing Nazi Germany much like the geneticist Ursula Philip), reviewed the *Chromosome Atlas* for the *Eugenics Review* (his choice of journal is interesting in itself). Grüneberg stated that classical taxonomy of plants based on purely morphological characters and desiccated museum specimens had long 'suffered an eclipse,' as it had neglected to take into account the chromosomal basis of speciation, and that diversity occurred through processes, which do not necessarily find their expression in those structural changes on which the classical systematist relies. For a genuine understanding of the evolutionary relationships of plants, both in the past and the latent genetic qualities to be explored in the future, depended on a thorough and detailed knowledge of the evolutionary pattern of each individual group. Such a pattern is revealed by the study of chromosome numbers in species, genera and the higher taxonomical groups, and 'forms one of the most important building stones,' which the *Chromosome Atlas* helped trace. Although the book did not provide a critical review of literature, it listed about seventy-five unpublished counts, mainly by Darlington's colleagues at the John Innes, and double that number, about one hundred and fifty unpublished counts, 'by the junior author,' noted Grüneberg, referring to Janaki.[34]

Notes

1 Darlington and Janaki Ammal, *Chromosome Atlas of Cultivated Plants*, p. 7.
2 Short for *Vergeltungswaffen*, which killed almost 18,000 people in the cities of London, Antwerp and Liège, the chief targets. The V1 was a pulse-jet powered cruise missile and the V2, a liquid-filled ballistic missile.
3 BHL, UoM: University Herbarium Records (University of Michigan) H. H. Bartlett series, Box 11, E. K. Janaki Ammal to H. H. Bartlett, letter dated 9 February 1945.
4 Incidentally, it was also in 1945 that the Tata Institute of Fundamental Research (TIFR) came into existence with physicist Homi J. Bhabha at its helm.
5 Harman, *The Man who Invented the Chromosome*, pp. 166–167.
6 Smocovitis, 'Unifying biology: The evolutionary synthesis and evolutionary biology', pp. 20–40.
7 Ibid., pp. 45–46.
8 Darlington and Janaki Ammal, *Chromosome Atlas of Cultivated Plants*, p. 9.
9 Vavilov's colleague at Leningrad, the brilliant Russian cytologist Karpechenko also met with a similar fate, when he was arrested by the NKVD for belonging to an alleged 'anti-Soviet' group, sentenced to death and executed in July 1941.
10 In 1924, Vavilov proposed three 'centres of origin', which became eight in 1935, and then amended to seven towards the end of his life, with some minor additions.
11 Published on the occasion of his 100th birth anniversary, in November 1987, this volume was inspired by *The World was my Garden* (1938), a book by the intrepid American plant hunter, David Fairchild, describing his worldwide travels in search of plants; parts of Vavilov's manuscript were lost or perhaps destroyed. In their place, the editors inserted abridged versions of Vavilov's published papers.
12 Vavilov, *Five Continents*, p. xliii.
13 Fowler and Mooney, *Shattering*.
14 Bonneuil, 'Seeing Nature as a "universal store of genes": How biological diversity became "genetic resources", 1890–1940'. Also see, Portin and Wilkins, 'The Evolving definition of the term "Gene"'.
15 Pistorius, *Scientist, Plants and Politics*, p. 7.
16 de Candolle, *Origin of Cultivated Plants*; Darwin, The *Variation of Plants and Animals under Domestication*.
17 Pickersgill, 'Taxonomy and the origin and evolution of cultivated plants in the New World'.
18 Darlington and Janaki Ammal, *Chromosome Atlas*, p. 7.
19 Ibid., pp. 7–8.
20 Ibid., p. 8.
21 In the nuclei of some plants and animals, besides the normal A chromosomes there might be present one or more accessory or B chromosomes called supernumeraries, which when present in large numbers could contribute to infertility and loss of vigour of the species.
22 Darlington and Janaki Ammal, *Chromosome Atlas*, p. 15.
23 Ibid., pp. 16–17.
24 Ibid., p. 17.
25 Ibid., p. 18.
26 Ibid., p. 19.
27 Anderson, *Plants, Man and Life*, p. 15.
28 Darlington and Janaki Ammal, *Chromosome Atlas*, pp. 23–25.

29 Ibid., pp. 27–28.
30 Anderson, *Plants, Man and Life*, p. 207.
31 Two parts of the chromosomes may be distinguished: the heterochromatin, which remains concentrated in the resting nucleus, and the euchromatin, which uncoils and engages in protein production. The proportion of euchromatin to heterochromatin, which is visible, varies across species and between individuals, and is subject to selection.
32 DP: C. 42 (E. 140), R. E. Cooper to E. K. Janaki Ammal, letter dated Royal Botanic Garden, Edinburgh, 25 Feb 1946.
33 Between 1921 and 1930, Cooper worked at the Maymyo Botanic Garden (today called the National Kandawgyi Botanic Gardens) in Mandalay, after which he returned to the RBGE to become the assistant curator and later curator. He remained attached to the Garden until his retirement in 1950, by which time Janaki had joined the RHS Wisley Garden as Cytologist.
34 Grüneberg, 'Review of *Chromosome Atlas of Cultivated Plants*', pp. 93–94.

13

WISLEY I
Maker of Tetraploids

I only long for Silence and Solitude and my microscope and to
be left in peace to work.
—E. K. Janaki Ammal (July 1948)[1]

In 1946, Janaki was working mainly with material cultivated at Kew and
investigating the cytology of several ornamental (and also economic plants),
especially Magnolias, bamboos and species of *Philadelphus* and *Morus*.[2]
The Jodrell Laboratory in the Gardens was her second home; she was
only earning a subsistence wage at this time. Some months after the war
ended, Edward Salisbury, the Director of Kew Gardens, wrote to Professor K. Shepherd of the Imperial College of Agriculture, Trinidad on behalf
of Janaki, who was anxious to have her tetraploid *Euchlaezia mertonensis*
tested under tropical conditions; her tetraploid plants were thriving in one
of the glasshouses at Kew. Salisbury hoped Shepherd would agree to test it
out on a small scale at his College and suggested he discuss the issue with
his colleague, the banana cytologist K. S. Dodds, who was familiar with
Janaki's work on the intergeneric fodder grass, having spent some time at
the John Innes and Kew, and had succeeded Ernest E. Cheesman as Professor of Botany at the Imperial College.[3]

At about the same time of year, Janaki visited the Caerhays Castle Gardens in Lanarth, Cornwall (owned by the Williams family, which generously
funded plant hunting expeditions to China, and owned the Burncoose Nurseries),[4] to collect Magnolia material; this was perhaps the first ever time she
was collecting material of an ornamental for cytological study.[5] She would
also begin to collect different species of the *Morus* (mulberry) available in
England. Further, plant identification queries sent by the Colonial Office
to the John Innes would often be re-routed to Janaki, if the plant in question was a tropical one, and in particular grass. In September for instance,
she was asked to report on the Columbus grass received by the Colonial
Office, from Argentina. Examining the plant with C. E. Hubbard of Kew,
Janaki identified it to be *Sorghum* x *almum* Parodi, which she believed

DOI: 10.4324/9781003267089-13

would serve as a valuable fodder grass in those parts of the British Commonwealth that had climatic conditions similar to Argentina, and therefore deserving of trial.[6] She was also often the contact point for botanists and agricultural experts, especially those based in Australia (a region actively involved in the introduction and acclimatisation of plants and animals since the nineteenth century), on the look-out for seeds or specimens of grasses. Janaki never failed to extract specimens in return for the ones despatched. In March 1947, H. N. Barber, Janaki's friend from her John Innes days, and now in Australia, sought her assistance in obtaining *Sorghum* specimens suitable for New South Wales.[7]

Appointment at Wisley

It was perhaps Darlington who first introduced Janaki to the Colonel. Col. Frederick Claude Stern (1884–1967), a merchant banker and former army colonel, a figure of authority in the horticultural circles of England in the twentieth century and later an important node in Janaki's network of friends. Stern was also Treasurer of the Linnean Society (1941–58), Chairman of the RHS Wisley Garden's Advisory Committee and of the John Innes (1947–61). She would have had several opportunities to meet Stern, particularly at the Wisley Garden, which she often visited to collect material for her researches, while at Kew.

At a meeting of the Wisley Advisory Committee convened on 15 July 1946, members discussed the possibility of someone carrying out cytological research in the laboratory, which hitherto had focused only on physiological, mycological and entomological research. They wished in particular to develop 'improved forms of ornamental plants by chromosome doubling with Colchicine and other reagents.' It was on the recommendation of Stern perhaps that the Committee suggested that Director Robert L. Harrow (who had just retired) approach Janaki, working at Kew, to enquire whether she was willing to accept a cytologist's post at the Wisley Laboratory 'to do work of this kind,' at a salary of about £300 per annum.[8] When Harrow spoke to her on the telephone, Janaki responded that she would not be free until the Spring (of 1947), but was pleased to discuss the matter further. However, three months later, when a Committee meeting was held (on 21 October 1946), it confirmed that Janaki had begun duties as cytologist from the 1st of that month (that is from 1 October 1946). She would be the first ever salaried woman scientist in the Garden's history; the plant pathologist, Dorothy Ashworth had worked as assistant mycologist (1935–44) and undertaken advisory and research work at the Garden, but we must deduce that she was not a salaried member of staff. Ashworth incidentally was a graduate of the Royal Holloway College, who went on to receive a doctorate in 1934 from University College, Nottingham, and published chiefly on the cytology and biology of plant rusts.

In June that year, John Scott Lennox Gilmour (1946–51) assumed charge as the new Director of the Garden replacing Harrow. Gilmour, previously Assistant Director at the Royal Botanic Gardens, Kew, had been deputed during the war to the Ministry of Fuel and Power. By 17 September 1946, the former Physiology Department in the Wisley Garden, under the charge of M. A. H. Tincker (who was about to retire), had been adapted for use as a cytological laboratory. Here, close to six years, until 1952, Janaki would busy herself manipulating plant material, using colchicine, for the RHS, besides working, when time permitted, on her own research problems. Among Janaki's other colleagues at Wisley were the entomologist G. Fox Wilson, botanist Norman K. Gould and mycologist D. E. Green, besides I. C. Enoch, the laboratory assistant. Stanley Pittman grew plants for her in the experimental glasshouse and generally assisted her in the work at the laboratory. Pittman remembered her fondly as 'a very colourful figure when one saw her out and about' but 'nevertheless quiet and unassuming in her disposition.' She disliked public attention of any kind, and on one occasion refused an interview by a national newspaper doing a series on career women. When Pittman departed from the Garden, Janaki gifted him a signed copy of *The Chromosome Atlas*. Several decades later, Mrs Molly Gilmour would recall that Janaki was 'a great asset to the staff' and 'an extremely kind and generous person, particularly fond of children [and] a very good and loyal friend.'[9]

Social Life at Wisley

Janaki rented a room in the sixteenth-century Cedar House, located at the corner of Rose Lane and the High Street in Ripley village, close to the Garden. 'The host/hostess treated N with great respect, which came down to me during my prolonged stays there,' recalled nephew Hari Krishnan. 'Often I was treated [by his aunt Janaki] as an immature child in need of protection from the wicked Western world. Sometimes she would be as tender as an equal to me & and would want an ice cream or a visit to a Cinema, which she would insist on paying [for] much to the embarrassment of a young captain,' he would add.[10] Darlington would often drive down from Merton to Wisley, a distance of 34 miles, accompanied sometimes by sons Oliver and Andy, to spend the day with Janaki. In fact, he visited her within a few days of her appointment at Wisley. An entry from his diary reads: 'To Wisley—all day 9 Oct 1946.'[11] Janaki adored the Darlington children (having significantly contributed to their care during the war years) and invariably pampered them with presents. She derived immense satisfaction playing hostess—cooking several Indian dishes for her guests and laying the table herself, which she did in the most elegant fashion. If for some reason she was unable to cook when they visited, she would treat them to a good meal at a local Indian restaurant.

As always, in late December, Janaki posted her Christmas and New Year greetings to dear friend Frieda Blanchard at Ann Arbor. She would recall the Christmas she had spent with the Blanchards, just after Frank Blanchard had returned from hospital (he would pass away in 1937). The Blanchards had a son called Nelson, born after Janaki had left Ann Arbor. 'What changes have taken place in the household. Your son I have not seen & the girls must be young ladies. As you see I am still in England,' she wrote to Frieda. Janaki mentioned that she was in charge of the nascent Cytology Department at the Royal Horticultural Society at Wisley. 'I am well-paid (first time since I came to this country) and the work is very interesting,' she added. She had been experimenting with colchicine, Janaki explained, 'to put out new horti plants,' as a result of which firms such as Suttons (Suttons Seeds, established in 1806) and Carters (Carters Tested Seeds, established in 1930) besides several other nursery-men had become interested in her work. Janaki in fact began to feel somewhat like a 'missionary' despatched 'for the introduction of cytology and genetics in Horticulture.' 'The members of the RHS are, some of them great gardeners,' she remarked to Frieda; she had been visiting some of the stately gardens of England owned by Lords and Ladies. She had also been planning a visit to India, but having just taken up her new position, decided to delay her trip a bit more. Janaki's profound longing was however for Ann Arbor, which she hoped to visit soon.[12]

Colchicining Ornamentals

Janaki began work in Wisley in full earnest in late 1946, and within a year produced tetraploids in a number of genera of garden plants, including the *Magnolia* and *Rhododendron*, using colchicine (perhaps material for this came from the Caerhays Gardens in Cornwall, which she had visited not long ago). However, it was not possible to ascertain the full extent of the value of the tetraploids unless the seeds from the treated plants could be grown under normal conditions. Colchicine could be applied to plants in various ways. Some breeders experimented with twigs immersed in solutions of varying concentrations, and for varying durations, while others applied a warm solution of the drug in agar to the growing buds, but in this case one had to ensure that the concentration was higher than when in an aqueous solution. Some breeders let the drug reach the growing bud through a capillary string, or by being dropped at regular intervals upon it. Some sprayed the colchicine solution on the growing tip of the plant, an act repeated at fixed intervals. There were also those who applied the drug as a lanolin-based paste. Some were convinced that the best results were obtained by using water-soaked seeds, soaked thereafter in an aqueous solution of colchicine. Through trial and error, breeders devised definitive methods by which to produce desired results, and of course this varied with each plant, and each practitioner. Janaki was considered particularly adroit in the use of colchicine, capable of inducing polyploidy even in woody plants such as the rhododendrons. Students, amateur gardeners and nurserymen flocked to her laboratory to learn her 'tricks of the trade'; her tried and trusted methods were even put together as a kind of colchicine vade mecum for use by all those interested in making 'plants to order.'

Colchicine triggered the doubling of chromosomes (polyploidy) by impacting on the spindle mechanism: during cell division, chromosomes appeared longitudinally split into thread-like strands called chromatids, which separating, moved towards the two poles at anaphase (or anaphase II during meiosis). With the injection of colchicine, the separation of the chromatids was arrested and instead of forming two daughter nuclei, a single nucleus with double the number of chromosomes was formed. Poyploidy resulted in morphological, physiological and genetic changes, with the polyploidal forms exhibiting a 'gigas' character (from the Greek meaning giant or huge, and a reference to de Vries' *Oenothera gigas*, the tetraploid mutant of evening primrose). That is, the doubling of chromosomes resulted in larger cells, and therefore a larger plant, with the stem thicker and stouter, leaves larger, broader, thicker and darker green, the hairs on the vegetative parts coarser and thicker, and the floral parts, fruits and seeds bigger and more abundant than in the diploid form. By mid-April 1947, it was decided that whenever favourable results were demonstrated by Janaki, polyploid plants would be exhibited side by side with the original control plants at the Society's shows at Vincent Square in Westminster.[13]

To Stern's Chalk Garden

A few days after an exhibition of her new tetraploids, on 19 April 1947, at London Victoria, Janaki boarded a train for Highdown, Sussex, to visit Col. Stern's Chalk Garden; Darlington had accompanied her. A man of many interests, including big game hunting (African trophies hung in his library) and amateur jockeying, Stern had begun shaping a garden out of a harsh chalk pit as early as 1909, a time when plant hunters, such as Reginald Farrer and Ernest H. Wilson, were being sent out to China and the Himalaya; Stern's 8.5-acre garden contained several of these exotics. He was however best known for his work on the genus *Paeonia*. The publication of his beautifully illustrated *A Study of the Genus Paeonia* (1946) had been delayed six years thanks to the war.

Work at the Cytology Laboratory

By this time, Janaki's colchicine treatment had led to an asparagus tetraploid, besides a suspect tetraploid in *Lilium*. She had also begun studying cytologically the species and cultivars (cultivated varieties) of garden

asparagus, besides conducting a survey on the evolution of sex chromosomes and distribution in the trial plants at Wisley. Further, the cytology
laboratory under her supervision had carried out cross-pollinations between
colchicine-induced tetraploids in flowering plants like *Clarkia* and between
various species of *Magnolia* and *Morus*. On 15 May 1947, she held a demonstration of the cytological work in progress at the laboratory, before fifty
members of the RHS and the visiting Gynaecological Society.[14] It was still
early days for the commercial distribution of new forms evolved by Janaki
at Wisley; a further two-year trial was necessary to identify new and stable
forms before they could be offered to the highest tender from a wholesale
nurseryman or seedsman.[15]

At Wisley, Janaki had been using a microscope borrowed from the John
Innes, but this had to be returned in early December that year; she was
in dire straits as the microscope that the RHS had ordered for her was to
take several more months to arrive. She approached her friend, Haldane, in
the hope that he would lend her one from the Biometry Department of the
University College London.[16] Haldane replied that they had 'only one microscope suitable for serious cytological work' in his department; the 'lesser'
ones had already been lent out.[17] This was disappointing news for her.

Visits to Kew

Janaki frequently visited the Kew Gardens during her tenure at RHS Wisley, not only in connection with her own research but also sometimes to
determine chromosome counts for Kew, making the Jodrell Laboratory as
usual her base.[18] Charles Russell Metcalfe was Keeper of the Laboratory;
the Experimental Officer was Miss C. I. Dickinson, assisted by a Miss E.
M. Slater. Among Janaki's friends at Kew was Gerald Atkinson, the official
artist and photographer, whose studio was a well-lit room with high ceilings, equipped with sky-lights, and situated at the canteen end of the Jodrell
Laboratory. When faced with issues of grass identification, Janaki turned
to the agrostologist Hubbard, who was Assistant Keeper of the Herbarium.
Janaki's friend N. L. Bor, who had retired from the Indian Forest Service
that year (he had served as the Director of Assam Relief Measures reha-
bilitating people in the Naga Hills and Manipur, and later as Conservator
of Forests), had been appointed Assistant Director but was not expected to
join before 5 May 1948.

Visits to the John Innes

While at Wisley, Janaki also visited the John Innes in the capacity of 'collaborator,' a term by which researchers from outside the institution were
referred. Her friends Stern and Eileen Macfarlane would also work as John
Innes collaborators at this time. Darlington paid generous tributes to some

of them, including of course, Janaki: 'I have been fortunate in having many gifted collaborators. Kenneth Mather who did most to join chromosome studies and breeding experiments, Margaret Upcott, Margaret Richardson and Pio Koller who did most to show the chromosomes by their excellent drawings, Leonard La Cour whose technique did most to develop their study; H. N. Barber who made the best of all experiments with chromosomes. Janaki Ammal who gave me her immense knowledge of tropical plants and also of Indian society.'[19] Notably, he referred to Janaki's work at a personal level, rather than as an important contribution to the field of cytotaxonomy. Indian visitors to the John Innes in 1947 (in the year of Indian independence), included cytogeneticist A. R. Gopal-Ayengar of the Tata Memorial Hospital, Bombay (but at this time based at the Barnard Free Skin and Cancer Hospital, Washington University School of Medicine, St Louis), Dontcho Kostoff of the Academy of Sciences, Sofia, Bulgaria (whose research notes/papers appeared frequently in *Current Science* in the 1930s–40s), N. R. Bhat and D. K. Mukherji of the Plant Breeding Institute, Cambridge, Moti Vachhani of the Agricultural Station, Sindh, and V. S. Sakhdeo of the Bombay Veterinary College.

That same year, Darlington with Ronald A. Fisher founded *Heredity*, an international journal of genetics bringing together for the first time a disparate range of scientific practitioners with interests in evolution and systematics, besides physiologists relying on cytology and experimental techniques, medical researchers who used genetics in diagnosis and treatment, plant and animal breeders, and physicists and chemists who worked on the borders of biology. With the intention of broadening the scope of genetics as a subject (or even better, as a border discipline), the journal also addressed social scientists studying the nature-nurture divide. Moreover, *Heredity* provided Darlington with the much-needed platform to publish his own researches having been rejected by journals such as Bateson's *Journal of Genetics* for being 'too cytological,' the *Journal of Experimental Biology* for not being experimental enough and the *Proceedings of the Royal Society* consequent to damning reviews from peers John Farmer and Ruggles Gates. Darlington was also keen on publishing opinions on a range of issues without the fear of being tempered down or censored, which *Heredity* made possible. This journal is also believed to have been a reply to former John Innes colleague Haldane's joining the Communist Party (despite what was happening in Soviet Russia) and taking the *Journal of Genetics* with him to the University College London.

Itching to Move On

By mid-1947, Janaki was already itching for change; the Wisley job had not turned out to be how she had imagined it. She toyed with the idea of a teaching post, as Professor of Botany at the University of Ceylon. A letter of recommendation from Darlington stated that 'her long experience of teaching and research admirably qualify her' but warned the authorities categorically that her 'only drawback from your point of view . . . would be that she would probably not stay for very long. She never stays anywhere very long.' He had judged her perfectly and it appears that they took his opinion seriously for she did not receive an offer. A born nomad, she valued her freedom above all else in the world. She was adventurous and enjoyed going on collecting expeditions to the wild and added more books on plant hunting, exploration and travel to her personal collection.

In London in the late 1940s, she bought O. M. Chapman's *Across Iceland: The Land of Frost and Fire* (1930), besides Russell Lesley's slim *Alpines I have Grown* (1940). She also received as a present F. Kingdon-Ward's *Modern Exploration* (1945), from the wife of her former Michigan teacher, the botanist Carl De la Rue. She often gifted her siblings, nieces and nephews, books on adventures and expeditions. From the Kew Gardens in 1946, she sent her elder brother Padmanabhan the *Voyage of Chelyuskin* (1935), the official account of the Russian exploration mission to the Arctic and the subsequent shipwreck and rescue operation. She would add a few more volumes to her collection of travel and botanical literature in the following year such as *The Carnivorous Plants* (1942) by F. E. Lloyd, with whom she occasionally corresponded. Despite his loss of vision owing to a serious illness, he had responded to her letter of mid-1945 requesting specimens of plants. Everything she needed, he had advised, could be had in the British Isles, and the most extensive collection including those of hybrids was in the possession of Smith and H. H. Dixon of Trinity College, Dublin. Lloyd was convinced that these 'two workers along with what Kew could give' could furnish her with all that she wanted.[20] Another important addition to her book collection from this period was Constance Helmericks' *We Live in Alaska* (1944), an account of an adventurous round trip by canoe undertaken by the author and her husband over a period of five months, taking them from Fairbanks via the Yukon and its tributaries, to the island settlements, the Russian Mission, Kuskokwim and Bethel; they would return to Fairbanks by plane.

Congress at Stockholm, 1947

Between 10 and 17 July 1947, Janaki attended the Sixth International Congress of Experimental Cytology at Stockholm; 400 members from 23 countries gathered at the Congress. During this visit, Janaki had the opportunity to go round the University of Lund and plant breeding stations in Southern Sweden like Svalöf and Alnarp, besides Uppsala outside Stockholm.[21]

The Indian Independence Day

For Janaki in England, 15 August 1947, went past like any other. She was at Kew engrossed in collecting material from the *Morus acidosa*, to be examined under the microscope at the Jodrell.[22]

Missing Ann Arbor

Mid-December, Janaki sent the Blanchard family in Ann Arbor the season's greetings; this time it was a card depicting Westminster Abbey, the Big Ben and St Margaret's. She longed to visit Ann Arbor and hoped to surprise them all by walking into the University Botanic Gardens someday to 'see the old place.' She hoped Frieda would write to her (she hadn't heard from her for quite some time now) enclosing a photograph of the Blanchard girls, Dorothy and Grace, whom she had seen last in the late 1920s. She had just received a parcel of *Darlingtonia californica* plants (California pitcher plant, or cobra plant, named after botanist, William Darlington 1782–1863, of Philadelphia) from the Carl Purdy Gardens (established 1879), Ukiah, California, as a gift from the Botanic Gardens, Michigan, courtesy Bartlett, who was at this time in Manila; she couldn't wait to go there 'just to see him!' There was only one plant of the *Darlingtonia* in the whole of Great Britain, and this was in Edinburgh, she explained to Frieda. Janaki had only been 'able to get a single small scale like leaf from it for chromosome count,' and to her delight discovered it was a tetraploid. Material being so scarce, she was desperate to obtain a collection for the RHS Garden.[23]

Exhibiting at the Chelsea Flower Show

The year 1947, saw the return of the Chelsea Flower Show (begun in 1913 at the Royal Chelsea Hospital) after a gap of seven years (the war was responsible for its closure). This would be Princess Elizabeth's first recorded visit to the Show, in the company of her mother and sister. Although the majority of the exhibitors wanted postponement because of low stocks and staff presence, the 2nd Lord Aberconway (Henry Duncan McLaren, 1879–1953), President of the RHS, who was responsible in appointing Janaki at Wisley, insisted that the show be resumed that year. Aberconway was also an industrialist-politician and a passionate horticulturist, who collected and bred magnolias and rhododendrons among other woody plants at his Bodnant Gardens in Wales. He had sponsored plant collectors like George Forrest, who had brought back exotics from Yunnan in southwest China. With time, Aberconway and Janaki would become good friends, exchanging plants and much horticultural information. She would also occasionally visit Bodnant. At the 1947 Show, members of the Wisley cytology laboratory under Janaki's direction staged exhibits in the Scientific Tent on the Chelsea Hospital grounds.

The Chelsea Flower Show of the following year had Edward Augustus Bowles (1865–1954) replacing Lord Aberconway as President of the RHS; Bowles kept an important garden at Myddleton House in Enfield, Middlesex, where he was born. The Scientific Tent as on the previous occasion included Janaki's contributions.[24] Being the key person responsible for the beautiful new cultivars produced using colchicine, she was introduced to personalities no less than King George VI and his wife Queen Elizabeth, besides several other royal dignitaries, each of who had a favourite ornamental. While the King loved rhododendrons and azaleas, the Queen's favourites were magnolias and camellias. An elated Janaki wrote to brother Raghavan: 'Did I tell you that both the King & Queen congratulated me on the work I produced since coming here [Wisley]. The Queen especially greeted me as an old friend when I was presented to them at Chelsea. So I am on the top of things here. Most of the patrons of Horticulture are Lords & Ladys—so I move only in the highest society! Ha Ha much I care!'[25]

The Hannyngtons and the Seligmans

It was through her connections with the British aristocracy that Janaki was able to meet with the family of J. C. Hannyngton (her Irish maternal grandfather, biologically speaking) living in London. Her first contact with her mother's Irish half-sister, who 'for N was so alike her mother in every way,' turned out to be extremely emotional and strange an experience for her. 'Her eager whispered voice of this encounter still rings in my ears!' recalled Hari Krishnan.[26] It appears however that the Hannyngtons were not too keen to acknowledge their link with the 'EK' family. Another of her close social contacts in Britain, who had visited Edam on several occasions, was a Miss Lowe of the Indian Educational Service, a Quaker. It was also during her tenure at Wisley that Janaki had the opportunity to befriend the activist Hilda Mary Seligman nee McDowell (1882–1964).

While her husband Richard Seligman (1878–1972) was a leading metallurgist and businessman, passionate about gardening, Hilda was a prominent campaigner, along with suffragette Sylvia Pankhurst, against Britain's pre-war appeasement of Mussolini and Hitler. The exiled Ethiopian Emperor Haile Selassie had taken refuge with the Seligmans in Wimbledon in 1936; Hilda was also a sculptor and her bust of Selassie made during the latter's residence at her house was displayed at Cannizaro Park in Wimbledon Common until sadly destroyed by protestors in June 2020. Post-war, Hilda became renowned for her humanitarian work in Ethiopia and was associated with the Royal African Society, London. She also wrote and published three books: *When Peacocks Called* (1940), *Skippo* of *Nonesuch* (1943) and *Asoka, Emperor of India* (1947). *When Peacocks Called* (the foreword to which was written by Rabindranath Tagore), a historical novel on the Maurya kings, was translated into Malayalam and would be published in 1952 at Shoranur as *Mayurakahalakalam* by the writer Malabar K. Sukumaran. Sukumaran, as we know, was married to his cousin Cousalya, Janaki's elder sister; the translation was in all probability, made at the behest of Janaki. Incidentally, Hilda donated a bust of Chandragupta Maurya that she had sculpted to be installed in the Indian Parliament complex.[27]

An Attic of One's Own

By the end of June 1948, the attic of the cytological laboratory in Wisley Garden was converted into a flat for Janaki by the Wisley authorities,[28] which she took pains to make warm and inviting. She enjoyed cooking South Indian dishes for her many guests, which included colleagues, occasional students, assistants and visiting friends. Len la Cour of the John Innes and his wife Anne visited a couple of months before she had moved into her new flat, but they were nevertheless received by her with great warmth and treated to a good meal. Janaki had not yet heard from Frieda Blanchard, but that did not stop her from writing to her again, a few weeks after the Chelsea Flower Show, sitting by the attic window late at night, and looking out into the darkness: 'Please do not forget me. You used to be my great luck with Michigan.' She spoke of her attic flat above the laboratory: 'It is very nice to live in the Gardens though a bit lonely at night. They say the building is haunted but so far I have not seen the ghost (who is supposed to have a long beard!). I am enjoying my work very much as it is mostly practical chiefly making new hybrids & Col. Tetraploids [colchicined tetraploids].' Janaki was worried, she told Frieda, that she would 'cease to be a cytologist & become only a "chromosome counter".' She also spoke of her plans to visit India, after what would be nine long years: 'I am going back this year. I am afraid to see the great changes at home. I am become quite a hermit & a recluse.'[29]

Yet again, Janaki expressed her longing to visit her alma mater, the University of Michigan, and see all her friends. 'I get quite thrilled when any Americans visit Wisley,' she wrote. The American plant physiologist N. W. Stuart had recently visited and was now at the American Embassy, Janaki noted. 'Sometimes I feel more American than English. You know what I mean—in thought.' Her work now revolved around lilies and she wrote to Frieda that she would be grateful for seeds, especially of the Chinese species.[30] Janaki was meant to go to Stockholm to attend the International Genetics Congress to be held in July 1948 (the first, after that fateful Edinburgh Congress of 1939) but having developed an allergy of some sort, cancelled her trip; she did not want 'to risk eating strange foods' and aggravate the condition, she told Frieda. It was at this Congress that the delegates took a decisive step to escalate their fight against Lysenkoism gaining strength in the Soviet Union; Darlington, Haldane and R. A. Fisher were among the 112 delegates from England. At year-end, Janaki would send Frieda her greetings for a merry Christmas and a happy new year; this time the postcard carried a sketch of the building in the Garden that housed her laboratory and her attic flat.

Colchicine Tutorials

Meanwhile, the Wisley Advisory Committee recommended the name of Miss Constance Margaret Eardley of the University of Adelaide, for training under Janaki for a year (1948–49); this was following a request from Eardley's teacher and Janaki's former colleague at the John Innes, the cytogeneticist H. N. Barber (who had recently moved from Adelaide to the University of Tasmania). The Committee believed this opportunity would benefit both Miss Eardley and the Wisley Garden.[31] Eardley joined Wisley on 19 April 1948, and two days later, Ann P. Wylie from the John Innes enrolled at the laboratory as well. One of the first tasks Janaki completed during this time was mapping the chromosome numbers of the *Narcissus*, assisted by Wylie.[32] In late April an exhibit of the polyploidy series in *Narcissus* would be staged at the Daffodil Show held in the Garden.

Janaki's work with colchicine took a toll on Janaki's health, recalled the late Gordon D. Rowley, an expert on succulents and xerophytes, who had joined the John Innes early in 1948. Although he had studied cytological techniques under La Cour at the John Innes, it was recommended that he also train under Janaki, who was not only an expert in the preparation of highly refined slides but also in the use of colchicine in breeding. Rowley spent five weeks training at Wisley's cytology laboratory, alongside his John Innes colleagues Miss R. Hurcombe and Miss Wylie, besides Miss Eardely.[33] There was also John Hamerton, who worked for two summers in Janaki's laboratory, making crosses under her supervision. In a letter to the author, worth citing, Rowley described his time at Wisley, as a student-friend of Janaki:

> I arrived at Wisley on May 24, 1948, and first met the staff with Director John Gilmour. Dr. Janaki was a popular and outstanding figure in traditional Indian dress: her colourful saris and outflowing personality immediately caught the attention. She excelled in cytology, with infinite skill and patience in manipulating material. My ham-fisted attempts at making the simplest squash preparation must have been a sore trial to her, but she persevered with perpetual optimism and good cheer. It had recently been discovered that the autumn crocus, Colchicum, contained a potent alkaloid that could affect cell division in other plants, leading to chromosome doubling

and the production of new polyploids. Janaki was a pioneer in using it, and taught me the technique. Unfortunately, she had already suffered a fate similar to that of Madam Curie in handling radium. Colchicine is highly poisonous, and even traces on the human flesh cause irritation and pain. She had suffered terribly before this was found out, although I heard about this only later. With her assistant Paul Town we quickly became a merry team socially with much laughter and badinage to lighten the serious moments. She loved entertaining, and we were often invited up to her flat at the top of the building to sample delicious Indian dishes she had prepared. But research came first before all human comforts. I well remember one occasion when we climbed the stairs to be greeted by a powerful smell of burning. The saucepan had boiled dry and been forgotten hours ago! However, we had a good laugh and I am sure that I didn't leave still hungry.[34]

Cytotaxonomy of Ornamentals

Janaki was rarely ever carried away by worldly success or fame. 'It is all very interesting,' she would remark referring to her encounters with the polite society of England, 'but I only long for Silence and Solitude and my microscope and to be left in peace to work.'[35] 'She used to work late on her own in the laboratory. Her work was very rewarding & enabled her to move in the upper echelons of scientists, researchers, writers and socialites in the UK,' Hari Krishnan would recall decades later.[36] Without doubt, she derived immense joy from chromosomal work, even when years of peering down the microscope severely strained her eyes, and had even begun to exhaust her physically. The experience of immersing oneself 'in the wondrous crystalline world of the microscope, where silence reigns, circumscribed by its own horizon, a blindingly white arena,' to borrow Nabokov's words, was 'so enticing' that it took away all her weariness.[37]

Garden plants differed from those in the wild, owing to the changes they underwent during cultivation. Darwin had shown that the changes were not the direct result of the better conditions of growth received in fields through ploughing and manuring but due to better and stronger plants emerging in the normal course of variation, which also then benefited from the richer soil provided by the cultivator. Although Darwin continued to be relevant, it was the chromosomal theory of inheritance of the 1920s that provided the knowledge of how and why plants were improved through a man-made selective process. From the number, shape and other morphological characters witnessed during meiosis, one was able to infer not only what was going on in the living plant but also in all its historical avatars, going back even a million years, but of more relevance in the last ten thousand years when plants began to be cultivated. To elucidate this, Janaki began with the example of the strawberry from the times of Shakespeare, her favourite author, in a non-academic lecture illustrated with lantern slides delivered before the members of the Royal Horticultural Society:

> In Shakespeare's play *Richard III* we all remember how the Bishop of Ely sent for strawberries for his guest the Duke of Gloucester. These strawberries, picked in his London garden, were not the kind we give our guests now. They were the small wood strawberry, *Fragaria vesca*, which still grows in grassy banks in the south of England. A little later, say when Sir Isaac Newton was eating straw-berries, they were of the species *Fragaria elatior*, introduced from Europe, with somewhat larger fruits, but before Newton died, two other species of strawberries had been brought into Europe from the New World. . . . Towards the end of the eighteenth century these two species were growing together in Brittany. . . . It was from the seeds of these fruits that our modern garden strawberries were descended.[38]

The modern varieties seen in the garden were thus the results of a man-made process of selection, and produced through the hybridisation of species, which if left simply to nature's course, would not have crossed with each other. More importantly, Janaki's strawberry example shed light on the chromosomal theory of inheritance: chromosomes show us 'the course of the events,' and 'the conditions that made them follow that course.'[39] From an examination of the cells of the different species of strawberries, it was possible to say with certainty how the story of mod-ern strawberries had unfolded. Species with different chromosome num-bers did not cross easily, and if they did, produced sterile hybrids with no fruit. Our modern strawberries therefore could only have descended, Janaki concluded, from the two American species, first brought together in Europe, as they were the only species with the same chromosome num-ber. Further, the size of the fruit was related to the number of chromo-somes. While the wild *F. vesca* was a diploid (2n = 14), *F. elatior* had six times that number (a hexaploid), and the two American species, eight times (octoploids), making them the largest of the fruits. The modern strawberry not unlike the sugarcane was an octopolyploid, as were flow-ering plants like the dahlia.

In the case of ornamentals, the grower selected those plants which were either larger or more striking in appearance. In Nerines, *Canna* and *Phil-adelphus* that Janaki studied, the larger-flowered plants had three times the basic number of chromosomes, while in Daffodils, *Iris* and *Primula malacoides*, they had four times as many.[40] In the case of triploidy, the odd set of chromosomes resulted in irregular pairing, leading to sterility, and therefore it was unsuitable in vigorous grasses such as rye, which depended on seeds for continuity. At the same time, triploidy was a desirable trait in plants which could be propagated through cuttings or were equipped with bulbs or corms because there was no waste in seed production. Several seedless fruits were also triploids. Thus triploidy, which also expressed itself in gigantism, was of utmost significance to growers of both ornamental and food plants.

Janaki's interest in the cytology of the flowering plants was however not limited to producing beautiful new cultivars for the ornamental garden, but to work out the true race history of the family, which she argued was the ultimate aim of taxonomy: The 'plants we see on the earth today and which we collect and classify into families, genera and species to preserve as dry specimens in herbaria or to grow in our gardens, are only part of the total numbers that have existed since plants appeared on the face of this earth.'[41] By mapping the 'true race history' of plants was meant tracing evolution from the most primitive to the most highly developed; that is charting the evolutionary history of a species by not only studying living plants but also their ancient ancestors, and predicting the missing links in the series or discovering the correct phylogenetic relations. Marking a new dimension in her thinking, this approach took her beyond the scope of the *Chromosome Atlas* and to the discipline of geology and palaeobotany and eventually to ethnobotany and ecology. The cytologist's task of mapping the origin and evolution of cultivated plants in space and time through chromosomal study was as thrilling to Janaki as the hunt for the missing link in the history of *Homo sapiens*.[42] She wished to know from what forms cultivated plants originated, where and when they were domesticated, and how these plants changed, travelled or spread across landscapes since they were first cultivated.

The Great Travellers

Some were great travellers like the *Morus nigra* or black mulberry (the example cited in the *Chromosome Atlas*), crossing borders (brought into domestication, that is) several times over, much like Janaki herself. We know that Janaki had been collecting material of the genus *Morus* at Kew and Wisley (from plants and seeds sent out from their native habitats) since 1947. The black mulberry was unique among flowering plants in having the highest chromosome number (2n = 308) for any known species, and posed an interesting problem to the cytogeneticist, as to the origin of the species. The chromosome numbers in different species of *Morus* ranged from 2n = 28 to 2n = 308 or, expressed in terms of the ploidy level, from x to 22x. Most of the *Morus* species were diploid, having 28 chromosomes (2n = 28), such as *Morus alba* (white mulberry). To this group of diploids, Janaki was able to add four other species: *M. rubra* and *M. microphylla* from North

America and *M. serrata* and *M. laevigata* from the Western Himalaya. Her discovery of four forms of *M. cathayana* (Hua sang, a deciduous mulberry tree, native of China, Japan and Korea) with 24, 48, 72 and 144 chromosomes helped to bridge the gap between the two economic groups, namely the diploid species and the 22-ploid (expressed as a series, the members of the family are 2x, 3x, 4x, 6x, 8 x, . . . 22x). She was hopeful that a further search for *M. cathayana* in Central China, or the allied species *M. notabilis*, would reveal even higher chromosome numbers than the ones at Kew, and provide a further connection with *M. nigra* and indicate its probable origin as China, rather than Persia as suggested by Vavilov.

In their *Chromosome Atlas*, we may recall that Janaki and Darlington had introduced the term 'centres of development' to inject dynamics and complexity to an otherwise static and simplistic Vavilovian deductive logic. Movement, they argued, had a redoubling (snowball) effect on variation, the snowball being the variants picked up by hybridisation and selection on the path of the migrant. For this reason, the species names that they provided represented a dynamic (rather than the fixed types known to systematics) combination of ecologically diverse systems (considering the movement of genetic material across larger populations and geographies) as yet known to genetics. The black mulberry had never been found in the wild, and had been cultivated from very early times. Janaki appended a cytogeographical map in her paper on the mulberry, an aid which she used to much effect, plotting the distribution of diploid and polypoid species of *Morus* across Asia, backed by evidence from palaeobotanical/archaeological sources. Much like tracing a songline, she would chart the migration of the mulberry along the once fertile regions of Central Asia, connecting Kashgar with the ancient buried sites of Niya (an archaeological site and formerly an oasis on the Southern branch of the Silk Road in the Southern Taklamakan Desert) and Lop Nor in the Tarim basin (an extinct salt lake in Xinjiang, China), all of which strongly pointed to its Chinese origin.[43]

Like the black mulberry, the black nightshade *Solanum nigrum* was also a 'great traveller.'[44] The species had travelled to almost every continent as a weed from its origin in South America. However, what took a very long time to evolve in nature could be duplicated in the 1940s in the laboratory in the shortest possible time, by the application of drugs such as colchicine (in the form of a solution: usually, 0.01% for herbaceous plants and 1% for shrubs; it could also be applied on seeds, tips of seedlings or the buds of old trees), which doubled the number of chromosomes in the diploid to produce tetraploids. The tetraploids were distinct for exhibiting thicker and hairier leaves, and flowers larger and more deeply coloured than in the diploid plant. In general, the impact of the drug was easier seen in herbaceous plants than the woody ones like magnolias or rhododendrons.

Magnolias

The manifestation of a tetraploid in a cultivated plant 'marked a turning point in the history of its cultivation,' and a eureka moment for the cytologist-breeder. Janaki would have several such epiphanic moments in her plant breeding career. At Wisley, Janaki treated young seedlings of *Magnolia stellata* and its close ally *Magnolia kobus* in a first ever experiment to break through genetic barriers that existed between some of the sections of the Magnoliaceae. Plants grown from seeds from colchicined tetraploids were planted out by mid-1948, together with diploid plants for comparison.[45] These tetraploids were then employed in producing new *Magnolia* cultivars such as the magnificent *Magnolia kobus* 'Norman Gould' (named after botanist Norman Gould of the Wisley Garden, an exact contemporary of Janaki),[46] which had thicker leaves, flowered for longer periods and produced a greater number of very large flowers (floriferous). One of these new hybrid varieties was also named after Janaki by the RHS: the *Magnolia kobus* 'Janaki Ammal.' Several of the hybrid magnolias sold by plant nurseries in England today, including a floriferous *Magnolia stellata* cultivar, indeed originate from Janaki's crossings at Wisley. Her new magnolias were planted on Battleston Hill (acquired by the RHS in 1936, it covered about 26 acres of land) in the Wisley Garden, which today stand tall and beautiful attracting numerous visitors during the flowering season in early spring.

Speaking of which, one is reminded of the magnolia blossom immortalised by her older contemporary, the American woman photographer Imogen Cunningham (1883–1976) in the 1920s. Interestingly, during her stay in England, perhaps in the late 1940s, Janaki acquired a beautiful Japanese wood-block print titled 'Magpie & White Magnolia,' perhaps by Ohara Koson (1877–1945), well-known for his 'birds and flowers' series (*kachō-e*) of paintings and prints.[47] Incidentally, at the same time as Janaki at Wisley, Charles Percival Raffill of Kew was breeding magnolias. Raffill was known as an authority on rhododendrons, lilies, irises and fuchsias. He was a friendly and generous person, who believed in the saying, '[if] you want a friend, you must be one yourself.' Janaki and Raffill often exchanged plants and seeds and enjoyed a close relationship.

Rhododendrons and Other Ornamentals

It was also in 1948 that Janaki treated the wild species of flowering annuals, *Godetia* and *Clarkia* collected in California (by the RHS) with colchicine, besides producing an asparagus with all its flowers hermaphrodite, by the same procedure. These flowers were of substantial importance in the production of true-breeding varieties.

Janaki made a major contribution to the cytology and breeding of the genus *Rhododendron* while at Wisley. The Rhododendron era in England is said to have begun with Joseph Hooker's collection of 30 spectacular species from the Sikkim Himalaya in 1850, which he brought back to Kew and distributed to various gardens in the country. By the early 1900s, there were about a hundred species growing in Britain, increasing fivefold in the years following, through the efforts of plant hunters like Ernest Henry 'Chinese' Wilson, Reginald Farrer, George Forrest, Frank Kingdon-Ward, George Sheriff and Frank Ludlow, and George Taylor, who collected in the Sino-Himalaya, 'the great home of Rhododendron.' More than half of the 800 species of *Rhododendron* in cultivation originated from that region, which extended from Sikkim and Bhutan eastwards into central China, northwards into Tibet and south-eastwards into Assam and Burma. To explain the classification of Rhododendrons, Janaki used the analogy of the solar system. Each of the forty-three series of this large genus, with several subseries, was centred round the oldest known species from which the series took its name: 'Around these "Master Species" revolve like the planets round the sun, one to as many as fifty species, and round some of these species may be gathered subspecies, revolving like moons.'[48]

The genus was a cause of annoyance to the regular systematic botanist, but not Janaki; it was a challenge to know where one species ended and the other began, and there was much family resemblance between the species within a series. Excited, she would comment: 'nowhere in nature is it possible to see so clearly a population of plants in which the minute gradations from one type to another are so well preserved as in Rhododendron. Nowhere too in the plant kingdom have we such a large assembly of species with such diverse morphological characters held together under one generic name.'[49] Given this botanical singularity/uniqueness, Janaki was certain that a chromosomal analysis of the species would shed light not only on the species problem but also on fundamental questions relating to plant migration, distribution and evolution. She painstakingly examined over 360 species of the family, chiefly taken from the established collections on Battleston Hill in Wisley Garden, besides the collections at Kew and Edinburgh, Lord Aberconway's garden at Bodnant, and J. B. Stevenson's garden at Tower Court in Ascot. With one exception (*R. diaprepes* from Tower Court), all the elepidotes (156 species) were diploid (2n = 26) and the 78 lepidotes, polyploid.[50] The polyploidy series for the Rhododendrons ranged from triploids (2n =39) to dodecaploids (2n = 156); importantly, she discovered that Rhododendrons growing at higher altitudes of the Himalaya were polyploid in nature.[51]

The Nerines

Among Janaki's students/assistants at Wisley (most of whom were women) was a young lady from Plymouth, Marion Nancarrow, who after

studying botany at the Royal Holloway College, London, was undertaking doctoral research. Tutored by Janaki in cytological techniques at
Wisley, Nancarrow was in complete awe of her teacher. In fact, Nancarrow's research involving the breeding of Nerines (of the family Amaryl-
lidaceae, with the basic number, 11) was much influenced by Janaki's
work on *Nerine* hybrids; Janaki had discovered these to be chiefly triploids. Nancarrow wanted to use pollen from the triploid *Nerine sarnien-
sis*, but triploids rarely occurred in nature, and it was this need that had
brought her back to Janaki.[52] At the Wisley cytological laboratory, under
Janaki's supervision, about 100 plants were examined by her small team,
which included her assistant Margery Bridgwater. In fact, Janaki would
develop a technique for counting chromosomes of the *Nerine* from young
ovules, the part of the ovary of seed plants that contains the female germ
cell, and after fertilisation becomes the seed. The majority of hybrids
were diploids, but what attracted breeders were the 16 triploids and
an equal number of aneuploids (plants with an abnormal number of
chromosomes in the haploid, that is, not an exact multiple of the usual
haploid number), which could be used to produce hybrids of the mag-
nificence of the single tetraploid, 'Inchmery Kate' ($2n = 44$) or its parent
'Alice' ($2n = 36$). To complete the series of cultivated Nerines (with their
description, parentage and chromosome number) for use in future breeding programmes, Janaki appealed to breeders for information and for 'a
single open flower' of each to examine the chromosomes in the ovule.[53]

Preparing to Leave for India

Janaki had once again embarked on a vegetarian diet just after the war
(during the war years, she had obviously found it difficult to subsist on
vegetables alone), when she began work at Wisley; she would be tempted to
eat some fish occasionally, only to end up ill. 'I have been having a wretched
time eating some fish that disagreed with me. Now I am again a *full vegetarian* and feel better,' she would write to brother Raghavan in July 1948.
Janaki had missed the recent wedding celebrations of her niece, late brother
Vasudevan's daughter, at Shoranur; she would wistfully remark: 'How
lovely it would have been if our dear brother was also with us. I dare say he
must be happy in the beyond seeing his children . . . I am waiting to hear of
a good match for your two daughters. I long to see everybody.'

Janaki was expecting to travel to India in a matter of few months, 'flying
all the way to save time.' She had planned to do 'quite a lot of collecting
in India' and was also hoping to visit Nepal for the purpose. She wanted
Raghavan to send her orchid seeds from his neck of the woods: 'I know
there are many in Malabar,' she wrote; she had herself collected orchids
in Tellicherry. Wild orchid pods from Shoranur were to be sent in an 'air

letter,' wrapped in a separate envelope, and plants from his garden were to be marked A, B, C, etc. 'so that I'll know them.' She had just begun working on orchids and was finding it 'tricky business germinating these seeds.'[54] Incidentally, at Wisley, Janaki was also growing such tropical plants as *Momordica charantia* (bitter gourd) and *Luffa* from seeds despatched to her from Bangalore by her forester brother Kittu.

On 17 September 1948, Janaki held a grand party to celebrate the second anniversary of the opening of her cytological laboratory. Fortunately for us, Rowley faithfully recorded this event in his journal:

> Weather: fair and bright
> Sept 17th Friday
>
> *Dr Janaki Ammal* phoned just as I was settling down to work and invited me so temptingly to join a party at Wisley that I could not resist. Packed up and came by bus, stopping for a little shop-gazing at Raynes Park and Kingston, and going direct for an early lunch at the Cedar House, Ripley. Then from 2.0 to 4.0 worked on the old record books, copying out data on roses, when the guests—over 20 in all—began to fill the cytology lab., and the Big Eat began. Dr Janaki laid the entire centre table with savoury sandwiches, cakes, sponges and in addition some typical vegetarian delicacies of her own which aroused keen interest. The celebration commemorated the 2nd anniversary of the opening of the Cytology lab., and departure of 5 members of the staff, including her trusted assistant Paul Town, going as a student to Reading. After the bun fight helped tidy up, and had the added pleasure of the hostess's company on the bus back as far as Raynes Park.[55]

In less than two months of this tea party, after an eventful nine years in Britain, Janaki boarded a plane for India.

From her friend Eileen, Janaki gathered that Bartlett had been collecting lily seeds and bulbs for her; he had just returned to Ann Arbor from the Philippines. 'It is indeed very kind of you to do this for me,' she wrote to him, before mentioning her plans to leave for India in early November, and of how 'troubled in mind' she was about whether she 'should stay on there and work for [her] own country. . . . Sometimes the conflict in my mind makes me feel crazy. How I wish I could have a talk with you.' Her work at Wisley was keeping her away from fulfilling her academic intentions. She would have been happier, she wrote to him, making 'better food plants for the starving millions in India' rather than making tetraploids 'by the dozen, for the amusement of rich Lords and Ladies of the Horti world.' Having said that, she was also 'afraid to go back and get into the nest of Brahmins who are in control of Agricultural Research in India,' obviously referring to men like T. S. Venkatraman of the Sugarcane Breeding Station. All she wanted was 'to be somewhere quietly working on some interesting cyto-logical problem—or collecting plants.' Her chief interest was in cytological research with respect to plant migration and polyploidy; and in fact, she had by now almost completed a survey of the Magnoliaceae. 'I am hoping to col-lect when I go to India and there is a possibility of my entering Nepal,' she revealed to him, and was curious about the American expedition to Nepal, led by ornithologist Sidney Dillon Ripley in late autumn that year. Janaki's yearning for Ann Arbor remained undiminished as ever: 'I look back on my Ann Arbor days and hope some day I will be in a position to visit my Alma Mater,' she told Bartlett.[56]

On Board with Nehru

Meanwhile, on 7 October 1948, Nehru arrived in London, where he was received by the Indian High Commissioner V. K. Krishna Menon. He was in the city to attend the third Commonwealth Prime Ministers' Conference, at which platform he reiterated that 'India wants to be completely sovereign and independent, at the same time being close to England.'[57] It was the first such meeting bringing together the prime ministers of the newly independ-ent countries, India, Pakistan and Ceylon. Nehru would leave for Paris on 15 October to attend the third session of the UN General Assembly, but briefly returned to London on 22 October, for the last day of the Common-wealth Conference. He was to leave Paris for Cairo on 2 November, having accepted the invitation of the Arab League, but the weather having played havoc, his journey was delayed by a day or two. Nehru would land in Cairo on 4 November and immediately after the meeting, board an Air India Inter-national flight to reach Bombay on 6 November, a little after midnight; Air India International had begun operations only that year.

The flight to Bombay had originated in London; on board were several Indians including Janaki (the day was special for her, it being her fifty-first birthday). At Santa Cruz airport, Bombay, a grand reception awaited the Prime Minister from the officials of the Bombay Provincial Congress Com-mittee including S. K. Patil, the ADC of the Governor of Bombay, Morarji Desai, the Home Minister of Bombay, Dr M. U. Mascarenhas, the Mayor of Bombay, M. D. Bhat, the Chief Secretary to Government, besides a large number of prominent citizens. It was among cheers of Jai Hind that the plane landed. 'Though nothing of this was meant for me and all of it was for Panditji,' Janaki remarked to Darlington, 'it was very gratifying to return to India in such good Company after my 9 years exile.' Dressed in a brown sherwani, Nehru appeared cheerful and brimming with energy as he walked down the gangway in the company of his nieces Nayantara and Chandrale-kha, despite his very busy month outside the country; on 14 November he would turn 59.

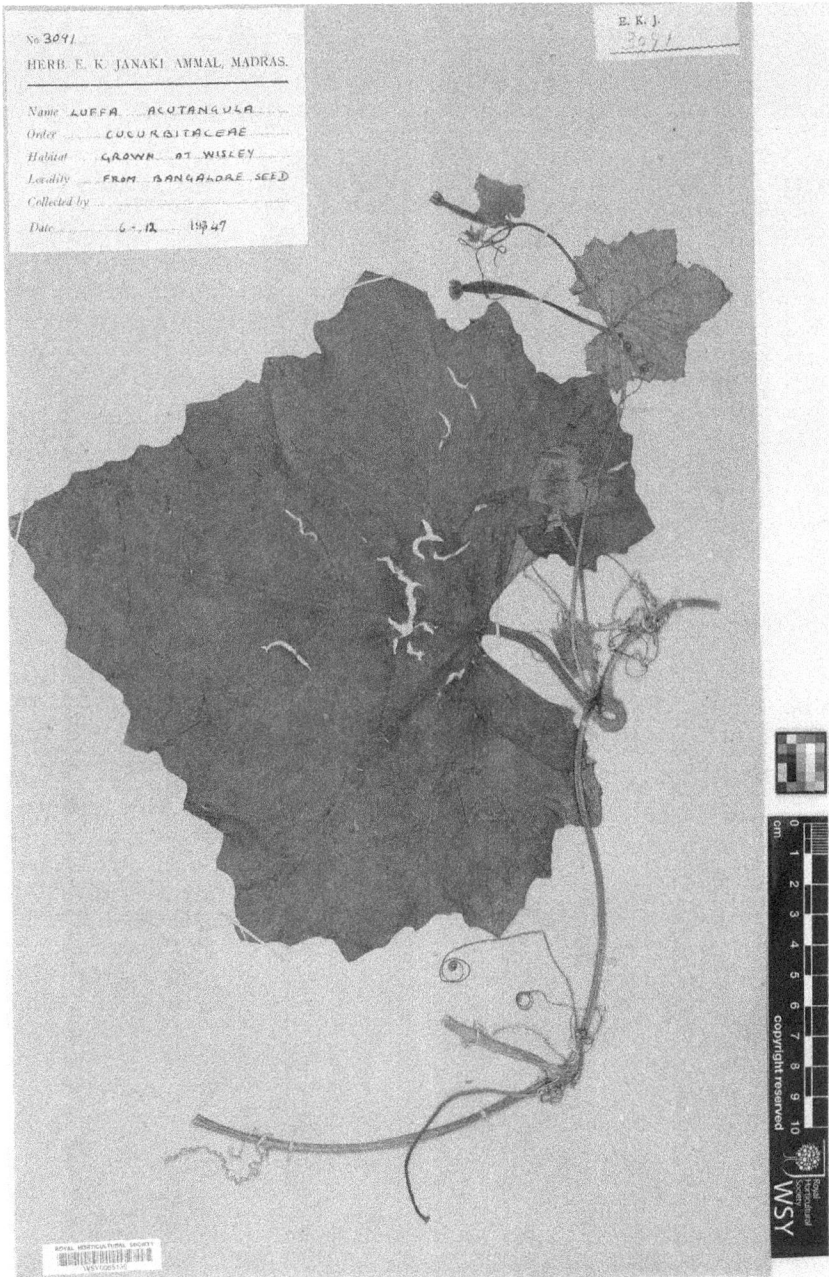

No 3041

HERB. E. K. JANAKI AMMAL, MADRAS.

Name *LUFFA* *ACUTANGULA*
Order *CUCURBITACEAE*
Habitat *GROWN* *AT* *WISLEY*
Locality *FROM* *BANGALORE* *SEED*
Collected by
Date 6 - 12 1947

E. K. J.
3041

Figure 13.1 *Momordica charantia*, grown by Janaki from Bangalore seeds at Wisley.
6 December 1947.

Source: Courtesy of the RHS Wisley Collections. WSY0065142.

267

Figure 13.2 Janaki. Brighton, 9 September 1948. Blanchard Family Papers c. 1835–
 c. 2000), Box 37.

Source: Courtesy of the Bentley Historical Library, University of Michigan.

Figure 13.3 The Seligmans, London, c. 1948.

Source: Courtesy of the EK family.

Notes

1 E. K. Janaki Ammal to E. K. Raghavan, letter dated RHS Garden, Wisley, Ripley, Surrey, 23 July 1948, personal collection.
2 'Review of the work of the Royal Botanic Gardens, Kew, during 1946', p. 5.
3 RBG Kew, Library & Archives: 1/RHS/10A, Edward Salisbury to Prof. Shephard, letter dated 24 May 1946.
4 The gardening staff had just returned to the estate, post-war, when Janaki visited.
5 RBG Kew, Library & Archives: Printed Notebooks, Herbarium E. K. Janaki Ammal, Madras.
6 DP: C. 42 (E 140), Letter from the Colonial Office to Darlington, dated 30 August 1946, Darlington to the Colonial Office, dated 27 Sep 1946, and Janaki's Report on the 'Columbus Grass', 25 Sep 1946.
7 DP: C. 42 (E 141), letter from H. N. Barber to E. K. Janaki Ammal, dated 13 March 1947.
8 RHS, Lindley Library: RHS/Minutes/WY/WY Garden Advisory Committee, Box No. 5, Wisley Advisory Committee Minute, 1945–47.
9 Excerpts from Mr Pittman's and Mrs Gilmour's letters sent to the John Innes in April 1984. See 'John Innes Visitor's Book' No. 1, Library & Archives, John Innes Centre, Norwich.
10 Hari Krishnan, 'Nachee', private collection, July 1993.
11 DP: g. 7, pocket diary, 1947.

12 BHL, UoM: Blanchard Family Papers c. 1835–c. 2000, E. K. Janaki Ammal to Frieda Blanchard, letter dated 27 December 1946.

13 RHS, Lindley Library: RHS/Minutes/WY/WY Garden Advisory Committee, Box No. 5, Director's Report of a Meeting of the Wisley Advisory Committee held on 14 April 1947.

14 Ibid., Director's Report of a Meeting of the Wisley Advisory Committee held on 2 June 1947.

15 Ibid., Report of a Meeting of the Wisley Advisory Committee held on 30 June 1947.

16 University College London (UCL), Special Collections: Haldane/5/2/2/153, from E. K. Janaki Ammal to J. B. S. Haldane, letter dated 17 December 1947.

17 UCL, Special Collections: Haldane/3/1/1/2/43, from J. B. S. Haldane to E. K. Janaki Ammal, letter dated 19 December 1947.

18 Serial numbers, 3000–3100 of her herbarium records/field notebook pertain to collections made at the Kew Gardens, from June 1946 to September 1947, chiefly of the family Moraceae (mulberry). See RBG Kew, Library & Archives: Printed Notebooks, Herbarium E. K. Janaki Ammal, Madras.

19 Lewis, 'Cyril Dean Darlington', p. 148.

20 Letter from A. R. Moore (representing F. E. Lloyd) to E. K. J., dated 2 September 1945, found in Janaki's copy of *The Carnivorous Plants*.

21 RHS, Lindley Library: RHS/Minutes/WY/WY Garden Advisory Committee, Box No. 5, Director's Report of a Meeting of the Wisley Advisory Committee held on 11 August 1947.

22 Serial No. 3074 (*Morus acidosa*), RBG Kew, Library & Archives: Printed Notebooks, Herbarium E. K. Janaki Ammal, Madras.

23 BHL, UoM: Blanchard Family Papers c. 1835–c. 2000, E. K. Janaki Ammal to Frieda Blanchard, letter dated 16 December 1947.

24 RHS, Lindley Library: RHS/Minutes/WY/WY Garden Advisory Committee, Box No. 5, Report of a Meeting of the Wisley Advisory Committee held on 19 April 1948.

25 E. K. Janaki Ammal to E. K. Raghavan, letter dated RHS Gardens, Wisley, Ripley, Surrey, 23 July 1948, personal collection.

26 Hari Krishnan, 'Nachee', private collection, July 1993.

27 It stands today in the courtyard near Gate No. 5, Parliament House, New Delhi, on a red sandstone pedestal, carrying the inscription, 'Shepherd boy— Chandragupta Maurya dreaming of India he was to create.' Hilda Seligman was also involved with the rural extension programme undertaken by the AIWC, in the post-war years, to improve woman and child welfare in rural areas. She donated to the cause the royalties received from the sales of her book *Skippo of Nonesuch*, and with the help of Lady Pethick-Lawrence and Lady Stafford Cripps founded a Skippo Fund in London 1945. The Fund helped custom-build mobile health vans, called Asoka-Akbar Mobile Health Vans, which were given to the AIWC to begin a rural health project; the President of AIWC, Lady (Dhanvanthi) Rama Rau and the Bombay-based lawyer and family planning activist Mrs Avabai Wadia designed a 'Skippo Project'. Later, both women would alternately hold the position of President of the Family Planning Association of India (founded in 1949).

28 RHS, Lindley Library: RHS/Minutes/WY/WY Garden Advisory Committee, Box No. 5, Report of a Meeting of the Wisley Advisory Committee held on 21 June 1948.

29 BHL, UoM: Blanchard Family Papers c. 1835–c. 2000, E. K. Janaki Ammal to Frieda Blanchard, letter dated 16 July 1948.

30 Ibid.
31 RHS, Lindley Library: RHS/Minutes/WY/WY Garden Advisory Committee, Box No. 5, Report of a Meeting of the Wisley Advisory Committee held on 30 June 1947.
32 Janaki Ammal and Wylie, 'Chromosome numbers of Cultivated Narcissi', 1949.
33 RHS, Lindley Library: RHS/Minutes/WY/WY Garden Advisory Committee, Box No. 5, Report of a Meeting of the Wisley Advisory Committee held on 20 June 1948.
34 Personal communication from the late Gordon D. Rowley, letter dated 30 November 2007, enclosing a page from his journal dated 24 May 1948.
35 Ibid.
36 Hari Krishnan, 'Nachee', private collection, July 1993.
37 Zimmer, *A Guide to Nabokov's Butterflies and Moths*.
38 Janaki Ammal, 'Chromosomes and Horticulture', p. 237.
39 Ibid.
40 Janaki Ammal and Wylie, 'Chromosome numbers of Cultivated Narcissi', 1949; Janaki Ammal, 'Chromosomes and the Evolution of Garden Philadelphus', 1951; 'The story of primula malacoides', 1952; 'Chromosome relationships in Cultivated Species of Camellia', 1952.
41 Janaki Ammal, 'The Race History of Magnolias', 1952.
42 Ibid., 'Chromosomes and Horticulture', 1951, p. 237.
43 Janaki Ammal, 'Contributions from the cytological department, RHS Garden, Wisley, I: The origin of the Black Mulberry', 1948.
44 Ibid., 'Chromosomes and Horticulture', 1951, p. 237.
45 Incidentally, *M. stellata rubra*, the first triploid *Magnolia* (2n = 57), had been found. Based on cytological evidence, Janaki suggested it to be a hybrid between *M. stellata* (2n = 38) and *M. liliflora* (2n = 76).
46 Norman Gould (1897–1960) joined the RHS Wisley Garden as a gardener in 1914 and later became plant recorder and lecturer in botany. Desmond, *Dictionary of British and Irish Botanists and Horticulturists*, p. 265.
47 When he began publishing with Watanabe Shozaburo, Koson began signing his work 'Shōson'. See Newland, Perree and Schaap, *Koson Ohara*.
48 Janaki Ammal, 'Polyploidy in the Genus Rhododendron', 1950.
49 Ibid.
50 Rhododendrons are divided into two categories, lepidotes and elipidotes; lepidotes have minute scales on the leaves and are usually small-leaved species, while elipidotes do not have scales and are usually large-leaved.
51 Janaki-Ammal, Enoch and Bridgwater, 'Chromosome numbers in Species of Rhododendron', 1950.
52 Wagstaff, 'Dr Marion Wood: Early life and career.'
53 Janaki Ammal and Bridgwater, 'Chromosome numbers in Hybrid Nerines', 1951.
54 E. K. Janaki Ammal to E. K. Raghavan, letter dated RHS Garden, Wisley, Ripley, Surrey, 23 July 1948.
55 Gordon D. Rowley to the author, letter dated 30 November 2007, enclosing a page from his journal dated 17 September 1948.
56 BHL, UoM: University Herbarium Records (University of Michigan) H. H. Bartlett series, Box 11, E. K. Janaki Ammal to H. H. Bartlett, letter dated 26 October 1948.
57 *The Times of India*, 7 Oct 1948.

14

NEPAL

A Pilgrim of Science

Nobody has had time for research since August 15th 1947.
—E. K. Janaki Ammal (1948)[1]

The first important task Janaki set out to accomplish on her return to India in November 1948 was a botanical expedition to Nepal to aid her research on phytogeography and the evolution of cultivated plants. The first woman ever to go on an expedition to Nepal, Janaki wished to enter the Himalayan country in the spirit of a 'Pilgrim of Science from India.'[2] It was to be a private undertaking, in no way supported by a scientific body in England or India for that matter, unlike for instance the Ripley expedition. Incidentally, Dillon Ripley was in Nepal in the company of the Peabody taxidermist Edward Migdalski, two Yale graduates and a couple of National Geographic photographers, in quest of the elusive *Ophrysia superciliosa* (the Himalayan quail) that was feared to be extinct. Nepal had been out of bounds for foreigners interested in collecting or even just trekking, for a very long time. Ripley, a close friend of the Indian ornithologist Salim Ali, had however managed to wangle permission to enter the country, but limited to the lowlands. A large and well-furnished bungalow had been placed at their disposal by Nepal's new Prime Minister Maharajah Mohun Shamsher Jang Bahadur Rana (Ripley would name a bird after the Rana). At a meeting with the Jang Bahadur, Ripley strategically mentioned that many of his friends, including Nehru, were eagerly awaiting the outcome of their collecting expedition to Nepal. The name dropping did the trick and, much as expected, they were allowed to go where they liked in Nepal. Nehru was however indignant when he came to know of this and it would take a great amount of tact and explanation on the part of Salim Ali to restore his own good relationship with Nehru. The Ripley expedition, which began in the fall of 1948, would end in the spring of 1949. The *Ophrysia superciliosa* would however continue to elude them.

To the question, why Nepal, Janaki explained: 'Nepal is the most unique part of Asia botanically, for it is in Nepal that the Flora of the Eastern

DOI: 10.4324/9781003267089-14

Himalayas with its Chinese affinities reaches its most Westerly limit and meets the Flora of the Western Himalayas with its European affinities. Any information regarding the composition of Nepalese plant population is therefore very important for the understanding of early migration of plants. They would also throw much light on the origin of some of the cultivated plants of Asia, a subject which is of special interest to me.'[3] Incidentally, from the point of view of avifauna, Ripley was attracted to Nepal for strikingly similar reasons. The 'bird picture in Nepal,' he explained 'is of extraordinary consequence for the topography is so plentifully and sharply ridged by great rivers and gorges, and has so great a variation between its highlands and lowlands, that somewhere in the country a dramatic break occurs in the fauna so that widely different forms of the same species are found.'[4]

Darlington wrote out a letter of recommendation for Janaki, addressed to the Indian High Commission in London. It stated: 'I should like to say that from my knowledge of her work during the last eighteen years, I think she is remarkably well-equipped for such an expedition. Her knowledge of the origins of cultivated plants based on her studies in India, in the United States and in this country (at Merton and Wisley) give her an unrivalled capacity for finding in the unexpected territory of Nepal what will be most useful for the development of both Horticulture and Agriculture in other parts of the world. Any facilities given to her would be an encouragement to science.'[5] The President of the Royal Horticultural Society, London, also admitted that he knew 'of no one whose ability and reputation render[ed] such research more useful and productive.'[6] Her acquaintance with the Indian Ambassador V. K. Krishna Menon (it is said that Janaki acted as guardian to his sister V. K. Janaki at the Presidency College, Madras), helped expedite the official arrangements from the Indian side. She was warned that travelling in Nepal was arduous as the communication network was far from robust, meaning she would have to 'rough it out in some places.' To help her with travel arrangements, Menon sent her a copy of 'Notes on journey to Nepal.'[7] This was the first time that the Nepalese government would be sponsoring the visit of a woman to Nepal, but Menon believed they would find it worthwhile in the case of a scientist like Janaki. Janaki had provided them with a list of her publications, including the *Chromosome Atlas*, and the names of four referees (two each from India and Britain): Darlington (Director, the John Innes, London), Birbal Sahni (Lucknow University), K.L. Moudgill (Director, University Research Laboratory, Travancore University) and Geoffrey Evans (Economic Botanist, Kew Gardens). The Indian government was hopeful if not sure that the Nepalese government would agree to its request.[8] Janaki wished to take one botanical assistant with her, chiefly for the preservation of specimens and materials in situ, and 'a dooly and a pony each' for travel between Amlekhganj and Kathmandu.[9]

The letter permitting her to travel to Nepal finally arrived at Janaki's doorstep, perhaps at Shoranur, where her siblings Raghavan, Vasudevan

273

and Cousalya lived with their families. Elated, she immediately set out for Delhi, where on arrival, she met V. M. M. Nair, Deputy Secretary of the Department (External Affairs) to present her itinerary in person. She hoped to reach Raxaul on the India-Nepal border (in the East Champaran district of Bihar) on 20 November (but she would eventually get there only two days later, on the 22nd) and then on the following day travel the 75 miles to Kathmandu, where she planned to spend ten days. Arrangements had been made with the Legation Overseer (part of the Indian diplomatic staff in Nepal) to receive and accommodate her in the Legation bungalow.

Predecessors

Janaki's visit to Nepal in the winter of 1948 coincided seasonally with the dates of visits to that country in 1820–21 by Nathaniel Wallich, the Danish surgeon-botanist attached to the East India Company, and nine decades after this (in 1907) by I. H. Burkill, the Assistant Reporter on Economic Products; both went in search of plants. If the visits of Wallich (he travelled from Raxaul in the south to Nuwakot or Niakot) and Burkill had lasted only a fortnight, Janaki's was even shorter, at ten days. The results of Wallich's researches would be published as *Tentamen Flora Nepalensis Illustratae* (1824–26) and Burkill's as 'Notes from a Journey to Nepal' (1910). Prior to Wallich's visit, Francis Buchanan-Hamilton had made extensive tours in Nepal (1802–03) collecting plants; his *An Account of the Kingdom of Nepaul* (1819) provided an outline of the history of Nepal and included chapters on its flora. The book also drew on Col. Kirkpatrick's account of Nepal (1811), which contained a description of the vegetation from the Terai at Birgunj to the Kathmandu valley. The British Resident in Nepal (1822–43), Brian Hodgson, who made the Himalayan kingdom his home for more than two decades, also contributed much to the knowledge of Nepalese flora and fauna.

Joseph Hooker was among those who visited Nepal in the first half of the nineteenth century. In November 1847, Hooker had set out on his three-year-long Himalayan journey; in late October 1848, Hooker and his local assistants left for Eastern Nepal, to Zongri, the spurs of Kanchenjunga and along Nepal's passes right into Tibet.[10] Thomas Thomson joined him in May 1850 on what was Hooker's final Himalayan tour, to Sylhet and the Khasi Hills in Assam, where they resided until the end of that year. Hooker intended to botanise in Nepal, a little-known region of the Himalaya, but his friends Lord Auckland and the Scottish geologist Hugh Falconer suggested Sikkim instead. He eventually entered Nepal through its eastern border and collected in the valley of Tamur and Arun rivers, with the help of surgeon Archibald Campbell, who had gained the friendship of the Jung Bahadur. Hooker's collection of Nepalese flora was described in his classic *Flora of British India* (vols. 1–7, 1875–97). Post-Hooker, in 1876 and later in 1880–84, botanists J. Scully and J. F. Duthie collected plants along the Mahakali river and the Doti and Baitadi districts of West Nepal. There is little doubt that Janaki had sufficiently acquainted herself with what was available of the published sources on the flora of Nepal before she set out on her expedition. Incidentally, a few months prior to Janaki's visit to Nepal, Mohan Lal Banerji (1916–2012), a doctoral student of the Spanish Jesuit botanist Hermenigild Santapau (1903–70), who worked extensively on Indian flora and was professor of botany at St Xavier's College, Bombay, was collecting in Nepal.[11] He was perhaps the first Indian to go on a scientific expedition to the Himalayan kingdom, which had in general practised an isolationist policy.

There was 'but one excuse for writing his report' even when based on just a fortnight's observations of the country, Burkill had observed in his 'Notes from a Journey to Nepal,' for it revealed 'the great want of knowledge of the Botany of that part of the chain.' By 'chain' was meant 'the succession of vegetation between Raxaul and the Himalaya of Central Nepal [Kathmandu valley] as far back as 35 miles in a straight line from the skirts of the plains and not higher than 7000 ft.'[12] Nepal from the floristic point of view still remained largely unknown. The vegetation had been altitudinally classified by Dietrich Brandis, the Inspector General of Forests in India, into four belts: the alpine belt, temperate forest belt, subtropical belt (which was the cultivation zone, clearly cutting off the upper forests from the lower forests), and finally the tropical forest belt.[13] It was this unique phytogeography that had attracted Burkill to Nepal, and later Janaki, a study of which she was sure would shine light on the early migration of plants and on the origin of some of the cultivated plants of Asia.

Simla

Before heading towards Nepal, Janaki visited Simla, located on the South-western ranges of the Himalaya, on a brief collecting excursion. She was a guest of botanist Kailash Nath Kaul (1905–83), and resided at 56 Grand Hotel, Simla. Only that year, Kaul had established the National Botanic Gardens, Lucknow (today called the National Botanical Research Institute). Incidentally, during the war years, 1939–44, when Janaki was at the John Innes, Kaul was researching at the herbaria of the Royal Botanic Gardens, Kew, and the Natural History Museum, formerly the British Museum (Natural History), besides lecturing at several universities in the United Kingdom, including the University of Cambridge.

Simla to Janaki was the loveliest hill station she had ever set foot in; she could hardly take her eyes away from the snowy peaks of the Himalaya. From Kalka, Janaki boarded the slow hill train to Simla, rather than the car that had been offered to her, because she could hop off at stations and do a little collecting 'while the thirsty engine was being watered.' 'I am having the greatest botanical treat of my life, and feel ashamed I did not come here before. Coming up from Kalka on the plain of Punjab to Simla you pass through a succession of vegetation that is almost like a demonstration in plant geography,' she described the route to Darlington. At Simla, she found her friends 'very kind botanically' speaking and accompanied them to 'many interesting beauty spots' and also along the 'Road to Tibet,' where she collected 'maize cobs and peas at 9000 feet.' First laid out by the British in the nineteenth century, the Hindustan-Tibet Road passed through Simla and connected with Tibet (a distance of 228 miles) for trade through the Shipki-La border post on the Indo-China border. It was a treacherous high-way connecting the trans-Himalayan Buddhist area of Kinnaur and neigh-bouring Spiti to the rest of Himachal, and travelling largely parallel to the Satluj River in Kinnaur district.

Among the several plants Janaki spotted or collected from Simla (on 11 and 12 November 1948) were the *Jasminum* with black fruit at Kandaghat on the Kalka-Simla National Highway, Acanthaceae from the 'way-side' (perhaps the *Justicia adhatoda*), Chenopodiaceae (the *Chenopodium album* 'weed can be seen everywhere'), Convolvulaceae (*Convolvulus arvensis* and *Cuscuta reflexa* were taking over hedges in Simla city and the Elysium Hill, she noted), several interesting plants near the Wild Flower Hall (a luxury three-storeyed hotel, which had come into the hands of the Indian Govern-ment post-independence) and Rosaceae (perhaps the Himalayan musk rose at Vincent Hill and in the open spaces at Kufriat 8000 ft above mean sea level). From Simla, she travelled by train and road to east Champaran in Bihar and on to Sugauli and then Raxaul—from where she would cross over to Nepal. On 22 November at Sugauli, she noted the Gramineae growing in the dry rice fields, and at Raxaul, *Erianthus* growing along the bunds bordering the rice fields, and *Utricularia* in a pond.[14]

Towards Nepal

From Raxaul, a narrow-gauge railway to Amlekhganj had opened in 1927, the railhead being 29 miles away. Shortly after crossing the frontier at Bir-gunj (the gateway to Nepal), the line passed through a forested area chiefly formed by sal trees (harvested heavily for use as railway sleepers), and entered the terai, or the lowlands (distinguished by tall grasslands, marshes, scrubs and sal forests), 12 to 20 miles wide, stretching almost continuously along the Southern border. From Amlekhganj, the journey was continued by road, and after some 27 miles, passed through a tunnel 300 yards long. Two passes, Sisagarhi (6,225 ft) and Chandragiri (7,200 ft), between Bhimphedi and Thankot, acted as natural ramparts protecting the Kathmandu valley. As one looked down upon the valley below from the Chandragiri pass, 'the most jaded traveller must feel his imagination stirred by its secluded posi-tion, its turbulent past, and by the mystery and sanctity attending this most ancient shrine of Hindu and Buddhist tradition.'[15] Usually a night was spent at a resthouse below the first pass, before proceeding to Thankot, some six or seven hours away. Kathmandu was a further 9 miles from Thankot.

Kathmandu and Surrounds

After three days of travel by road, Janaki entered the valley of Kathmandu, braving the cold, carried by eight men on a massive 'Tamdan—or a chair of nobility . . . This Tamdan is a most humiliating vehicle,' she would remark. Janaki would walk a great deal of the distance and use the Tamdan mostly as a vasculum, or collecting container.[16] Surjit Singh Majithia, the Indian Ambassador to Nepal (1947–49), accommodated Janaki and her assistant (whose name is unknown) at his official residence in Kathmandu as promised.[17] 'What a change from the simple life at Wisley to be surrounded by liveried peons, mukhias, subedars, havildars and kukried gurkawallas,' she wrote to Darlington towards the end of her stay in Nepal.[18] She had been provided with a state car and an assistant but after the austerity in England (the war and its aftermath) she found it bad taste to live a life so well-provided. Although Janaki had found it unnerving to return to India after nine long years, 'to come to Nepal and see a land where time seems to have stopped 1000 years ago . . . was fantastic . . . I feel like one who has completely lost her bearings,' she wrote to Darlington. However, Kathmandu was bristling with 'militarism, like Roman Britain,' and this she found abhorrent. It made her 'blood boil' to see the Ranas and Jung Bahadurs living in large palaces 'walled in from public gaze,' while 'the Buddhist peasant worked philosophically in his fields.'[19]

On 23 November 1948, she collected Solanaceae from the dry forests near Simara in the Bara district in the Southeast part of Nepal, Compositae and grasses from the roadside near Bhimphedi (3000 ft), spotted *Strobilanthes* in the undergrowth on the Chandragiri Hills (7000 ft), collected dried bamboo from Sisagarhi (6000 ft), Gramineae from the side of the hill at Kulekhani (Indrasarobar) and *Eugenia* from Chitlang in the southwest, an ancient and salubrious Newari settlement only a few hours away from Kathmandu. Four days later, at Godavari she sighted several orchids and lilies. From 29 November until 4 December 1948, Janaki botanised in the surrounds of the hilly terrain of Sundarijal (at 6000 ft), 15 km northeast of Kathmandu, of temperate climate and dense forest cover of pine, oak and rhododendron besides being the habitat of animals like the Himalayan black bear, leopards and monkeys and several hundred species of birds and butterflies. Between the rock crevices, from the wettest spots, she recorded several plants, and from one place, wild tea plants of 'the Chinese variety.'[20]

One of Janaki's chief reasons for visiting Nepal was collecting information on the cultivated plants of the country and seeds of 'queer vegetables like the rat tail raddish [sic]—with roots 2 ft long' to take back with her to England.

In her letter to Darlington, she described the Nepalese *Luffa* as being 'taller than Oliver [Darlington's elder son, Oliver Darlington] and gourds as tall as Andy [his younger son, Andrew Darlington]' and pumpkins 'big enough to make Cinderella her Coach.' The Newars she observed were expert agriculturists, growing crops on cliffs 'where even a goat would hesitate to go.' She was thrilled to discover wild tea plants growing on one of the hills.[21] On 29 November, a day after she had despatched the letter to Darlington, she collected from the Royal Gardens of Kathmandu, *Amaryllis* bulbs and later, on 16 December just before she left the valley, more *Luffa* seeds.

Among the other plants or seeds she collected in Nepal in the first week of December were the green and red varieties of Amaranth, Capsicum, Peas, Lapsi (Nepali hog-plum, eaten pickled), *Sechium edule* and other Cucurbitae, Pomelo (*Citrus maxima*), edible *Dioscorea* from Markhu village and soybean from Godavari (at 5000 ft), both places rich in biodiversity. Incidentally, her interest in soybean had begun as early as 1943. At Kulekhani, she collected more seeds of Amaranth. Janaki sometimes also collected from the village shops; she bought mixed beans from a shop in Sundarijal, and ginger and *Colacasia* from the bazaar in the Markhu valley. In a dump heap in Markhu village, she chanced upon *Nicandra physalodes*, the subject of her doctoral dissertation.[22]

Janaki also went botanising in the *shola* jungle near the Kathmandu Aerodrome. On 30 November, she collected Mistletoe and plants of the Myrtaceae from near there, and specimens of the Labiatae (or Lamiaceae) family, commonly called the mint or dead nettle, from the mud walls of Kathmandu. Among flowering plants collected or recorded in the first week of December were those that belonged to the Verbenaceae family near Markhu village, rhododendron and Rosaceae on the mountainside of the Sisagiri Pass, and on the hillside of Chandragiri, plants of almost all the flowering families were recorded: Acanthaceae, Campanulaceae, Compositae, Ranunculaceae, Boraginaceae, Rubiaceae, Rhododendron and *Buddleia*. There, she also spotted certain plants with red berries, hydrangea and a kind of 'shrub with white fruit.' Again, in the Markhu valley, she noted Leguminaceae and in Chitlang village vegetable plants belonging to the Cruciferaceae (whose leaves were used as a vegetable), Solanaceae and Cucurbitaceae ('called Karella in Nepal') families. Along the Sisagiri-Thankot road, on the hillside she found Andropogoneae and on the wayside, *Saccharum spontaneum*. In Sundarijal, she observed or collected a large number of plants of the Myrtaceae family, besides several ferns and grasses.[23]

Successors

The only exception the Jung Bahadurs ever made were to men of science, so much so that mountain climbing expeditions included at least one scientist to ensure access to the region. In 1950, owing to a favourable change in the political situation, following a popular uprising which marked an end to the autocratic rule of the Ranas, the Himalayan kingdom began to see a greater number of visitors, including botanists and mountaineers. In all, seven botanical expeditions, including a couple by the British Museum (Natural History), were organised to East Nepal between the years 1948 and 1957.[24] Banerji had focused on the entire region studied by Burkill and Wallich and would continue to visit Nepal throughout the 1950s.[25] Taking cue from Burkill, and corroborated by his extensive fieldwork, Banerji would argue that it was in Nepal 'that the differing floras of the Eastern and Western Himalayas merged'; the outstanding result was 'an expression of the easternness of the vegetation of East Nepal, while in West Nepal the West Himalayan elements abound,' making Nepal a botanical transition zone, and a region of enduring interest to botanists.[26] For some reason, Janaki did not choose to write a travelogue of her journey to Nepal, or even a report, which, had she done, would have been an extremely useful historical and scientific document.

Four years after Janaki's visit to Nepal, in 1952, William Russell Sykes, a botanist from New Zealand, who had studied for a diploma in horticulture at the Wisley Garden (1949–51) went on an expedition to Western Nepal, exploring an area of about 1000 sq miles lying between the Karnali and Kali Gandaki rivers, accompanied by botanists, Oleg Polunin, Adam Stainton and John Williams. Their expedition was supported by the British Museum (Natural History) and the RHS Wisley. At Wisley, Sykes would have had ample opportunity to meet Janaki and learn about her expedition to Nepal and perhaps it was Janaki's independently funded visit which had prompted Wisley to co-sponsor a plant collecting expedition to Nepal. During the team's first expedition (1952), which lasted eight months, they collected 5000 specimens, and two years later (1954), over a similar duration, a further 9500, which was eventually shared between the two institutions.

Visit to the IARI, Delhi

As soon as she returned to Delhi from her adventurous collecting expedition to Nepal, Janaki paid a flying visit to the Imperial Agricultural Research Institute to meet friends such as B. P. Pal; en route to the IARI, she collected Amaranth growing in a wasteland in Mehrauli. Benjamin Peary Pal (1906–89) was a foremost wheat geneticist and breeder and an expert on roses and bougainvillea. In 1965, he would become the first Director General of the IARI (Indian Agricultural Research Institute). Pal was educated in Burma, from which place (the University of Rangoon) he completed his BSc and MSc degrees (1924–29). In 1929, Pal was awarded a scholarship by the Government of Burma to pursue a doctoral programme at the University of Cambridge under the wheat breeding experts Rowland Biffen and Frank Engledow. He returned to Burma in 1932 on completion of his PhD dissertation (one year after Janaki), which demonstrated that the hybrid vigour in wheat could be commercially tapped, if sound methods for the large-scale production of F1 seeds could be developed and applied. A year later, Pal was appointed Second Economic Botanist at (what was then called) the IARI in Pusa.[27] Janaki had a special affinity for Pal, not the least because of her strong connections with Burma through brother Raghavan and sister Devayani; she had herself visited the place in the late 1920s.

Post-independence, science policy (applying scientific knowledge to the development of public policies) in India had taken on an aggressive outlook, something that Janaki was hardly prepared for. She was appalled at the state of research at the IARI. 'They are all busy planning great things' a sceptical Janaki wrote to Darlington. 'Nobody has had time for research since August 15th 1947,' she would comment.[28] She had also sent him a postcard, with a picture of the 'Bull Capital' in polished sandstone, from the Indian Museum collection, wishing him 'Best Birthday' and to his family 'A Very Jolly Xmas and a Prosperous New Year.'[29] She would receive from him, as Christmas present that year, L. Dudley Stamp's *Britain's Structure and Scenery* (New Naturalist Series, 1946), which attempted to explain evolution of the earth to lay readers.

Making improvements to wheat continued to be the chief focus of the IARI at this time, and B. P. Pal was at the helm of it, as he had been since the 1930s. In his Presidential address delivered at the eighth meeting of the Indian Society of Genetics and Plant Breeding in Patna in January that year (1948), he had spoken on aspects of wheat improvement. The origin of wheat, 'one of the world's most important foodgrains was lost in antiquity,' would be his opening remark.[30] Vavilov's expeditions to about 60 countries had led to the identification of eight independent centres of distribution of the chief cultivated plants in the world. These centres containing wild forms (genetically dominant) of the cultivated plants were of utmost use to breeders in the selection of desirable characteristics; out of the eight centres identified, at least four were believed to include the various wheat species and subspecies. The Central Asian Centre, which included Northeast India, Afghanistan and parts of Russia, was considered the birthplace of the so-called bread-wheat and of much interest to the IARI as it was here that the great potential of varietal material of *Triticum vulgare*, 'the most important breadstuff of the world,' was located. Inspired by Vavilov and his Institute of Plant Industry, the IARI had over time accumulated the largest collection of wheat species and varieties (genetic resources), the result of years of collecting not only by the staff but also through contributions made by the Provincial and State Departments of Agriculture within the country (which included both the hill and plains varieties; the Institute of Plant Industry, Indore, had a collection of *T. durum*, and the Cereal Botanist of Lyallpur in Punjab had an important collection of the dying species *T. sphaerococcum*) and those imported from abroad.

An Indian Textbook on Cytogenetics

To address the nation's concerns regarding food production in the post-war period, it was crucial, the plant geneticist Ramanujam had stressed, to execute a 'vastly expanded and co-ordinated plan of active research in pure and applied plant genetics whereby the breeding material and methods at . . . disposal could be enriched for exploitation by the breeder.'[31] His suggestions for the future organisation of genetic research in India included, firstly, long-range research for discovering material, principles and techniques for breeding and for training geneticists; secondly, breeding for yield, and other qualities in crops; and thirdly, testing, multiplication, certification and distribution of bred varieties. It dawned on Indian scientists that the Western countries had been successful in plant breeding because they focused on two lines of activity: on the one hand, making progress in fundamental research (genetics and allied sciences), or strengthening the scientific basis of plant breeding, and secondly, the practical application of available knowledge and techniques of plant engineering. It was felt that in India, 'fundamental research of the right type' was a serious desideratum: 'Genetics which is basic to plant breeding and is being hailed elsewhere as the science of the future with untold possibilities for the improvement of both animals and plants is not receiving the attention it should in this country . . . the teaching of this important subject is neglected and in India not a single university has a chair of genetics.'[32]

One of the highlights of the year 1948, as far as academic plant genetics in India was concerned, was the publication of the book, *Cytogenetics and Plant Breeding* by S. N. Chandrasekharan and S. V. Parthasarathy, lecturers at the Agricultural College, Coimbatore; the foreword was written by the rice breeder K. Ramiah, who was at this time Director, Central Rice Research Institute, Cuttack. The authors stated at the very outset that plant breeding was no more an art but a science, and that, perhaps with the exception of USSR, the chromosome basis of heredity had come to be universally accepted, and it was important for a plant breeder to have a sound knowledge of a set of allied sciences including genetics, cytology, taxonomy and statistics. While taxonomy was taught very effectively in the Indian universities, its unification with the study of the wild and cultivated species of agricultural crops had never been attempted earlier. Where genetics was taught, it was limited to Mendelism, and the classic examples of peas, maize and *Drosophila* drawn from textbooks published abroad. In general, the teaching of genetics and plant breeding was more successful in the country's agricultural colleges (affiliated to the universities) rather than the universities themselves, where it was usually offered as part of a subject such as agricultural botany. The chief reason being that agricultural colleges had a greater extent of live segregating material growing in the experimental fields, which students could observe for themselves. With this publication, the first of its kind in the country, a book specifically designed as a textbook of genetics and plant breeding, suited to the needs of Indian students, and with specific reference to tropical crops, had come into being. It would have undoubtedly made Janaki happy. The book would be the only publication from India to find a place in the bibliography of Darlington and Mather's *Genes, Plants and People: Essays on Genetics* (1950).

It was only when the book was going to press in December 1947 that the authors had access to the *Chromosome Atlas of Cultivated Plants*; in haste, they would cross-check some of their chromosome counts against the list provided in the *Atlas*.[33] The authors interestingly enough included a section on eugenics in the book, taking a position in favour of it. They remarked that the study of eugenics was useful in improving society and that by the study of family traits, 'marriage between persons that are likely to beget defective progenies due to genetic causes may be avoided.' In extreme cases, they suggested most insensitively that 'sterilization laws' could be 'promulgated to prevent undue and rapid multiplication of the defective in society.'[34]

Lysenko and Vernalization in India

It may be noted that vernalization experiments in India began in the 1930s and would continue into the mid and late 1940s. In an attempt to verify Lysenko's claims, B. P. Pal of the IARI (with G. Narayana Murthy) conducted preliminary experiments on the effect of vernalization on Indian crop plants namely, gram, wheat, chilli and sorghum. He observed that vernalization experiments in the different provinces of India until 1935 on crop plants such as wheat, barley, oats, cotton, millets and rice had failed to give positive results, but these had not been undertaken in a systematic manner and therefore needed further verification. Based on scientifically designed experiments undertaken in the chemical section of the IARI, Pal concluded that out of the four crop plants, only gram and wheat demonstrated a clear-response to pre-sowing temperature treatment.

The fields attached to Boshi Sen's Vivekananda Laboratory in Almora were also devoted to vernalization experiments. These experiments were funded by the Elmgrant Trust, Dartington (in Totnes, Devon, established in 1936 by Leonard and Dorothy Elmhirst; Leonard was an agronomist, who worked extensively in India and was a close friend of Rabindranath Tagore), and the ICAR. While vernalization experiments were on the one side being carried out in India (with no conclusive results to either prove or disprove Lysenko), oblivious of the tragic happenings in the USSR and the threat to genetics in that country, there were some in India like Panchanan Maheshwari, who were doing everything possible to fight this dangerous trend.[35] When in 1949, Maheshwari was invited by Maurice Gwyer, the Vice-Chancellor of the Delhi University, to head the new Botany Department, he discovered that a member of the teaching staff 'with a formidable personality' (perhaps this was Girija Prasanna Majumdar, a doctorate in botany from the University of Leeds) had begun teaching Lamarckist Michurin's biology (which argued that the inheritance of acquired characters in the genotype of the plant could be altered by a change in the environmental conditions, a view fiercely supported by Michurin's student Lysenko). Lysenko's discoveries had also fired public imagination in India through newspapers, which carried stories about his miracles on the agricultural fields of Soviet Russia. Maheshwari strongly objected to teaching Lysenkoism to young research students as if it was 'good science' and published Lysenkoist critiques in journals like *Nature*.[36]

Making a Stronger Presence

In the 1940s, the *Current Science* and the *Proceedings of the Indian Academy of Sciences* not only received a far greater number of communications from Indian women scientists (as a single author or co-author) when compared to the previous decade, but it also revealed a growing trend towards biochemistry. Further, none of those whose names appeared in the two journals during this period were doctorates by qualification, except Janaki (1931) and Asima Chatterjee (1944); Chatterjee had earned a doctoral degree from the University of Calcutta in organic chemistry, the first Indian woman to do so in any one of the sciences, from a University in India.

Among the physicists who were making their presence known at this time via research publications were Anna Mani, K. Sunanda Bai, C. Shantakumari (students of C. V. Raman at the Indian Institute of Science, Bangalore), besides K. Savithri (Annamalai University) and the mathematician-physicist, Aleyamma George (Trivandrum). The chemists, besides Asima Chatterjee, included Miss K. G. Gupte, K. S. Radha, Rashmi Bala Pandya, Miss V. S. Vakil, Roshan J. Irani, K. V. Kantak, Sunita Inderjit Singh (human physiology), Miss K. D. Paranjape, Alamela Venkatraman, A. Kameswaramma, Miss Vimala Puri, Miss K. Padmasini, Miriam George and Miss M. Prema Bai (several of them were attached to the Fermentation Technology Section of the Indian Institute of Science). Zoologists outdid botanists in number; Miss G. Mahadevan, Indira M. Gajjar (later Bhatt), Miss Meera Dey, C. K. Rathnavathy, Violet De Souza (later Bajaj),[37] Mary Samuel, H. Sunanda Kamath, T. S. Sarojini, Miss L. Yogeswaran, and Mrs Mubarika Shah. Only three botanists (excluding Janaki) communicated an original note or paper to either of these journals: Shanti Khosla, P. R. Bhagavathi Kutty Amma and C. K. Soumini.

Incidentally, Janaki's arrival in India in late 1948, after a nine-year self-imposed exile in Britain, coincided with the return of the physicist Anna Modayil Mani (1918–2001) to the country, to a scientific career at the Instruments division of the India Meteorological Department (IMD) in Pune. A much younger contemporary, Anna was a former research student (between 1940 and 1945) of C. V. Raman at the Indian Institute of Science (at this time working on fluorescence and absorption patterns and spectra of diamonds); she was not however awarded a doctorate by the University of Madras for her work despite five research publications because she did not hold a master's degree in the subject. In 1945, Anna enrolled at the Imperial College, London (Janaki was at the John Innes at this time, and the *Chromosome Atlas* had just been published) for a master's in Physics, but had ended up training at the Meteorological Office in Harrow, England, backed by a Government of India scholarship, for almost three years. Her chief job at the IMD involved arranging the meteorological instruments imported from Britain and gradually replacing these by instruments manufactured in India.

Edam, at Last

By late December 1948, Janaki reached Edam, physically exhausted after her Nepal expedition but elated nevertheless. It was a creditable achievement (she was by herself), leave alone for a woman, given the trying terrain of the country and its poor communication network. As for Edam, it was no more the haven of solace she had known it to be, after her mother's departure but her joy in wandering about its neighbourhood collecting plants (to carry back to Wisley) remained supreme. Her entries in the field notebook from this time record plants of the Annonaceae family growing on the mud wall at Illikunnu near Edam, a collecting location, she invariably visited. On 30th of that month, she visited Wayanad, collecting grasses en route from the wetlands along the Lakkadi-Pookkode Lake Road. Rich in biodiversity, serpent-groves (*sarpakavukal*) would be among her favourite hunting grounds. This time, she visited the Chirakkakavu (one of the oldest shrines in Malabar, located within a sacred grove of trees) and the Thayarkavu in Tellicherry, and walked along the river's edge at Koduvalli, the seaside, and the grounds of the Andalurkavu at Dharmadom, in search of plants. Janaki discovered wonderful specimens of the *Gnetum* (a liana or woody climber, perhaps *Gnetum edule*) at both the Chirakkakavu and the Thayarkavu. From the botanically rich Edam homestead, she gathered *Morus laevigata*, *Morus acidosa* and *Caesalpinia*.

On 17 January 1949, Janaki was in Shoranur with her elder siblings—Raghavan, Vasudevan and Cousalya, and their families. Plant hunting excursions were an integral part of her life, and more so when in Shoranur, where she had the wonderful company of Raghavan, a nature-lover exemplar. She went about collecting ground orchids, from the borders of fields, jasmine, and the cultivated *Clerodendrum* (of the family Lamiaceae, which she had seen in Nepal) from the family gardens. Janaki was particularly drawn to a *Citrus medica* growing in sister Cousalya's garden, which would figure in a later publication on the cytologenetics of the Citrus. After a week's break in Shoranur, Janaki would head back to Edam, where she would remain for almost a month, and during which time she indulged in more collecting: *kakkapoo* or *Utricularia reticulata* (carnivorous bladderworts) from rice fields in Tellicherry, wild *Canna indica*, water lilies from ponds and Amaranth from her brother Varad's homestead at Dharmadom. She would also pay a brief visit to Coimbatore, keeping Shoranur as base; her field notebook recorded a rare pink *Crossandra* (the endemic fire cracker flower of the family Acanthaceae) from this place.

In early February that year, she travelled to Bangalore, where her forester brother Kittu was now stationed, and visited the Lal Bagh Gardens in his company. From Lal Bagh, she collected several varieties of bougainvillea and Paspalum grass.[38] Kittu we know frequently sent her collections of seeds and plants to aid her researches.[39] Janaki would spend more time with sisters Lekshmi, Parvathi and Sumithra at Edam, before finally making her way back to England by the end of March 1949. She also appears to have visited Madras during this trip to India; her field notebook records a *Nyctanthes arbor* (night-flowering jasmine) from the place. Perhaps she flew out from Madras, and the plant was recorded during her short time in the city.

Back at Wisley

When Janaki settled into her attic flat in the Wisley Garden, after almost five months of being away, she was filled with a sense of foreboding about Edam. She was hardly prepared for what she saw during her visit. Edam, her refuge, that great house by the sea, stood before her like a ship that had run aground, in the absence of her mother and a steady source of income. Moved by the experience, the first thing she did was to prepare a detailed missive on the management of Edam. Among her various suggestions were the partition of the family properties (landholdings), sale of the Kuruvey House (which had belonged to her grandmother Kunhi Kurumbi, and acceded to Devi Amma, after her death in April 1918) and so on.[40] By mid-year, Janaki had her proposal circulated among members of the family. While elder sisters Lekshmi and Sumithra vehemently disapproved of it, Raghavan found it to be a beneficial intervention in the affairs of Edam, as expected from a responsible sibling like Janaki. He felt that the proposal was 'broad and will help in binding all members together and have a common interest in the Taravad.'[41] Sumithra on the other hand took it personal, interpreting it as Janaki's lack of appreciation for all she had done, and was doing, for Edam.

Lekshmi's objection (she had been residing at Edam while her graduate daughters Shantha and Savithri were employed in Madras) was of a more serious nature. She accused Janaki, who wanted the properties to be divided among members of the family, of being a radical, attempting to change the very system of inheritance on which a Thiya family was based (that is *marumakkathayam* or matrilineal inheritance within a joint family setup, to *makkathayam*, where sons and daughters inherit the family's property, implying a nuclear family), 'which cannot be done.' On the other hand, to Janaki's credit, she had only suggested a more reformist, mixed system of inheritance, combining certain features of *marumakkathayam* (for instance, she had suggested the properties be divided equally among the females) and *makkathayam*. The consequence of the discord that ensued was such that Janaki entirely stopped her remittances to Edam.

Figure 14.1 Janaki plant hunting in Nepal. Seen here with her Nepali assistant, who helped her preserve specimens in situ. Late 1948.

Source: Courtesy of the EK family.

NEPAL

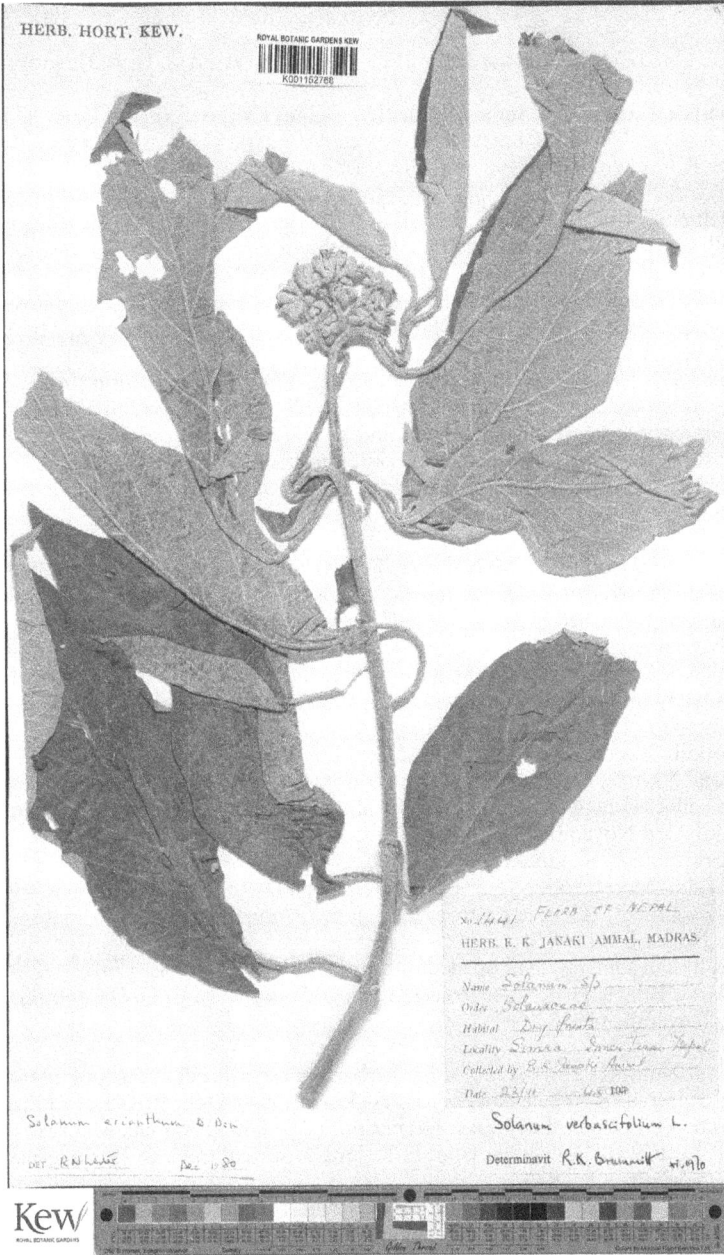

Figure 14.2 Solanum collected by Janaki from the dry forests of Simra, Terai (Nepal), November 1948.

Source: © copyright of the Board of Trustees of the Royal Botanic Gardens. http://specimens. kew.org/herbarium/K001152958

287

Notes

1 DP: C. 109 (J. 113), letter from E. K. Janaki Ammal to C. D. Darlington, dated 14 November 1948.

2 National Archives of India (NAI), New Delhi: 'Proposed visit of Dr E. K. Janaki Ammal to Nepal for Botanical Research': File No. 511-C.A., Secret Collection, 1948, Government of India, Ministry of EA & CR C. A. Branch.

3 Ibid.

4 Hellman, 'Curator getting around'.

5 DP: C. 109 (J 113), letter from C. D. Darlington to The High Commissioner for India, India House, London, dated 26 August 1948.

6 NAI: 'Proposed visit of Dr E. K. Janaki Ammal to Nepal for Botanical Research': File No. 511- C.A., Secret Collection, 1948, Government of India, Ministry of EA & CR C. A. Branch.

7 Ibid., File No. 511- C.A., Secret Collection, 1948, Government of India, Ministry of EA & CR C. A. Branch. The booklet, *Notes on journey to Nepal*, was not however filed with this document and so could not be accessed. However, it is possible that this was only a copy of 'Notes from a Journey to Nepal', by I. H. Burkill published in the *Records of the Botanical Survey of India*, 1913, Vol. IV, no. 4, pp. 59–140, which Janaki was already acquainted with.

8 Ibid.

9 Ibid., letter from V. M. M. Nair to E. K. Janaki Ammal, dated 10 November 1948.

10 Joseph Hooker's *Himalayan Journals*, dedicated to his good friend Charles Darwin was published in 1854 by the Calcutta Trigonometrical Survey Office.

11 That same year, an Indian group led by J. Banerji had visited the Tamur river valley in connection with the Kosi project, but otherwise Indian visitors to that Himalayan kingdom were few and far between.

12 Santapau, 'I. H. Burkill in India', p. 344.

13 Burkill, 'Notes from a journey to Nepal', p. 60.

14 Serial Numbers 1401–1498 of her herbarium records/field notebook plants seen or collected between Simla and Nepal and some parts of the latter, from 11th until 23rd November, 1948. RBG Kew, Library & Archives: Printed Notebooks, Herbarium E. K. Janaki Ammal, Madras

15 Burkill, 'Notes from a Journey to Nepal', p. 11.

16 DP: C. 109 (J. 113), letter from E. K. Janaki Ammal to C. D. Darlington, dated 28 November 1948.

17 An Indian Embassy had opened in Kathmandu in December 1947.

18 DP: C. 109 (J. 113), letter from E. K. Janaki Ammal to C. D. Darlington, dated 28 November 1948.

19 Ibid.

20 Serial Numbers 1601–1700 of her herbarium records/field notebook plants seen or collected in Nepal, November 29–December 4, 1948. RBG Kew, Library & Archives: Printed Notebooks, Herbarium E. K. Janaki Ammal, Madras.

21 DP: C. 109 (J. 113), letter from E. K. Janaki Ammal to C. D. Darlington, dated 28 November 1948.

22 Serial Numbers 1701–1800 of her herbarium records/field notebook also deal with plants in Nepal, but pertain chiefly to the first week of December 1948. RBG Kew, Library & Archives: Printed Notebooks, Herbarium E. K. Janaki Ammal, Madras.

23 Ibid., plants observed or collected in Sundarijal, Nepal, between 29 November and 4 December, 1948.

24 Rajbhandari, 'History of botanical exploration in Nepal'.

25 M. L. Banerji was awarded a doctorate by the University of Punjab for his dissertation, 'Flora of East Nepal' in 1958.
26 Banerji, 'Outline of Nepal phytogeography'.
27 For a short biography, see Swaminathan, 'Benjamin Peary Pal'.
28 DP: C. 109 (J. 113), letter from E. K. Janaki Ammal to C. D. Darlington, dated 14 November 1948.
29 Private Collection.
30 Pal, 'Some Aspects of Wheat Improvement in India', p. 59.
31 Ramanujam, 'Genetical Research as Applied to Plant Breeding in Post-War India', p. 65.
32 Ibid.
33 Chandrasekharan and Parthasarathy, *Cytogenetics and Plant Breeding*, pp. 184, 321.
34 Ibid., p. 443.
35 Maheshwari, 'Remembering Panchanan Maheshwari'.
36 See for example, Maheshwari, 'Lysenko's Latest Discovery—The Conversion of Wheat into Rye, Barley and Oats.'
37 For a short biography of Violet de Souza, see Deepika. S. 'The Colourful Life of Violet', http://connect.iisc.ac.in/2018/06/the-colourful-life-of-violet/
38 Serial Numbers 1801–1900 of her herbarium records, deal with collecting in Wynad, Tellicherry, Shoranur and Coimbatore between 30 December 1948 and 4 March 1949. RBG Kew, Library & Archives: Printed Notebooks, Herbarium E. K. Janaki Ammal, Madras.
39 In early April (after she had returned to Wisley), he would send her seeds of the asparagus and of the tree *Aleurites fordii* (Chinese wood-oil tree), both sourced from Lal Bagh.
40 The Kuruvey House, in the old part of Tellicherry town, would eventually be sold off in 1990 and every member of Devi Amma's family would receive a share.
41 E. K. Raghavan to E. K. Sumithra, letter dated Shoranur, 9 August 1949, private collection.

15

WISLEY II

Craze for Chromosome Counts

All this is good for the propagation of science amongst the
nobility but this sort of fragmentary work is very unsatisfying.
I am trying to strike a balance.

—E. K. Janaki Ammal (1949)[1]

On return to Wisley in March 1949, Janaki began working on her Nepal collections, besides growing vegetables and ornamentals from the seeds she had brought back with her. It was a particularly dry year; the Garden had also suffered much damage when a freak snow storm struck in the early hours one morning. An important event in the Garden Calendar that year was the private visit of the King and Queen on 11 May; they were shown around the Garden by Gilmour, the Director. They were so pleased with what they saw that they expressed a wish to visit again and this time include Janaki's cytology laboratory in their inspection.[2]

Visits to Kew

As usual, Janaki made regular trips to the Kew Gardens to collect plant material and to determine chromosomal counts, at the Jodrell Laboratory. She would perhaps have run into cytogeneticist Irene Manton of the University of Leeds (an expert on Pteridophyta, she had recently turned to algology), who would turn up from time to time at the Jodrell, to fix material; Manton had by this time already realised the usefulness of electron micrographs, a tool that would revolutionise her practice, leading to her election as Fellow of the Royal Society.[3] In the same period, another woman cytogeneticist, the Polish Maria Skalińska (1890–1977) was also visiting the Jodrell to conduct research; as a peer, at least on one occasion, Janaki reviewed Skalińska's research paper for the Linnean Society journal.[4]

Janaki's friend, N. L. Bor had only recently joined Kew as Assistant Director. William Macdonald Campbell was Curator of the Gardens and his team included Arthur Osborn, the Deputy Curator in charge of the Arboretum,

290

DOI: 10.4324/9781003267089-15

Charles P. Raffill of the Temperate House, Lewis Stenning of the Tropical House, Sydney A. Pearce of the Decorative Department and R. Holder in charge of the Herbaceous and Alpine collections. Friendly, warm and generous by disposition, Janaki had several friends in the Gardens, including Charles Raffill, who on one occasion sent her seeds of the 'Blue-purple Wallflower-like plant.'[5] When she needed help with identification or naming, she would turn to W. B. Turrill, Keeper of the Herbarium and Library. On one occasion, when Janaki found a *Rubus* growing in Wisley Common, which to her appeared like a natural cross between the blackberry and raspberry, she approached Turrill for confirmation, on behalf of Wisley, after working out its chromosome number: 'If . . . it is a true species we would be very grateful to know its name; chromosome no. is 2n=28.'[6]

Janaki did not fail to put up a small exhibit at the Chelsea Flower Show this year too (1949), on the theme 'The Ancestors of Cultivated Plants.' She would request Edward Salisbury, Director of the Kew Gardens, for 'some ancestors.'[7] On a later occasion, she would send cuttings to Kew of a very primitive type of sugarcane [Saccharum spontaneum] she had found around some hamlets in the Marku valley in Nepal. 'This should prove interesting as an ancestral form of the indigenous cane of India. I thought you would like to include it in the collection of *Saccharum* at Kew Gardens,' she suggested.[8] Sometimes, Gilmour would write to Salisbury seeking permission, on behalf of Janaki, to collect material from Kew. In December, he sent one such: 'Dr Janaki is at present working on a cytological survey of the Rhododendron and is very anxious to examine material of some of the species growing in the Temperate House at Kew. All she would require from a plant would be a single bud and I wonder therefore, if you would be kind enough to let her come with her Assistant to collect material, under the supervision of Mr Raffill or one of his men?'[9]

Requests for Her Expertise

People like Edward A. Bowles, President of the RHS, occasionally sought Janaki's assistance in working out chromosome counts of plants in his collection. In early November 1949, she wrote to him confirming his doubts that the Crocus distributed as *C. karduchorum* was only an unspotted variety of *C. zonatus*; the bulb had the same number of chromosomes (2n = 8) as the *C. zonatus*, which he had despatched to her for comparison, while the true *C. karduchorum* had 2n = 20. She had been away in India when the Crocus flowered and was thus late in replying to him.[10] Earlier, Bowles had sent her the Plymouth Strawberry, but she needed more time to examine it cytologically. Meanwhile, she had him send her seeds or plants of the *Fragaria chiloensis* (coastal strawberry), *F. elatior* (musk strawberry) and *F. virginiana* (wild strawberry): 'I would like to have all these in my little collection,' she wrote to him.[11]

From Frieda at Last

On her return to Wisley from India, Janaki found a letter from Frieda Blan-chard awaiting her, enclosing photographs of her family; she hadn't heard from her in a very long time and was simply overjoyed. 'The children all look grand! And you have not changed much. How much I wish to see you all.' Janaki thought the Blanchard girls already looked like biologists. Nelson's age (Frieda's son) marked the years she had not seen Michigan. 'I had a glorious 5 mths in India. I wish I could have stayed longer. The prodigal son [Jawaharlal Nehru] was welcomed home as I was. It was lovely to be with the family after nearly 10 years. Now I am very homesick. While in India I entered Nepal and collected somewhat,' Janaki told her. 'I did all my collecting on my own [she had employed an assistant only to fix material in situ). The trip to Nepal was a very arduous and expensive one but it was worth every penny I spent on it,' she added. Janaki was also more than happy to send seeds 'if the University of Michigan Gardens cared to have some,' but in return wanted Frieda to 'induce the Bot Dept to make [her] a collaborator in Asiatic Research,' an honorary position, that friend Eileen had held previously. 'It will give me the link I want with my University and America,' she explained. As always, she remembered Bartlett: 'Please give him my salaams and to Dr Davis [Bradley Davis, her doctoral supervisor] and others who know me.' Janaki at this time had no clue as to how long she planned to remain in England but felt quite strongly that she must work for her own country before she became 'too old.' She was also beginning to 'like systematic botany in relation to chromosome studies more and more' and wished 'to investigate the distribution of some of the less known genera of flowering plants of India.'[12]

Signs that she had become somewhat disgruntled with her work at Wisley were loud and clear. 'At Wisley, I am a sort of maker of tetraploids by colchicine. I have made dozens for the RHS. Also since the publication of the *Chromosome Atlas*, there is a craze for knowing chromosome numbers of each member's pet plant & all the Lords & Ladies send me their Rhododendron hybrids to count chromosomes. All this is good for the propagation of science amongst the nobility but this sort of fragmentary work is very unsatisfying. I am trying to strike a balance,' she remarked to Frieda. At the same time, she found the Wisley Garden 'very beautiful' and enjoyed living in her little attic-flat on top of her laboratory. 'We are right in the country in beautiful Surrey,' Janaki wrote, urging Frieda to visit England while she was still there. Post-war, 'England is recovering marvellously thanks I think to Marshall aid. Meat is scarce—but it does not trouble me as I am a vegetarian,' Janaki added.[13]

Visitors to the Cytology Laboratory

The Seligmans, Hilda and Richard, visited Janaki at the laboratory on 14 April 1949, about a month after she had returned from Nepal. To celebrate their friendship, and acknowledge the use of his collection, Janaki included Richard Seligman's name as co-author of a paper on the cytology of the *Dianthus*.[14] During a visit to Storo in Trentino in Northern Italy in June 1932, Richard had chanced upon a *Dianthus* growing at about 500 m altitude, identified later as *D. monspessulanus*. He had brought a plant home to Wimbledon, where for many years it survived beautifully in limestone scree, but by the time Janaki saw it in the mid-1940s, it had begun to show signs of 'exhaustion.' Richard returned to Storo in 1948 but was unable to find any *Dianthus* at that altitude or even higher at 600 m. However, he discovered some growing luxuriously at about 680 m on the banks of the Lake Garda, a few of which he brought back home and planted at several spots in his five-acre garden. Further collections were made in 1949 and 1950 from a number of locations around the lake. Janaki observed that the plants of 1932 were conspicuously different from the ones collected by Richard in 1948–50. In 1951, Janaki determined the chromosome numbers of the plants (grown from seeds brought back from Storo and distributed to a number of gardens); the ones with larger flowers and longer stems (collected in 1948–50) were chiefly tetraploids (only a few were diploids), while the ones collected in 1932, despite being hexaploid, were small-flowered. There were thus three races of the *Dianthus* growing in this region, she concluded, of which the diploids and hexaploids were small-flowered, while the tetraploids were always large-flowered. This, Janaki found to be similar to the cultivated Daffodil, in which the bigger flowers were all tetraploids, with the hexaploids hardly remarkable in their appearance.

Janaki received several other guests at her laboratory at Wisley in 1949. At the end of April, her friend Pio Koller from the Chester Beatty Research Institute visited; he would visit again, this time in the company of his wife. From the Indian subcontinent, M. I. Khan, from the office of the Chief Conservator of Forests, West Punjab (Lahore), visited in June, followed by H. C. Mirchandani from the University of Illinois (based at the Paddy Research Station, Nagina, United Provinces/Uttar Pradesh). The other visitors included Kamala Nair (c/o the India House) and chemist K. Padmasini from Bangalore (of the Fermentation Technology Section, IISc),[15] besides Janaki's friend, the phycologist Margaret A. P. Madge (or Rita Madge of the Royal Holloway College). Darlington's wife Margaret Upcott, with whom Janaki enjoyed a close relationship, visited in September. It was in March that year that Darlington had announced he was leaving Upcott, having begun a relationship with Gwendolen Harvey (babysitter to his children and wife of his cousin Jack Harvey). In deep anguish, he had once driven down to Wisley, to be comforted by Janaki. 'Dr Darlington was having a wretched time when I met him last. Margaret picked the lock of his Deed Box and took away important papers! It all sounds so fantastic and unbelievably complicated. What a sad thing for a scientific genius to be so throttled by domestic worries,' she had remarked to Col. Stern, their common friend.[16]

Looking Out for Kate Pinsdorf

Sometime during the year, Janaki contacted her old friend, the pioneering woman mathematician, Mary Evelyn Wells (1881–1965) of Maine, USA, a 1915 doctorate from the University of Chicago; Wells headed the Mathematics Department at the WCC, Madras, in 1926–27, where Janaki taught in the 1920s. Janaki requested Wells to help her ferret out the postal address of Darlington's first wife, Kate Pinsdorf, who taught history at Vassar, but why she wanted to do this remains a mystery. Having located Pinsdorf's last known address, Wells wrote to Janaki to share the same and also to, enquire whether she was still determined to carry out her earlier plan of travelling to the United States to visit her alma mater, the University of Michigan, before returning for good to India via Hawaii.[17]

Visit to Paris

In October 1949, Janaki went on a holiday to Paris, where niece Padma (daughter of forester brother, Kittu) and husband Padmanabhan, Diplomatic Secretary to the Indian Ambassador to France, currently resided. Padmanabhan had brought with him to Paris, their cook Ramunni, who would often indulge them with delicious Thiya fare. While in Paris, Janaki would hear from good friends, Col. and Lady Stern, who were also holidaying in France at this time. They were stationed at the hillside city of Blois in Centre-Val de Loire. 'I am glad to hear you and the "chauffeuse" [Lady Stern] are having a good holiday—so I am, even though I am in Paris and rather in the centre of things. How I wish you were both passing this way to have a real Indian meal with us!,' Janaki replied. It was in this letter that she expressed hope that Stern would nominate her for a membership of the Royal Geographical Society, 'if you feel I can be a member.' In one of his, Stern promised to lend her maps demonstrating the distribution of Magnolias: 'I am feeling very refreshed and ready for hard work of any type,' she replied.[18] Of course Stern was most happy to support her candidacy to the Royal Geographical Society; his friend, the botanist and plant hunter, Major Patrick Millington Synge (editor of the *Horticultural Journal*, who had gone on expeditions to Nepal, including once in the company of Colville Herbert Sanford Barclay) was also glad to support her nomination.[19]

Janaki Meets Nehru

Prime Minister Jawaharlal Nehru was at this time scouting the world for accomplished and highly educated Indians to contribute to the nation-building project. V. K. Krishna Menon, then Indian High Commissioner to the United Kingdom, would introduce Janaki to Nehru on 8 November 1949 at a reception held in the PM's honour at the India House. Although Janaki happened to travel with Nehru on the same flight to India a year ago, this was the first time they were meeting each other. A second and longer meeting would take place on the day following, the venue being Nehru's private room in the Waldorf on The Strand, near India House, where they had an hour-long 'fireside chat ranging from food and the agricultural problems of India to the origin of maize and mango' and on how her work could be made useful to India. Incidentally, only a few days earlier, Nehru had met with the physicist Albert Einstein at the latter's Princeton office.[20] 'It is rather nice to have a PM who is also a keen scientist,' Janaki would remark. 'He knows his Himalayan Flora very well. His home used to be in Kashmir,' she enlightened friend Col. F. C. Stern.[21] Nehru after all had to his credit a Tripos in the natural sciences from Cambridge. Eight years her senior (they were both born in the same month), Nehru struck an instant friendship with Janaki and invited her to India as an advisor to the Ministry of Agriculture.

As a New Year gift, Janaki presented Nehru with a copy of Darlington's Conway Memorial Lecture of 1948, *Dead Hand on Discovery* (the published version was titled *The Conflict of Science and Society*), and insisted he read it, even playfully threatening him with non-cooperation if he did not. However, nobody in power could have understood the import of Darlington's lecture better than Nehru himself, for only a month earlier he had addressed a student gathering at the Indian Institute of Science, wherein he stated that the quality and output of scientific work in India was not proportionate with the amount of capital spent on research nor with the potential scientific talent that the country could mobilise. In fact, one of the leading agricultural scientists in India would state that in America the career of an agricultural research worker began with the completion of his PhD degree, while in India it ended with it.[22] This undesirable situation, Nehru observed, was chiefly due to the dearth of a congenial 'research climate,' but this could be achieved only if a 'dynamic and inspiring leadership and enlightened administration free from pettiness and red tape' was ensured. 'If we are to make any worthwhile contribution to world's science and technology we should not hesitate to draw freely from the world's pool of scientific and administrative talent with a view to make up our deficiencies,' he added.[23]

In fact, two years later, in October 1951, at the annual General Meeting of the National Institute of Sciences of India, attention would be drawn to the promotion and safeguarding of the interests of scientists in India. It acknowledged that although the number of scientists had increased in number, 'the quality of an average Indian scientist' had reduced, owing to the absence of a 'scientific atmosphere.' Social and economic conditions and bureaucratic retapes were believed to have contributed to this situation. 'As a matter of fact, when science is receiving great attention in India, the scientist, as an individual, is perhaps being neglected,' Nehru observed perceptively.[24]

A few days after her meeting with Nehru, Janaki wrote to her friend, N. L. Bor of the Kew Gardens: 'Dear Dr Bor, I was overcome by the fog and turned tail from Portsmouth Road [in Ripley, near Wisley, the whole journey might have taken her close to two hours by train] on my way to Kew! I was thinking of the return journey. May I come when it is finer?' She was eager to tell him all about her meeting with Nehru: 'My Prime Minister (Nehru) sent for me when he was here the other day—he wants me to work for India. Again, I said I thought I could work better from this end—much as I love to be back home with my people—my psyche will not consent. I am too frightened of Venkatraman [the sugarcane expert, T. S. Venkatraman] and the other Raman [C. V. Raman]. What would you do? With Salaams from Janaki.'[25] In the same letter, she requested Bor to source some material for her of the *Magnolia pterocarpa*, through his contact in Assam, namely, the Chief Conservator of Forests, Shillong (perhaps C. Mackarness). She asked that it be packed in damp moss and sent the very day it was collected to her friend the cytologist A. R. Gopal-Ayengar of the Tata Memorial Hospital, Bombay, for if it dried out it would be impossible to obtain a chromosome count.[26]

The Conway Lecture

The choice of Darlington as Conway Memorial lecturer for the year 1948 was not a surprise, as there were several eminent biologists in the thirty-eight-year history of the series, and all without exception emphasised the need for 'humanistic values of scientific thought and discovery.' In recognition of his original contribution in the field of cytology and 'of his co-ordination of a diversified body of apparently disconnected facts,' Darlington had been awarded a medal by the Royal Society only a couple of years previously (1946).[27] Moreover, his life and work exemplified freedom of thought and expression, beyond the control of any authority or moral institution, even to the point of being labelled arrogant, brash and eccentric. It was precisely this sense of freedom and irreverence combined with intellectual brilliance which Janaki admired in him. Whether Janaki attended the lecture (delivered on 20 April 1948 and chaired by Richard Gregory at the Conway Hall at 25 Red Lion Square, London), we have no clue to, but there was no reason why she would have given it a miss. Whatever the case, in late January 1950, Janaki was eager to have a copy of the lecture and requested Darlington to send one to her Delhi address because she wanted Nehru to read it.

Vavilov's murder in 1942 on the orders of Stalin, and the rise of Lysenko had been aggressively exposed by Darlington in articles published in scientific journals and weekly papers and through radio talks (from 1946 to 1969), as not only the elimination of one man but the undoing of the whole field of genetics and its numerous practitioners worldwide. He not only expressed concern at this exceptional instance of suppression for ideological reasons, of science, but also at instances where there was a 'dead hand on discovery,' and where 'the conflict between science and society' were played out. For example, he criticised the animal breeding societies, the Ministry of Agriculture and the Forestry Commission in Britain for neglecting to use the modern science of genetics for the improvement of live stocks and forest trees.[28] In his Conway Lecture, Darlington argued forcefully that the need for security and stability often outweighed the desire for discovery; even scientific institutions that began with radical objectives later turned volte face to become suppressors of originality. This was also true, he said, of government departments, which ought to be supportive of new knowledge but for several reasons were simply the opposite, like the Soviet Union which had debunked genetics because it went against its political ideology. Darlington spoke persuasively on the need to reduce conflict between science and society through change in the conservative outlook of academic and social institutions and expanding the collaboration between science and the humanities.

Prompted by recent events, in particular the rise of Lysenkoism, Darlington and Mather had published *The Elements of Genetics* in 1949, in which they attempted to present the whole field of genetics, which they said had acted as an 'isthmus' linking continents that had hitherto been separate from one another.[29] The book cited the *Chromosome Atlas* at least four times, and credited the term 'triplopolyploid' to Janaki; we might recall that she had introduced it in her publication on *Saccharum* in 1941.[30] By the time *The Elements* was released, Darlington and Mather were already working towards the completion of *Genes, Plants and People: Essays on Genetics* (1950). A majority of the essays in the book were familiar to readers of journals like *Nature* and to those who had read Darlington's *Recent Advances in Cytology*, *The Evolution of Genetic Systems*, and Darlington and Mather's *The Elements*. The majority felt that the essays in their last book, although stimulating when they first appeared, had now lost their edge.[31]

Additions to the Library

Additions to Janaki's personal library in the year 1949 included a copy of the *Journal of the Royal Asiatic Society* for the year, Herbert Butterfield's *The Origins of Modern Science* (1949) and *Burma's Icy Mountains* (edition 1949, first published 1946) by her favourite explorer-nature writer, the botanist Francis Kingdon Ward. Ward's tome brought alive the extreme suffering and tenacity involved in exploring unknown lands, making it inspiring reading for Janaki who loved travel, exploration and adventure. Moreover, she had a deep love for Burma, a place she had visited in 1927. Ward's book provided a vivid account of Burma prior to the outbreak of World War

II, with descriptions of its terrain, its bountiful botanical diversity, and the fauna and people encountered during his expedition. In 1948, Ward's paper, 'Botanical Exploration of Manipur' had been published in the *Journal of the Royal Horticultural Society*.[32]

Science News from India

In May 1949, the former sugarcane expert, T. S. Venkatraman, enjoying his retirement in Madras, sent a letter to the editor of *Current Science* on the inter-generic sugarcane-bamboo hybrid, his greatest achievement of 1936, in response to a recent piece of research by the Russian cytogeneticist, N. P. Avdulov, the French agrostologist H. Prat and the Cornell agronomist C. A. Taylor, indicating a close relationship between the bamboo and the Panicoid grasses to which the sugarcane belonged. Venkatraman noted that 'there were certain Botanists who doubted the possibility' when the hybrid was first produced but that subsequent work on the chromosome numbers (by Janaki) and the morphological and the histological characters of the F1 hybrids (by Ekambaram and Kutty Amma) had confirmed the genuineness of the hybrid, and now the recent researches of Avdulov et al. had reaffirmed this fact.[33] Two months later, he sent 'a sequel' to the letter, in which he stated that C. A. Taylor believed the systematic positions of sugarcane and bamboo in the family of Panicoid grasses needed revision, in the light of the successful hybridisation between sugarcane and bamboo. Taylor was of the opinion that the floral morphology and anatomy of the genera *Arundo* and *Ochlandra* (the reed bamboo endemic to the Western Ghats) could throw light on the problem, and wished to get in touch with researchers in India who had worked on the two genera. On the other hand, if the success of the Sugarcane x Bamboo cross was because the two genera were not as distant from each other as was hitherto believed, did that make Venkatraman's intergeneric cross a less 'revolutionary step' in cane breeding?[34] He did not however choose to dwell on this.

The year 1949 was also the year that Janaki lost her friend Birbal Sahni, who only a few days earlier had inaugurated his new institute of palaeobotany. Besides Sahni, Ramunni Menon, the former Vice-Chancellor of the Madras University, responsible for establishing the University laboratories such as the UBL, passed away that year, as did botanist M. S. Sabhesan of the Madras University, who had been a fellow member on the Board of Studies for Botany, with Janaki. In August, on the second anniversary of India's Independence, the Government confirmed the decision to establish eleven National Laboratories, under the auspices of the Council of Scientific and Industrial Research (CSIR), to aid the nation-building process. That same year, the Institute of Nuclear Physics was established (growing out of the Palit Research Laboratory) for teaching and cultivating nuclear physics, with Meghnad Saha as its director.[35]

Accepts Nehru's Invitation

Janaki decided to take up Nehru's offer of the post of Director of Agriculture (as a special scientific advisor to the agriculture ministry). Ten days before she departed for India, she wrote to Frieda Blanchard that she was at Wisley working on Rhododendrons: 'Very difficult material cytologically.' She was still 'however only counting chromosomes to map out the distribution,' rather than doing anything of a more serious nature. 'I am making this sort of study a hobby of mine and wish I could do more genera but at Wisley we have to satisfy the curiosity of 36,000 fellows of the RHS and now that "tetraploidy" is become a topic of the dinner table of the Lords & Ladies of England—every great squire or Lord or a Giles Loder wants to know if the Rhododendron his grandfather "made" is a tetraploid—this is a distraction from one's major work.' The only bright spot, she commented, was having an American postgraduate student from the Missouri Botanical Gardens (perhaps F. G. Meyer) working with her at the laboratory, on the genus *Cotoneaster* (which had a strong presence in the Sino-Himalaya, a border zone of immense interest to her). Janaki was curious to know if Frieda was planning to attend the International Botanical Congress in Stockholm the next year. 'Do persuade Prof. Bartlett to come to it. I want to see him so much,' she begged. Janaki mentioned to Frieda that she had been keeping in touch 'with work in India' since her last visit home (in late 1948).[36]

On 31 December 1949, on the eve of her departure for India, Janaki enclosed a map of the distribution of *Morus* that she had prepared, in the post, to dear friend Stern, which she hoped he could improve upon and perhaps even use in his lecture; she handed over the lantern slides of the *Magnolia* with a copy of the *Morus* map to Gilmour, the Director of RHS Wisley, for safe-keeping.[37] She had just discovered two more *Nerine* triploids in the Kew collection, the news of which she eagerly shared with Stern. 'Nehru has given me the status of a Director of Agriculture while I am in India!' she announced to him, full of excitement and a certain pride.[38] Nehru's personality and charisma had charmed her without doubt into accepting the post, but she was still unsure of what awaited her, and whether she could cope with the bureaucracy and patriarchy she would be faced with. She had at this time completed her investigations on the *Cyclamen* received from Messrs. Frampton's Nurseries, chromosome counts of the early interspecific hybrids of the *Nerine* discovered at Kew and of the *Rhododendron loderi* forms received from Giles Loder at Leonardslee. In her absence, her assistant Enoch was given charge of the Cytology Department.[39] On the eve of her departure, Janaki gave away thirty-nine packets of seeds of plants of horticultural, economic and botanical interest, collected by her from Nepal and South India in 1948–49 to the Kew Gardens.[40]

Figure 15.1 Staff members outside the Laboratory, RHS Garden, Wisley, including J. M. S. Porter, R. E. Adams, J. S. L. Gilmour (the Director), I. C. Enoch, G. Fox Wilson, N. K. Gould, Janaki and D. E. Green. October 1950.

Source: Courtesy of the RHS Wisley Collections.

Notes

1 BHL, UoM: Blanchard Family Papers c. 1835–c. 2000, E. K. Janaki Ammal to Frieda Blanchard, letter dated 8 May 1949.
2 RHS, Lindley Library: RHS/Minutes/WY/WY Garden Advisory Committee, Box No. 5, Report of a Meeting of the Wisley Advisory Committee held on 30 May 1949.
3 'Review of the work of the Royal Botanic Gardens, Kew, during 1949'. p. 13.
4 The Linnean Society: LL/10; this document contains a fragment of the crossed out version of Janaki's review of Skalińska's paper titled, 'Polyploidy in *Valeriana officinalis* Lin. In relation to its ecology and distribution', which eventually appeared in the *Journal of the Linnean Society of London* 53, no. 350 (1947): 159–186.
5 C. P. Raffill to E. K. Janaki Ammal, letter dated 193, Kew Road, Richmond, 30 August 1949, private collection; Raffill would be no more by 1951. He had just retired after a five-decade long career at Kew.
6 RBG Kew, Library & Archives: 1/RHS/10A, E. K. Janaki Ammal to W. B. Turrill, 16 July 1948.
7 Ibid, E. K. Janaki Ammal to Edward Salisbury, letter dated 16 May 1949.

8 Ibid, letter dated 19 September 1949.
9 Ibid, J. S. L. Gilmour to Edward Salisbury, letter dated 22 December 1949.
10 RHS, Lindley Library: EAB/2/6/2/5, E. K. Janaki Ammal to E. A. Bowles, Myddelton House, Enfield, letter dated 7 November 1949.
11 Ibid.: EAB/2/4/1, E. K. Janaki Ammal to E. A. Bowles, letter dated 27 May (no year given, but perhaps 1949).
12 BHL, UoM: Blanchard Family Papers c. 1835–c. 2000, E. K. Janaki Ammal to Frieda Blanchard, letter dated 8 May 1949.
13 Ibid. The Marshall Plan of 1948, also known as the European Recovery Program, and put into force by the 80th United States Congress, provided post-war reconstruction aid to Western Europe.
14 Janaki Ammal and Seligman, 'Notes on the Occurrence of Chromosome Races in *Dianthus monspessulanus* in Northern Italy', 1952.
15 See Padmasini, 'Antiseptic Culture of Ragi Seedlings—and Their Response to Vitamins.'
16 RBG Kew, Library & Archives: F. C. Stern and Lady Stern correspondence, E. K. Janaki Ammal to F. C. Stern, letter dated 10 October 1949.
17 DP: C 109 (J. 113), Mary Evelyn Wells to E. K. Janaki Ammal, letter undated 1949.
18 RBG Kew, Library & Archives: F. C. Stern and Lady Stern correspondence, E. K. Janaki Ammal to F. C. Stern, letter dated 10 October 1949.
19 Ibid., letter dated 1 December 1949.
20 *The Times of India*, 11 November 1949.
21 DP: C 109 (J. 113), undated 1949.
22 Kadam, 'The light of agricultural research in India'.
23 'Maximization of scientific effort'. *Current Science* 18, no. 10 (October 1949): 365.
24 'Promotion and safeguarding of the interests of scientists in India'. *Current Science* 1, no. 7 (July 1952): 180.
25 RBG Kew, Library & Archives: 2/BUR/2A, E. K. Janaki Ammal to N. L. Bor, letter dated 15 November 1949.
26 Ibid., 17 November 1949.
27 Darlington, *The Conflict of Science and Society*, p. vii.
28 Lewis, 'Cyril Dean Darlington 1903–1981', p. 144.
29 Preface to Darlington and Mather, *The Elements of Genetics*, pp. 7–8.
30 *The Elements of Genetics*, p. 423.
31 See for example, a review by Kenneth W. Cooper of the Department of Biology, Princeton University, *Science* 113, no. 2938, 20 April 1951, p. 456.
32 Vol. LXXIII, Part 2, February 1948, pp. 37–43.
33 *Current Science* 18, no. 5 (June 1949): 218.
34 Parthasarathy, 'Chromosome Numbers in Bambusae'.
35 In 1958, it would be renamed the Saha Institute of Nuclear Physics, in memory of Saha who had died two years before.
36 BHL, UoM: Blanchard Family Papers c. 1835–c. 2000, E. K. Janaki Ammal to Frieda Blanchard, letter dated 20 December 1949.
37 Ibid., E. K. Janaki Ammal to F. C. Stern, letter dated 31 December 1949.
38 Ibid.
39 RHS, Lindley Library: RHS/Minutes/WY/WY Garden Advisory Committee, Box No. 5, Report from the Cytological Dept, 18 November 1949–5 May 1950.
40 Ibid, E. K. Janaki Ammal to Richmond, letter dated 29 December 1950.

16

DELHI

Director of Agriculture

I am yearning for the silence of my little attic flat in Wisley and
the quiet of the Garden.

—E. K. Janaki Ammal (1950)[1]

Janaki was back in India early in the first week of January 1950 having
accepted Nehru's offer of appointment as a Director of Agriculture (Special
Scientific Advisor to the Ministry of Agriculture, Government of India), but
she would remain attached to the RHS Wisley as cytologist. She was on a
short-term contract (she would not have it any other way) with the government, which was keen on using her expertise to tackle the agricultural
issues and food scarcity facing the nation. Janaki had been only a few days
in India, when she received the happy news of her being elected member of
the Royal Geographical Society (the certificate was dated 4 January 1950).
The Society considered her a suitable candidate for election, on the grounds
that she had travelled in the United States, Canada, Japan, China, Malaya,
Burma, Nepal, Ceylon, Europe and the United Kingdom and extensively in
India and that she was interested 'in races of mankind and plant Geography
in relation to origins of cultivated plants and the migration of peoples also
speciation.'[2] About her travels to Canada, Japan, China and Malaya we
know nothing (and it is doubtful she travelled to these countries), but we
do know the last three countries interested her very much from the cytogeographical perspective.

The Grow More Food Campaign

When Janaki arrived in India, activities revolving around the Grow More
Food campaign (begun in 1946) were escalating, with even students being
mobilised to grow more food.[3] Public campaigns had been launched, using
government institutions such as the All India Radio, informing people that the
government had made all the necessary arrangements for cultivators to adopt
improved methods of agriculture. Bombay's minister for Agriculture and

DOI: 10.4324/9781003267089-16

Forests in a Marathi broadcast on All India Radio explained the new schemes for increased food production in the province, adding that much depended on the cultivator himself. Over the next couple of years, the food deficit was to be drastically reduced. The Minister stated that the increased production of food depended on such factors as improved seeds, more effective manures, irrigation, soil-conservation and so on, which the government was prepared to make available. It would undertake bunding operations to deal with soil erosion.[4] A 35 lakh plan for Bombay was announced towards this.[5] At a meeting at Vile Parle, Balasaheb Gangadhar Kher, the Chief Minister of Bombay, observed that the most important factor of the Grow More Food campaign was to bring about a change in the psychology of the cultivator, who would be thus empowered to play a vital role in the drive to produce more food.[6]

The introduction of advanced farming technology was another objective, and Pandit D. K. Mehta, the Central Provinces Agriculture minister, attended the demonstration of a Fordson Major tractor (manufactured by the Ford Motor Company Ltd, UK) in Nagpur.[7] Massey Ferguson, the British company manufacturing farm machinery, advertised their tractor alongside Nehru's appeal: 'We must produce enough food for all the millions who inhabit India by 1951.'[8] It added: 'Grow More Food—More cheaply with Ferguson.' Further, a 'Waste No Food and Save More Food' Exhibition was opened by Raja Maharaja Singh, the Governor of Bombay, who said: 'My Government and I are extremely keen that the "Grow More Food" campaign and its corollary "Waste No Food" should be taken to the heart of this city and province.' The exhibition displayed hundreds of posters, charts and cartoons illustrating the enormous wastage of food, and the measures to be adopted to increase agricultural output.[9] To save cereals and return them to the general pool, the Union Minister of Health Rajkumari Amrit Kaur initiated the slogan 'Miss a Meal a Week.'[10] In an effort to make citizens food-conscious two forms of pledges to be taken at public gatherings were released by the Director of Publicity, Bombay.[11]

Money was also sanctioned for improving the quality of manure and a subsidy of two rupees granted for every new manure pit dug. Progressive farmers would begin to receive official recognition, with their farms becoming agricultural demonstration centres. During a visit to the United State of Travancore-Cochin in early December, the Union Minister for Food Supply Jairamdas Daulatram noted with appreciation the state's efforts towards achieving self-sufficiency in food. No part of India was more highly deficient in cereals than Travancore-Cochin at this point in time and fish served as an added supplementary diet for the population. At the behest of the state's Minister of Agriculture, Mr E. Ikkanda Warrier, Daulatram participated in the christening ceremony of three deep-sea fishing boats belonging to the West Coast Fisheries Ltd., besides the inauguration of the Chalakudi irrigation project and the inspection of the huge land reclamation scheme launched by the Travancore-Cochin state at Valasserikandam, which contemplated the conversion of 25,000 acres of forest land adjoining the Anamalai Road and Parambikulam and Sholayar valleys of the Anamalai ghats; incidentally, the Union Minister himself occupied a seat on one of the bulldozers and witnessed with 'delight the quick felling of a huge tree.'[12]

The state hoped the public limited company would in due course make a substantial contribution to the Grow More Food campaign; in this context Daulatram cited the example of the United States, which alongside boosting rice production was focused on developing the potential wealth of 'a food as cheap and nutritious as fish.' In addition, the ICAR approved of a scheme for research on tapioca and sweet potato in the state. A District Rural Development Board had also come into being by this time to aid the Grow More Food campaign. Further, the Department of Fisheries, Bombay, began to promote consumption of vitamins as a supplement to food; a typical advertisement read: 'Why be afraid of Food Scarcity when vitamins can enrich your diet and help tide over the crisis? Shark-liver oil Elasmin Pearls fortify your diet with extra vitamin A & D from fresh liver oil of Indian sharks.'[13] The Grow More Food campaign launched by the United State of Travancore and Cochin (a short-lived state, created in 1949 and lasting until 1956, its first Chief Minister being Parur T. K. Narayana Pillai of the Congress) went a step further and suggested the abolition of landlordism.

The committee appointed by the Government of Travancore and Cochin a few months before integration (which took place on 1 July 1949), when T. K. Narayana Pillai was Premier of Travancore, to report on the question of investing the proprietary rights of land with agricultural tenants, had finalised its report; the unanimous decision was that landlordism should be abolished without delay, and landlords should be suitably compensated. The proposal was to reclaim 50,000 acres of forest land and allot plots of land up to ten acres and also to provide financial help to those willing to cultivate these plots. By this time, about 200,000 acres of land in Travancore and Cochin had already been brought under cultivation. The major and minor irrigation schemes had been given a boost, and the Government of India allotted seed grants to initiate food schemes; it also provided fifteen tractors. Jute cultivation experiments made in some parts of north Travancore over an area of 300 acres were proving to be a success.

Doubts were however expressed as to the feasibility of achieving self-sufficiency in food by the end of 1951.[14] There were several setbacks. It was expected that India would raise an additional 2,30,000 tons of food grains by transferring the acreage under sugarcane to food crops, but sugar scarcity led to this plan being abandoned.[15] In several provinces, tube well irrigation had been planned but received a setback because the International Bank of Reconstruction and Development had rejected the Indian request to fund the project. Similarly, plans for intensive cultivation which required massive imports of fertilisers also suffered a setback due to the Bank's refusal of a prospective loan; not much progress had been made in the development of compost manuring either. There was also no money to import agricultural machinery. Moreover, acreage under food crops had been diverted to cash crops like cotton and jute, with the Commerce Ministry adopting a strong line in favour of this move.[16]

By the end of 1948, it had become quite evident that the Grow More Food campaign was far from successful, with the country continuing to import grains from abroad and in ever-increasing quantities. Although the acreage under cultivation of food crops had substantially increased, there was no corresponding increase in yield. The reason for this, the Government reasoned was the poor quality of soil, which required better quality fertilisers for improvement. It was hoped that with better and more intensive methods of cultivation, combined with adequate manuring, even with the existing acreage, a major contribution could be made towards reducing the food deficit. What was worse, the extension of acreage had been accomplished at the expense of pasture lands and forests, and this was detrimental to livestock, forest resources and rainfall. The need for a close working relationship between agricultural policy and forest policy began to dawn on at least some policy makers.

At Dehra Dun

In Delhi, on 15 February 1950, Janaki recorded a Hibiscus on Tuglaq Road in her field notebook. A week later, she would leave for the Forest Research Institute, Dehra Dun; on 21 January, she recorded a *Prunus* from the place. At the Institute, she managed to persuade its President, C. R. Ranganathan (the first President of the Institute, post-independence, 1947–54), to begin a tree cytology and breeding laboratory there. 'There is a man trained at Minnesota, a PhD doing what they called "Documentation." I think it means filing newspaper cuttings! He has been moved to the Systematic Dept. and a Small Cytology lab rigged up in one of the rooms,' Janaki wrote to Darlington. She had made the first smear that day, and had advised the man to begin work on *Shorea robusta* (Sal) and *Tectonia grandis* (Teak), 'our most important timbers.'[17] Incidentally, Janaki's friend, N. L. Bor had worked as forest botanist at the Institute between 1937 and 1942. Janaki would send Darlington a sample of her work at the Institute, but this has not been traceable.

With the Bureaucrats

Janaki suddenly found herself extremely busy, spending close to a month, 'studying the work of the Ministry with its dozens of Joint, Deputy and Under Secretaries' and touring the provinces. 'The Commissioner for Food Production likes to hold my hand—I mean figuratively—not literally—when he goes out to see how the Grow More Food schemes of the Government are taking shape,' she joked in her letter to Darlington. She had also been deputed to Universities and Research Stations 'to goad them to activity' and discovered that she was highly influential. 'What surprises me is that they are ready to do what I ask them to do,' she would remark to Darlington. Janaki was about to leave for Assam and Orissa via Calcutta. At Cuttack, she would stay at botanist Prankrushna Parija's house, 'where we stayed in 1937,' she reminded Darlington. Parija was a doctoral student of F. F. Blackman (as were Indian botanists T. Ekambaram, R. S. Inamdar and Shri Ranjan), and made his name with the Blackman-Parija paper(s) on the respiration of apples published in the *Proceedings of the Royal Society* (1928), and considered a classic in the field of plant physiology. He was also an eminent educationist and a humanitarian.[18] Parija's daughter, whom Darlington and Janaki had met as a young girl in 1937, was 'now a full blown Doctor of Medicine,' and Parija himself had become Pro Vice-Chancellor of the Banaras Hindu University, where Janaki's niece Leela Desai was based (at the Botany Department of the Mahila Mahavidyalaya).[19]

'Various good posts are being dangled in front of me to tempt me to stay on in India. I have said No to all. I am yearning for the silence of my little attic flat in Wisley and the quiet of the Garden. I wonder if I am doing the right thing in refusing,' she wrote to Darlington, a remark strikingly reminiscent of McClintock's comment when she began to receive public attention: 'I don't like publicity at all. All I want to do is retire to a quiet place in the laboratory.'[20] Janaki wondered how Darlington was faring, his wife having walked away with the keys to his document box; Janaki was particularly concerned about the fate of her 'letters etc.' She hoped he was still able to see his children; in fact, Janaki had been collecting 'some quaint Indian toys' for them.[21] In another letter, she told him how much Somerset Maugham's *Of Human Bondage* (1915), which she had just finished reading, had made her think about him. 'C [Cyril]—why did you not trust Pio [Koller] and me with the truth—We suffered so much with you,' she wrote, obviously referring to Upcott, but what that 'truth' itself was we do not know.[22]

Collecting Tour to Assam

On 6 March 1950, Janaki returned from her trip to Assam, her friend N. L. Bor's home turf for several years, loaded with specimens including live orchids (from the Khasi and Jaintia Hills) and magnolias, besides tubes of fixed material of michelias, rhododendrons and magnolias endemic to the region. Incidentally, Bor's doctoral thesis submitted to the University of Edinburgh in 1930 was on the Synecology of the Naga Hills forests in Assam. Synecology referred to the study of whole plant or animal communities (or whole ecosystems) as opposed to autoecology, which focused on individual plants and their relationship with the environment. In fact, from the Naga Hills, the forest department had sent Janaki a *Coix* (belonging to the grass family, like the *Coix lachryma* or Job's Tear, which she had earlier examined cytologically). She was hoping to include her findings in the paper she was preparing for the Seventh International Botanical Congress in Stockholm in July that year. However, she also had some distressing news to share with Darlington. In Assam, she had seen 'valuable forests' being destroyed in 'the name of the Grow More Food campaign.' She had 'gone trekking 37 miles from Shillong in search of the only tree of *Magnolia grif-fithi* [perhaps it was Bor who had mentioned the tree to her] in that part of Assam but found it had been burnt down.' She was deeply pained by what she saw: 'It is terrible to see so much done in India without any scientific principle behind it. I find myself losing my temper with Ministers and Food Commissioners. The Govt. of India wants me to stay on. Very handsome salary is being offered. That does not tempt me. I hate to be a public figure and I do not like Congress men—Nehru is the only exception,' she would remark to Darlington.[23]

Visit to Patna

Janaki also called on the Central Potato Research Station at Patna, established in August 1949, and the brain-child of S. Ramanujam, B. P. Pal, Push-karnath and R. S. Vasudeva. Ramanujam, the Second Economic Botanist at the IARI, who had been appointed Officer on Special Duty at the Station 'was doing very good work,' she informed Darlington, and very soon it would be 'possible to buy seeds of pure strains of potatoes as we [the British] do mustard and radish seed . . . It is quite possible to do it in India.'[24]

Planning a UNESCO Conference

In early March 1950, Janaki had written to Darlington, to let him know that Alexander Wolsky, the Hungarian biologist of the UNESCO Science Co-operation Office for South Asia (Delhi), was intending to contact him regarding a 'Symposium on the Cultivated Plants of South Asia,' which he was organising under the aegis of the Indian Society of Genetics and Plant Breeding. Incidentally, it was when Janaki was away in Britain that an Indian Society of Genetics and Plant Breeding had come into existence. The war and the post-war reconstruction work would give a major boost to genetical research and plant breeding in India, as it did elsewhere. In January 1941, at the Indian Science Congress session held at the Banaras Hindu University, Varanasi, the idea of founding a society and a journal exclusively devoted to genetics and plant breeding had been first proposed; this was accepted unanimously as it was thought both 'desirable and natural.' As a result of this, the Indian Society for Genetics and Plant Breeding and the *Journal for Genetics and Plant Breeding* came into existence by the end of that year (1941). At the Banaras meeting, the rice expert Ramiah expressed hope that more crop plants would begin to be described in the journal from the genetical point of view so as to expose the breeder to the material available with his [her] colleagues in other parts of

India. The wheat geneticist B. P. Pal of the IARI had been elected the Society's first secretary and the journal's first editor. The journal was thought to be a welcome addition to publications on genetics and plant breeding 'providing a valuable outlet and source of reference for the ever-increasing work of the research institutes of India.'[25]

Wolsky wanted Darlington to chair the Symposium. It was Janaki who had suggested the theme, when members of the Society gathered in Delhi early that year, to discuss the idea of holding a symposium. Janaki was eager to have Darlington accept the invitation because she wanted him to inspire botanists in India to take up research on the lines of 'shall I say *our* book,' the *Chromosome Atlas of Cultivated Plants*. It 'is very vital for India that you visit it now. I shall be having a talk with Nehru on the dangers of pseudoscience springing up in India under Congress leadership,' she told him. 'I am afraid some people are lending support to Lysenkoish projects. There was actually a question in our Parliament about the inadvisability of using improved seed as it would impoverish the soil!' Janaki wrote in exasperation. She was 'torn between staying on and returning to England. . . . Perhaps I am missing the chance to serve my country. I could not have a better position in India as this one in the Ministry. All the same I want to return to England!'[26] Darlington would however decline the invitation. In his place S. C. Harland, Edgar Anderson and Arne Müntzing would be brought in as expert consultants. Delegates from the South Asian countries of Ceylon, China, Malaya and Pakistan would also attend.

Founded in November 1946, UNESCO had devoted one of its sections to the exact and natural sciences and its applied forms (agriculture, medicine, technology), on the understanding that it was no more possible for science to develop freely without an international collaboration of scientists and scientific institutions. Science could no longer be studied individually; it demanded teamwork and a sharing of the results of investigations, as had been done during the war years, which resulted in the atom bomb, penicillin drugs and radar. Some of the democratic governments had set up Science Co-Operation Offices in their capitals during the war, the largest being the British Commonwealth Scientific Office in Washington and the United States Scientific Mission in London. There was also the Sino-British Scientific Office in China, headed by Joseph Needham, the head of the Natural Sciences Section of UNESCO, which dealt not only with the war sciences but also the pure and applied sciences, useful for post-war reconstruction. Needham proposed a scheme to establish similar offices on an international scale, devoted to peaceful purposes. S. S. Bhatnagar, President of the National Institute of Sciences of India, had also proposed a similar idea, of 'regional scientific co-operation stations.' In 1947, UNESCO began to establish Science Co-Operation Offices in various regions, beginning with Rio de Janeiro (for Latin America), Cairo (for the Middle East) and Nanking (for the Far East). The fourth office was opened in Delhi (for South Asia), in the Science Buildings of the Delhi University. Janaki was already by this time toying with the idea of applying for a post with UNESCO South East Asia, as she was keen on exploring Indonesia, the Philippines, Siam (Thailand) and also China so as 'to develop Cytogeography on a large scale.' She had become bored with the research atmosphere, or more, the lack of it, at Wisley. 'I have a feeling RHS is not interested in research. [James] Gilmour is not anyhow,' she would comment in a letter to Darlington.[27]

The symposium on 'The Origin and Distribution of Cultivated Plants in South Asia' would be held in Delhi in January 1951, funded by the UNESCO South Asia Science Co-operation Office. Organising symposia periodically, the Society believed, was a way of exchanging information, sharing knowledge and discussing various issues relating to plant breeding and cytogenetics; this was the first time a symposium was being organised with respect to tropical crop plants. From India, about twenty-seven members of the Indian Society of Genetics and Plant Breeding either contributed to or were present at the Symposium, besides delegates from various scientific institutions. K. Ramiah, B. P. Pal, S. K. Mukherjee, R. D. Asana, S. Kedharnath, S. M. Sikka, T. K. Koshy, P. Maheshwari, N. Parthasarathy, S. Ramanujam, P. N. Bhaduri and S. P. Agharkar were among the more prominent names. The papers presented at the Symposium ranged from the origin and distribution of rice, wheat, banana, mango, citrus, sugarcane, millets and brinjal to orchids, cotton, jute, sesame, spices and palms. Müntzing of the University of Lund would later travel to Hyderabad to speak at the meeting of the Indian Academy of Sciences (Section B) on genetics in relation to plant breeding.

The Society chose this opportunity to express its gratitude to Vavilov, who had stressed the importance of collecting and using wild relatives of cultivated plants in breeding improved crop plants. As we know, these wild relatives exhibited a wide genetic diversity across specific landscapes, which Vavilov called the 'centres' of origin of cultivated plants. From these primary centres, species migrated across time and space, adapting to different environmental conditions (through polyploidy for example) and evolving new and improved forms. The collection of plants with desirable characters from the centres of origin and the study and utilisation of such plant material to breed improved plants for cultivation demanded in-depth knowledge of several branches of botanical science including taxonomy, morphology, cytology and genetics, which the Society believed could only be accomplished through collective effort.

Congress at Stockholm, 1950

The Seventh International Botanical Congress (12–20 July 1950) held in Stockholm was attended by an unprecedented number of botanists (about 1,450 from over 50 countries). Janaki and P. Maheshwari represented the Department of Scientific Research, Government of India at the Congress.

She was still attached to the Wisley Garden at this time, and it was the second time she was visiting Stockholm. Significantly, she was the sole person from India to present a paper at the Congress. Among the Indians present at the Congress were also K. C. Mehta of Agra University, J. Venkateswarlu of Andhra University, R. V. Sitholey of the Birbal Sahni Institute of Palaeobotany, K. A. Chowdhury of the Forest Research Institute, Dehra Dun, F. R. Bharucha, representing the Indian Ecological Society and the University of Bombay, and T. S. Sadasivan of the University of Madras; they would all be entertained by the diplomat R. K. Nehru (a close relative of Jawaharlal Nehru) and his wife at the Indian Embassy in Stockholm.

The President of the section on Genetics (General) was Øjvind Winge, the Danish yeast geneticist, and among the Vice-Presidents were such eminent biologists/geneticists as Edgar Anderson, G. W. Beadle, D. G. Catcheside, R. E. Clausen, K. Mather and Elisabeth Schiemann of the German Research University. Schiemann was one of the earliest among women to hold a doctorate in genetics (1912), and came to be recognised for her work on cultivated plants,[28] just as Janaki would by this time. It was as part of the session, 'Induced and Natural Polyploidy,' chaired by Schiemann and L. F. Randolph, that Janaki presented her work on polyploidy and migration (or 'wanderings' as she called them) in the genus *Magnolia*. In her paper, she discussed the significance of chromosome numbers in determining the taxonomic position of the family Magnoliaceae. She demonstrated how the migration of Magnolias before and after the Ice Ages was related to the present distribution of diploid and polyploid species in Asia and North America.[29] Fellow presenters included Irene Manton and Tischler; R. R. Gates and P. N. Bhaduri acted as discussants. Darlington was the President of the Cytology section (and among the Vice-Presidents were McClintock, K. Sax, C. L. Huskins and C. A. Jörgensen), and also delivered a 'general evening lecture' (chaired by Arne Müntzing), titled 'The Study of the Cell in the Understanding of Life,' in which he alluded to Janaki's research on the *Magnolia*: 'The woody flowering plants retain a conservative chromosome structure which reflects their phylogeny often as far back as their origin from non-woody plants . . . in the primary and secondary geological periods. It can therefore be readily related, as Janaki Ammal has suggested in *Magnolia*, to the greatest geological and climate changes of the tertiary period.' He also referred to his friend Col. Stern's work on the *Paeonia*, which demonstrated that annuals and herbaceous plants (in contrast to woody plants) often could not be traced even to a common chromosome origin for each genus; at best they reflected the movements of the Ice Age.[30]

Science and Society

In Britain, in the early 1930s, a social relations of science movement arose in response to several events and developments such as the economic depression, the rise of the new Soviet State and its active encouragement of applied science, the advance of fascism and its abuse of science (as in the use of mustard gas by Fascist Italy against Ethiopians in 1935 during the Second Italo-Abyssinian War) and, finally, World War II. The movement was led by scientists who were not only interested in investigating the natural world but also deeply concerned about science and its relationship to society, influencing thereby such early organisations as the Royal Society, the British Association (by 1934, it was encouraging its members to make presentations on the social impact of science) and the British Science Guild and the Association of Scientific Workers; the last two organisations, the latter dominated by left-wing scientists like Bernal, were of the view that scientists should play an active role in contributing to discussions on science in the British Parliament.[31]

While on the one hand the social relations of science movement had reformists like Julian Huxley and Richard Gregory (the editor of *Nature*), on the other, it was constituted by radicals like Haldane, the X-ray crystallographer Bernal, biochemist and historian Joseph Needham, mathematician Hyman Levy, zoologist Lancelot Hogben and physicist Patrick Blackett. For the reformists, the zealous nationalism in Germany and Soviet Union was a threat to the spirit of international science, while the radicals believed that the most ethical use of science could only be accomplished within a socialist order. In 1938, Haldane would present a series of lectures, published as *The Marxist Philosophy and the Sciences*, which would become one of the most complete and authoritative affirmation of the application of the dialectical principles to science. A year later, Bernal would publish his *The Social Function of Science* (1939), which would become the British left's manifesto for the reorganisation of society along scientific lines; the reformists fiercely opposed the ideas in the book and became increasingly suspicious of the state's intervention in scientific practice. However, compelled by the adverse global situation, the radicals and reformists united to support planning for science. Ever since his visit to Russia, Darlington had been a strong supporter of the social relations of science movement and a staunch advocate of planning. Like Huxley, Hogben, Haldane, Levy, Bernal and science journalist James G. Crowther, Darlington was a scientific humanist and internationalist, a supporter of the view that science if pursued well was mankind's greatest saviour.

Between 1935 and 1937, the Social Relations of Science movement began to attract international interest, especially the activities of the British Association, which in 1938, thanks to the science journalist Richie Calder's persuasion, founded its Division for the Social and International Relations of Science, on the belief that the study of social relations was as important as investigations of the natural world. Under Calder's chairmanship, the division dealt with the impact of scientific discovery on human affairs; the 'Calder Plan' for a world association of science, in a way, anticipated the 'S' in UNESCO, an organisation founded a few years later.[32] If the social relations of science movement emphasised the importance of science to society and the utilisation of science for nation-building, as realised by Soviet-style planning, a second group argued for freedom of science from the influence of any political ideology, for they were convinced that radical central planning and totalitarian forms of government were detrimental to science. For instance, Michael Polanyi, who had left the Kaiser Wilhelm Institute in 1933 in protest against the Nazis, believed in a mutualistic relationship between scientific progress and liberal democracy. The Society for Freedom in Science that he founded demanded that scientists be given the freedom to pursue their own investigations, without any interference from the state, a position that was far from attractive at a time of war.

In 1938, the Indian Science News Association would suggest that the Indian Science Congress discuss the subject of science and its social relationships in a plenary session at the forthcoming meeting in Lahore and a year later, coinciding with the appearance of Bernal's *The Social Function of Science*, *Current Science* welcomed the creation of the 'New Division' (for the Social and International Relations of Science) of the British Association, which believed that by studying society on lines similar to that of natural science, some of the paradoxes of modern life could be resolved. The formation of a department of scientific investigation under the patronage of the Indian Science Congress was thus found relevant and urgent. The *Current Science* editorial observed: 'Science has too long been divorced from society, because of the idea that the province of science is matter, and the human sciences like biology, sociology and economics had not acquired the status and importance of the physical sciences. The consequence has led to a dreadful state of affairs where the physical and the moral are indistinguishably mixed up in the social conditions. It becomes increasingly clear how hopeless it is to disentangle them.'[33]

Nehru and Blackett

Blackett's book *Fear, War and the Bomb: Military and Political Consequences of Atomic Energy* was published, first in Great Britain (1948); that year, Blackett would win the Nobel Prize in physics for his investigation of cosmic rays using his invention of the counter-controlled cloud chamber method. Blackett's book was virulently censured by several groups, including the British State, but there were several others who heartily welcomed it, chiefly pacifists. An exact contemporary of Janaki, Blackett would visit India on a regular basis over the next two decades. Incidentally, three years before he won the Nobel, Bibha Chowdhuri (1913–91), an Indian woman physicist from the University of Calcutta, enrolled with Blackett at his cosmic-ray research laboratory for research leading to a doctorate, but is believed not to have been part of the team that brought him his Nobel.[34]

Blackett's left-wing ideology and internationalism was particularly attractive to third world development, and thus it was no surprise that it brought him in contact with Nehru, whom he met in 1947. Nehru had sought his advice on the research and developmental needs of the Indian armed forces and it was in this capacity that he had first travelled to India. Nehru would approve of Blackett's advice to found a new research capability within the Ministry of Defence. At this time, technical development establishments under the Indian Army that could inspect and undertake quality control in ordnance factories did not exist; neither were scientists and engineers involved with defence establishments.[35] In 1948, Blackett recommended the name of physicist Daulat Singh Kothari as scientific advisor to the Ministry of Defence, a post Kothari would continue to hold until 1961; he was the head of the Defence Science Organization, later called the Defence Research and Development Organization (DRDO). It was again Blackett who suggested the name of his close friend Homi Bhabha as leader of the Indian Atomic Energy Commission, founded by the Constituent Assembly that same year. Vikram Sarabhai, who succeeded Bhabha as head of the Commission in 1966, was also a close friend of Blackett.

Even while Blackett and Nehru recommended the use of atomic energy as a source of electric power in third world countries, they were united in thinking that atomic weapons should not be allowed to proliferate. Homi Bhabha, S. S. Bhatnagar and K. S. Krishnan (all Fellows of the Royal Society) were appointed members of the new Scientific Advisory Committee to the Defence Ministry on the advice of Blackett. They held talks at India House in London on atomic energy; the meeting was attended by Charles Darwin, Director of the National Physical Laboratory and A. V. Hill, Scientific Adviser to the Indian government (1943–44). Nobel prize–winning scientists would visit India on the invitation of Bhatnagar, who was the chief of Government of India's Department of Scientific and Industrial Research, and the person behind (with Arcot Ramasami Mudaliar) the constitution of an autonomous body, the CSIR in September 1942.[36]

Nobel Scientists Visit India

Accepting Bhatnagar's invitation, in late December 1949, Jean Frédéric Joliot-Curie and Irene Joliot-Curie, the Nobel Prize–winning couple (in chemistry, 1935) arrived in India to attend the Indian Science Congress at Poona (in January 1950) along with such men as J. D. Bernal, who had already visited once before; this time it was as the Vice-President of the World Federation of Scientific Workers. Addressing the plenary session of the Indian Science Congress in Poona, January 1950, Nehru reiterated his unflinching trust in the potential ability and resourcefulness of science and technology in solving the economic and sociological problems confronting the nation. He stated: 'The problems are to be tackled with the spirit of a man who does things himself and not sit in the office ordering people about.' He was of course addressing both men and women of science; of the latter there were quite a few at the Congress. He asked them to inculcate a spirit of internationalism in the youth (rather than drill them with nationalist rhetoric).[37]

Only three months before his arrival in India, on 20 September 1949 to be precise, Bernal had delivered a speech at a peace conference in Moscow fiercely critical of Western countries. In an interview published in an Ipswich newspaper, he had wholeheartedly supported Soviet style agriculture, namely, Lysenko's proletarian science. As a result of Bernal's pro-Soviet position, in November that year, the British Association removed him from membership of its council. To the journalists waiting at the Bombay airport, Bernal was outspoken: there were 'many great men of science in India,' but there was 'a lack of intermediation necessary to scientific research,' he commented.[38]

That same month, in connection with Stalin's seventieth birthday celebrations, a decision was taken to award an annual Stalin Peace Prize; the first prize went to Joliot-Curie (in 1951) and the third to Bernal and ten others (in 1953); his photograph would appear on the cover of *Pravda*, which described him as a 'progressive scientist who wants the achievements of science to serve human progress and peace.'[39] In 1950, Bernal's book, *The Freedom of Necessity*, on the relationship between science and society, was reviewed in *The Times of India*. In the book, Bernal had argued that the USSR provided the most evolved kind of social organisation, and Marx and Engels were the only prophets worth heeding to. The reviewer was however of the opinion that 'even such an able writer as Prof. Bernal' would find it difficult to convince people, given the 'record of Soviet Russia, its ruthless dictatorship, its suppression of freedom and of truth, the vast inhuman cruelties of its slave labour system.'[40] This opinion was not shared by journals like *Current Science*, which remained tight-lipped about the excesses of Stalin and the setbacks to science in Soviet Russia, in particular genetics.

'All About Rice'

While Janaki was in India, the writer R.K. Narayan, working on his novel *The Financial Expert* at this time, commented on the Grow More Food campaign in a piece titled 'All About Rice':

> And they attempted to offset Nature's unhelpful attitude with the slogan 'Grow More Food.' The earth ran with the cry 'Grow More Food.' There were hundreds of Grow More Food Conferences all over the country; people flew long distances to attend these meetings. At least half the tonnage of paper marked Food, which at the moment chokes our building were GMF files. It seemed at first improbable that such a simple notion as growing more food (especially when people were hungry) should need all this complex

machinery and expenditure. It might as well have been necessary to start a country-wide campaign to tell people to breathe in order to avoid suffocation.[41]

Even as widespread doubts persisted that self-sufficiency could be achieved, the Food Minister Daulatram reiterated that the Government was determined not to import any food after 1951, 'except in case of an unforeseen emergency.' He tried to reassure people that the food machinery was on the move and difficulties would be overcome soon. At the debate in Parliament on the issue of increased output, the Andhra political leader Dr Pattabhi Sitaramayya was of the opinion 'that the so-called shortage' was due to rationing and that if rationing was given up, everything would be alright. Another representative, from Madras, suggested that a high-power committee be constituted to enquire into the genuineness of the food deficit in the country and that the government should remedy the issues on the procurement system in the various states, under which the 'rich landlords went "scot free," while the poor had to surrender more grains than they could afford to.'[42] Nothing would come out of all this, a fact well reflected in Werner Bischof's photographs of the famine in Bihar taken in April 1951.

The subject of Soviet biology would be raked up once again in India in early 1950, by *The Times of India*, which carried a piece by the British botanist and educationist Eric Ashby (at this time the Chair of Botany, University of Manchester) titled, 'Soviet Biology: The "Prestige" of Lysenko,' published previously in the *Manchester Guardian*.[43] As the *chargé d'affaires* at the Australian legation in Moscow in the early 1940s, Ashby had had the opportunity to meet several leading personalities in the USSR including Stalin and was the author of the *Scientist in Russia* (1947). In his article in *The Times of India*, Ashby referred to a heavily attended meeting on the subject of 'Soviet Genetics' at the Linnean Society, while he discussed three books published in 1949, which were 'certain to be much more widely read than books on ordinary genetics.' He was referring to Julian Huxley's *Soviet Genetics and World Science*, I. V. Michurin's *Selected Works* and Gennadi Fish's *A People's Academy*. Huxley's book provided a lucid summary of the facts already known about the controversies in Soviet biology since 1948, of which Ashby wrote that Huxley very logically separated the ideological issue from the scientific one: 'His statements are accurate and documented, and he puts Soviet genetics in its proper perspective beside the genetics of the rest of the world.' The book was such that it could 'be warmly recommended to the man in the street who wants to know what all the fuss has been about.'[44]

Janaki Returns to Wisley

Janaki meanwhile was growing more fretful as time passed. She left India for England on 4 April 1950, at the end of her contract, and ostensibly, without submitting a report to the government. She must definitely have visited Edam and Shoranur before she departed, for she would later gift the Kew Gardens a collection reportedly 'put together in South India' in 1950.[45]

Notes

1 DP: C. 109 (J 113) E. K. Janaki Ammal to C. D. Darlington, letter dated 22 February 1950.
2 Royal Geographical Society, London: E. K. Janaki Ammal, Fellowship Certificate.
3 *The Times of India*, 5 November 1949.
4 Ibid., 10 November 1949.
5 Ibid., 11 November 1949.
6 Ibid., 14 November 1949.
7 Ibid., 11 November 1949.
8 Ibid., 28 December 1949
9 Ibid., 12 November 1949.
10 Ibid., 19 November 1949.
11 Pledge 1: I will do my utmost to produce more food crops in the land under my control and I pledge myself to sell the entire surplus to Government after leaving just enough for the use of myself and the members of my family and for seed and other agricultural requisites. Pledge 2: I will make every attempt to reduce my consumption of food grains by using non-cereal foods and to avoid wastage of food in the kitchen, and on the table, and I pledge myself to support the drive for increasing food production in the country by doing everything in my power, like growing vegetables and other food crops in my backyard or in pots, helping the cultivators to increase the supply of manure, etc. *The Times of India*, 23 November 1949.
12 *The Times of India*, 1 December 1949.
13 Ibid., December 1949.
14 Ibid., 24 December 1949.
15 Ibid., 18 November 1949.
16 Ibid.
17 DP: C. 109 (J 113) E. K. Janaki Ammal to C. D. Darlington, letter dated 21 January 1950.
18 Subramanian, 'Plant physiologist and humanist: A birth centenary tribute to Prankrushna Parija'.
19 DP: C. 109 (J 113) E. K. Janaki Ammal to C. D. Darlington, letter dated 22 February 1950.
20 Keller, *A Feeling for the Organism: The Life and Work of Barbara McClintock*.
21 DP: C. 109 (J 113) E. K. Janaki Ammal to C. D. Darlington, letter dated 22 February 1950.
22 Ibid., letter dated 6 March [1950].
23 Ibid.
24 Ibid.
25 Burns, 'Some ideas and opportunities for plant geneticists in India'.
26 Ibid., letter dated 12 March 1950.
27 Ibid.
28 The German geneticist and crop researcher, Elisabeth Schiemann (1881–1972), was best known for her book *Entstehung der Kulturpflanzen* (*Origin of Culti-vated Plants*) published in 1932. Also present at the Congress was the Russian

cytologist activist and feminist, Sophia Satina (1879–1975), who worked as research assistant to Albert Blakeslee.

29 Osvald and Aberg (eds.), *Proceedings of the Seventh International Botanical Congress*, p. 333.
30 Ibid., p. 97.
31 For a detailed analysis of the links between the social relations of science movement and these bodies, see McGucken, *Scientists, Society and State*.
32 UNESCO came into existence in November 1945, but would begin operations only a year later.
33 'Science and Society'. *Current Science* 8, no. 1 (January 1939): 1–3, 2.
34 For the story of her life and work as a physicist in Manchester, see Singh and Roy, *A Jewel Unearthed: Bibha Chowdhuri*.
35 Anderson, 'Empire's setting sun? Patrick Blackett and military and scientific development of India', pp. 3703–05+3707–20.
36 For a history of the formative years of the CSIR, see Krishna, 'Organization of industrial research: The early history of CSIR, 1937–1947'.
37 *Current Science* 21, no. 1 (January 1950): 1–2.
38 *The Times of India* (Bombay), 28 December 1949.
39 Brown, *J. D. Bernal: The Sage of Science*, p. 341.
40 *The Times of India*, March 1950.
41 *The Hindu*, 25 February 1950.
42 *The Times of India*, 14 March 1950.
43 Ibid., 2 February 1950
44 Ibid.
45 RBG Kew, Library & Archives: QG 169, Correspondence with Janaki Ammal, 1945–1970, E. K. Janaki Ammal to N. L. Bor, letter dated 20 January 1952.

17

WISLEY III

The 'Wanderings' of Flowering Plants

On 12 April 1950, within a week of her return to Wisley from India, she attended the meeting of the Cytological Sub-Committee of the RHS Wisley, with President F. C. Stern in the Chair, and F. T. Brooks, M. B. Crane, and Director Gilmour, as members. The group discussed and assessed the work done in the cytology department over the past months under Janaki's direction, which included colchicine treatment, breeding, and chromosome counting undertaken for the Fellows of the RHS, and advisory work. Janaki had by this time incorporated her cytological findings on the genus *Nerine* in the form of two forthcoming papers in the RHS Journal.[1] The paper dealt with the evolutionary history or deep phylogenies of thirteen species of the genus in cultivation from 1659 to 1949, when the first tetraploid 'Inchmary Gate' came into existence; this was the only tetraploid in the genus, and it had taken nearly three centuries of cultivation to produce it. The second paper listed the chromosome counts of 112 garden hybrids, together with their known parentage, the result of a laborious task undertaken by Janaki and her team; sixteen triploids, useful for the production of new hybrids, had also been identified.

Janaki had also written a paper on the cytology and evolution of the garden *Philadelphus*, a work she had undertaken on behalf of Gilmour, which would again be published in the society's journal in 1951. All the species of *Philadelphus* in nature were diploids, but when species from a range of geographies were brought together in cultivation, it resulted in large triploids, of which three beautiful ones were selected for breeding. The first tetraploid had been produced, when Lewis Palmer, a breeder and a Vice-President of the RHS, crossed the triploid 'Sybille' with *P. Burfordiensis* to create the beautiful 'Beauclerk'; it had received the RHS Award of Merit in 1947.

Janaki's study of the cytology of the *Viburnum* species had revealed that it could be classified into three chromosome groups, with polyploids in each, but her results were yet to published. Some of her work on the *Rhododendron* had already appeared in the *Rhododendron Year Book* (1950),[2] but there was more work to be done on the well-known garden hybrids in the genus, which she now turned her attention to.

DOI: 10.4324/9781003267089-17

Colchicine treatment was restricted to *Rhododendron, Magnolia, Salvia* and *Narcissus*, as the Committee dictated that instructions regarding colchicine treatment were to be given only at the Director's discretion. Janaki's breeding work in the previous season chiefly involved making selections from the tetraploid *Salvia splendens* to ensure high fertility and having *Narcissus* bulbs X-rayed at the John Innes with the aim of intensifying the pink colour, in some of the Guy Wilson's pink varieties. Janaki and her team had also undertaken more than 100 chromosome counts in the *Philadelphus, Cyclamen, Nerine, Camellia, Paeonia* and *Primula*. Following discussions, the Committee now decided to abandon the work on the *Cyclamen*, but thought it important to continue work on the *Camellia* and its hybrids; this work had been undertaken at the request of the horticulturist George Horace Johnstone of the Trewithen Gardens, Cornwall.[3]

A Colchicine School

Janaki resumed work at Wisley on 16 April 1950: her time from now on would be chiefly devoted to a survey of the genus *Rhododendron*, besides undertaking chromosome counting for the RHS Fellows, W. B. Turrill of Kew, and amateur gardeners and nursery men. Between May and October that year, she generated over 500 counts involving 360 species; she covered all but one of the forty-three series into which the genus was classified.[4] A collection of *Rhododendron* hybrids and their parents selected for exhibition at the forthcoming Chelsea Flower Show (1951) were analysed by the Committee; almost 130 species of *Rhododendron* of known chromosome number had been colchicined at the seedling stage, but they were still too small to be submitted to chromosome counting. Further, if the last time *Narcissus* bulbs were X-rayed as an experiment, this time they were injected with colchicine, to be later examined cytologically. As far as breeding work was concerned, attempts were made to cross tetraploid asparagus with diploid, to produce triploids; the *Delphinium macostachym* was crossed with garden forms (but seed setting was poor); and the tetraploid Salvias and Antirrhinums were grown for seed multiplication. One of Janaki's assistants, Miss Janet Wadsworth was at this time involved in a comparative study of sterility in the diploid and tetraploid *Salvia*.[5]

The cytology department undertook the task of making chromosome counts, as per the orders of the Director, for some of the RHS Fellows and amateur gardeners such as A. C. Herrick (*Iris*) and A. G. Rait (*Narcissus*), and commercial plant nurseries like Messrs. Frampton & Co (*Cyclamen*). However, advice on colchicine methods was now offered only to those who were in a position to execute the work on a scientific basis; Dr Dark of the Wye College (on Hops specifically), a Messrs. Bywaters, and John Moreland of Australia were given lessons by Janaki in the general techniques used in colchicine treatment. A note on colchicine techniques evolved at Wisley by her was supplied to W. J. C. Lawrence, Curator of the John Innes Gardens, for his book on plant breeding. In his third edition of *Practical Plant Breeding* (1951), Lawrence acknowledged Janaki's 'advice on the uses of colchicine.' She was easily one of the best colchicine-technicians of Britain (and India) at the time, and the one who had used the drug most effectively in plant breeding. There were also those who were referred to as Visiting Workers, like Meyer from the Missouri Botanic Gardens, Miss Margaret Fox, Nancarrow and later Rita Madge (all postgraduate students at Royal Holloway College), who spent some time at Wisley to train in cytological techniques, and John McLeish, who wished to study the chromosome structure of X-rayed annuals. In addition, visitors came to the laboratory from a range of locations as varied as Versailles, Wageningen, Coimbra, Ottawa and Ghent.[6]

British Association Meeting at Birmingham, 1950

In early September 1950, Janaki travelled to Birmingham to attend the meeting of the British Association and to present a paper in the session 'Cytology and Genetics in Relation to the Classification of Plants and Animals' (arranged by the Zoology and Botany Sections). Janaki's paper dealt with the correlation between morphology, chromosome number and geographical distribution, through a study of polyploidy in *Rhododendron*. The session also had Col. F. C. Stern speaking on the use of chromosome number and other cytological characteristics in solving taxonomic and phylogenetic problems (illustrated with excellent maps), and David H. Valentine of the University of Durham on the use of interspecific compatibility as a taxonomic criterion, and the evaluation of the term comparium, defined 'as a group comparable in size with a genus or subgenus, composed of all those species which can be united, directly or indirectly, by hybridization.'[7]

Less than a year before this meeting, Janaki had drawn Stern's attention to an article by Valentine titled 'The Units of Experimental Taxonomy' where he had introduced certain descriptive phrases to refer to interspecific relationships and evolution within groups.[8] She had taken a dislike for Valentine's terms 'g-ecospecies' and 'a-ecospecies,' derived by combining the chromosome composition or 'ploidy' of a species with the ecological phraseology used by the Swedish evolutionary botanist Göte Turesson (the term 'ecotype' was coined by Turesson in 1922 to describe locally adapted plant populations) because she did not find the terms helpful in differentiating between groups. In fact, Janaki was of the opinion that 'Darlington should be called in to coin more descriptive terms than *a* and *g* which will not be easy to differentiate except by memorizing.' As regards Valentine's use of the terms 'autoploidy' and 'alloploids,' she reacted somewhat disapprovingly with, 'Is this American, I wonder.'[9]

At the British Association meeting, Janaki purchased Alex Du Toit's *Our Wandering Continents: An Hypothesis of Continental Drifting* (London, 1937). She had bought her copy of Darlington's *Genes, Plants and People*

(1950) on 14 August that year, just after she had purchased C. E. P. Brooks' *Climate through the Ages* (1950). Francis Kingdon Ward, whose books Janaki enjoyed reading, was at this time on an expedition in Assam with his young wife Jean, and very close to the epicentre of an earthquake, one among several catastrophes he would encounter during his travels.

Exciting Discoveries

In early January 1951, Janaki communicated to Stern on the chromosome counts of the *Cyclamen* she had borrowed from his garden. A month later she had examined more material and found that there was 'a very peculiar large lagging chromosome in the complex which divided precociously' and sometimes gave rise to daughter nuclei with 9 and 11 chromosomes. Having seen something extraordinary going on in the nucleus, she felt the genus deserved a detailed study by somebody else: 'It might help to clear up the doubt and uncertainty around the chromosome number of Cyclamens. I am beginning to doubt my own counts!' She had also been writing up a note on *Viburnum* hybrids for the RHS journal, when she discovered among the Wisley Viburnums, on a 'lightning examination,' a white form of *V. fragrans*, which was a tetraploid. She had never before found a single American species of *Viburnum* with a basic number other than x = 9. She announced to Stern that she was pleased her theory of the origin of the American *Viburnum* was being supported by the chromosome numbers. All the Himalayan species and a few from Japan were x = 8 and she had an American hexaploid, which was 2n = 54. 'All very exciting from the cytogeographical angle,' she remarked. She would send Stern her paper on the *Viburnum* for his comments, accompanied by a map and diagram, before sending it away for publication.[10] Janaki had only recently attended his RHS lecture, 'Snow-drops and Snow-flakes,' and which she enjoyed thoroughly.

She would write again to Stern to share her great excitement at having broken through the diploidy of the non-lepidote *Rhododendron* (the non-scaly kind of *Rhododendron*, identified by close observation with a magnifying glass of the underside of the leaf) using colchicine. Janaki had with her a population of 130 colchicined seedlings (having selected 12 seedlings of each species to work on) and *R. wardii* was the first group she had examined. To her utter delight, she discovered 'the doubling of chromosomes in two to become tetraploid! I am too thrilled to do anything else,' she would comment. Janaki decided to dedicate the *R. wardii* tetraploid to Koichiro Wada, the Japanese nurseryman, who in the 1930s had sent his finest selection of the *Rhododendron yakushimanum* to Lionel de Rothschild in England; she gave it the name, *Rhododendron yakushimanum* 'Koichiro Wada.'[11]

It was when Francis Hanger joined the Wisley Garden as Curator in 1946 that a new era in rhododendron cultivation and introduction had begun; he planted at Wisley specimens collected from the gardens of gentlemen horticulturists such as Lord Aberconway (Bodnant), Edmund de Rothschild (Exbury),

J. C. Williams (Caerhays), Giles Loder (Leonardslee) and J. B. Stevenson (Tower Court), besides the Kew Gardens. Hawk-eyed, Hanger chose those inter-specific crosses among them which were pregnant with possibilities. Importantly, he recognised the usefulness of *R. yakushimanum* as a seed parent, given its excellent qualities: its dwarf compact habit and ability to flower early. In 1945, Hanger had brought to Wisley one of the two *R. yakushimanum* plants introduced by Rothschild to Exbury in 1934, and planted it, feeding it periodically with generous amounts of spent hops and water. Two years later, it would be awarded a First Class Certificate (FCC) at the Chelsea Flower Show.[12]

Meanwhile, Janaki had also written to Edward Salisbury, Director of the Kew Gardens for a selection of *Buddleia*, 'cutting or without flowers.' She assured him that Kew (and himself) would be duly acknowledged in any publication that resulted from her cytological investigations of the genus.[13] Later, she would donate Hitchcock's *Manual of the Grasses of the United States* (2 vols., 1935, and 1950 with Chase) to Kew.[14]

More Visitors to the Cytology Laboratory

In the period, October 1950–April 1951, there was an even higher demand for literature on colchicine techniques at Wisley, and a list of names would be prepared for the distribution of the printed notes prepared by Janaki.[15] During this period, Janaki was busy selecting and potting out those that she thought were hopeful among the 130 *Rhododendron* species treated with colchicine in the previous year. Among the Magnoliaceae, young seedlings of *Magnolia* x *highdownensis* (the Highdown Magnolia) and *M. wilsonii* were treated with 1% colchicine in lanolin out of doors. Janaki was also giving six species of *Salvia* the colchicine treatment.

In March 1951, Barbara J. Saunders joined the laboratory; she would train under Janaki, and work with her on the breeding and cytology of the *Lonicera*. Among the visitors that year were M. B. Crane from the John Innes, Hilda Seligman, Carl Olsen and his wife, and those from India such as K. Damodaran from Varkala (a doctor with his own medical practice in England), botanist Shanti Batra from Calcutta, the Finance Secretary K. R. K. Menon and his wife Saraswathi, one K. Rukmani, one Mrs C. K. Nair, A. M. Menon, Dr K. Parvathi from Bangalore, and Janaki's friend, A. R. Gopal-Ayengar of Trombay, who signed the register, 'from London & the World' beside his name, accompanied Shanta V. Iyengar of Bangalore, who was en route to Bloomington to enrol for a doctoral degree under geneticist Hermann Muller at the University of Indiana.[16]

Falling Out with Darlington

Darlington had intimated Janaki of his plan for a second edition of the *Chromosome Atlas* in late 1950. She had agreed with him at once that it ought to be a thoroughly revised new edition rather than a reprint and promptly began work to 'get together' her notes and list of new chromosome numbers. 'Dr [John] Hutchinson is revising his classification of *Flowering Plants* [*The Families of Flowering Plants*, arranged according to a new system based on their probable phylogeny, 2 Vols.] and we shall have to rearrange our families,' she had suggested. She was also planning a meeting with Hutchinson (who had worked for six years in the early part of his career as 'Assistant for India' at the Kew Gardens and had retired as Keeper of the Museums of Botany at the Gardens in late May 1948), before sending Darlington a report on the rearrangement.[17]

However, a meeting with Darlington on 2 November 1950 would turn out to be deeply unpleasant, so much so that Janaki chose to opt out of the new edition. Darlington sent her a very formal note on the following day: 'Confirming our conversation of yesterday, I understand that you do not wish to take any part in the production of the second edition of the *Chromosome Atlas of Cultivated Plants*, which I have been asked to prepare. Also, I gather that you are indifferent as to whether your name appears on it or not. Further that you will not have any unpublished chromosome counts of your own to supply for publication in it.'[18] Darlington's letter made Janaki seethe with anger but she cautiously decided to mull over it before responding. She also wanted to speak to Stern, Chairman of the Cytology Sub-Committee, before she did anything further. Somehow, talking to him always worked like an emollient for her.

More than two weeks later, Janaki would respond to Darlington, with an apology for the delay. She cited Darlington's unrealistic deadline for revision as the chief reason for opting out of the project. 'In my opinion there is a whole year's work to be done before I can be ready for a new edition and you have asked for my amendments for next January!' she wrote in utter indignation. She was not ready for a hasty revision, come what may. Once made up, Janaki's mind was difficult to change, nobody, not even Darlington, could influence her decision. In no uncertain terms, she would state: 'Placed as we are geographically and, if I may be allowed to put it frankly, psychologically, collaboration will be most difficult and rather than see it break down I believe it is best not to attempt it.' She challenged him on the remaining points as well: 'You are wrong if you have gathered from my talk that I am "indifferent" as to whether my name appears on the *Atlas* or not. Far from it. I am positive that my name shall not appear on the book without my collaboration. This is why I suggested that you might alter the name of the book so that you can make it your very own, or take in another collaborator should you so desire.' Janaki assured him that she would be charitable and give up the right to four years of work she put into the *Atlas*, 'as a free gift,' if this would enable him to follow her suggestion. As to the third point in his note to her, that of supplying him with her unpublished counts, she said she was willing to send him a list of those chromosome counts made by herself and her team at the Wisley cytology laboratory, with the permission of the

Cytological Sub-Committee, on the condition that 'premature publication' of them 'will not take away from the value of the thesis for which the chromosome counts were made.'[19] Janaki would never forgive Darlington for this.

The New Chromosome Atlas

The new version, called the *Chromosome Atlas of Flowering Plants*, would be co-authored by Darlington and the young Ann Phillipa Wylie (from New Zealand) of the John Innes (who had trained at Wisley under Janaki) and would take five years before it saw the light of day. The new *Atlas* was extended to cover all the gymnosperms and angiosperms and included about 18,000 species in 2,500 genera, giving the somatic number and the chief uses of each. In the making of the new version, the whole of the introductory chapter that appeared in the old *Atlas* was removed (to make space for new additions). This introduction was expanded and brought out as a separate volume titled *Chromosome Botany and the Origins of Cultivated Plants* (1956). It would be dedicated to Otto Renner of the University of Jena who unravelled the nature of hybrid species (but would still hold on to a static approach to genetics, in contrast to the evolutionary approach adopted by the proponents of the modern evolutionary synthesis).

At least some reviewers of the new *Atlas*, like the cytobotanist Valentine, were of the opinion that the removal of the introductory essay was not a good move, but conceded that it was in general far superior in format and appearance.[20] Aimed at students and experts alike, the new *Atlas* was arranged as per Hutchinson's system, but with minor modifications; the first chapter dealt with the chromosomes themselves, and the observable differences in their form and behaviour, followed by a chapter on the uses of chromosome studies in addressing the problem of systematic groups, informed by a dynamic view of the concept of species using suitable examples. The third chapter demonstrated the immense value to be derived from combining cytological study with biogeographical (or ecogeographical, to use Valentine's term) investigations of groups of related plants such as in the *Rhododendron* and *Oenothera*; the fourth chapter aimed to show how chromosomal study shed light on evolution, or on phylogeny and the temporal and spatial parameters of the evolutionary process; and the last two focused on the cultivated and ornamental plants, which were of interest not just from the evolutionary point of view but because they shed light on the very origin of human civilisation.

Disgruntled

Janaki was becoming increasingly despondent with the lack of a research environment at Wisley. She deeply resented the fact that she could not focus on specific research problems, because all her time was spent on undertaking tasks for the Fellows of the RHS and commercial establishments at the

324

orders of the Director. She felt more like a technician of sorts, a maker of tetraploids, or a chromosome counter rather than an academic researcher. And now, with Darlington having let her down, she desperately longed for a change of scene. She was ready to move on.

On 30 September 1951, Janaki resigned from Wisley and moved to Paris to work at the genetical laboratory at the Institut National Agronomique on Rue Claude-Bernard. The French government had offered her a liberal grant-in-aid for research. On the eve of her departure from Wisley, her friend Hilda Seligman presented her with Percy Sykes' *A History of Exploration* (1935), and among the books she purchased that year were some of the most recent and diverse of publications such as the *Uttermost Part of the Earth* by E. Lucas Bridges (1951); *Everything has a History* (1951) by her friend, Haldane; *Hevea: Thirty Years of Research in the Far East* by M. J. Dijkman (1951); and the RHS publication, *The Fruit Garden Displayed* (1951).

Figure 17.1 Janaki's *Rhododendron yakushimanum 'Koichiro Wada.'* 1950.

Source: Courtesy of the RHS Wisley Collections. WSY0013311.

Figure 17.2 Buddleia collected by Janaki with the help of M. Bridgwater, from Cornwall. 29 June 1951.

Source: Courtesy of the RHS Wisley Collections. WSY0060441.

Notes

1 Janaki Ammal, 'The Chromosome history of Cultivated Nerines', and Janaki Ammal and Bridgwater, 'Chromosome numbers in Hybrid Nerines', both 1951.

2 Janaki Ammal, 'Polyploidy in the Genus Rhododendron', 1950.

3 RHS, Lindley Library: RHS/Minutes/WY/WY Garden Advisory Committee, Box No. 5, Meeting of the Cytological Sub-Committee, 12 April 1950.

4 Janaki Ammal, Enoch and Bridgwater, 'Chromosome numbers in species of Rhododendron', 1950.

5 RHS, Lindley Library: RHS/Minutes/WY/WY Garden Advisory Committee, Box No. 5, Meeting of the Cytological Sub-Committee, 12 April 1950.

6 Ibid.

7 For an overview of the papers presented at the session see, Muldal and Valentine, 'Cytology, genetics and classification.'

8 *Acta Biotheoretica* 9, nos. 1–2 (November 1949): 75–88.

9 RBG Kew, Library & Archives: Letters of E. K. Janaki Ammal from F. C. Stern & Lady Stern, letter dated 13 November 1949.

10 Ibid., letter dated 9 February 1951; Janaki Ammal, 'Chromosomes and the Species Problem in the Genus Viburnum', 1953.

11 Ibid., letter dated 4 May 1951.

12 See Gardiner, 'The "Rhododendron" Horticultural Society', especially p. 86, which contains a note on Janaki's work on the cytogenetics of the *Rhododendron*.

13 RBG Kew, Library & Archives: QG 169, Correspondence with Janaki Ammal, 1945–1970, E. K. Janaki Ammal to Edward Salisbury, letter dated 22 June 1951.

14 Ibid., letter of thanks from Edward Salisbury to E. K. Janaki Ammal, dated 17 September 1951.

15 RHS, Lindley Library: RHS/Minutes/WY/WY Garden Advisory Committee, Box No. 5, Report of the Cytology Department, October 1950–April 1951.

16 Ibid.: Visitor's Book, Cytology Department, RHS Wisley, maintained by E. K. Janaki Ammal, 1 October 1946–30 September 1951.

17 DP: C. 42 (E 138), E. K. Janaki Ammal to C. D. Darlington, letter dated 14 October 1950.

18 Ibid., C. D. Darlington to E. K. Janaki Ammal, letter dated 3 November 1950.

19 Ibid., E. K. Janaki Ammal to C. D. Darlington, letter dated 20 November 1950.

20 For a review see D. H. Valentine in the *Botanical Society of Britain and Ireland*, http://archive.bsbi.org.uk/Wats4p49.pdf accessed 11 December 2018.

18

PARIS–LONDON

On the Camellia Trail

> Paris is interesting but Ann Arbor is even better for developing
> my new found science—Cyto-geography of Flowering Plants . . .
> I could work on Camellias.
>
> —E. K. Janaki Ammal (1952)[1]

At Paris, Janaki would reside with her niece and family, never failing to take them presents, especially books on fairies or plants, for the children. As soon as she settled down, she wrote to her mentor Bartlett, enclosing a few reprints of her papers published in the Journal of the RHS. He had not corresponded with her for quite some now but was quick to respond to this one. 'I'm a poor letter writer, but must acknowledge at once your greetings from France and the three very interesting reprints from the JRHS. You do not state how long you intend to be in France,' he wrote to her. Bartlett wanted Janaki to send him 'a sort of biographical article about your numerous activities and experiences after leaving us, and even before on a rather intimate and personal level' for publication in his 'little lithoprinted magazine,' *The Asa Gray Bulletin*, jointly edited with Rogers McVaugh, and which carried 'Michigan botanical news.' It was a huge challenge for Janaki to sit down and write about herself; a regular curriculum vitae was all that she could accomplish on that front. In his letter, Bartlett assured her that he had been 'trying to figure out ways and means of getting [her] back for a visit sometime,' but nothing 'practicable' had 'ever developed.'[2]

Clearing the Air with Kew

Sometime earlier, when she was at Wisley, Janaki had sent a collection of Indian and Nepalese plants to Kew, and wanted them named. Edward Salisbury, the Director of the Garden, replied that he would not be able to help her in the matter of naming because of a shortage of staff. He used the opportunity to say that he hoped that the Indian Government would realise before long that 'it is to their own interest to appoint a botanist to work at

DOI: 10.4324/9781003267089-18

Kew specifically on the Indian Flora.'[3] In a subsequent letter, he reminded Janaki that the dried specimens which she had left behind at Kew for identification could be taken back, surprising her, for these had been gifted by her to the Herbarium, and about which she had already alerted friends, N. L. Bor (the Assistant Director) and W. B. Turrill, quite a while ago. Vexed and somewhat hurt by the rejection, Janaki wrote to Salisbury: 'I am very sorry that the collection of plants I presented as a gift to the Kew Herbarium is proving a trouble for the staff. I was unable to devote any time to them while I was on the staff of the RHS Garden at Wisley. These plants represent my 1949 collection from Nepal and besides those I made in South India in 1950. I shall at the very first opportunity relieve Kew of this burden.'[4] She next dashed off a letter to Bor, with a mild threat: 'if they are not acceptable, I shall remove them when I am next in England probably in Spring and pack the whole lot to my Old University in the U.S.A. [University of Michigan]. Until then, I hope you will bear with them. I had hoped to identify the Nepal specimens to enumerate the species collected for a note on my trip to Nepal.' Such a note would never see the light of day; not even a manuscript copy has been discovered this far.

In the same letter, Janaki mentioned to Bor that she had begun to learn French and was finding Paris 'very interesting.'[5] Within a week of this, Salisbury would write to her clearing the air. He explained that it was a misunderstanding and that Kew was very glad to have those valuable specimens for the Herbarium.[6] Incidentally, Eleanor Bor's book titled *Adventures of a Botanist's Wife* (1952), on her life with her husband in India in the 1930 and 1940s, containing illustrations by herself, had just been published.

Falling in Love with the Camellia

A few years earlier, during a visit to the renowned gardens of Colonel George Johnstone at Trewithen (still home to some of the first-rate specimens of tree Magnolia) in Cornwall, Janaki had fallen in love with the genus *Camellia*. She had several large-leaved seedlings in her collection at Wisley, of the *C. saluenensis*, grown from seeds brought back from Yunnan by the plant hunter George Forrest. She later discovered that the plants of the variety *C. saluenensis*, like *C. japonica* and *C. sinensis*, were diploids (2n = 30), while the large-leaved seedlings were either natural tetraploids or hybrids between *C. saluenensis* (2n = 30) and a close relative, *C. reticulata* (2n = 90). Janaki had noticed similar large-leaved and large-flowered specimens at the Herbarium of the RBG Kew, among specimens collected by Forrest from Yunnan.

Of particular interest were her findings with respect to the section *Thea* of the genus *Camellia*, to which belonged the diploids *C. sinensis* and *C. taliensis*. Most of the cultivated plants were results of the duplication of whole sets of their ancestral chromosomes, displaying the phenomenon called polyploidy. The case of tea was different. In China, the tea plant had been in cultivation from very early times, and the best tea there was what was called Yeh Ch'a—'the wild tea.' Every time a tetraploid was produced, it would be removed from cultivation, owing to its increased tannin content, or at best, grown as an ornament. For this reason, the cultivated tea always remained a diploid 'through all its thousands of years of cultivation. The conservative tastes of the Chinese and Japanese had worked towards keeping Chinese tea diploid. Until the nineteenth century, China was the world's leading supplier of tea, but when tea was discovered in Assam, the centre of production shifted to India, Ceylon and Java. Assam tea was different from Chinese tea in that it was large-leaved and thinner in texture than the latter. It was only a different botanical variety, for, chromosomally speaking, it was still a diploid like Chinese tea. 'Whether the taste of Western tea drinkers for strong tea will eventually result in the cultivation of polyploid tea remains to be seen,' wrote Janaki. She had discovered a triploid among the six tea plants she had examined in the Kew collection of Chinese and Assamese tea and had suggested that it could be exploited by tea planters who produced tea for the Western palate, the added advantages being its sterility, vegetative vigour and increased leaf production, traits associated with triploidy. Her work in Paris would continue to be on the *Thea*, in particular on the Assam teas.

British Association Meeting at Edinburgh, 1951

In August 1951, the British Association's annual meeting took place in Edinburgh, where Janaki presented a paper titled, 'Polyploidy as a factor of active speciation in S. E. Asia.'[7] This was the first time, she was presenting her pioneering ideas on cytogeography on a public platform, based entirely on her researches at Wisley, especially on the *Rhododendron* and the *Camellia*. It also clearly marked an archipelagic turn in her thinking, which involved conceiving the world as a collection of interconnected islands rather than as closed continental forms (a scatter versus consolidation). Janaki's interest was in mapping on a planetary scale (moving beyond regional, national or local boundaries) the migratory patterns or wanderings of plants as they occurred in the footsteps of man. There has been a rising interest in thinking with the archipelago, but the focus has largely been anthropocentric (studies on human migration in the recent past for instance), whereas in Janaki's cytogeographical project, we have a robust example of how thinking with the archipelago provided a new ecological perspective to the cytogeneticist studying plant migration, speciation and evolution. The cytogeographical approach allowed the probing of the entangled histories of human and plant migration across time and space (to shine light on the evolutionary development of old and new crops and of ornamentals such as the *Rhododendron*, *Magnolia* and *Camellia*) in terms of relations and processes, and thus breaking through the staticity and centripetality of the Vavilovian centre (of origin)-periphery framework.

Connecting with Ralph S. Peer

Meanwhile, Ralph Sylvester Peer (1892–1960), American businessman (a pioneering entrepreneur, he founded the Southern Music Publishing for recording country and blues music in the 1920s–30s) and President of the Camellia Society of America, had expressed interest in Janaki's work and hoped she would undertake a cytosystematic study of the *Camellia* for the Society. It was in 1950 that Peer had first heard 'about a very learned Indian woman (she comes from Madras) who was doing brilliant work with chromosome counts of Camellias, Rhododendrons, Magnolias etc.'[8] The two had not yet met, when Janaki wrote to Ralph Peer enquiring whether he could include her paper on the *Camellia* in the American Camellia Society's Yearbook of 1951.

Peer replied (from the Westchester Country Club at Rye, New York) in the negative, owing to a time limitation; he however assured her that he would be happy to include it in the next Yearbook. He was eager to meet her, either at Wisley or in London, in the coming October. It was perhaps in reply to a query from her that Peer had written in one of his earliest letters, that, as far as he was aware, there were only two sources of *Camellia pitardii* in America, one of which was in the Park Hill estate—the permanent home of the Peers located on 3 acres of land in Hollywood Boulevard, California. He promised to send her the *C. pitardii* scion for propagation. In fact, it was one of his 'pet projects,' he observed, to establish *C. pitardii* in England. 'Mrs Peer and I certainly look forward to making your acquaintance,' he had added.[9] Janaki would eventually meet with the Peers in Paris in late January (1952) and they would learn excitedly that she had obtained 'her Doctor's degree' from America. At this time, Peer had (wrongly) assumed that she had 'ample financial resources as she [was] connected with one of the richest families in India.' He would soon gather that Janaki required financial support if she were to live and work in the United States.[10]

Immediately after their first meeting, Janaki sent a letter to Peer saying that Paris was interesting but Ann Arbor was even better to develop her 'new found science—Cyto-geography of Flowering Plants (I am hopeless at learning French).' She wrote: 'I have a feeling you can help me to get there. You know so many people in the scientific and Horticultural world. I could work on Camellias.' She had enclosed Bartlett's letter of recommendation for his perusal and hoped Peer would get in touch with her. For Peer's information, Janaki mentioned that her University (Michigan) had actually sent for her when the war had broken out in late 1939, but she 'did not want to run away from the bombs' falling on London. 'I am ready now,' she told him. 'It was lovely meeting you both. I cannot tell you how lovely. You have revived all that spirit of adventure and adventurous dreaming once again,' she told him.[11] Peer replied that he would be in Paris again, two months later, in March. 'I hope we can meet again,' Janaki responded. She was planning to go to London to see her niece and family 'off on the boat to India' and to visit Kew 'to get some material of Camellia.' She also wished to take a good look at the hybrid C. *saluenensis* x C. *reticulata* of Francis Hanger, the Curator at Wisley. With Peer's help, Janaki believed it would be possible for her to return to America 'for a couple of years.' She was planning to write to Bartlett 'for ideas' on how to go about it. 'There must be an assurance from a University,' she enlightened Peer. 'We Asians are not allowed to take up jobs in America! I want to do research only,' she clarified.[12]

Janaki wasted no time in requesting Bartlett to approach Peer from his side with a view to arranging a research grant for her: 'He wants me to write a paper on Camellias for the American Camellia Society and got me some cuttings from California. I wish you would contact this gentleman by letter perhaps—I want to return to Ann Arbor for 2 years to do cytosystematics. I have a great deal of data and a few original ideas about cytogeography especially of the plants of SE Asia.' She wanted to explore 'more of the relationship between genera found in China and America—especially C. America' which was Bartlett's region of expertise. In fact, she had just completed a paper, 'Race History of Magnolias' for a 'book on Magnolias by Mr Johnstone' [*Asiatic Magnolias in Cultivation*, 1955], but she much wished it could be published from America instead; she would eventually succeed.[13] Bartlett acted quickly and wrote to Peer, strongly recommending Janaki's name for a possible research grant: 'We could arrange for her to have lab. facilities here as a guest investigator, and would gladly do so by having her appointed to an honorary position as Research Associate at the Bot. Gardens of the University. She is one of the most highly esteemed and distinguished botanical graduates, and we should be delighted to grant her guest privileges.' However, Janaki had no financial resources herself, he noted, 'and in these difficult times of keeping up with the inflation we have no proposal for a stipend for her unless it should come as a gift [that is untaxed].' Bartlett wanted Peer to induce the Camellia Society to float a grant for the Fellowship, which would then render everything 'plain sailing.' His calculation was based on the surmise that 'special societies in the horticultural fields' generally comprised wealthy members, therefore 'possibly one of them, or a group' might be willing to 'contribute to such a worthy end as a cyto-systematic review of the genus Camellia from a modern standpoint.' Janaki had 'attained international distinction' in the field, Bartlett noted. 'Certainly we would very greatly appreciate anything you could do to enable her to come to undertake the proposed work,' he added.[14]

Janaki was however unable to make it to Paris in time to meet the Peers (in March), because she had 'no place to stay in Paris,' her niece and family having left the city. In London, she resided at West Acton and was making the best use of her time in the city by working on the Camellias at Kew. 'My paper I hope will be ready soon. I am including a map showing the distribution of Polyploid Camellias,' she informed Peer.[15] In her map, she had demonstrated how the area of endemic distribution of the two Yunnan species overlapped. After Paris, the Peers did travel to London, where Janaki was, but this turned out to be a very short visit, to present a plant or two of the yellow Camellia (which Peer had rediscovered) to the Kew Gardens. William Campbell, the Curator at the Gardens would be 'thrilled' to get back the supposedly 'lost' yellow Camellia of Robert Fortune. 'We came across this item in a very old nursery in Porto, Portugal,' Peer had revealed to Janaki. 'This is a sasanqua (a Japanese Camellia with fragrant flowers), which can be propagated only by grafting. I suspect that a study of the chromosomes would disclose great irregularities. The two plants, which I sent to Kew should have some growth by this time next year,' he explained, writing from aboard the luxury cruise liner, *Queen Elizabeth*.[16]

Peer also shared with Janaki the intelligence that a batch of fifteen 'Kunming reticulatas' (*C. reticulata* from the Kunming Botanical Garden, Yunnan established in 1938) had just been received by Lord Aberconway, for the exclusive use of the Bodnant; these were hybrids, mostly *C. pitardii* x *C. reticulata*, he hinted. He wanted Janaki to persuade Lord Aberconway in the following year to let her have enough material for cytological study, but as far as her research grant was concerned, stated that there was 'no interesting reaction' as yet, but that he was still at it.[17] One of the people he was in touch with was the millionaire Calder W. Siebels of Columbia, South California. Inheritor of the 738-acre Lark Hill Plantation, Siebels would write the foreword to *The American Camellia Yearbook*, 1952, in which Janaki's paper would finally appear. Siebels was also President of the South Carolina Camellia Society.

Janaki was prompt in replying to Peer's letter sent to her in early April. She was deeply disappointed she could not meet them at Kew, where she usually spent most of the week. She had been detained in London, Janaki explained, due to a 'hitch in the grant' she was to receive from the Ministry of Agriculture, France. 'We are allowed only £25 when going abroad, so I cannot earn francs. I must remain here. All this makes me wish so much to be in USA,' she wrote. Janaki informed Peer that Bartlett had assured her that the University of Michigan was ready to provide all facilities to work on *Camellia* if a grant could be had from the American Camellia Society. 'I think he has written to you. I hope you will be able to help. I want to study the related genera to get at the evolution of the genus Camellia,' she appealed. At Kew, she had had 'a look at the rare yellow Camellia of Robert Fortune' (which Peer had rediscovered at Porto), but found it 'looking a little unhappy.' However, Campbell had allowed her to 'one tiny bud' for counting chromosomes, for which she was grateful.[18]

Janaki had finished a preliminary paper (a map showing the distribution of polyploids was also included) on cultivated Camellias to be published by the Southern Camellia Society (California) but was unable to 'add some data' as her papers had been left behind in Paris; in addition, she had come to learn of a *C. tsaii* (Forrest 25252) at Carharp Castle, Cornwall, and wished to examine it for its count and include that as well. Janaki mentioned to Peer that Joseph Robert Sealy of Kew, who was working on revising the genus (his 'Revision of the genus Camellia,' 1958, went on to win the Veitch Memorial Medal in silver of the RHS), was 'very interested' in some of her counts. She listed them out for Peer: *C. pitardii* (which Peer had sent her) was a diploid (2n = 30), *C. pitardii* var. *yunnanica*, a hexaploid (2n = 90), 'just the same relationship as between *C. saluenensis* (2n=30) and *C. reticulata* (2n=60)!' Janaki titled her paper 'Chromosome Relationships in Cultivated Species of Camellia.'[19] She planned to send Bartlett a copy of her paper, 'to see how much there is to do.' 'As far as I know there has been no work on the other 16 genera of Theaceae (Ternstroemiaceae) and the chromosome relationship between these & Camellia would be an interesting problem in cytogeography. The number of Camellia species in cultivation is also very small,' she explained.[20] Janaki incidentally had collected a *C. kissi* from Nepal in 1948.

'I am hoping some good luck will be on my way very soon to visit USA—otherwise I must return to India and grow tapioca & yams. No chance of earning Francs in Paris!!', she wrote hopefully in a letter sent from the Herbarium at the Kew Gardens.[21] At Kew, Janaki's research was by no means restricted to the *Camellia* collection. In fact, she would present seeds of asparagus and *Michelia neilgherrensis* (*Michelia nilgirica*), which she had received from India, perhaps from brother Kittu, to Kew's Campbell, with whom she discussed her keen interest in the study of asparagus cytology.[22]

Back to Paris

By late April 1952, Janaki returned to Paris. From the Laboratoire de Genetique (she had only arrived in Paris that day), Janaki wrote to Kew's Edward Salisbury, enclosing some seeds of two Teosintes (or *Euchlaena perennis*, any of the four species of tall and stout grasses in the genus *Zea*, of the family Poaceae), which she had received from the American agronomist Paul C. Mangelsdorf (best known for his studies on the origins and hybridisation of maize). She asked him for these to be back-crossed with her intergeneric fodder grass hybrid *Euchlaezea mertonensis*, which Kew had been taking care of for her. Janaki also wanted it back-crossed with *Zea*, but as the hybrid flowered very late, the maize had to be grown rather late in the season. The letter was also a request to Salisbury to permit her to complete her researches at Kew.[23]

By this time, her paper on the cytology of the *Lonicera* had appeared in the *Kew Bulletin* and, as always, the name of her assistant was included as co-author. She had begun work on its cytology towards the end of her stay at Wisley, in 1951, assisted by Barbara Saunders, who had joined the laboratory that year. Of the one hundred species known to horticulture, Janaki and Saunders examined fifty species growing in Kew and the Wisley Garden, besides twenty-two garden varieties and hybrids. As with most of her papers since the publication of the *Chromosome Atlas*, Janaki included a phytogeographical distribution map for the genus, but now in addition there was one mapping polyploidy in the Asian *Lonicera*,[24] as with the *Camellia*, her chief interest being the study of cytogeography of flowering plants on a global scale.

Fair Version of the Camellia Paper

In the first week of May, from the genetical laboratory in Paris, Janaki sent Peer the final copy of her paper on the Camellia, accompanied by a letter: 'I have somewhere amongst my papers a letter from the Camellia Society of America asking me for this paper. I hope you will be so good as to communicate it for me to them. You will find the paper waiting for you at Westchester Country Club, Rye, New York.' In her great desperation to leave Europe, for America and given the sorry state of her finances, she was unashamed to appeal to him again: 'I hope you will soon find some way of opening up a chance for me to come to America. I am waiting to hear if you have heard from Prof. H. H. Bartlett, Director of the University of Michigan, Bot. Lab & Gardens, who is prepared to give me all facilities and assemble Camellias at the Gardens.' Janaki had been looking around for grants herself and mentioned to Peer that she had started 'preliminary negotiations for securing a USA Fulbright Senior Travelling Grant to come to America,' but the issue was that they needed an official document to say that 'some institution would guarantee [her] expenses in the States.' 'I am eking out an existence in Paris because I have so far not been able to earn francs—and you know how difficult it is to get Sterling! The sooner I get away to USA—the better for me,' she added.[25]

Looking Up to Bartlett

Meanwhile, Janaki had also sent Bartlett her paper on the Camellia; she warned him that the paper had been written 'for the layman hence the elementary cytology in it for which I apologize.' She told him that she was dejected that Peer was unable to find her a funding opportunity in America yet. 'I think I shall have to return to India and do the best I can there but before I go I would like to visit my old Alma Mater in USA and see old friends.' She wanted to apply for a Fulbright Travelling Grant if the University Botanic Gardens at Ann Arbor could give an assurance that she would be provided with dollars while in America. 'I am ready to give a course of lectures on the work I have been doing on Sugarcanes in India and Horti plants in England, if that will help. So that I can call it a lecture tour,' she begged. She was in touch with the United States Educational Commission in the United Kingdom, Janaki informed Bartlett. She held a British passport but when that expired in June, she was to be given an Indian one. 'Everything points to a return to India!!, which is I suppose as it should be,' she remarked on a resigned note.[26] In a later letter, with the Royal Geographical Society as her address, she wrote to Bartlett: 'Here is another stumbling block to "Westward Ho". I am not yet a citizen of India because I still have a British Passport. Of course, when the time to change comes, I shall take up an Indian Passport and that will be soon.'[27] She knew a return to India was ineluctable.

One Last Attempt

Impatient, Janaki wrote again to Peer on 16 May 1952 to enquire whether he had received her paper; she had sent it by airmail, along with the original maps, and was anxious to know if it had reached him. She had also despatched 'a slightly modified version of the paper as the final one for publication from France.' If possible, I would like the paper to be published by the Camellia Society of America in its Yearbook 1952,' she requested Peer. She used the occasion to let him know that the yellow Camellia from Portugal, which he had presented to Kew's Campbell was in 'a sorry way.' 'He [Campbell] is arranging for cuttings by air [from Portugal] at the proper season—Such a pity! But I am certain there is nothing extraordinary about its chromosomes—It is a hexaploid, 2n=90,' she informed Peer. Once again, she would appeal to him: 'I hope it will be possible for me to come to America for at least a little while to work on the Camellia family at the University of Michigan—I am very keen to do it.'[28] Their letters crossed each other, and on 19 May she would hear from him. He had read the revised version of her paper but was surprised that she was not 'familiar with the theory (arrived at separately) held by both Te Tsun Yu and Walter E. Lammerts (1904–96) that the so-called "Kunming reticulatas" are all hybrids.'[29] Te-Tsun Yu (1908–86) was a Chinese botanist, who was the co-founder of the Kunming Institute of Botany in Yunnan and as for Lammerts, he was a controversial geneticist and rose breeder, who vehemently attacked the theory of evolution, to promote creation science.

Sometimes, hybrids, especially of garden plants, were mere claims by nursery-men or elite horticulturists. This was also the case with certain species of Camellia. Peer's correspondence with Tsai Hsi-Tao, the protégé of Te-Tsun Yu, had revealed that 'since time immemorial it [had] been customary to cross Camellia C. reticulata (presently the species form) with the local Camellia C. pitardii in order to secure attractive garden forms;

as these hybrids cannot ordinarily be reproduced from cuttings it would be entirely normal that all of them would gradually disappear, excepting only certain varieties which were considered exceptionally beautiful garden forms.' Peer noted that the 'varieties, which had been imported from Yunnan seem to be presently available because of some old tree growing in a sheltered position.' This was reconfirmed, he informed Janaki, by a 'new hybrid *reticulata* x *pitardii* (so described by Tsai), which was received by Dr Lammerts direct from Kunming during 1949, and which has just flowered.' The flowers were magnificent and 'the resulting plants in all respects appear[ed] to fit in to the series known as "Kunming reticulatas". This variety has been named Budha,' he added. Peer was completely convinced by the theory on the strength of his own 'amateur observations.' He observed that the ordinary garden *reticulata* and the rare double-form ('florepleno'), both with large and very beautiful flowers, displayed 'a lack of viability,' by which was meant that they could only be propagated by grafting. As for the chromosome count provided by a cytologist, the garden form revealed irregularities, which had forced him to think this was indeed the case with the double-form. Peer was suggesting that these plants were mutations of *reticulata*, whereas the weaknesses of the Yunnan group could be attributed to its hybrid ancestry. Lord Aberconway, Peer added, had thirteen varieties of the Yunnan group growing successfully in his garden at Bodnant, of which he had already alerted Janaki.[30]

On the 'Kunming reticulatas' that Peer had first mentioned (in a letter from early April), Janaki the cytogeneticist (writing from the Herbarium at Kew), had commented that she could not pass a verdict until she had examined their chromosomes. 'I shall reserve that for a later date,' she had told him. She used this opportunity to ask him categorically whether he really wanted her to work on Camellias or whether there were people in United States already on the job, in which case she would not be needed. She remarked that the 'Royal Horticultural Society of England [had] given up cytological research at Wisley ... [which was a] great pity.' It made her wonder if Lord Aberconway's thirteen varieties would ever get analysed in England. Whether Janaki actually responded to Peer's question on the 'Kunming reticulatas' on a later occasion is not clear, but she was outspoken about the possibility of a 'species inflation' within Camellia taxonomy as it existed.

According to the revised classification of Camellias (Linneaus had recognised only two: *Thea* and *Camellia*), all native of East Asia and adopted by Joseph Sealy, they were placed in five subgeneric groups: *Camelliosis*, represented in cultivation by *C. salicifolia* of Hong Kong and Formosa (Taiwan); *Theopsis*, represented by *C. cuspidate* of East and Central China; *Thea*, represented by the tea plant, *C. sinensis*, its variety *assamica* and *C. taliensis* of Yunnan; *Camellia*, which included the large flowers and beautiful *C. japonica*, *C. reticulata*, *C. saluenensis* of the garden variety, and the oil-producing C. *oleifera* of China and C. *sasanqua* of Japan; and lastly the genus *Calpandria*, found in the East Indies and the Philippines, the representative species being C. *lanceolata*. Janaki did not approve of T. Nakai's classification of the *Camellia* (he was based at the Tokyo Botanical Garden) because there was 'too much of splitting of the Genus,' and as for Sealy's ordering, it was not very different from Adolf Engler's.[31] In all her researches and publications, she had followed the Kew classification; 'I should be consistent,' she would remark to Peer.[32] For the regular taxonomist, Janaki observed, it was sufficient if species differed from each other in one or more morphological characters to give them specific names; the concept of species as per this view would appear as if it was only 'a reflection of the mind of the botanist,' and not one based on scientific reasoning. To the cytotaxonomist on the other hand, permanent variations in external morphology, inherited from the parent by the offspring and so on, were reflected in the chromosomal profile of the plant, making cytology an important tool for the taxonomist. The *Chromosome Atlas* had driven this point home very clearly. Janaki believed that if she were given a funded opportunity in America, she would be able to work out the cytotaxonomy of the *Camellia*.

Peer replied to Janaki's letter in late June conveying the good news that the American Camellia Society was 'quite pleased' to receive her article and that it would indeed appear in the Year Book (1952), which would be ready for distribution that Fall.[33] To her query, he assured her that 'nobody [was] presently engaged in special botanical studies of Camellias as such.' Also, that the 'research work carried on in the Los Angeles area, over a period of years was completed two years ago and that it had not been possible to develop a new research programme.' 'If therefore, it can be arranged to have you come to the United States, you will be the sole investigator in a field where public interest exceeds scientific knowledge,' he put it succinctly. The Peers were planning to spend the whole of the summer at their Park Hill estate in Los Angeles, so that Peer could work in his Camellia garden and 'to spend time with the many Camellia enthusiasts living in that area.' He continued to be hopeful that he could 'develop some plan,' which would fulfil Janaki's desire to migrate to America to research on the Camellia. Peer also mentioned the names of two books published in Paris (he had come across these during his search for information on the *Camellia*), containing descriptions of several species of *Thea* which grew in Indo-China. These were the *Flore Generale de L' Indo-China* (1910) and another, which was actually a supplement to it (1943). 'It would seem that this section of the world contains more different sorts of Camellias growing in a wild state than any other. It is only because of the war, which now disturbs that country, that I have not investigated this region horticulturally,' he would remark, perhaps suggesting that Janaki explore that option as well.[34]

Regrets

When Janaki finally heard from Bartlett, she was overjoyed, but had nothing great to tell him. She only told him that she was 'really an exile in France' and hardly knew how long she would remain here and rather unexpectedly for us, that she regretted missing 'a great opportunity' to serve in India on the Grow More Food campaign. Though she had accepted Nehru's invitation, she confessed to Bartlett, she 'did not have the courage needed to battle with the problem.' Nehru's idea to make India self-sufficient in food by 1951 was 'so immense,' and so unrealistic that it hastened her return to England, after delivering 'some advice' to the Ministry of Agriculture. Before she had 'seen the condition India was in,' she had thought it was possible and had even written a note 'as to how it could be done.' When however, she saw 'the machinery of Govt. in Delhi—a huge Jagannath! [she] felt [she] didn't have the spiritual strength to reform it.' As a result, she felt, she had lost a chance to serve the nation 'and be an important person in the Ministry of Agriculture.' In her 'spare time' at the Agronomy Institute in Paris, she confessed, she was 'working out a solution for all the mistakes I saw in India and this while [she] was hard at work on [her] own research.' At the same time, Janaki was glad she had not stayed on in India because the job was purely 'administrative' and would have tolled the death knell as far as her research was concerned.[35]

For now, however, it was proving difficult to survive in Paris without a proper stipend, even though Janaki was enjoying her work; she was desperate to go to America and hoped Bartlett would be her saviour. She had 'worked very hard at RHS Gardens' and had 'discovered some rather exciting things regarding plants in that region of SE Asia, the Sino Himalayas where so many of Garden plants of Europe came from.' In every genus she worked on she had found high polyploidy, and it was to the problem of cytogeography that she wanted to devote the rest of her life. 'It means working in a place where I can get facilities for cytological work + a good herbarium + living plants I can work on + money to live. France has given me all the 3 first' but she had been living with her niece and family, and they had now left. 'I want to come to USA and you must help me to get there. I have faith you can,' she appealed to him, clearly revealing her desperation. In a postscript, Janaki promised Bartlett 'an article' on her 'collection trip to Nepal' (as per his request of January 1952) for the *Asa Gray Bulletin*, but warned that her life was 'too much of a failure to write about.'[36] She would never send it.

Life and Work in Paris

Her niece and family having left Paris for good, Janaki found herself homeless. She was forced to move into the International House for Women (inaugurated in 1936); similar ones had been established in New York (1924),

Berkeley (1930) and Chicago (1932). The French Ministry of Agriculture, luckily for her, made her a very liberal grant, and she was thus comfortably off but extremely lonely. 'Paris is empty without Padma & family—for the same reason, I am getting to know it better as I have to do everything myself & manage with broken French,' she would write to brother Raghavan. She kept busy with her work at the genetical laboratory, chiefly on the cultivated species of the genus *Camellia*. In a few days, she would set out on a journey to the South of France, to the Antibes in Provence to see the breeding of scented plants at the Agronomy and Plant Biochemistry Station (a daughter institution of the plant genetics laboratory she was at in Paris) and study the perfumery industry there.[37] She was keen on extending this industry in India. On the Pyrenees, she observed, farmers grew miles and miles of lavender, and she suggested to Raghavan they do something similar and on a large scale somewhere in Malabar: 'we in Malabar could make fortune on such things as Lemon Grass, Patchouli etc.'[38] This business proposition would however never take off.

Nehru's Second Offer

In a letter to Bartlett in August 1952 Janaki revealed that Nehru had offered her 'the post of Director of the Botanical Survey of India' under the Ministry of Natural Resources and Scientific Research. 'After much thought I have said Yes,' she told him. Provisionally, the contract was for a year, but with the possibility of an extension. 'The Botanical Survey of India Directorship has been long held in abeyance and I shall have to organise the Department. It will not be easy! But I shall do my best to modernize Botany in India.' Ever hopeful of going to America, she remarked to Bartlett: 'I was looking forward to visiting Michigan somehow in 1953. This will have to be postponed. Perhaps I will be able to have a flying visit some time in the not too distant future.' She wanted Bartlett's help in publishing 'one or two papers in an American Journal of Botany,' as the papers she had submitted to the RHS journal had been returned to her by its editor. 'I want to see them in print as they deal with the Cyto-geography of Sino-Himalayan plants. I may have to rewrite some to make them more fitted for a scientific journal (RHS journal is for the layman!).'[39]

On the Family Front

Meanwhile, Janaki had heard from brother Raghavan, that the Malayalam translation of Hilda Seligman's *When Peacocks Called* by Malabar K. Sukumaran, her cousin (and elder sister Cousalya's husband), and titled *Mayurakahalakalam*, had been printed in Shoranur; she was yet to hear about the proceeds of its sales though. 'I am beginning to think it had better all go to Sumthi [sister Sumithra],' she wrote to elder brother Raghavan, 'who is a splendid business woman.' After all, Sumithra had invested money in it. Janaki was hoping she would be able to get Indian universities to include the translation in their curriculum, but it doesn't appear much came of this family project, monetarily or otherwise.

In her letter to Raghavan, Janaki disclosed the news of her being offered the post of Director of the Botanical Survey of India; she was yet to receive the appointment letter in her hands though. 'I have not heard the details but the offer came directly from the PM. This will combine Research with Economic Problems in Indian Plants,' she explained to Raghavan, obviously pleased that her research would not be put on the backburner, if she accepted the offer. She was longing to be back with her family: 'It was high time I returned to the fold,' she told him, 'Edam wants a little pulling up.' Janaki was particularly worried about youngest sister Devayani's illness (she had been admitted to the Madras General Hospital at this time), as also the fragile state of health of a niece and of Raghavan's finances. Her concern extended to her grand-nephews and nieces, some of whom often assisted her in collecting and drying plants and packing them up: 'that boy has to be settled in life and I think a strong hand is needed,' she wrote referring to a grand-nephew.[40] Incidentally, when it came to her nieces, Janaki preferred marriage to a career for them, unless they were particularly bright like sister Cousalya's daughter, Leela (later Desai). Speaking of Raghavan's youngest daughter, Uma, Janaki said she would like to see her 'either getting on with her studies—or married—preferably the latter. She is such a charmer!'

Janaki was hoping to be back in India in November (when she would turn fifty-five) and much looked forward to 'wandering in the forests with [Raghavan].' She was not keeping too well however, as her left arm was bothering her (despite the surgery in Edinburgh in 1939), and she had painful shoulders, thanks to the years of slouching over the microscope to examine chromosomes. 'I have fibrositis [fibromyalgia] and I suppose what we call *katachal*, but painful. I would have liked some good Malabar oil. When I come I must take a course of oil bath.' Janaki enjoyed medicated oil baths (much like her father and elder brother Kittu), especially during the auspicious month of *karkidakam*. She enclosed for Raghavan seeds of the new red *Salvia*, the variety she had produced at Wisley 'by treating the ordinary one with Colchicine.' 'I am calling it Lekshmi after our elder sister, and the goddess,' she would remark[41]; Lekshmi, the eldest female member of the Edathil household, an elegant and accomplished person, had only recently passed.

Departing for India

All her attempts at migrating to America, even if for only a couple of years, had not met with success. Janaki would leave Paris for India by the end of September 1952, earlier than planned. She had decided to accept the offer, but little did she imagine her return to India would be for ever. Towards the end of her time in Paris, Janaki bought the Brazilian physician-geographer Josue de Castro's newly published book, *Geography of Hunger* (London, 1952), a ground-breaking work in ecology relating to the politics of hunger in Brazil. It was translated into twenty-six languages and came to be regarded as a classic study on food and population. Contrary to the Neo-Malthusian worldview, de Castro had argued that hunger was a man-made phenomenon.

Meanwhile, Cyril Jones, an Assistant Under-Secretary at the Foreign Office, had despatched to Janaki, care of Kew, some freshly ripened tea seeds he had received from India. Janaki as we know had been working on Indian teas (from Assam) among other species of the *Camellia* at the genetical laboratory in Paris.[42] A couple of months later, in January 1953, Gopalswamy Doraiswamy Naidu (G. D. Naidu, 1893–1974), the ingenious inventor and amateur plant breeder from Coimbatore, whom Janaki knew (in what context is not clear, but he lived in the same locality in Coimbatore as she did), also sent her, to her Kew address, 'by sample post three packets containing tea seeds (Betjam, Rajghur and Dangri Manipuri varieties) collected in the 1952 season in Assam.' Naidu had packed them up in 'carbon, so that they may not be spoiled. You may continue your experiment with these seeds also,' he wrote to her. 'Is there any chance of your coming to our country in the near future?' he queried. He was planning a visit to England that year to attend the British Industries Fair, and hoped to meet her, little knowing that she had already returned to India and was in Calcutta not too far away from where the tea seeds came.[43] Ever enterprising, Naidu would continue to provide her with vegetables, fruits (such as the seedless and extraordinarily large papaya, perhaps a triploid) and seeds from his farms for her research, which sometimes were the results of her experiments but mostly his own.

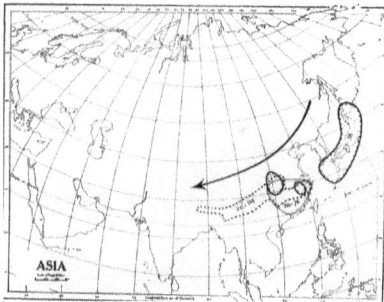

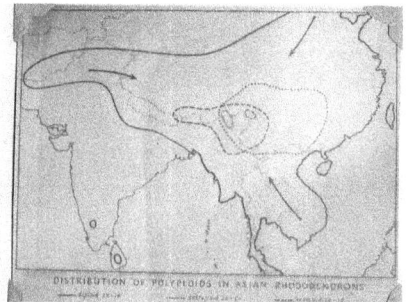

Figure 18.1 Janaki's maps showing polyploid distribution in Asia of the *Magnolia* (left) and the *Rhododendron*.

Source: Private Collection.

Notes

1 Huntington Library, San Marino, California (hereafter HL): Mss Peer, Box 2, Folder 6, E. K. Janaki Ammal to Ralph S. Peer, letter dated 28 January 1952.

2 Ibid., H. H. Bartlett to E. K. Janaki Ammal, letter dated 3 January 1952.

3 RBG Kew, Library & Archives: QG 169, Correspondence with Janaki Ammal, 1945–1970, Edward Salisbury to E. K. Janaki Ammal, letter dated 11 January 1952.

4 Ibid., E. K. Janaki Ammal to Edward Salisbury, letter dated 20 January 1952.

5 Ibid., E. K. Janaki Ammal to N. L. Bor, letter dated 20 January 1952.

6 Ibid., Edward Salisbury to E. K. Janaki Ammal, letter dated 23 January 1952.

7 Kedharnath, 'Edavaleth Kakkat Janaki Ammal (1897–1984)', p. 99.

8 HL: Mss Peer, Box 2, Folder 6, Ralph S. Peer to Calder W. Siebels, letter dated 5 April 1952.

9 Ibid., Ralph S. Peer to E. K. Janaki Ammal, letter dated 14 August 1951.

10 Ibid., Ralph S. Peer to Calder W. Siebels, letter dated 5 April 1952.

11 Ibid., E. K. Janaki Ammal to Ralph S. Peer, letter dated 28 January 1952.

12 Ibid., letter dated 15 February 1952.

13 BHL, UoM: University Herbarium Records (University of Michigan) H. H. Bartlett series, Box 11, E. K. Janaki Ammal to H. H. Bartlett, letter dated 19 February 1952.

14 Ibid., H. H. Bartlett to Ralph S. Peer, letter dated 27 February 1952.

15 HL: Mss Peer, Box 2, Folder 6, E. K. Janaki Ammal to Ralph S. Peer, letter dated 18 March 1952.

16 Ibid., Ralph S. Peer to E. K. Janaki Ammal, letter dated 5 April 1952.

17 Ibid.

18 Ibid., E. K. Janaki Ammal to Ralph S. Peer, letter dated 16 April 1952.

19 Ibid.

20 BHL, UoM: University Herbarium Records (University of Michigan) H. H. Bartlett series, Box 11, E. K. Janaki Ammal to H. H. Bartlett, letter dated 17 April 1952.

21 Ibid.

22 RBG Kew, Library & Archives: QG 169, Correspondence with Janaki Ammal, 1945–1970, E. K. Janaki Ammal to W. M. Campbell, letter dated 29 April 1952.

23 Ibid., E. K. Janaki Ammal to Edward Salisbury, letter dated 30 April 1952.

24 Janaki Ammal and Saunders, 'Chromosome Numbers in Species of *Lonicera*', 1952.

25 HL: Mss Peer, Box 2, Folder 6, E. K. Janaki Ammal to Ralph S. Peer, letter dated 7 May 1952.

26 BHL, UoM: University Herbarium Records (University of Michigan) H. H. Bartlett series, Box 11, E. K. Janaki Ammal to H. H. Bartlett, letter dated 9 May 1952.

27 Ibid, letter dated 16 May 1952.

28 Ibid, letter dated 19 May 1952.

29 HL: Mss Peer, Box 2, Folder 6, Ralph S. Peer to E. K. Janaki Ammal, letter dated 19 May 1952.

30 Ibid.

31 Adolf Engler (1844–1930), with Karl A. E. Prantl (1849–93), published the twenty-volume work *Die Natürlichen Pflanzenfamilien* (1887–91).

32 HL: Mss Peer, Box 2, Folder 6, E. K. Janaki Ammal to Ralph S. Peer, letter dated 27 May 1952.

33 Janaki Ammal, 'Chromosome relationships in cultivated species of *Camellia*', 1952.
34 HL: Mss Peer, Box 2, Folder 6, Ralph S. Peer to E. K. Janaki Ammal, letter dated 23 June 1952.
35 BHL, UoM: University Herbarium Records (University of Michigan) H. H. Bartlett series, Box 11, E. K. Janaki Ammal to H. H. Bartlett, letter dated 15 June 1952.
36 Ibid.
37 Ibid., letter dated 8 August 1952.
38 E. K. Janaki Ammal to E. K. Raghavan, letter dated 31 July 1952, private collection.
39 BHL, UoM: University Herbarium Records (University of Michigan) H. H. Bartlett series, Box 11, E. K. Janaki Ammal to H. H. Bartlett, letter dated 8 August 1952.
40 E. K. Janaki Ammal to E. K. Raghavan, letter dated 31 July 1952, private collection.
41 Ibid.
42 RBG Kew, Library & Archives: QG 169, Correspondence with Janaki Ammal, 1945–1970, Sir Cyril Jones to Director of Botanic Gardens, Kew, 16 November 1952.
43 Ibid., G. D. Naidu to E. K. Janaki Ammal, letter dated 1 January 1953.

19

CALCUTTA

Modernising Botany in India

Here in India I am struggling to reorganize the Botanical Survey on a broader basis than the past Survey.
—E. K. Janaki Ammal (1953)[1]

Janaki had to leave Paris earlier than planned, to reach Calcutta by 14 October 1952 and take up the appointment of Officer-on-Special Duty, Botanical Survey of India (BSI), Chowringhee, in the heart of the city. A small office had been set up for her, with the permission of the Trustees of the Indian Museum, in the Museum House wing, where she settled down to the onerous task of drafting a scheme for the reorganisation of the BSI.[2] For about a year, without once visiting her family at Edam or Shoranur, she worked single-mindedly on it.

Correspondence with Friends

Correspondence with friends in England, particularly Stern, Darlington, and Bor, would however be kept up, sharing with them the tribulations and challenges she faced as a government officer in India. Peer would also continue to correspond with her, after her move to Calcutta. In late January 1953, he wrote to let her know that the Year Book carrying her article had been published. 'It turned out very well,' he would remark. Peer observed that he would be pleased to receive 'scions of any unusual Camellias' which, she may come across during her official tours, 'especially of the species . . . indigenous to India.' 'Please remember I use the cleft grafting method and therefore, two very short scions will normally be sufficient to reproduce a species or variety,' he explained. Seeds were also welcome, he added. Peer wished to know if Janaki had facilities for making chromosome counts at her new place. 'If so, do you want me to send you any of the odd items, which come to my attention,' he queried. He was also more than happy to send her scions if she was interested in introducing Camellia 'to the garden of [her] friends in Northern India. They should certainly grow in the

DOI: 10.4324/9781003267089-19

region of Darjeeling, that is, the garden variety.' Peer was curious to know if Janaki had been in touch with the horticulturist Sydney Percy Lancaster (1886–1972), who lived at 1 Alipore Road and was famous for his hybrid Alipore Canna collection among others; it was in Calcutta that he had met Lancaster first and ever since had kept up a horticultural correspondence with him. Peer did not forget to tell Janaki that his efforts 'to find a spot' for her in America continued.[3] He would write again, when his plans to visit India were finalised; the Peers were to spend a few days in Bombay in mid-February 1954, before flying out to Colombo, en route to Australia and finally, Japan. It is not known whether Janaki and the Peers met up on this occasion or ever again.[4] In any event, we know for certain that her pursuit of the Camellia would run out of steam for want of funds, with her paper in the American Camellia Society's Year Book (1952), remaining her only publication on the genus.

Publication on the Viburnum

Janaki would also write up papers for publication during this time, the research for which had already been completed in England. Her first publication after joining the BSI was her paper on the cytogeography of the genus *Viburnum*, which she had been collecting and studying since 1950. This was published in the *Current Science* in early 1953; she had planned to publish it in the RHS journal, but having resigned from Wisley, felt no more compelled to do so.[5] The paper was the result of her survey of the species in the live collections of Wisley, the RBG Kew and the Jardin des Plantes, Paris. The white form of *V. fragrans* was the only tetraploid found among them, which she suggested had its origin in cultivation in China.

O. J. Eigsti Visits

Famous for his seedless watermelon, a product of colchicine-induced tetraploidy, the cytogeneticist O. J. Eigsti visited India (and Pakistan) in 1952 as a Fulbright Fellow. He would call on Janaki at the BSI, and together they would visit the Banaras Hindu University (where niece Leela Desai taught), among other institutions of interest. Several Indian cytogeneticists including Janaki supplied him with unpublished data from their colchicine-induced research,[6] which would be incorporated into his colchicine vade mecum, *Colchicine in Agriculture, Medicine, Biology and Chemistry* (1955), co-authored with Pierre Dustin.

Symposium in Delhi

In early March 1953, Janaki travelled to New Delhi to present a paper at the symposium on 'Organic Evolution.' Organised by S. S. Bhatnagar, representing the National Institute of Sciences of India, the conference was an opportunity to celebrate the presence of the German biologist Bernhard Rensch in the country. The conference was convened by ichthyologist Sunder Lal Hora, an almost exact contemporary of Janaki, who had by now become well-known for his Satpura hypothesis, a biogeographical theory on the affinities of the fauna of the Western Ghats to the Indo-Malayan forms. The idea of species adaptation was the key note of the symposium. Almost half the papers delivered or discussed at the gathering were by non-Indian contributors, and the list included such big names as Bernal, Haldane and his partner Helen Spurway, entomologist Sidnie M. Manton (sister of the plant cytogeneticist, Irene Manton), Joseph Needham and C. F. A. Pantin. Indian biologists were not well represented at the symposium, but papers 'of wide general interest' were presented by people like Janaki, the star delegate on the Indian side, besides Sunder Lal Hora of course. Others included C. P. Gnanamuthu, A. K. Ghosh and A. Bose, K. Jacob and wife Chinna from the Birbal Sahni Institute, besides H. K. Mukherjee, Sivatosh Mookerjee and M. L. Roonwal.[7] Janaki also visited Bombay during this trip, where, following her friend Bor's suggestion, met up with the Spanish Jesuit botanist H. Santapau, teaching at the Bombay University; he impressed her very much. 'I was sure you would be impressed by Santapau's work. He really is keen, knows the right way to do things, and does not spare himself in the field,' remarked Bor in his letter, thanking her for the seeds of the *Chenopodium album*, a commonly cultivated plant of Northern India, which she had despatched to him.[8]

Against Conservatives

In early April, Janaki received news of Darlington's appointment as Sherardian Professor of Botany at Magdalen College, University of Oxford. 'I wonder if this is a promotion. I thought there could be nothing bigger in the world of Biological Sciences than being the Director of JIHI. Why has this changed?' she queried impatiently. 'At least Oliver & Andy can be Oxon now!' she remarked. 'Here in India I am struggling to reorganize the Botanical Survey on a broader basis than the past Survey but it looks as if the greatest objectors to change are the old fashioned systematists of England who by direct & indirect methods are putting stumbling blocks to reform. They want one of their kind to take charge who will not contaminate Taxonomy with cytology,' Janaki complained to Darlington. She wished he could help in some way. Janaki had heard that Edward Salisbury of Kew was arriving in the country to advise the Government of India on the BSI. 'How I wish it could be you!' she wrote to him. She enclosed C. V. Raman's letter to her for Darlington to read, with the words: 'He is one of the few here with a broad outlook. I am beset with politics—Bengal politics is a very virulent kind.'[9]

Darlington's reply was prompt, because he was himself in a similar situation: 'I am sorry you are having trouble with the Bourbons of systematics. My reason for going to Oxford is that I shall be able in a few years, to overthrow them in this country which from this Institution [John Innes] is unfortunately impossible since here I am in fact (as anyone who surveys our Governing Body can see) directly subject to the Bourbons themselves.' He had news to share; Wylie and he were on the verge of completing their edition of the *Chromosome Atlas*. 'I have written a much longer Introduction, which challenges all the traditional opinions of systematic botany in an uncompromising and explicit,' manner, he explained. Darlington was eager to have any new chromosome numbers Janaki might have for inclusion in the book, 'especially on Magnolia and Viburnum.'[10]

Leaps in Genetics

In what has been referred to as the classical period of genetics, the gene was understood as a dimensionless point on the chromosome. From the 1940s however the gene began to be ascribed an 'unambiguous spatial dimension,' namely length, and (later) a 'linear chemical identity' in the shape of the DNA molecule.[11] As early as 1944, it had come to dawn on scientists that genetic change in bacteria owed itself to the DNA and not protein and that it was the DNA molecule that determined heredity in genes and chromosomes. The molecular structure of the DNA became a matter of great curiosity. In 1950, Erwin Chargaff demonstrated that in DNA, the amount of adenine was equal to the quantity of thymine, and the amount of guanine was same as that of the cytosine, a major clue that helped Watson and Crick develop their model of the molecular structure of the DNA, which in turn shed light on how DNA replicated and how hereditary information was coded on it, thus setting the stage for the rapid development of the field of molecular biology. The American James Dewey Watson and the Englishman Francis Crick of the Cavendish Laboratory, Cambridge, would go on to share the Nobel Prize in medicine (in 1962), with Maurice Wilkins, for their modelling of the DNA double helix in 1953 (the news of their discovery was published in April that year), based on the X-ray diffraction studies by Rosalind Franklin (and Wilkins) at King's College, London.

Incidentally, McClintock's microscopic studies of the maize between the 1940s and 1950s would lead to the path-breaking discovery that genes were dynamic in nature, with the ability to move from one position to another, thereby switching physical traits on or off, a discovery that would bring her a much-belated Nobel, in 1983. Her finding shook up the world of classical genetics, which assumed that the positions of genes on chromosomes were fixed, and for which reason it failed to find immediate acceptance. It had to wait until the late 1960s before the biological community gave it serious attention, thanks to sophisticated molecular methods, which aided the discovery of similar jumping genes (later called transposons) in the bacterium *Escherichia coli*.

Studies on Citrus

Within a month of the symposium on 'Organic Evolution,' Janaki was ready with a short paper on a new Citrus hybrid, material for which she had begun collecting in Nepal in 1948 (the Eastern Himalaya was suspected to be the origin of the Citron, as natural hybrids displaying Citron and Lemon characters were abundant there). To this material she would add an interesting hybrid species, which bred true, from sister Cousalya's house in Shoranur in the Western Ghats, besides some species from the Kew Gardens such as the giant 'Mitford Lemon.'

Among citrus fruits, *Citrus medica* (Citron) and *C. aurantifolium* (acid Lime) had been in cultivation in India from very ancient times. *C. limonia* (Lemon) was closely related to the Citron; it was considered a variety of *C. medica* by Linnaeus, and also by Company botanists like William Roxburgh and those like Hooker and of course Brandis, who worked on India's forest flora. Janaki suggested that the Lemon might have come into existence as a hybrid between the Citron and the Lime in Western India, and spread to the Mediterranean countries in the early years after Christ. Interspecific hybridisation was a very common trait in the genus and there were intermediate forms between the Lime and the Lemon, one of which was the Malta Lemon. A Maltese surgeon attached to the Indian Medical Service, Emanuel W. Bonavia (1829–1908), a famed horticulturist and Superintendent of the Government Horticultural Garden in Lucknow, had introduced the Malta Lemon to this princely state in 1863. He was also the author of an important monograph on the *Citrus* published in 1886.[12]

The Malta Lemon was introduced to Malabar from Burma in 1931 by Janaki's elder brother E. K. Raghavan, among several other plants including the bougainvillea. This lemon was a heavy yielder, producing fruits all year round, often in bunches of four or five, and ideal for the small gardens of Malabar, Janaki observed, but these were mostly seedless, meaning they were triploids. An interesting fact about Citrus hybrids was that seeds produced true plants, owing to a phenomenon called vegetative embryony (production of an embryo without fertilisation, or apomixis, typical of the family Rutaceae, where the offspring is genetically identical to the parent plant). Further, where embryos were produced sexually, it provided useful segregates, from which novel varieties were constantly evolving for selection by the Citrus breeder. One of the natural hybrids was *C. karna* (rough lemon), a hybrid between *C. aurantium* (sour orange) and the Malta Lemon, and the universal stock plant for the Citrus industry. Janaki's paper pointed to the immense potential of the genus for interspecific hybridisation, without doubt a boon to the plant breeder.

Already, there existed artificial hybrids like the 'Citrange,' a hardy hybrid between the sweet orange and the trifoliate orange; the 'Tangelo,' a cross between the *C. reticulata* variety like tangerine and the *C. maxima* variety such as the pomelo (other examples include the grapefruit and the 'Tangor,' a hybrid between tangerine and orange). To this list Janaki added one more, the 'Citromelon,' a natural hybrid between Citron and Malta Lemon, which she had found growing in sister Cousalya's garden in Shoranur in 1949, among its parents. The fruit of the hybrid was intermediate in size and when exhibited at the local horticultural show, received an award of merit. Cytological examination showed that both parents and the hybrid were diploids (2n = 18). Janaki developed seedlings from the abundant seeds the hybrid fruit produced, to be treated with colchicine at her makeshift laboratory at the BSI. This useful Citrus hybrid she named 'Sukumari' after her cousin and brother-in-law, Sukumaran (Cousalya's husband), but gave it a gender twist, by altering the declension (from 'an' to 'i').[13]

Janaki's Memorandum

On 25 June 1953, Janaki submitted a *Preliminary Memorandum on the Reorganization of the Botanical Survey of India*.[14] It had taken her almost seven months of intense work to complete the task and contained the germs of all her future projects. The thoroughly researched piece involved sifting through a mountain of primary material, such as official correspondence and other documents, besides the extensive published literature on Indian botany that was available, to provide context and rationale to her scheme; several drafts were written out before a final version worthy of submission to the Government was produced. Janaki's interest in the ancient world, including Ayurveda and Sanskrit learning, which came easily to her as a descendant on her paternal side, of the famed Oracheri *Gurunathanmar* of Tellicherry, and of course her father's eclectic erudition, found clear articulation in the *Memorandum*, aspects of her worldview that had seldom found public expression before.

In Part I of her *Memorandum* on the history of botanical explorations in India, rather than begin with the contributions of colonial botanists to the study of Indian flora, she harked back to India's ancient past and wrote of the earliest known 'Botanical Survey to be recorded in India,' that of Jivaka, physician to King Bimbisara (late fifth century BC):

> who as a student at the University of Taxila, was asked by his great teacher, the Bikshu Atreya, to collect, describe, identify and mention the properties of all the plants that grew within the four *Yojanas* of the University Town. One 'Yojana' being a little over nine miles, this must have been a considerable task imposed by the Master on his pupil. It is also recorded that Jivaka did this to the

entire satisfaction of his teacher. Thus Botany or *Vrikshayurveda*, as it was called in ancient India, was part of the curriculum in seats of learning. The duties of a *Vrikshayurvedajna* or applied botanist, according to *Arthasastra* (350 BC) [were] to learn the art of collection, the selection of seeds and the treatment of plants in health and disease, as well as to identify and classify them. Classification was based not only on external morphological characters, but on medicinal properties and environmental association (ecology). Two names were generally given for a plant, one for the ordinary man in the street and the other for the Physician. Thus the castor plant was popularly known as 'Chitravija' (with painted seeds) and 'Vatahari' (enemy of rheumatism) by the physicians.

In Sanskrit nomenclature, the genus is indicated by a name while the species is described by a prefix, denoting its distinctive character. Thus *Bala* is the modern genus *Sida*, and Mahabala = *Sida rhomboidea*, Ati Bala= *Sida rhombifolia*, Naga Bala= *Sida spinosa* and so on.[15]

Janaki would cite the name of the Orientalist William Jones, who admitted that even Linnaeus would have adopted this system of nomenclature had he the knowledge of Sanskrit. In direct contrast to Jones was the colonial botanist George King, she remarked, who when Director of the BSI referred to the literature of Indian flora in Sanskrit works as 'vague and obscure' (in his Presidential Address, 'The History of Indian Botany,' before the Section on Botany of the British Association in 1899). King had reduced two thousand years of the history of recording plants in India to nothing, by tracing the beginnings of Indian botany to the arrival of the first colonialists, she observed critically. To her an exhaustive study of the ancient Indian sources was as important as the colonial compilations and treatises on Indian flora, in shining light 'on the history of the evolution and the migration of many of the commonly cultivated plants of India in particular, and of Asia in general,' her life's quest as a cytogeographer and evolutionary biologist.[16] This ought to be one of the core purposes of the BSI, she stressed very early in her *Memorandum*. Indeed, what Janaki truly envisioned was a Botanical Survey of Asia rather than one limited to India, for it was imperative to look beyond national boundaries and explore border zones, to better understand speciation. This approach also tied in well with her ideas for an Asian sorority, we may recall, that she put forth during her stint at the University of Michigan as a Barbour Fellow. The need of the times was to break out of colonial habits of thinking (chiefly the Kew legacy), she warned, and shape the BSI along modern lines and beyond national frontiers, but one also rooted in an indigenous episteme.

Janaki's introductory essay provided a gist of the botanical contributions of the Portuguese, the Dutch and the Danes (she left out the French however, though there were such men of science as Jean-Baptiste Leschenault de la Tour who founded a botanic garden in Pondicherry in the early nineteenth century and whose Indian collections aided the study of botanists like Antoine-Laurent Jussieu). Among the Dutch achievements was of course the monumental *Hortus Malabaricus*, a compilation by Malayali physicians under the direction of Hendrik Adriaan van Rheede. The botanical activities of the East India Company such as the founding of botanical gardens and formation of regional floras by the surgeon-naturalists Koenig, Roxburgh, Wallich, Griffith, Thomson and so on, she discussed in some detail, but her chief focus was the compilation of the flora of British India under the Crown (post-1857), which marked the beginning of a new era of botanical survey, best exemplified by Joseph Hooker's *Himalayan Journals* and *Flora of British India* (1872–97); incidentally, the preface to the last volume of Hooker's *Flora* was written in November 1897, exactly when Janaki was born. By Hooker's own admittance, his *Flora* was not exhaustive, even if it was a pioneering attempt at re-ordering the Indian flora:

> I must remind those who may use it that it has no pretentions [sic] to give full characters of the general and species contained in it. It aims at no more than being an attempt to sweep together and sys-tematise within a reasonable time and compass, a century of hith-erto undigested materials scattered through a library of botanical books and monographs, and preserved in vast collections, many of which later had lain unexamined for half a century in the cellars of the India House and in public and private herbaria. It is a pioneer work, which besides enabling botanists to name with some accu-racy a host of Indian plants, may, I hope, serve two higher purposes, to facilitate the compilation of local Indian Floras and monographs of the large Indian genera; and to enable the phytogeographer to discuss the problems of the distribution of plants from the point of view of what is perhaps the richest . . . and most varied botanical area on the surface of the globe, and one which, in a greater degree than any other, contains representatives of the floras of both East-ern and Western Hemispheres.[17]

Janaki was clearly driving home the point that Hooker had shown the way, but it was the turn of the BSI to undertake a revision of the *Flora*, not only because several new plants had been discovered since Hooker but also because there was much that had changed in nomenclature arising from the progress in botanical science, in particular the rise of cytosystematics.

In the period under the Crown, there was a general decline in the qual-ity of work done at the Indian institutions such as the botanic gardens and the herbaria attached to them. When an assessment was made of the botanical establishment in the Bombay Presidency in the 1880s, for instance, it was discovered that the work was mainly related to the marketing of vegetable seeds rather than anything of a scientific nature. The general lack of focus and apathy was apparent in the other botanical departments as well. It dawned on the Government of India rather late that a central authority to direct and coordinate botanical work in India was urgently needed. Accordingly, in 1887, a scheme for the organisation of a botanical survey was approved, and about three years later, in February 1890, it was formally sanctioned. The post of Director of the BSI was instituted in March, and the Survey itself came into existence in 1891, with a two-pronged objective: to coordinate the botanical work carried out in the different institutions of the Central and Provincial Governments; and to promote the exploration of regions that had not been explored by the four botanical departments of Calcutta, Bombay, Saharanpur and Madras. Calcutta was considered the head office of the BSI, and George King, who was Superintendent of the Sibpur Botanic Gardens, was appointed Director of the Survey, on an honorary basis. J. Cook (Principal of the Science College, Poona), J. F. Duthie (Superintendent of the Saharanpur Gardens) and M. A. Lawson (Government Botanist, Madras and Director of the Government Cinchona Plantations in the Nilgiris) were made Directors of the regional surveys at Poona, Saharanpur and Madras, respectively.

Janaki spoke highly of the fine work done by the Survey as reflected in the first five volumes of the *Records of the Botanical Survey of India*, which included explorations by Gage in Assam and Burma, Prain in the Andamans and Sundarbans and Burkill in the Abor Hills. However, much like a shooting star, after a brilliant start, the BSI began a downward spiral thanks to several unforeseen difficulties but chiefly the lack of dedicated staff. Further, coordination by the central authority was limited to demanding periodical reports from the Directors of the regional surveys, to be forwarded to the Government. Neither were these regular, with the Southern Regional Survey (following the death of government botanist M. A. Lawson), filing only three reports in the first decade of its existence; the situation would be revived somewhat with the appointment of C. A. Barber, the sugarcane expert (who would go on to head the Coimbatore Sugarcane Breeding Institute) as Government Botanist in 1900, but within a decade, the Southern Regional Survey would cut away entirely from the BSI. The Western Region followed suit, entrusting its botanical work, like Madras would do, to the Economic Botanist, while the Northern Regional Survey handed over control to the Government Botanist of the United Provinces, with the retirement of Duthie in 1902. The Eastern Survey based at the Sibpur Botanic Gardens now remained the only surviving unit. However, with the establishment of the ICAR in the first decade of the twentieth century, the tasks of plant breeding, improvement and introduction, 'the fundamental key-note of the earlier days of the Botanical Survey, viz., plant introduction, field survey and commercial exploitation of plant species,' were moved out of its fold. The BSI thus became redundant, or almost so, and in 1939 the arrangement under which the Superintendent of the Sibpur Gardens also acted as the Director of the BSI ended, with C. C. Calder being the last to hold the joint post. The position of Director remained vacant for several years until mid-October 1952, when Janaki was appointed Officer-on-Special Duty, a temporary designation created by the Government of India at the recommendation of Nehru, to draft a suitable plan for the reorganisation of the BSI, one deserving of the new-born nation.[18]

If Part I of Janaki's *Memorandum* chronicled the historical developments that led to her appointment as Officer-on-Special Duty, Part II was an exegesis of her scheme for the reorganisation of the BSI. She could not emphasise enough the importance of scientific taxonomy (or the new natural history) to the other plant sciences such as morphology, physiology, ecology, genetics or cytology: 'the conclusion reached by the plant geographer on the distribution of species and the origin and relation of Floras depend for their reliability on the accurate identification of the species concerned. Thus in all lines of botanical work the identity of a species is a fundamental consideration,' she stated. This point logically led to her second very important, Humboldtian one on the interconnectedness of nature:[19] the need to be conscious of the connection between botany and the other sciences such as geography and geology, zoology, public health and agriculture, or simply put, the importance of border-crossing. To demonstrate the interconnectedness of nature, she spoke of how the knowledge of the distribution and periodicity of growth of algae was crucial to understand the life-history and migration of fishes, and how plants and the life-histories of insects and pests like the mosquitoes were linked, and had a direct impact on public health. Janaki warned that 'the identification of Indian plant fossils, especially of microfossils involving pollen analysis,' could be accomplished only through comparison with those species in existence. It was of utmost importance, she argued, that ecological studies be conducted jointly by the Botanical and Zoological Surveys. As for agriculture, its relationship with botanical science was very obvious. Further, it had come to be widely accepted that the wild species of cultivated plants provided genetic material for improvement through hybridisation, especially in the production of varieties that were disease or drought-resistant. Janaki cited the example of her favourite grass, the sugarcane, and the success met with in breeding better varieties in both Java and India using the wild species of the cultivated crop plant. The cytogenetic study of wild populations, which was to be one of the chief objectives of the reorganised BSI, would have 'a direct bearing on plant breeding programmes in this country . . . We have yet to discover the wild ancestors of some of our cultivated plants like the turmeric, betel, Citrus species, and this work will have to be done by the Botanical Survey,' she observed.[20]

An all-India survey of the whole group of Indian grasses (the Gramineae being her own research focus) with respect to their nutritive and edaphic (relating to the soil) characteristics was to be an important function of the BSI, as far as Janaki was concerned. She was also strongly in favour of the BSI undertaking the study of the botany of medicinal plants, with special reference to their ecology, cytogenetics and geographical distribution, which knowledge could be made available to pharmacologists for the correct identification and utilisation of drug plants in the making of indigenous drugs: 'It will be the function of the Botanical Survey to make available to the Central Drug Research Laboratory and other institutions botanical data regarding the species most useful to medicine,' she stated.[21] According to Janaki, the functions of the BSI were twofold: first, to act as the Keeper of the botanical collections of India, on which the identification of the flora of India was based, and second, to secure the 'fullest possible knowledge' of the flora of India, which could be put to use in boosting the nation's wealth.

Janaki stressed on the need to establish a Central National Herbarium (CNH) in India, where plant material could be compared and checked against type specimens, to confirm their identification. At present all new specimens were being compared with those at Kew, she noted, and 'This should not be so. If it is not possible to secure the "type" sheets, at least the co-types of all Indian plants should be deposited in India in a Central National Herbarium.' We may recall that such a demand had been put forward by Birbal Sahni, Agharkar, Parija and Janaki as early as the 1930s. The lack of an institution like the CNH (the Herbarium at Sibpur had proved inadequate for the purpose), Janaki observed, had badly hampered the teaching of botany and the correct identification of plants, so crucial to the fields of agriculture, horticulture, medicine and forestry. For systematists, the type specimen (the plant employed in the description of the species for the first time but which could not be very different from the other members of the species), preserved as a dried and pressed specimen in the herbarium of an institution, was the final go-to source to settle issues regarding the characteristics of a species. Being crucial to the systematist's practice, it was invariably preserved and kept under lock and key in a safe location.

As to the choice of location of the CNH, a survey was conducted through a questionnaire sent to 160 Fellows of the Indian Botanical Society. Of the ninety-four botanists who replied, only twenty-one were in favour of Calcutta being the seat of the CNH. The end result was that the majority had chosen Delhi over Calcutta as the location for the CNH, and 'as the capital city of the Indian Union [Delhi] rightly deserves this honour,' Janaki would remark. She was convinced that a change of location would give the reorganised BSI added strength and motivation.[22] Although the tropics were the source of a great number of type specimens, most were taken away and stored at European botanical institutions such as museums, gardens and universities; the Kew Gardens Herbarium represented the best example of this colonial injustice. The centre of the botanical world in the nineteenth and twentieth centuries, the Royal Botanic Gardens at Kew enjoyed the ultimate say when it came to plant identification and nomenclature, particularly under the Hooker regime represented by William Hooker, his son Joseph Hooker and son-in-law William Turner Thistleton-Dyer. Hooker 'insistently and persistently prioritised global over the local taxonomies' and clearly belonged to the tribe of 'lumpers' rather than the 'splitters.'[23] Janaki drew attention to the need to re-establish the post of 'Indian Botanist at the Kew Herbarium' but warned that the botanist sent to Kew should have 'a good working knowledge of the Kew Herbarium' as 'there was no question of wasting time training a new person at Kew.'[24]

Secondly, the reorganised BSI was to undertake the task of exploring unknown territories of India in a systematic fashion as part of the revision of Hooker's *Flora of British India*. Janaki insisted that handbooks of the flora of every province in India, even every district, be prepared by the BSI. She stated in no uncertain terms that botany had suffered a major setback in India due to the dormant nature of the BSI in the past decades and more importantly because taxonomic studies in the country had not gained from the new sciences of cytology and genetics. 'The recognition of the importance of chromosome studies as an aid to taxonomy and the genetical and cytological analysis of wild populations undertaken in different parts of the world for the understanding of evolutionary tendencies in plant life is once again bringing Systematic botany to the forefront of Plant Sciences. The reorganization of the Botanical Survey of India will do much to enhance the position of Systematic Botany in India,' she stated.[25] Janaki also insisted that there be a close connection between the BSI and the universities, with research students taking up problems relating to the flora of India. She suggested that the BSI also take on board a handful of university students when they went on plant collection tours and that the BSI could offer a training course in collection and preservation of specimens to university students.

On Administration

On the administrative front, Janaki urged the creation of a post of Director General of the BSI, a salaried position, assisted by a Joint Director-General, under whom were to be six Directorates or Circles—Northern, Eastern, Central, Southeastern, Southern and Western—each controlled by a Director. The divisions were not based on geographical, but on phytogeographical considerations, with the result that the boundaries were not fixed, but subject to revision when phytogeographical affinities came to light. Further, the headquarters of each Circle would maintain a Herbarium under the care of a Curator, and attached to each of the six Regional National Herbariums would be the National Botanical Gardens (growing regional flora as well as acclimatised exotic ones from other parts of India and abroad), besides the cytotaxonomic laboratories, where plant investigations specific to the region would be undertaken by the cytogeneticist.

In Janaki's opinion, the headquarters of the reorganised BSI ought to also be the location of a National Botanical Laboratory (NBL) of India, from where all botanical studies would be coordinated. She saw the NBL as a necessary adjunct to the CNH, fitted with state-of-the-art technology for the investigation of the living plant, 'manned' by a team of botanists who were experts in the different branches of botany and the related sciences. Besides the above institutions, she suggested the establishment of a Botanical Museum, which would also be a museum of evolution of plant life in general and India in particular.[26]

Hopes and Disappointments

Janaki would often send collections from India to her friends at Kew, including Bor, who was researching on tropical plants, especially grasses. Located in Calcutta, she acted as a nodal point of a collecting network that spread far and wide. In July that year, she sent Bor yet another packet of seeds collected from the Chandra valley by the Norwegian palaeobotanist, Ove Fredrik Arbo Høeg, Director of the Birbal Sahni Institute of Palaeobotany, Lucknow.[27]

In September (1953), Janaki would hear of the death of her friend, Lord Aberconway who had passed away on 23 May that year. Janaki knew him very well, not least because he was the President of the RHS and Chairman of the Wisley Advisory Committee, and had visited his Garden several times during her residence in England. 'The news saddened me greatly. He was a good friend to me at Wisley, and truly a great man,' she wrote in her reply to F. C. Stern who had given her the belated piece of bad news.[28] The letter to Stern had been drafted while she was in New Delhi (c/o The Ministry of Natural Resources and Scientific Research), to submit her report to the government. 'There were many discussions & Committees to sit over it and now it has been accepted almost in toto by the Ministry,' she told Stern with an obvious sense of pride. 'For purposes of Survey I have subdivided India in six phytogeographic regions (it used to be four earlier). . . . I have asked for the creation of a Botanical Lab attached to the Survey for cyto-taxonomic and plant geographic studies. This was well received and I have been asked to go ahead with organizing it. You will I know be pleased to hear of this,' she commented. At Delhi, she found herself busy trying to work out the logistics of her scheme through discussions with the Finance Ministry and the Planning Commission. 'Just think of it—The 5-year plan [India's First Five-Year Plan, 1951–56] made no mention of the Bot. or Zoo, Surveys! I have to argue my case with the Planning Commission myself. I am getting quite good at writing documents and less and less at cytology. This worries me lot,' she would add.[29]

The Primitive Magnolia

Janaki's paper on the Magnolia had just been published in the *Indian Journal of Genetics and Plant Breeding*, a copy of which she would enclose in her letter to Stern.[30] The *Magnolia pterocarpa* material which she had sourced from Assam through Bor also figured in the paper as one of the Magnoliales, some of the most primitive genera of flowering plants. Incidentally, J. Hutchinson considered the *Magnolia pterocarpa* of Assam and Burma, as 'the oldest flowering plant . . . a living fossil.'[31]

By the time Janaki's letter and copy of her paper reached Stern, he was holidaying in Italy with his wife. They had spent a delightful weekend at the Villa Taranto gardens at Pallanza on Lake Maggiore belonging to a Scottish Captain McEachern, where the latest introductions of magnolias and rhododendrons were thriving. Ever her well-wisher, Stern was very happy at Janaki having wangled a botanical laboratory for conducting 'cyto-taxonomic and geographical studies' from the government. 'I am glad you are going on with that which you are so expert at and about which there is so much to be done.' Darlington's new *Chromosome Atlas* had not yet been published, he assured Janaki. 'I know nothing about it except that sometime ago I gave Darlington at his request some of the maps I did on the distribution of *Leucojum*, *Primula* and *Iris* which he wanted to publish in the *Atlas*. I don't know if he is publishing them or not,' he added. Stern and Darlington had had a row before the latter left Bayfordbury (to which place the John Innes had relocated from Merton) for Oxford, and so were 'not now on [good] terms.'[32]

After he had read her paper, Stern would write again just to let her know that the Magnolia paper was 'a very fine bit of work.' He had grown keen on her cytogeographical research and sincerely hoped she would write a book on the 'Chromosome History of Flowering Plants.' 'You are welcome to any of my maps if they are any use to you,' he offered kindly. That weekend, at Highdown, the Sterns had had the delightful company of the legendary plant hunter, the Austrian-born Joseph Francis Rock, from whose seeds of the *Paeonia suffruticosa* Stern had raised plants in his Chalk Garden. 'He is a most interesting person with all his stories of his 27 years of living in Yunnan. His photographs of plants and the country in China are magnificent,' Stern would enlighten Janaki. As to her invitation to visit India, he responded: 'Your suggestion of our coming to India sounds wonderful but I wonder if I shall ever get away from all my engagements here.'[33] He never would. Janaki had about this time applied to the Linnean Society for a re-election to a fellowship, on this occasion, as Officer on Special Duty at the BSI, and as 'attached to the study of Natural History, especially Cytology and Plant Geography in relation to Systematics.' Her application was recommended besides others by Stern, George Taylor of Kew, and J. E. Dandy and W. T. Stearn of the British Museum (Nat. Hist.); she was re-elected FLS on 19 November 1953.[34]

Writes to Darlington

Janaki was still smarting at how Darlington eased her out of the second edition of the *Atlas* without showing the least remorse. 'I wonder if the *Chromosome Atlas* II ed. is out. I wanted to make it "out of date" in a year. In fact the skeleton of the rival *Atlas* is already set up...' (she had already let Stern know this).[35] However, keeping aside her differences with Darlington, she updated him of the developments at the BSI. '[The Ministry of Natural Resources and Scientific Research] has accepted my scheme for the reorganization of the Botanical Survey of India with a few amendments,' she informed him: instead of the six phytogeographical zones she had suggested in her draft plan (and which she had written to Stern about), the Government had accepted five, she explained, cutting off the Central one (which she then cleverly saved by merging it with the Madras Region), 'each with its own herbarium and a cytotaxonomic laboratory!!' There was to be a Central or National Herbarium either at Calcutta or 'if W. Bengal fails to give us the Sibpur Herb. [Herbarium] at Dehra Dun.' The idea of having it at Delhi seems to have been shot down despite a majority vote in its favour among members of the Indian Botanical Society. Janaki disclosed to Darlington that K. P. Biswas (the first Indian Superintendent of the Sibpur Botanic Gardens, Calcutta) had been 'fighting tooth and nail to keep the Herbarium at Sibpur with himself as the Director of the BSI.'[36]

In Janaki's scheme there was a provision for a Central Botanical Laboratory, to be under the control of the Director of the BSI, where the 'living' Flora of India would be studied 'from an evolutionary and cytotaxonomic angle.' However, she would receive news that the Laboratory was to be under a separate Director, of equal standing and salary as the Director of the BSI, and that they were planning to offer her the post. Whatever that decision, she was determined to 'try and get the Universities to give up the study of "macrosporogenesis" and begin cytological and genetic survey of plant populations.'[37] Janaki hoped students would begin to study variation in plant populations to shed light on speciation (the evolutionary process by which populations evolve to become distinct species) rather than narrow their focus to the changes in the cell during meiosis. 'I want to tell you that the "old fogies" of England were partly responsible for this. They did not wish me to be the Director of the Botanical Survey fearing I would be too unorthodox. So Nehru in his usual way has given me even better scope to practise my unorthodoxy in Systematic botany. The P.M. is head of Scientific Research in India! He is a gem!' she would remark in her letter to Darlington. Janaki was terribly mistaken however for not even Nehru's blessings could see her scheme to fruition. The last line of her letter to Darlington was a tongue in cheek response to his move to Oxford: 'I hope you are liking Oxford—Will Oxford like you?'[38]

Science and Freedom

The Congress on *Science and Freedom* held in Hamburg in 1953 (the Proceedings of which were published later, in 1955) had as the Chairman of its Organizing Committee none other than the Hungarian-born British polymath and scientist Michael Polanyi of the Society for Freedom in Science, founded in the late 1930s. The Committee demanded that scientists be given the freedom to pursue their own investigations, without any interference from the state. However, it had now become increasingly evident that modern science was now "big science," demanding huge amounts of capital investment, which only the state could provide. This was true of America, the United Kingdom and Germany, and, *Current Science* speculated, perhaps of India too. The Congress critically examined the organisations of scientific activity in various countries, with a view to suggest reform measures and give extensive coverage to the nature and extent of suppression of intellectual freedom under totalitarianism. It also intended to explain clearly the philosophical foundations of the idea of freedom as related to science.[39]

Interestingly, in December 1953, the Indian journal *Science and Culture* carried the presidential address delivered by the Nobel Prize–winning radio-physicist Edward Victor Appleton (1947) to the British Association at Liverpool, on the theme 'Science for Its Own Sake.' Appleton stated that it was not possible for a scientist to make any progress in his work unless he/she examined his/her problem 'with an unprejudiced and dispassionate mind and conscientiously record[ed] his observations with a scrupulous regard for truth.' As he noted, 'even the great body of scientists . . . are known to be as dogmatic, as prejudiced, as jealous, as sentimental and as self-seeking as anybody else. In every land there are many among the scientists who do not hesitate to sell the freedom of their conscience for the sake of paltry reward from the State, or for the applause of popular passion or out of fear for public opinion and social pressure.'[40] Appleton's words could not ring truer than in Janaki's case (and in today's world); the fate of her plan for the reorganisation of the BSI would rest in the hands of men who cared little about cytotaxonomy or the advanced methods for classifying flora.

Janaki Versus Santapau

The January 1954 issue of *Science and Culture* carried an outline of Janaki's plan for the reorganisation of the BSI, followed in the next number by Hermenegild Santapau's thoughts on the subject. The Spanish Jesuit botanist, who had worked extensively on Indian flora and had a large student following and clout in India, was at this time professor of botany at St Xavier's College, Bombay. He couldn't agree more that a revival of such an institution as the BSI was 'long overdue' and even found it irrational that India was prepared to spend large sums on a network of National Laboratories, but none at all to revive an institution like the BSI. In his article, Santapau voiced his views on what services the BSI could render to the country and on how it should be developed to make it efficient and indispensable to the progress of the nation. This was Santapau putting forth a plan to rival Janaki's for the reorganisation of the BSI, thanks to the platform provided by the editors of *Science and Culture*, who wished to make it a subject of public debate.

The truth was that Santapau was no cytotaxonomist; he was but a conventional plant taxonomist in the classical/Kew tradition, unlike Janaki (five years his senior), who was trained in the most advanced institutions of the world in cytogenetics and had through the *Chromosome Atlas* made a major contribution to cytosystematics. Cytological investigations were central to Janaki's proposal, but this was not the case with Santapau's. He was willing to admit that 'cytological notes are an added advantage' of any flora, but believed that this line of research ought to be conducted with care in a National Botanical Laboratory.[41] Cytological details, as far as he was concerned, merely aided the identification of the plant in question, or in differentiating it from similar species, clearly revealing a position in favour of morphology as the basis of classification, much in line with the colonial tradition with its centre at Kew, rather than with the evolutionary approach based on chromosomal study represented by cytosystematists like Janaki.

Desperate to Resume Research

Two months later, in March 1954, Janaki shared with Stern her plans for the Central Botanical Laboratory. She was 'thinking of Microscopes and other equipment for the Laboratory,' and could not wait to 'sit down to some solid piece of research.' Janaki had chosen the genus *Nymphaea* as her new subject for a cyto-taxonomic survey. 'If you have any friends in any part of the world who can send seeds of *Nymphaea*, I should be most grateful to get them,' she wrote to Stern.[42] In June, Janaki's paper, 'Cytogeography of the Genus *Buddleia* in Asia,' was published in *Science and Culture*, the last of her papers on the cytology of cultivated flowering plants, her subject of focus while at Wisley and Paris. She had begun collecting specimens of *Buddleia* on her trip to Nepal in 1948–49; more would be added over the next couple of years, sourced chiefly from the Kew Gardens. In May 1952, she had worked on the *Buddleias* of Asia (part of her larger project on the cytogeography of flowering plants), which she found 'very exciting.'[43] Of the 150 species included in the genus *Buddleia*, about half belonged to the Americas, a third to Asia and the rest to Africa. In all, Janaki had examined twenty-eight species of the Asian *Buddleia* at the Wisley laboratory, the results of which were published in her paper, along with a map showing the distribution of polyploid *Buddleia* in Asia and another on the distribution of *Buddleia* in comparison to the subfamilies of *Nicodemia* and *Chilianthus*.[44]

Of Assistance to Friends

The Central Botanical Laboratory (CBL) came into existence on 13 April 1954, and Janaki was appointed its first Director. The laboratory began functioning in the Indian Museum, Calcutta, in the same rooms where cinchona and other plant products had been studied and classified by the old BSI. It was a long hall, which had to be put to use for both scientific and office purposes. 'Thus the continuity of research which had snapped as a result of stagnation of the survey, was once again revived in the very same building where Watt [George Watt] worked on the Economic Products of India,' she would remark a few years later.[45] Stationed at the Indian Museum, Janaki enjoyed access to huge amounts of plant material, both live and preserved, and as earlier noted, was in a position to help botanist friends like Bor in sourcing research material.

In late April, she sent Bor a bulb of what she believed was the *Pancratium triflorum*; the specimen had been collected from the sea coast in Tellicherry (this must have been sent to her). On occasion, she would use the expertise available at the Jodrell Laboratory of the Kew Gardens, for information on economic plant products.

Incidentally, that same month, Janaki had a letter from Bartlett with regard to one Dr Howard Gentry, a student of Carl Erlanson (the botanist and former husband of her good friend, Eileen Macfarlane) of the US Department of Agriculture in Washington, who had sent home from India cuttings of pepper for propagation, 'to the great annoyance of Indian officials.' India had by this time drastically restricted botanical and agricultural exploration in the country by foreigners, but Gentry had been unaware of this fact. Bartlett wished to know if Janaki was in any position to resolve the issue. She certainly was, but we do not know how she responded to this request. Bartlett also used the opportunity to introduce her to Thakur Rup Chand, associate of Walter Koelz involved in 'biological survey work' for the University of Michigan. Bartlett warned her about a possible request from Rup Chand for *Coix* seeds to be delivered to cytogeneticist Miss Nalini Nirody (daughter of the acclaimed plantsman and gardener, B. S. Nirody), based at the Washington University at this time (and later at the IARI, Pusa).[46]

Field Excursions

Janaki was making exploratory trips to various parts of the country during this period. In late January that year (1954), she was in Shillong, from where she received a piece of wood of the *Aquilaria agallocha* (the agarwood tree) from the Forest Department; this she would despatch to C. R. Metcalfe of the Jodrell a few months later. 'The dark regions are where the resin "agar" is secreted. It is used as perfume in India. . . . It would be interesting to know how 'agar' is formed. Some think it is the result of fungal attack. . . . Dr Bor will be able to tell you how nuver (?) trees are being felled down in search of those with agar. As this is of economic value, any information will be interesting,' she wrote. In her letter, Janaki also expressed excitement at having learnt from the *Kew Bulletin* that the cytology of the *Dianthus* (which she had worked on) was being investigated at the Jodrell. Neither did she forget to convey to Metcalfe news of her recent appointment: 'You will be pleased to hear I am now the Director of the Central Botanical Laboratory of the BSI.'[47]

At Shillong, on 1 February 1954, Janaki bought a copy of Jean Kingdon-Ward's, *My Hill so Strong* (1952). Born to a High Court Judge in Bombay, Jean Macklin at twenty-six had married sixty-two-year-old Frank Kingdon-Ward, a plant hunter of renown. Many of Frank's books on his adventures, we know, had found their way into Janaki's personal library. Frank and Jean's first expedition together, to Manipur, had turned out to be very productive; they made a vast collection of 1,400 herbarium specimens, of about 1,000 species, and several with seeds, which included a small lily endemic to the region, named after Jean, *Lilium mackliniae*. In 1950, supported by the RHS, Frank and Jean had trekked up the Lohit river valley on the Assam/Tibet border, when they found themselves (in August) very close to the epicentre of the terrible Assam earthquake, measuring 8.7 on the Richter Scale. They were lucky not to lose their lives but were unable to bring back a substantial collection. Their fateful expedition was described by Frank in an article published in the journal of the RHS (1952), while Jean provided her more graphic version (without the botany) in her only book, *My Hill so Strong*.[48]

Scope of the CBL

The CBL aimed to address a wide span of botanical problems—physiological, cytological, genetical, ecological and economic—relating to India, and provide information on the live flora, a knowledge of which could be used in the fields of agriculture, horticulture, forestry, trade, industry and medicine to boost the wealth of the country. The laboratory also aimed to study the development of the taxonomic hierarchy—family, genus, species—of Indian plants in its different regional variations and the role played by human beings and animals in altering the flora of India. It was also concerned with how deserts and wastelands could be revived. The CBL was comprised of four departments: cytogenetics, ecology, economic botany and plant physiology, complete with a radioisotope laboratory attached to it. In a report on the CBL, which Janaki submitted on 16 March 1954, she described in great detail the type of investigations to be undertaken by each department. Of the four, only the cytogenetic laboratory had been active since the CBL came into being. Its chief functions included undertaking a cyto-taxonomic survey of the Indian flora with special focus on medicinal and other economic plants and to maintain a cyto-taxonomic herbarium for reference by botanists and the creation of new plants (timber, fruit and medicinal) by inducing polyploidy using drugs such as colchicine, X-rays or radioisotopes, and hybridisation.

Under Janaki's directorship, a department of economic botany was initiated at the CBL, with a section devoted to ethnobotany; although an economic botanist had not yet been appointed, the core of the department had been developed by Janaki. For the first time ever, an attempt had been made to investigate the plants grown and cultivated by the primitive tribes of India, such as *Dioscorea*, *Colocasia* and *Curcuma*, which, it was hoped, would not only provide genetic material for the improvement of cultivated plants but also new material for introduction.

Disillusionment

Janaki was excited and happy with how the CBL was progressing and how the BSI was metamorphosing in the way she had imagined, when like a bolt from the blue she received news that Santapau had been appointed to the post of Chief Botanist of the BSI, a definitive indication that her plans for putting the BSI on a modern/decolonial footing had come crashing down. To make things worse for her, the Government decided to enhance the powers of the Chief Botanist, making him the supreme head of the institution, while her own position as Director of the CBL was reduced. In this extreme moment of distress, and despite the growing emotional distance between them, she turned to Darlington for solace, but would write to him only after regaining some composure. Her immediate response on hearing the news was to run away to Edam for some sanity and to where the Irulas of Malabar lived, looking for primitive cultivars, a new research interest she had been developing at the CBL.

On her return to Calcutta from Malabar, a still despondent Janaki wrote to Darlington: 'My dear Cyril, I wish I could span with my thoughts the thousands of miles that lie between you and me—both geographically & in other ways—and stand before you just now! I bring you news of a major defeat for Bot[anical] Science in India. The Government of India has appointed as the Chief Botanist of India—a man with the Kew tradition and I the Director of the Central Botanical Laboratory must now take orders from him. Kew has won a decisive victory—and the news has been jubilantly received there. I am very angry.' Janaki introduced Santapau to Darlington as a Jesuit-Spaniard, who had become an Indian National 'as all Jesuits in India have done.' She had nothing against him as a systematic botanist (after all she had met him recently in Bombay at the behest of Bor and had been impressed), but she had fervently hoped the BSI after its reorganisation 'would be something different to what it was in 1856—when Hooker wrote his *Flora of British India*.' She remarked wistfully, 'Kew has won, Sir Edward [Salisbury] has won and we have lost.' She felt two years of her work had gone down the drain, and that thought made her 'feel sick.'

When she heard the news, she ran away 'to the wild of Malabar to collect wild yams that our aboriginal tribes dig up,' she told him.[49]

Transfer to Lucknow

Santapau was formally appointed on 1 December, but would not last long as Chief Botanist. In less than a year, on 1 October 1955, he resigned and J. C. Sengupta formerly, Principal of the Presidency College, Calcutta, took his place. As the accommodation in the Indian Museum premises was found to be inadequate for the proper functioning of the CBL, on 2 December, the Government of India moved the institution to the Chhattar Manzil in Lucknow. Built by that French man of Enlightenment, Major General Claude Martin, in the late eighteenth century, the Chhattar Manzil was occupied by the Central Drug Research Institute, under the administration of the CSIR, Delhi.[50] Janaki was attracted to danger, like a moth to the flame. She had bought herself a car (one that a relative had left behind in Calcutta), and drove it all the way to Lucknow, with her nephew for company; he would recall this adventurous journey several years later: 'N [Nachee] & I drove it all the way to Lucknow . . . This three day (& some nights) drive along the National Highway was most educational for me. We stopped at "dhabas" for hot tea & vegetable "bhajjis" & fruits, a kind of camping life that she adored. She moved easily with the villagers. We drove through Man Singh's (a notorious dacoit) country at night. She often wished she were captured!'[51]

Figure 19.1 Janaki at her office at the Botanical Survey of India, Calcutta. c. 1952.
Source: Courtesy of the RRL, Jammu.

Figure 19.2 Janaki with O. J. Eigsti and Mrs Eigsti, Banaras Hindu University. 1952.
Source: Courtesy of Dr E. R. S. Talpasai.

From left to right: Dr. S.K.Mukerjee, Dr. R.S.Rao, Dr. E.K.Janaki Ammal, Dr. J.C.Sengupta, Dr. G.S.Puri, Mr. K.S.Srinivasan, Dr. K. Subramanyam and Dr. M.A.Rau.

Figure 19.3 Janaki with the Botanical Survey of India, Central and Regional Survey
        Directors. c. 1955.
Source: Courtesy of the BSI.

Notes

1 DP: C 49 (E 268), E. K. Janaki Ammal to C. D. Darlington, letter dated 11 April 1953.

2 Incidentally, the Botanical Survey of India, Calcutta, organized a year-long exhibition on Janaki in 2016, in the same premises that she had occupied (albeit for a brief time).

3 HL: Mss Peer, Box 2, Folder 6, Ralph S. Peer to E. K. Janaki Ammal, letter dated 26 January 1953.

4 Ibid., letter dated 16 December 1953.

5 Janaki Ammal, 'Chromosomes and the Species Problem in the Genus Viburnum', 1953.

6 The names included B. P. Pal, P. N. Bhaduri, H. Chowdhury, S. Lodhi, P. Maheshwari, G. P. Majumdar, A. Mohajir and S. Ramanujam.

7 'Conference on Organic Evolution', *Nature*, 11 April 1953, p. 637.

8 RBG Kew, Library & Archives: QG 169, Correspondence with Janaki Ammal, 1945–1970, N. L. Bor to E. K. Janaki Ammal, letter dated 20 March 1953.

9 DP: C 49 (E 268), E. K. Janaki Ammal to C. D. Darlington, letter dated 11 April 1953; of the content of C. V. Raman's letter, we have no clue, unfortunately; the letter has not been traceable.

10 Ibid., C. D. Darlington to E. K. Janaki Ammal, letter dated 15 April 1953.

11 See Portin and Wilkins, 'The Evolving Definition of the term "Gene"'.

12 Bonavia, 'On the probable wild source of the whole group of cultivated true limes'. Also see his, *The Cultivated Oranges and Lemons of India and Ceylon*.

13 Janaki Ammal, 'A New Interspecific Citrus Hybrid', 1953.

14 Botanical Survey of India, Herbarium and Library (Sibpur): Official file on the Reorganization of the BSI by E. K. Janaki Ammal, June 1953.

15 Ibid., Part I, p. 1.

16 Ibid.

17 Ibid., pp. 4–5.

18 Sengupta, 'Botanical Survey of India: Its Past, Present and Future', p. 21.

19 After the German polymath Alexander von Humboldt (1769-1859) who relied on accurate observations of real phenomena to unravel patterns in nature.

20 Botanical Survey of India, Herbarium and Library (Sibpur): Official file on the Reorganization of the BSI by E. K. Janaki Ammal, June 1953, Part II, p. 2.

21 Ibid.

22 Ibid., p. 6.

23 Endersby, 'Lumpers and splitters: Darwin, Hooker, and the search for order'.

24 Botanical Survey of India, Herbarium and Library (Sibpur): Official file on the Reorganization of the BSI by E. K. Janaki Ammal, June 1953, Part II, p. 6.

25 Ibid., pp. 3–4.

26 Ibid., p. 7.

27 RBG Kew, Library & Archives: QG 169, Correspondence with Janaki Ammal, 1945–1970, E. K. Janaki Ammal to N. L. Bor, letter dated 3 July 1953.

28 Ibid., F. C. Stern and Lady Stern correspondence, E. K. Janaki Ammal to F. C. Stern, letter dated 16 September 1953.

29 Ibid.

30 Janaki Ammal, 'The Race History of Magnolias', 1952.

31 Ibid., p. 83.

32 F. C. Stern to E. K. Janaki Ammal, letter dated 3 September 1953, private collection.

33 Ibid., letter dated 26 October 1953, private collection.

34 The Linnean Society, London: CR 166 (1953).

35 RBG Kew, Library & Archives: F. C. Stern and Lady Stern correspondence, E. K. Janaki Ammal to F. C. Stern, letter dated 16 September 1953.

36 DP: C 109 (J 114), E. K. Janaki Ammal to C. D. Darlington, letter dated 25 September 1953.

37 Meiosis or reduction division in the cell is a process of micro- and macrosporogenesis occurring in plant stamens and ovaries, respectively; microsporogenesis involves the development of pollen grains or male gametes in the anther, while macrosporogenesis involves the development of embryos or female gametes in the embryo sac.

38 DP: C 109 (J 114), E. K. Janaki Ammal to C. D. Darlington, letter dated 25 September 1953.

39 'State Support and Research'. *Current Science* 24, no. 3 (March 1955): 71–72.

40 *Science and Culture* 19, no. 6 (December 1953): 271–273, p. 272.

41 Santapau, 'The Revival of the Botanical Survey of India'.

42 RBG Kew, Library & Archives: F.C. Stern, Purchase & Exchange of Plants, Vol. I, f. 13, E. K. Janaki Ammal to F. C. Stern, letter dated 8 March 1954.

43 HL: Mss Peer, Box 2, Folder 6, E. K. Janaki Ammal to Ralph S. Peer, letter dated 27 May 1952.

44 Janaki Ammal, 'The Cyto-Geography of the Genus *Buddleia* in Asia', 1954.

45 Janaki Ammal, 'Central Botanical Laboratory', 1959, p. 33.

46 BHL, UoM: University Herbarium Records (University of Michigan) H. H. Bartlett series, Box 11, H. H. Bartlett to E. K. Janaki Ammal, letter dated 27 April 1954. In 1922, gardener and landscape architect B. S. Nirody completed an MS in Horticulture from the University of Massachusetts Amherst on Avocado Breeding and is best known for his book, *Flower Gardening in South India: A Practical Guide for Amateurs*, Madras: 1927. Nalini Nirody would marry the aeronautical engineer, Satish Dhawan.

47 RBG Kew, Library & Archives: QG 169, Correspondence with Janaki Ammal, 1945–1970, E. K. Janaki Ammal to C. R. Metcalfe, letter dated 21 July 1954.

48 Carnaghan, 'Jean Rasmussen', pp. 546–547.

49 DP: C 109 (J 114), E. K. Janaki Ammal to C. D. Darlington, letter dated 4 October 1954.

50 Lucknow boasted of a Government Horticultural Garden, known as Sikandra Bagh, the brainchild of botanist K. N. Kaul, Janaki's friend and Nehru's brother-in-law. Established in the early 1930s, the Garden evolved under the supervision of a committee, which included Birbal Sahni, S. K. Mukherji and Kaul. During the seven years that the committee functioned, a great number of medicinal plants were collected and planted in the Garden. Work on the Garden came to a complete halt during the war years, 1939–44, chiefly because Mukherji had passed away and Sahni was too caught up with his own research and teaching. Moreover, Kaul was based at this time at the Kew Gardens as Botanical Assistant for India. On his return to India in December 1944, Kaul mobilised the Government to resume work on the Horticultural Garden and extend its scope. Two years later, he presented to the UP Government a scheme for the Garden's reorganization; this was accepted in 1948, and funds were sanctioned. The institution was renamed the National Botanic Gardens, and Kaul was appointed Director in an honorary capacity. In April 1953 (by which time the CBL under the Directorship of Janaki had moved to Lucknow), the Gardens were transferred to the CSIR, thanks to Kaul's untiring efforts; he took charge as their first Director.

51 Hari Krishnan, 'Nachee', private collection, July 1993.

20

OAK RIDGE–ANN ARBOR–PRINCETON
Tracer Atoms and Agriculture

I feel quite unworthy of all this and even a bit worried at the publicity.

—E. K. Janaki Ammal (1955)[1]

In 1955, Janaki was deputed by the Indian Government to attend a four-week course, beginning early May, in radioisotope or tracer atom techniques at the Oak Ridge National Laboratory (ORNL), in Tennessee. The *Current Science* had published a short note on the uses of the tracer technique in agriculture three years earlier: 'With the aid of radio-isotopes it has become possible to trace nutrients through the soil, into roots, and thence through plants, to measure the extent and spread of their movement, to determine at what stage in its growing cycle the plant needs fertiliser most; to know where and how fertiliser should be placed to give plants the maximum benefit, to establish what kinds of fertilisers work best in the country's varied soils,' it announced.[2] Radiation genetics for the development of new varieties of plants will soon become, the journal would state, 'one of the most important events in the history of Agriculture.'[3] Using radiation, new disease-resistant as well as high-yielding plant varieties could be 'made to order' not only within a short span of time but also cheaply. In a few years' time, however, scientists would raise concern over the excessive enthusiasm for radiation genetics shown by some plant breeders, who were little aware of its limitations.[4]

Recommendation from Bartlett

It was the Under-Secretary to the Government of India, M. R. Kalyanaraman, who conveyed to Janaki the news that her deputation to Oak Ridge, had been sanctioned by the President of India. She was in Bombay at this time 'to discuss with the Atomic Energy Committee about using Radio-Isotopes' in the laboratory work at the CBL. She was quick to convey the news to Bartlett: 'My Ministry wishes me to spend a month at Oakridge

DOI: 10.4324/9781003267089-20

Institute of Nuclear Studies when I am in America. They are deputing me to this.' Janaki needed a supporting letter from him despatched directly to the Education Secretary, Indian Embassy in Washington, USA, recommending her for the programme. 'I hope you will be so good as to say that I am a desirable person—academically and otherwise,' she wrote, a department for inducing mutations 'by X-rays and etc' had been planned and she was required to 'know something about Atomic energy too.'[5]

Bartlett would promptly send copies of his recommendation not only to the Indian Embassy but also directly to the Director, Institute of Nuclear Research, Oak Ridge. His covering letter to Oak Ridge expressed his contempt for 'Hindus' (by which he meant Indians) in general: 'I never like to send letters of recommendation through intermediaries, although that is regular procedure in India, which is a country notable for its devious ways. For most Hindus I have profound distrust. Dr Janaki is a notable exception for whose integrity I have the highest opinion. So I send you the copy of my recommendation direct in order to assure you that I really mean it in spite of the detoured routing of the original.'[6] His recommendation letter had some fine things to say about Janaki, whom he had known since the mid-1920s:

> Dr Janaki is one of our most highly regarded graduates. She took her doctorate at this University and will be here in June to receive an honorary degree and to lecture on her work as Director during the organization period of the Central Botanical Laboratory of India at Lucknow. She is here in this country to participate in a conference sponsored by the Wenner Gren Foundation and the National Research Foundation, to be held at Princeton in June on Man's influence in Modifying the Face of the Earth. She will be the only participant from India.
>
> Dr Janaki is eminent in the field of cytology and plant breeding and has been requested by the Government to set up a project for studying the effect of atomic radiation in the induction of mutations and chromosomal aberrations in plants. In this field she is inexperienced and wishes to acquire a proper background for handling radioactive isotopes etc. She will of course see what is going on here.
>
> I strongly recommend that Dr Janaki be granted the facilities for study that are being requested. She is a person of genuine eminence in science. We consider her the leading Indian botanist. I regard her as a thoroughly reliable person, of excellent character, and one who shall be received in the most cooperative spirit.[7]

She was permitted to leave the country on 24 April 1955 so as to reach Oak Ridge five days later; incidentally, Janaki's friend, Boshi Sen, the plant physiologist and agricultural scientist, would be deputed to attend the course this year, but as part of a different batch. Janaki was also permitted to visit the University of Michigan and reside on the campus from 1 to 15 June, and to accept the invitation of the Wenner Gren Foundation for Anthropological Research (New York), to participate in the landmark symposium at Princeton, New Jersey on 'Man's Role in Changing the Face of the Earth,' between 16 and 22 June. She was allowed to spend a week in Britain from 25 June to 1 July on her way back from America, 'to study world Distribution of certain genera of plants at Kew Gardens,' and to collect plants and bulbs from the Royal Horticultural Society's Garden at Wisley. The Government was willing to continue her pay and allowances during her absence from the country. Additionally, the Indian Embassy in the United States had been ordered to grant her a daily allowance of 8 dollars and provide her free lodging from 29 April until 15 June in the country. Similar arrangements were made with the Indian High Commissioner to the United Kingdom, who was to provide her with a daily allowance of fifteen shillings besides a paid-for bread and breakfast lodging in London.[8]

Radioactive Isotopes

The 'Atomic City' of Oak Ridge was founded in 1942 as a production site for the Manhattan Project, the huge American, Canadian and British mission that developed the atomic bomb. It was the site of the X-10 Graphite Reactor, which was used to show that plutonium could be created from enriched uranium. It was also where the physicist Enrico Fermi and his team of scientists and technicians worked in bare and make-do conditions, and with an urgency shrouded in secrecy, to beat the Nazis in developing a dreadful uranium fission weapon, already demonstrated in Germany before the war had broken out; the Manhattan Project was a success. It was following this that the laboratory evolved into a full-fledged research institution (referred to as the Clinton Laboratories) with the aim to apply atomic power to peaceful purposes. Some years after the war it was renamed the Oak Ridge Nuclear Laboratory (ORNL) and had by this time become a world leader in evolving nuclear reactor technology among other applications. Incidentally, the Union Carbide and Carbon Corporation helped the American government operate the huge nuclear materials plants at Oak Ridge, by locating, mining (chiefly from parts of Colorado, Utah, New Mexico and Arizona) and refining uranium ore.

The peace-time production of radioisotopes at the Graphite Reactor began in 1946 for industrial, agricultural and research applications. That year, ORNL produced the first medical radioisotopes, triggering a revolution in the life and medical sciences by shedding new light on metabolic and genetic processes. In August, the first shipment of reactor-produced radioisotopes (a container of carbon-14) was despatched to the Director of the Barnard Free Skin and Cancer Hospital of St. Louis, Missouri (where Janaki's friend A. R. Gopal-Ayengar was based); St. Louis was also the home of the Monsanto

Chemical Company, which provided skilled crafts-people for the machine shops that aided the various research projects. In the first year of their production more than 1,000 shipments of 60 different radioisotopes, chiefly iodine-131, phosphorus-32 and carbon-14, were produced for use in cancer treatment as part of the nascent field of nuclear medicine, and as tracers for academic, industrial and agricultural research.

The need to understand the impact of radiation on human health and environment led to an expansion of the biology and health physics divisions at ORNL. The Health Division was eventually split into two new research sections and a medical department. In October, physical chemist Alexander Hollaender was appointed to form and head the Biology Division. By the time Janaki visited ORNL in May 1955, to receive training in the use of radioisotopes in plant breeding,[9] its Biology Division had gained world-renown in the fields of radiation genetics, biochemistry, and later radiation carcinogenesis and molecular biology, and had become the largest biological laboratory in the world.[10] Thirty-one scientists and technicians from across the world (from twenty-two nations and twelve language groups) graduated on 27 May 1955, after a four-week special course in radioisotope techniques at ORNL, sponsored by the Atomic Energy Commission 'as one of several projects launched in furtherance of the President's atoms-for-peace program.'[11] The special session was the first such, exclusively devoted to non-Americans and the trainees were accommodated at the Alexander Inn, in double rooms each provided with a window air-conditioner (it being a very hot time of the year). At the same as Janaki, Shanta V. Iyengar and K. T. Jacob attended the 'atom school.'[12]

The course was structured in such a manner as to provide the students with a foundation in the physics of radioactivity and the use of radioactive isotopes through hands-on laboratory experience. They were all provided with a printed text titled 'Experimental Nucleonics,' and were required to attend a series of lectures, including one by William G. Pollard, the executive Director of the Institute, on Heisenberg's Uncertainty Principle, besides touring various facilities such as the X-10 and Y-12. The students toured the facility that prepared and shipped isotopes and 'watched technicians using remote robotic hands putting vials of isotopes into lead pigs,' after which they were invited to do it themselves but without real isotopes in the vials. 'The graduation was a fun event, conducted by Dr Pollard, who wore his formal gown for the ceremony.' Each student was called up by name and awarded the honorary degree of DRIP (Dabbler in Radioisotope Procedures), and the Graduating Class number.[13] On the occasion, Pollard spoke on 'Origin of Chemical Elements.'[14] Janaki attended the course to learn how plants could be irradiated to trigger mutation, and how plants subjected to background radiation such as in the monazite-rich sands of the west coast of India containing thorium, in particular Travancore, behaved cytologically, a field of investigation she hoped to develop at the CBL.[15]

Doctor of Laws

After completing her training at ORNL, Janaki travelled north to Ann Arbor, Michigan, perhaps by air, where in early June, she was to receive a *Legum Doctor* (Doctor of Laws) from her alma mater, the University of Michigan, in recognition of her seminal contributions to botanical science. It was Bartlett who had suggested her name to the committee on honorary degrees. He also had his department make a 'reservation of Non-Resident Lecturer funds' for a lecture by Janaki during her stay at Ann Arbor.[16] In late March that year, Janaki had received a letter from President Hatcher of the University inviting her to be the official guest of the University of Michigan at its commencement exercises on 11 June, and also to receive an honorary Doctor of Laws degree. She was 'thrilled' to be so recognised by her alma mater, but wrote to Bartlett that she felt 'quite unworthy of all this and even a bit worried at the publicity' this might bring her. *The Michigan Daily* reported the news, and introduced her to the readers thus:

> Edavaleth Kakkat Janaki, Director of the Central Botanical Laboratory, Government of India . . . has made herself an international reputation as a botanist of high order by her work in India, by her extensive research and writing and by her splendid contributions to important scientific conferences. . . . Her *Chromosome Atlas of Cultivated Plants* co-authored with C. D. Darlington, has become one of the great standard sources for cytological workers. Her extensive publications in the field of cytogenetics are widely respected. Blessed with the ability to make painstaking and accurate observations she and her patient endeavours stand as a model for serious and dedicated scientific workers.[17]

Janaki loathed publicity of any kind, much like her coeval McClintock; her excitement stemmed more from the thought of reuniting with old friends at Ann Arbor and walking about the Botanic Gardens than anything else.[18] Bartlett was 'pleased about this honorary degree' and hoped 'that the effect in India' would be favourable to her, but noted (engaging in his usual India-bashing) that there was 'such childish and silly antipathy to the United States' in India 'that the result can hardly be expected to be altogether helpful. Too much politics and too little realistic thinking prevent good relations.' He was sure however that her friends and botanists 'generally [were] going to view this degree as a well merited honour and . . . delighted' that it had gone to her. In his letter, Bartlett also mentioned that the 'great new Phoenix Laboratory' (it housed the Ford Nuclear Reactor, which was to open in 1957 on the North Campus to explore peacetime possibilities of nuclear science) at the University of Michigan, which she had not seen, was now nearing completion.[19]

Janaki would reside, in the University campus, spending two weeks (1–15 June 1955) with old friends Frieda Blanchard, Katherine Fellows, Eileen MacFarlane and, of course, Bartlett. Together, the women friends would visit the Botanic Gardens and their other usual haunts on campus and have a great time catching on news (in a matter of two years from this time, the University Botanic Gardens would relocate to a 200-acre parcel of land, donated by Frederick C. and Mildred H. Matthaei, about three miles away from Ann Arbor). It was after more than two decades that Janaki was meeting the Blanchard children, and she could not contain her joy; five decades later, Dorothy, the eldest of the children, would vividly recall seeing Janaki in 1955. As someone who enjoyed reading books on adventurous expeditions, Janaki bought French mountaineer Maurice Herzog's *Annapurna* (1953 edition) at Ann Arbor; Herzog was the leader of the successful French expedition that had climbed the extremely challenging 8,091-m peak, for the first ever time, in 1950.

A Landmark Symposium

Leaving Ann Arbor, perhaps on 15 June 1955, Janaki travelled east to Princeton, New Jersey (possibly with Bartlett) to attend the landmark international symposium on environmental history under the aegis of Wenner Gren. The symposium had literally taken three years to develop and organise. The theme, 'Man's Role in Changing the Face of the Earth,' was reflective of the ideas of the Russian geographer Alexander Ivanovich Woeikof and was the first interdisciplinary attempt to understand the impact of human activity on the environment. It attempted to address the question: 'What has been, and is, happening to the earth's surface as a result of man's having been on it for a long time, increasing in numbers and skills unevenly, at different places and times?' It had no political agenda; its purpose was simply to bring together the most eminent scholars in the world to assess the scope and rate of environmental change caused by human activity, evaluate the state of knowledge of human environmental impact, and identify areas that needed further study. About 40% of the participants were in the geosciences, 28% in the biological sciences, 12% in the social sciences and 20% in applied fields like administration and city planning. Six days of presentation and wide-ranging discussions covered topics such as anthropogenic climate change, the ecology of waste, urban and industrial demands on land, and the history of human modifications of the environment.

The American cultural geographer, Carl O. Sauer (1889–1975), as Chairman of the symposium made it clear that each contributor was free to select his/her own spot to state his/her topic within the limits of the conference; the object was to find persons with something to state and let them state it. The urban historian, Lewis Mumford's role was to add, delete and generally reorganise and regroup the potential contributions and their authors. When it came to choosing participants, Sauer had cautioned against 'salesmen' or the engineers who had future triumphs to push, and government men who had politics to defend. Sauer wanted to avoid quantitative methods and trend lines and instead consider geographical distribution maps as a tool to trace dynamic patterns. He felt it was important to have historically minded persons with a common interest in 'curiosity about what Man (Cultures) has been doing to and with his habitat.' To Sauer, a historical account was essential to counter the reductionist tendencies of the social scientists, who tended to theorise and universalise. He believed that science must continue to work humbly within the natural order and its limits and that the most rigorous participation would come from 'historically minded' persons.[20] Accordingly, the past was granted a separate section, chaired by Sauer and called 'Retrospect.'

In October 1954, Janaki had gladly accepted the invitation and promptly notified the organising committee of a change of address (the CBL having moved from Calcutta to Lucknow in early December) and selected the method for payment of funds to her: 'a statement or bill covering my transportation to be sent . . . for payment by check in USA currency by the Airline Company arranging my itinerary,' she stated. She had received her passport, but the visa for entering America, as per the information given her by the American Embassy at Delhi, would only be issued three months before the actual date of travel. She was also pleased to 'accept the honour of being one of the Guests of honour at the dinner of the Symposium at Princeton on 15 June.' At this time, on reserve were the names of Carl L. Hubbs (ichthyologist) and Rachel Carson; the consensus was in favour of the former.[21] The letter stated that the invitation was among 'those addressed to fifty other prominent scholars in the United States. The theme of the Conference is nothing less than "What has man done, and is he doing, to the earth as a result of his occupance?" with emphasis to be placed on the stimulation of cross-disciplinary thought and the delineation of the most profitable lines of future research.' The letter also stated that the eminent geographer Carl C. Sauer, Professor of Geography of the University of California Berkeley, had consented to act as Conference Chairman. For Janaki's background paper, the organising committee suggested, 'The Subsistence Economy of India,' but she found 'the title a bit ambiguous.' 'Will you please explain what you mean by Subsistence Economy?' she queried. 'An early reply will be appreciated,' she would add.[22]

The Modern Evolutionary Synthesis

Carl Sauer replied to say that he was forwarding her query to botanist Edgar Anderson (an exact contemporary of Janaki, born 9 November 1897), who had in the first place chosen the theme for her. If not Bartlett, it might have been Anderson (or both) who had suggested her name as a potential invitee, for Janaki and he shared the same social and intellectual world—had the John Innes as a common factor, and both adopted an evolutionary approach to systematics. Anderson's researches ranged from the genetics of hybridisation to economic botany, and he was considered a major figure in the development of what has been referred to as the modern evolutionary synthesis. Anderson and his colleagues aimed to make taxonomy or systematics a biological science (biosystematics) rather than a descriptive practice merely based on morphology; they advocated the application of biological tools to the questions of taxonomy. Having worked on the development and naturalisation of cultivated plants, Anderson was interested in studying the environmental conditions under which plants grew. We may recall that Anderson had visited India in 1950 as an expert consultant at the UNESCO symposium on the origin and distribution of tropical crop plants.

Graduating from the Michigan State University in 1918, Anderson had moved to Harvard to complete a Masters and then a DSc in botany in 1922. In 1929, he received a National Research Fellowship to study at the John Innes with Darlington, R. A. Fisher (it was Anderson's data set on three related species of irises that Fisher used to demonstrate his statistical methods of classification) and Haldane, after which he worked for four years at the Arnold Arboretum, with geneticist Karl Sax. This was followed by years of teaching at the Washington University. In 1941, Anderson was invited to present the prestigious Jesup Lecture series at the Columbia University on the role of genetics in plant systematics, alongside the evolutionary biologist Ernst Walter Mayr, who gave the lectures in zoology. While Mayr's lectures resulted in his substantial *Systematics and the Origin of Species* (1942), a concise discussion of species and species formation, Anderson did not produce anything of the kind. Anderson would instead work on introgressive hybridisation, publishing an important book on the subject in 1949; *Introgressive Hybridization* was considered a major contribution to the field. His later publication, *Plants, Man and Life* (1952), attained the position of a classic, in which he examined the long history of human and plant interactions, putting forth his ideas on hybridisation as an evolutionary tool in simple terms and stressing on the need for a greater focus on cultivated plants. In his view, man was a key evolutionary factor (Carl Sauer's 'ecological dominant') to be taken on board if one were to comprehend the changing floras of the world. Like Janaki, Anderson's border practice was a creative synthesis of the disciplines of morphology, taxonomy, genetics, geography, anthropology, archaeology and agronomy.[23] In the second half of her professional life, rather than with Darlington, it was with the American evolutionary biologists like Anderson, and the geographer Sauer (who was highly influenced by Vavilov himself), that Janaki found intellectual affinity.

The Invitees

Invitees included the German-American sinologist Karl Wittfogel (whose *Oriental Despotism: A Comparative Study of Social Power* was published a couple of years after the symposium), the British physicist Charles Darwin, Berkeley geographer and environmental historian Clarence Glacken and the botanist-anthropologist Bartlett. Several of the participants at the symposium either went on to produce seminal works in environmental history or were already well-known ecological thinkers. Participants were 'selected for their qualities as individuals' and hailed from America, Europe, the Middle East and Asia, with specialisations ranging across twenty conventionally defined disciplines.[24] As a participant, Janaki fitted the requirements of the symposium perfectly: her substantial work in the field of cytosystematics was historical (evolutionary) and geographical in its approach and matched Sauer's own approach to landscape. Sauer, who studied the impact of human interventions on landscapes was a fierce critic of environmental determinism and defined geography as a study of cultural landscapes rather than physical or natural ones. And as for Janaki, she had chosen as her life's aim the study of the changing relationship between man and plants (in Anderson's words, man and his transported landscapes), extensively employing cytogenetics, geology, history and anthropology to demonstrate how species travelled across space and time. Not only was Janaki the sole invitee from India, but it also appears she was the sole woman participant at this landmark international symposium. Rachel Carson, the marine biologist and conservationist (she had not yet published *Silent Spring*) and Aldous Huxley were on the suggested list of invitees but did not make it to the final round. Radhakamal Mukerjee, Director of the Institute of Sociology, Ecology, and Human Relations and Vice-Chancellor of the University of Lucknow, was invited to contribute a background paper but not required to be physically present.

Within the section 'Retrospect,' the idea was to have presentations on the first day by Sauer, Clarence J. Glacken and F. Fraser Darling on the 'over-all human time' or 'man's tenure of the earth' (besides a couple of other contributions), followed on the next day by a 'start down the corridors of time' with Omer Stewart speaking on fire as the first human tool, Wittfogel on hydraulic civilisations, Estyn Evans on peasant life in Western Europe, Janaki on the subsistence economy of India, Pierre Gourou on crops and animals and Mumford on a natural history of urbanisation, among other papers.

Reflecting on Primitive Agriculture

Within a couple of days of writing to the organising committee, asking for a clarification on the title of her paper that they had suggested to her, Janaki wrote to Bartlett from the new home of the CBL, the Chhattar Manzil in Lucknow, wishing him a happy new year and apologising for her long silence. 'I have been going through some difficult times over the sudden shifting of our laboratory from Calcutta to Lucknow and that in the middle of the growing season! All for the whims of one person—the [Joint] Secretary of our Ministry. It has upset my whole year's work on *Nymphaea and Dioscorea*. Now we have just 4 walls & no garden. However, one has to be prepared for all sorts of things in India,' she would comment. She told him that she had been 'doing some thinking' on 'Primitive Agriculture' and had gone *Dioscorea* hunting to the Anaimalai Hills and the Ghats, for those eaten by the food-hunting tribes, but was feeling 'less certain' about her background paper for the Symposium, 'Subsistence Economy of India.' 'I am not an economist but a botanist—and the new Jargon of Sociological Studies is not familiar to me,' she complained, while promising to send him a 'short note' on her life, and a photograph of hers that was long overdue for publication in the *Asa Gray Bulletin* that he edited.[25] This would again not happen.

Meanwhile as Chairman of the Symposium, Sauer offered to explain (on the basis of discussions with the organising committee) to Janaki what they meant by her title, and he did in a very perceptive and gentle manner: 'As I recall the discussion, we wanted someone who knew land, plants, and folk intimately—not merely the commercial surpluses and deficiencies. Our agricultural scientists have taken a supercilious attitude toward self-supporting farming everywhere. The crops are inferior, the methods primitive, the yield improper.' This was certainly the experience with the Grow More Food campaign that Janaki had had. Sauer stated that he had some 'strong convictions' with respect to Latin America 'that the small native farmer makes sense of his own and community's ends, that he grows things is some sort of an ecologic balance of which he is a part, and that we can do a lot of damage by knocking plants that do not interest us out of the pattern and also do damage to the surface and the soil by our immediate commercial ends. That's what I like to see you do an essay on . . . the cultural fit (or misfit) of people, animals, plants in Indian communities (plurality of cultures and also environments) as living for themselves rather than for world markets. If you will bend your great knowledge of plants that man made and your ample insight into Indian country life to some such . . . we shall be wisely and well instructed and benefitted.'[26]

Only a week after receiving Sauer's letter, Janaki had one from Bartlett, voicing his thoughts on both their papers. He told her that he had been taking his paper very 'seriously' and had travelled 'to Washington before Thanksgiving to look up a couple of hundred references not available' in Michigan but had ended up falling ill. As a result, he had failed to source anything from Washington until the previous week (luckily for him there was an interlibrary loan system in operation) and given his limited access to secondary material, felt he would have to alter his title to something like, 'Fire, in relation to Primitive Agriculture and Grazing in the Tropics,' (thus excluding Europe in his study), in contrast to her position, which in his eyes was a far better one; she was not 'limited by reference to material in European languages' (as he certainly was), and could simply rely on her 'wide experience,' to 'make a unique contribution.' Bartlett assured Janaki that the committee would not have expected her 'subject to be approached from the standpoint of technical or theoretical economics, but rather from the agronomic, geographic and ethnographic standpoints . . . If I were you I would discuss the chief types of agriculture of each province with consideration of adequacy for local subsistence, actuality or possibility of excess production for interprovincial commerce, possibility of extension of production without destruction of natural resources, and the geographic distribution of types of agriculture and crop plants as related to climate.'[27] Even if Bartlett had not hit it quite close to where Sauer was coming from, it gave her much food for thought.

She would write to Bartlett again in late March, when she was 'going through very strenuous times' reorganising the CBL at Lucknow and was always on the move: 'We were growing like a normal organization at Calcutta. When we were ordered to shift to Lucknow at almost a moment's notice. I had all my plants in the field and my Nymphaea from all over India were growing in huge mud pots. They are all at Calcutta and I keep moving from Lucknow and Calcutta.'[28] She was having to work on her paper amidst all this travel and disruption, making her very anxious. The papers were to be written and reached to the organising committee in time for printing and circulation among participants before they left their homes. This would, it was hoped, give ample time for fruitful discussions instead of unnecessary formal presentations of the papers.

Janaki's Symposium Paper

Janaki began her symposium paper, 'Introduction to the Subsistence Economy of India,'[29] by describing the physical geography of the Indian subcontinent, including its geology; incidentally her second major at the Madras University was geology and, from the 1950s, would be added to her toolkit to map the evolutionary process of speciation, vis-a-vis polyploidy and plant migration. Her paper was not just an enunciation of her cytogeographical ideas, but a reflection of her archipelagic thinking; the archipelago is a cultural and imagined space, and not simply a physical entity located at the intersection of land and water. It was also an epistemological concept that allowed the exploration of connections and similitudes on a global scale and involved creative interdisciplinary exchanges; fundamentally, it demanded getting out of disciplinary silos of thinking.

In Janaki's case, the focus was on flowering plants, many of which were world travellers, not unlike herself. Peninsular India, Janaki observed, was geologically very old and was a piece of the ancient continent of Gondwana, which split up to form the continents of Australia, South America and Africa. This break-up had occurred after flowering plants had evolved on the earth, for which reason, she stated, Indian flora had several genera in common with Africa and South America and, to a more conservative extent, Australia. She further observed that India became part of Asia, resulting in the Himalayan uplift, and that like Arabia it was an annexation to Asia. The evolution of man was very closely linked with this annexation, she noted: 'The close affinity between the flora and fauna of Sind and that of Africa and Arabia, the continuation of the Sahara into India as the Great Rajputana Desert, and the movement of this desert from west to east are factors that have to be taken into consideration when studying the movement of primitive man into and within India.' Further, the presence of temperate Himalayan plants like the *Viburnum* and *Rhododendron* on isolated hill ranges of the Nilgiris in southern India was indicative of their migration during periods when the climate of southern India was cooler and wetter than it was today. Through her cytological investigations, Janaki had been able to confirm that these Nilgiris species were identical to those of the Himalaya. The case of the wild goat, *Capra hypocrius*, was another case in point; it was found not only to inhabit the Nilgiri and Anaimalai ranges but also the temperate regions of the Himalaya from Kashmir to Bhutan.[30]

After the plants, Janaki turned to the history of human evolution and migration in India, with particular focus on the genesis of tribal economies. She observed that in the worldview of the primitive man, 'all aspects of life are harmonized into a whole,' that economy in primitive societies remained a static one until it came into contact with a different culture, and that for the aborigine, 'every natural process is a manifestation of magical power' resulting from certain ritual acts. Janaki was making an original observation when she stated that to the tribal man, magic was 'a way of organizing intensive economic results—it is his way of concentration.' Diversity in mankind and cultures arose from the encounter between the old and the new, she argued, by 'contact and fusion,' analogous to how 'the origin of new forms in plants and animals was brought about by hybridization, resulting in new combinations of parental characters (as well as by isolation in which mutational changes may occur).' Cattle-keeping, for instance, evolved from hunting; a mystical connection was established between man and cattle, as was seen among the Todas of the Nilgiri Hills, whose culture was bound up with their buffalo herds. Janaki also noted that female infanticide was ordinarily practised among the Todas because the females of the community were not of much use in tending to the herds. 'The inordinate reverence paid to cows in India, especially on the Gangetic Plain, is undoubtedly a relic of this mystic relationship of an earlier pastoral people with their herds of cattle . . . cow protection is a political issue in India . . . the antagonism of the Hindus to beef-eaters is based chiefly on this ancient mystic link with the cow,' she explained.[31]

In order to classify the different tribal economies, it was important, Janaki argued, to first discuss the types of primitive agriculture found in India: to understand 'the agricultural soul' of a place, as Vavilov would phrase it. The earliest example of subsistence economy was of the food-gatherers, as in the

Paniyars of Wayanad who used the wooden digging stick (the tip of which was replaced later by iron), similar to the one used by the Kadars of the Anaimalai Hills; the stick would be replaced firstly by the hoe, and then the basic plough, initially operated by women. The use of oxen for ploughing indicated, she observed, 'a synthesis of a cattle-keeping tribe with one having an agricultural economy.' The ox-drawn plough was a 'great landmark in the history of agriculture and human culture,' Janaki noted, because it had aided food production on a much larger scale than was necessary for subsistence. When population increased, it led to conflict and disaster, with tribes fighting over food reserves. She cited the example of the migration of the 'fast-multiplying and aggressive Syrian Christians of Travancore and Cochin' to Wayanad in Malabar, which had had a disastrous impact on the Paniyars, altering their agricultural practice and the economy itself: 'with their jungles under the plow of the new settlers, . . . a tribe that once was independent in its habitat [had] now become "depressed".' She also drew attention to the Kadars of Cochin, who once freely foraged honey, yams and roots from the forests, but had now come under the care of the Forest Department, and were paid in rice for gathering minor forest products; this had resulted in Kadar women taking to 'a life of idleness' and 'turning degenerate' (promiscuity had become a serious issue). 'Thus the "noble savage" in contact with civilization,' she noted, 'generally finds himself in an awkward state.'[32]

Janaki's discussion then geared towards agriculture in relation to climate in India, a country dictated by the monsoon cycle—its vegetation was very much influenced by the amount and seasonal distribution of rainfall received. 'The vegetative climax of India is the dense and tall tropical evergreen forest, found on the Western Ghats and in Assam,' the very regions Janaki had explored and botanised over the decades. The most highly evolved type of *jhumming* or shifting cultivation is seen in Assam, Janaki observed, 'where a regular rotation of forest clearing takes place once every eight years, so that sufficient time elapses for regeneration of the forest,' but unfortunately, this had a negative effect, 'for the flora that come up after clearance are not always the same as those previously destroyed.' The loss of genetic diversity associated with *jhumming* was for Janaki, an alarming issue. On the other hand, the practice of declaring parts of the forests sacred had contributed much to conserve vegetation, as people would not dare cut the trees for fear of committing sacrilege. Indeed, during the symposium discussion, Edgar Anderson drew attention to what Janaki's friend, N. L. Bor, had witnessed in the forests of Upper Burma and Assam. Anderson noted: 'Since unrecorded time the peoples there have had a simple nature worship; although they had a *Brandwirtschaft* [slash and burn agriculture]—they burned and moved on—their hilltops were not burned over. The hilltops were sacred; as placed of worship, they were not cut over or pastured.'[33] Sacred forests or *kavus* were also a feature of Malabar, and frequented by Janaki in search of rare botanical specimens; the *kavus*, she enlightened her listeners, 'serve as sanctuaries for snakes and wild life and often become the nuclei of sylvan temples.'[34]

As man evolved from being nomadic to sedentary, he strived to attain a certain equilibrium with his environment, best seen in Malabar, she stated, where mixed plantings followed *jhumming*, to which she gave the name *paramba* (homestead or subsistence) cultivation. In this kind of agricultural practice, the forest patch was cleared only partially, with useful trees such as the *Artocarpus*, wild mango, nutmeg, *Strychnos* and *Cycas* palm left standing for either timber, firewood, training the pepper vine, and in the case of Cycas, also as famine food. 'This *paramba* form of cultivation . . . does not upset the ecological balance of the natural vegetation,' she observed.[35] We may recall that even as early as 1932, Janaki had been inspired by a model farm that she saw at the Nedumpoyil village, not far from Tellicherry. The farm was a fine example of the *paramba* form of cultivation, employing *poonam* or shifting cultivation, and growing multiple crops including fruits and vegetables, besides rice paddy. The farm's successful application of the ideas of sustainability and polycropping, so typical of local agricultural practice, had struck her instantly as an example of how ecological balance in nature might be attained; she had thought of it as an example worth emulating and had even recommended it to her elder brother, Raghavan. Also, during her expedition to Nepal in 1948/49, she had not failed to notice how the Newari peasant worked 'philosophically in his fields.'

Janaki's symposium paper ended with a discussion on the intimate relationship between plants and man in India. The region including India, Pakistan, and further west the countries of Afghanistan and Iran had been Vavilovian centres of origin of a range of cultivated plants including rice, moth bean, horse gram, eggplant, rat-tailed radish, mango, hemp, pepper and indigo. Several economic plants such as the banana and rice had been domesticated concurrently in Southeast Asia, where several species had been recorded. Incidentally, to pursue her interest in the origins of the banana, she had recently purchased a copy of Cherian Jacob's monograph, *Madras Bananas* (Madras, 1952). 'The wild banana with its stony fruit must have attracted, as it still attracts, primitive tribes,' she observed. The wild banana had evolved into cultivated forms through the accidental outcome of sterile forms produced either by hybridity or by triploidy. The wild tribes would instantly notice the stoneless banana varieties (i.e. sterile bananas) wherever they occurred and collect and grow them near their huts. In Assam, Janaki had observed the gradation from seeded to seedless varieties in the local wild form. She also stated that with every one of the major food crops cultivated in the world—wheat, rice and maize—there was one of 'great local or regional importance,' such as the coconut palm (in combination with rice) on the west coast of India, used as food, timber and thatching. All parts of *Caryota urens*, or the toddy palm, for example were used by tribal communities such as the Paniyars of Wayanad, as food (the pith of the palm being full of starch), as building material (leaves as thatching), in ceremonies and in making their digging sticks (using the outer rind of the palm). Janaki also cited the example of palms such as the *Borassus*, the wild date *Phoenix sylvestris* and *Cycas circinalis*, which played important roles in the subsistence economy of the tribal people of the Western Ghats, while the tribal communities of peninsular India relied heavily on *Madhuca latifolia* (Mahua).

We might recall that one of the aims of the CBL of which she was Director was the search for new economic plants such as those cultivated by indigenous tribes, but had not included in its scope the nature of their agricultural practices. What the Wenner Gren symposium did for Janaki was to sharpen her awareness of folk agricultural practices in India and taught her to become politically alert to the power of market economies to destroy the environment; she had already a glimpse of this 'loot' (plundering precious resources) when she was a director of agriculture during the Grow More Food campaign in India. Her focus would no more be limited to the cytogenetics of cultivated plants, their breeding in controlled environments, or even a study of primitive cultivars, but would broaden to include a study of the origin and dispersal of agricultural practices. We know that she collected several books during this period, some of them recent publications, to help her write up the symposium paper. Among them were Indrajit Singh's *The Gondwana and the Gonds* (1944), S. C. Dube's *The Kamar* (1951), William Howell's *Mankind so Far* (1948), Alan H. Brodrick's *Early Man: A Survey of Human Origins* (1948), Arthur Keith's *A New Theory of Human Evolution* (1948), Sethumadhava Rao's *Among the Gonds of Adilabad* (1949), C. E. P. Brooks' *Climate through the Ages* (1950), V. Gordon Childe's *New Light on the Most Ancient East* (1952), bought at Lucknow, B. G. Gokhale's *Ancient India: History and Culture* (1954), and R. Linton's *The Tree of Culture* (1955).

Perspectives on Population

The Symposium Discussion of the section 'Retrospect,' under which Janaki's paper figured, was organised around four themes: 'Man's Tenure of the Earth,' 'Subsistence Economies,' 'Commercial Economies' and 'Industrial Revolution and Urban Dominance.' She participated in discussions on 'Man's Tenure of the Earth,' under the subsection, 'The Modern Era of Exchange Economy.' In fact, she opened it, with insights on the impact of market economy on the population of India: 'the industrial revolution meant the clearing of land used chiefly for a subsistence economy of village farming, cattle-grazing, and hunting and the shifting to a money economy, using the land for the growing of cotton, oilseeds, and other export crops.' Janaki argued that 'the result has been that for the last one hundred and fifty years the Indians have lived hedged in, as it were, by an artificial environment, with all the necessary facilities of security and safety to increase their population. Any kind of medical aid only makes them live longer and increases demands on food supply; any method of curtailing population must deal with people who have not yet assimilated the scientific attitude.'[36]

Janaki, in fact, shared some of Bartlett's controversial views on eugenics and population control—particularly, his belief that over-population was a bomb waiting to explode and had to be contained at all cost. She was though not as ruthless as he was. In a letter to Janaki, ahead of the Symposium, the unmarried Bartlett had remarked that he was convinced that 'there are altogether too many people' in the world; 'improvement in agriculture, manufacture, commerce, and every other temporary alleviation of human economy will be futile so long as people breed like that many rabbits.' Rather heartlessly he would argue that famine was the only effective remedy for over-population: 'War no longer helps because it has come to be too expensive to kill a man and the genetically inferior population left at home increases and leaves no economic niches for the expandable but left-over surplus of fighters to retreat to.' A diehard eugenicist, Bartlett observed:

> As I see it eugenic selection and birth-control, by force if necessary, are all that can save the world from overpopulation and wholesale starvation. I am not a good democrat—don't believe in human equality. The progress of medicine simply complicated matters by saddling society with expensive maintenance of improvident or unfortunate non-productive population. A well-ordered and regulated, really civilized economy can't continue to exist without eugenic control. And what a howl that will raise—anywhere! Yet I wonder how long Michigan, for example, will tolerate a situation in which the public support of each incompetent, insane, feeble-minded, or criminal costs as much as the income of the average family![37]

Janaki raised the issue of population in India once again, at the symposium discussion of the section, 'Process,' under the subsection, 'Changes in Biological Communities,' which dealt with 'man in relation to ecological processes.' Providing statistical data to demonstrate how population in India had risen rather alarmingly over the six-decade period, 1891–1951, she asked for suggestions 'in which a remedy might lie and expressed her willingness to answer any questions on difficulties that India might have in putting into action any remedy proposed.'[38] India's population problem would continue to interest her, and also make her exceedingly anxious.

Planning a Project at the Bishop Museum

His views on 'Hindus' and eugenics aside, Bartlett was an exemplary networker and a true mentor, untiring in his efforts to find Janaki a research position in America. At the Princeton Conference, he managed a private meeting with American anthropologist Alexander Spoehr (1913–93), Chairman of the section 'Subsistence Economies' and Director of the Bishop Museum, Honolulu. This meeting had to do with discussing a plan to 'somehow' get Janaki involved in a study of the yams (*Dioscorea*) of the Old World, 'especially the ethnobotany and geographical distribution and cytotaxonomy of the old cultivated types.' Bartlett hoped Spoehr would invite her to 'bring an active project' (that is a live collection of yams) 'that would result in a good Memoir or Bulletin in the Bishop Museum Series.' 'She could give some very interesting lectures on primitive agriculture in India,' he suggested to Spoehr. However, there were a few issues that had to be resolved first. It was illegal to introduce yams into America except by the agency of the US Department of Agriculture; the Department had begun to display an interest in yams as a source of cortisone and had been collecting them for biochemical study.

Bartlett was hopeful that Carl Erlanson, in charge of the Office of Plant Introduction, would arrange for the Bishop Museum to have 'a study collection' at Honolulu. 'Dr Janaki was tremendously enthusiastic over the prospect of a visit to Honolulu, but would want to stay long enough to make a real contribution. She has no personal financial resources and would have to receive something in the way of a salary enough to live upon. She is by all means the most distinguished Indian botanist and it will enable her to get back into personal scientific productivity if you can arrange for her to have a good long visit in Honolulu with enough of a stipend to cover her living expenses,' Bartlett appealed to Spoehr. He added that in his opinion, Janaki cared little for her present administrative position and 'would welcome a way to get out of it into personal research . . . The proposed Honolulu invitation would be just the thing for her, for she could take a leave of absence and then resign if anything in the way of a position turned up.'[39]

Bartlett also informed Spoehr that the American Camellia Society, which had 'two or three millionaire members' had been considering inviting Janaki 'to do a cytotaxonomic monograph of Camellia and its allies.' If that line of research could be taken further, he explained, it was hoped it would give her 'a chance to work on the variation of tea as well as ornamentals, if material could be found in cultivation.' 'That would not be as simple as the yam job,' Bartlett observed, because the *Dioscorea* could be propagated rapidly by vegetative means and would quickly flower unlike the Camellias; Janaki could 'make extensive herbarium collections and fixations for cytological study of Indian types' before travelling to Honolulu, he commented. He fervently hoped that Spoehr would invite Janaki 'for a period of several months or a year,' in the anticipation that this might offer her 'a little respite from politicians' and she would begin to 'feel differently about her position in India.' Bartlett hit the nail right on the head when he remarked, with immense empathy for his former student, that the 'trouble there [India] is that any number of make-believe scientists resent the idea of a woman being their official superior, and it makes her life difficult.' He was however quick to add that Janaki was no doubt 'patriotic and would wish to continue in public service if her leadership were willingly accepted, but too many big shots feel that they should have high-sounding positions without qualifications to fill them.'[40] Bartlett wasted no time in intimating Janaki of the proposed study of yams at Bishop Museum and assured her that 'something will surely come of it.' Spoehr, he informed her, was even thinking of expanding the project 'to cover other tropical primitive crop plants . . . in his attempt to secure funds.' He was also hopeful that the National Research Foundation would support the work to be carried out in Honolulu, which would 'surely be the best place for such a programme . . . Bishop Museum auspices will be excellent, because the institutional interests are largely in anthropology.'[41]

Ralph Peer Tries Again

In late November, Bartlett received a letter from the Camellia connoisseur Ralph Peer, which conveyed that the Peers had been 'personally friendly with Dr E. K. Janaki Ammal . . . for a number of years' and that in their garden at Park Hill, Florida, were a very large collection of 'Camellia varieties and species.' 'Dr Janaki while in England made several Camellia chromosome counts which were highly useful and since then I have done whatever I could to encourage her interest in Camellias and related items. As you doubtless know, she is most anxious to return to this country and to pursue her scientific work here,' Peer informed Bartlett. He wished to know if Bartlett had a place for her at the University and in that case 'what would be involved in the way of financial assistance to make this possible.'[42]

In reply Bartlett, who was now close to retirement (he had already stepped down from being Chairman of the Department of Botany way back in May 1947), wrote that he recommended $1800 as a half-time fellowship for a year, the same amount that the plant collector for the University of Michigan, Thakur Rup Chand was receiving. Bartlett promised to do all that he could to 'promote her coming to work on the cytotaxonomy of Camellia and Dioscorea, either "in parallel" or "in series".' 'Personally, I can't imagine a better plan to have her come for she is without question the most distinguished botanist in India, and a very productive worker,' he added.[43] Incidentally, about this time, Bartlett would send Janaki a whole set of representative Indian plants, including grasses, from the University of Michigan collection curated by Rup Chand.[44]

A few days later, Bartlett was able to confirm to Janaki that Carl Erlanson had expressed tremendous interest in the *Dioscorea* project and that for projects of a year's duration, he had a free hand in making appointments in the Department of Agriculture. 'If something of the sort could bridge the time until you were an accepted candidate for citizenship, I cannot believe that you would not find a permanent position,' he wrote to Janaki, under the misunderstanding that she was keen on permanently migrating to America.[45] All she wanted was a salaried position in America, which would allow her to get back to independent research. That very day, he sent away a letter to the Visa Office, Washington (to Roland Welch, the Director) stating that Peer was ready to provide financial assistance if Janaki were to come to America, for a research project on Camellia, in which he was keenly interested. 'She is the most distinguished botanist of India, best known for her work on the races and hybrids of economic plants,' Bartlett introduced her. He also informed the Visa Office that she had been in the country recently to attend the landmark Wenner Gren Conference at Princeton and a course in atomic energy for foreign scientists at Oak Ridge and that she had been awarded an honorary Doctor of Laws by her alma mater, the University of Michigan.[46] As soon as Janaki received Bartlett's letter, she responded, to clarify that although she 'would like to very much spend some years in America before [she] die[d],' she wished 'to do that only as an Indian Scientist working abroad.'[47] In the new year (1956), Bartlett finally heard from Spoehr, and in the negative; he had not met with any success in finding funds for Janaki at Honolulu.[48]

Missing the Haldanes

When Janaki passed through England after the symposium, en route to India, she missed meeting Haldane, but sent him a copy of her paper on subsistence economy anyway, along with his lighter which he had left behind at her office in Calcutta. Haldane wrote back that he was pleased Janaki had settled into a good job: 'I did what I could to help you, but I don't suppose it had any influence.' He and Helen were hoping to be back in Calcutta in July the next (1956), if all went to plan; they were also aiming to visit Lucknow and Izzatnagar on the same trip.[49] By way of a quick comment on her paper, Haldane remarked that he disagreed with her statement that the Namboothiris (the Malayali Brahmans) were a 'pure line.' On the other hand, Janaki had stated that a pure line of Namboothiris was preserved by the insistence that the eldest son marry a Namboothiri woman, while the other sons were 'free to take on Dravidian partners.' He had a second reservation as well, and perhaps this was a valid one, when he wrote: 'I also wonder whether p. 8 [page 328 in the published version of the essay] is the whole story. The people described in the *Mahabharata* occasionally ate beef, so it is not clear to me how the present prohibition of beef originated.' He was questioning Janaki's simplistic claim that the antagonism of Hindus to beef-eaters stemmed from an ancient mystic link with the cow, as if there were no exceptions to it but one can't help remarking on how Janaki's statement, 'Today, cow protection is a political issue in India,' resonates so strongly with today's India![50]

387

Figure 20.1 Janaki at the Radioisotope facility, Oak Ridge. 30 April 1955.
Source: Courtesy of the ORNL.

Figure 20.2 Commencement 1955, University of Michigan. 'Eleven Honorary
Degree [Doctor of Laws] Recipients, including Edavaleth K. Janaki,
director, Central Botanical Laboratory, Government of India.' 11
June 1955. News and Information Services. Box A-3.

Source: Courtesy of the Bentley Historical Library, University of Michigan.

388

Figure 20.3 Janaki with friends: Frieda Blanchard, Eileen Macfarlane, and Katherine Fellows. University Botanic Gardens, Ann Arbor. June 1955.

Source: Courtesy of the late Dorothy Blanchard.

Notes

1 BHL, UoM: University Herbarium Records (University of Michigan) H. H. Bartlett series, Box 11, E. K. Janaki Ammal to H. H. Bartlett, letter dated 23 March 1955.
2 Anon, 'Tracer Techniques in Agriculture', p. 119.
3 Anon, 'Radiation Genetics', p. 391.
4 See for instance, Myers, 'Some limitations of radiation genetics and plant breeding'.
5 BHL, UoM: University Herbarium Records (University of Michigan) H. H. Bartlett series, Box 11, E. K. Janaki Ammal to H. H. Bartlett, letter dated 23 March 1955.
6 Ibid., H. H. Bartlett to The Director, Institute of Nuclear Research, Oak Ridge, letter undated.
7 Ibid.

8 Copy of a letter from the Government of India, Ministry of Natural Resources and Scientific Research to E. K. Janaki Ammal, dated 19 April 1955, private collection.

9 'Radioisotopes and health: Trace of hope'. *Oak Ridge National Laboratory Review* 25, nos. 3–4 (1992): 36–38; also see Curry, 'Atoms in agriculture.'

10 'Alexander Hollaender: A Radiant Biologist'. *Oak Ridge National Laboratory Review* 25, nos. 3–4 (1992): 39–41.

11 Archives of the ORNL, Oak Ridge (AORNL): 'Students complete radioisotope study', *ORNL-The News* 7, no. 47 (3 June 1955).

12 'Indians for atom school', *The Times of India*, 20 March 1955, p. 8.

13 AORNL: Letter from Russell Hilf (class of 1954), Professor Emeritus, Department of Biochemistry/Biophysics, School of Medicine and Dentistry, University of Rochester, 2014.

14 Ibid., 'Students complete radioisotope study', *ORNL-The News* 7, no. 47 (3 June 1955).

15 Monazite was discovered by a German prospector C. W. Schomberg in Travancore as early as 1909, but it was geologist D. N. Wadia (appointed in 1949 to survey the country for rare earths by nuclear physicist Homi J. Bhabha, founding director of the Atomic Energy Establishment, Trombay, immediately after the Indian Atomic Energy Act was passed by the Indian Parliament) and his team which first surveyed the extensive monazite sands of Travancore-Cochin, later brought under the control of the Government of India. Indian Rare Earths Limited began processing monazite sands in Travancore in 1952, separating the rare earth minerals from thoria and phosphates.

16 BHL, UoM: University Herbarium Records (University of Michigan) H. H. Bartlett series, Box 11, H. H. Bartlett to E. K. Janaki Ammal, letter dated 8 July 1954.

17 University of Michigan, Proceedings of the Board of Regents (1954–57), p. 533, https://quod.lib.umich.edu/u/umregproc/acw7513.1954.001/551?rgn=full+text; view=image;q1=janaki accessed on 2 June 2019.

18 BHL, UoM: University Herbarium Records (University of Michigan) H. H. Bartlett series, Box 11, E. K. Janaki Ammal to H. H. Bartlett, letter dated 23 March 1955.

19 Ibid, H. H. Bartlett to E. K. Janaki Ammal, letter dated 1 April 1955.

20 Williams, 'Sauer and Man's Role in Changing the Face of the Earth', p. 221.

21 The Bancroft Library, University of California Berkeley, California: Sauer Papers, W. L. Thomas to Carl Sauer, letter dated 14 October 1954.

22 Ibid., E. K. Janaki Ammal to W. L. Thomas, letter dated 3 January 1955.

23 Kleinman, 'His own synthesis: Corn, Edgar Anderson, and evolutionary theory in the 1940s'.

24 Thomas (ed.), *Man's Role in Changing the Face of the Earth*, p. xxvi.

25 BHL, UoM: University Herbarium Records (University of Michigan) H. H. Bartlett series, Box 11, E. K. Janaki Ammal to H. H. Bartlett, letter dated 7 January 1955.

26 The Bancroft Library, University of California Berkeley, California: Sauer Papers, Carl O. Sauer to E. K. Janaki Ammal, letter dated 21 January 1955.

27 BHL, UoM: University Herbarium Records (University of Michigan) H. H. Bartlett series, Box 11, H. H. Bartlett to E. K. Janaki Ammal, letter dated 29 January 1955.

28 Ibid., E. K. Janaki Ammal to H. H. Bartlett, letter dated 23 March 1955.

29 Janaki Ammal, 'Introduction to the Subsistence Economy of India', 1956.

30 Ibid., p. 325.

31 Ibid., p. 328.
32 Ibid., pp. 328–329.
33 Thomas (ed.), *Man's Role in Changing the Face of the Earth*, p. 407.
34 Janaki Ammal, 'Introduction to the Subsistence Economy of India', p. 330.
35 Ibid.
36 Thomas (ed.), *Man's Role in Changing the Face of the Earth*, p. 405.
37 BHL, UoM: University Herbarium Records (University of Michigan) H. H. Bartlett series, Box 11, H. H. Bartlett to E. K. Janaki Ammal, letter dated 23 March 1955.
38 Thomas (ed.), *Man's Role in Changing the Face of the Earth*, pp. 931–932.
39 BHL, UoM: University Herbarium Records (University of Michigan) H. H. Bartlett series, Box 11, H. H. Bartlett to Alexander Spoehr, letter dated 13 July 1955.
40 Ibid.
41 Ibid., H. H. Bartlett to E. K. Janaki Ammal, letter dated 7 September 1955.
42 Ibid., Ralph S. Peer to H. H. Bartlett, letter dated 25 November 1955.
43 Ibid., H. H. Bartlett to Ralph S. Peer, letter dated 1 December 1955.
44 Ibid., H. H. Bartlett to E. K. Janaki Ammal, letter dated 10 April 1956.
45 Ibid., letter dated 6 December 1955.
46 Ibid., H. H. Bartlett to Roland Welch, The Director, Visa Office, Washington, letter dated 6 December 1955.
47 Ibid., E. K. Janaki Ammal to H. H. Bartlett, letter dated 15 December 1955.
48 Ibid., A. Spoehr to H. H. Bartlett, letter dated 13 January 1956.
49 Wellcome Library: Haldane Papers, 5/1/4/127, J. B. S. Haldane to E. K. Janaki Ammal, letter dated 3 October 1955.
50 Janaki Ammal, 'Introduction to the subsistence economy of India', p. 328.

21

KANDY

The Humid Tropics

Man has taken over the work of nature and can now produce changes in the genetic make-up of plants by inducing
polyploidy through such drugs as colchicine and by X-rays.
He will find in the humid forests of South Asia plants that
have remained static for 100 million years with which to
experiment.

—E. K. Janaki Ammal (1956)[1]

The eighth session of the General Conference of UNESCO held at Montevi-
deo in 1954 had taken a decision 'to promote the co-ordination of research
on scientific problems relating *inter alia* to the humid tropical zone and to
promote international or regional measures to expand such research.' As a
consequence of this, a symposium on the 'Study of Tropical Vegetation' was
jointly organised at Kandy in 1956 by UNESCO's Science Co-Operation
Offices in South Asia (New Delhi) and Southeast Asia (Jakarta), in collaboration with the National Commission of Ceylon for UNESCO. In all
twenty-six specialists participated from Australia, North Borneo, Ceylon,
France, Indonesia, Malaya, the Netherlands, Pakistan, Sarawak, the United
Kingdom, the United States, Vietnam and India, Janaki was again the sole
woman delegate. At the inauguration of the symposium on 19 March 1956
in the ballroom of the Queen's Hotel in Kandy, H. Jinadasa, Permanent
Secretary to the Ministry of Education, Government of Ceylon, invited the
attention of the experts to a couple of issues facing Ceylon, namely the
threatened extinction of the elephant and the havoc caused by the 'cancerous weed, Salvinia [molesta]' in the waterbodies of that country.[2]

The papers contributed were of two sorts: general reports containing fac-
tual data as per the UNESCO directive, and special papers on subjects of the
authors' choice. Like the Wenner Gren symposium, the papers were circulated and read beforehand, with presenters permitted time only to introduce
the subjects and review them; this was followed by questions and discussions. The subsequent part of the symposium was devoted to certain themes

392

DOI: 10.4324/9781003267089-21

selected by the Steering Committee, with each theme introduced, discussed and summed up by a selected expert. The organising committee claimed that the symposium 'demonstrated yet again the unifying influence of science, which admits of no distinction on the basis of colour, race, language or political views—an object lesson of no mean value.'[3] From India, besides Janaki, those who presented papers included Bharucha, Director of the Institute of Science, Bombay, D. Chatterjee, Superintendent, Indian Botanic Garden, Calcutta, R. Misra, Head of the Department of Botany, Banaras Hindu University, and G. S. Puri, Regional Botanist, BSI (Poona). At Kandy, were also present George Kuriyan, Professor of Geography, University of Madras, representing the International Geographical Union, besides the UNESCO coordinators, P. A. Varughese of its Department of Natural Sciences in Paris (of which Joseph Needham, the British biologist, was head until April 1948) and V. P. Kundra of the Delhi Office.

Cytotaxonomy of Humid Forest Flora

Janaki's symposium paper titled 'An Introduction to the Genetical Analysis of the Humid Forest Flora of South Asia,' fell into the special-papers category, where themes were chosen by the authors themselves. Her paper was in a way a response to the conclusions of the Fourth World Forestry Congress held at Dehra Dun in December 1954, wherein the future policy with respect to the conservation of natural resources, in particular forests, was discussed, the summary report of which was published in the *Indian Forester*.[4] At that meeting, resolutions on conservation, forest management and education, and a system of classifying forest land had also been passed. Janaki's major critique of the policy was that but for suggesting the use of improved strains by selection in breeding, there was no emphasis on the study of forest flora from a genetic or an evolutionary perspective. 'This seems rather surprising in an age when departments of horticulture and agriculture are utilizing the latest discoveries in genetics and cytology and even atomic energy [after all she had just been to the Oak Ridge National Laboratory to study tracer atom techniques for use in plant breeding] for the understanding of plant life processes and for the improving of crops by breeding,' she remarked. The genetics of forest trees had by this time attracted much attention in America and the Scandinavian countries.[5]

Janaki observed that the studies of rain forests in South Asia were being led to a good extent by 'principles and policies' relevant to the temperate regions of the world, where the flora (i.e. the coming together of individual species) exhibited far less diversity than in the tropics. In fact, foresters in Assam were promoting a plantation approach to cultivation, by conscientiously doing away with mixed cropping that was so natural to the region, 'instead of learning to perfect the technique of co-existence in different species which is normal for the type of forests in this region,' she stated, citing a recent paper by the Assam forester P. D. Stracey in the *Indian Forester*.[6] These reasons drove her to choose as her topic for the Kandy symposium a critical study of the genetic analysis of the flora of South Asia, clearly revealing once again her archipelagic thinking: the flora of South Asia she observed was really 'part of the great Indo-Malayan rain forests, which extend from Ceylon and Western India, Assam, Burma to Thailand, Indo-China, the Philippines, Sumatra, Java and Borneo to New Guinea,' the humid forests of the tropics. Janaki considered H. G. Champion's classification of Indian forest types 'a great milestone in the study of Indian forests'[7] but her paper was a plea for the consideration of humid forest studies as a special discipline, important not only for an understanding of the origin and evolution of tropical forests but also for their future development and management.[8]

Janaki argued for a 'genecology of forest flora,' which was a nod to Göte Turesson, who made a distinction between the study of species-ecology and the ecology of an individual organism. A quintessential border practice, genecology was a synthesis of the genetical, ecological and taxonomical approaches, in which a species was understood as 'a genetically complex community, the distribution and the composition of which is largely determined by the ecological factors and the genotypical constitution of the individuals composing the species community.'[9] Ecology and genetics had received much attention at the Fifth British Empire Forestry Conference (1947) held in London, at which her friend N. L. Bor had presented a paper titled 'Tropical Ecology and Research.' It doesn't appear that Janaki attended the conference, although she was in England at this time. The session on genetics had been organised with the assistance of M. B. Crane of the John Innes, P. S. Hudson of the Imperial Bureau of Plant Breeding and Genetics and Richards of Cambridge University; incidentally, it may be recalled that Hudson had visited Coimbatore in 1933 and spoken about Vavilov under the aegis of the Association of Economic Biologists. The overall argument was that genetic and silviculture were not opposed to each other and ought to be carried out along parallel lines.

A flora was more than a coming together of individual species; it was an association deeply rooted in the physiology and the soil requirements of the plants. They not only influenced the place they inhabited but also influenced each other. The association was also contingent on the migration of plants, which in turn was determined by reproductive efficiency and the adaptability to travel and withstand competition. Janaki argued that a study of the genetic composition of a given flora would not only shed light on all these characteristics but explain diversity in plants. 'An analysis of a mixed forest becomes even more significant when the plants composing an association are analysed cytologically . . . The present flora of South and South-East Tropical Asia is therefore worthy of closer examination to study the processes whereby species, genera and families have changed in the course of ages,' she stated. The aim of taxonomy, she reiterated, was to unravel the evolution or the true race history of plants in the Darwinian sense, beginning with the primitive and moving on to the more developed; a species in other words was not a fixed entity. Janaki was dismayed to see several systematic botanists still resorting only to morphological characters to distinguish between species even a century after the *Origin of Species* was published. Quoting Darlington, she stated, as the carriers of genes, the chromosomes 'mark many of the steps by which species, genera and even families have diverged.'[10]

The study of phenomena like polyploidy was of particular significance in this context. Diploid plants were in most cases parents of the polyploids and situated at a lower rung of the evolutionary ladder. The living relic species of Magnolias such as *M. nitida* (with several fossil forms), found in the evergreen forests of Southeast Asia, were diploids. Adopting a Humboldtian perspective, and the use of conceptual tools such as vegetation maps that revealed the interconnectedness, Janaki argued that an increase in the knowledge of chromosome numbers of tropical forest trees would not only help map the extent of diploidy in the evergreen forest flora but also throw up useful information on the chromosomal relationship between different families of woody plants, which formed a major part of the tropical forest flora. Janaki observed that in the associations present in humid forests, there was a conspicuous gradation in the families that composed the different tiers of the flora: in evolutionary terms, the highest canopy comprised of the 'primitive' genera such as the Magnoliaceae and Moraceae (she had studied the two families very closely), while the middle tier chiefly of climbers and lianas like the Apocynaceae assumed an intermediary position on the evolutionary scale, and the bottom-most tier of such plants as *Strobilanthes* were the most evolved of those belonging to the Gamopetalae.[11]

Humid forests had at one time, according to Janaki, been 'great centres of speciation by hybridization,' but had now become somewhat static, with a population comprised of secondary diploids: drastic climatic changes caused by the genesis of the Indian Ocean and the resultant southwest monsoon had given rise to extraordinary climatic variations, which in turn had led to discontinuity in the distribution of humid forest types in India. She suggested that there was an affinity between the African forest flora and that of India, by virtue of being connected in an earlier geological epoch. Their breaking away from each other was more recent, she speculated, as several of the commonly found plants in Western India were also found in Africa, but despite the interesting scenario hardly any cytogenetical work on a comparative scale had yet been undertaken. Yet again, this was her archipelagic thinking at work. Janaki hoped the BSI would take this up seriously, and that UNESCO would support it wholeheartedly, in particular the study of the flora of the Mascarene Islands (or Mascarenhas Archipelago), Janaki's own Galapagos, a group of islands in the Indian Ocean east of Madagascar and consisting of Mauritius, Réunion and Rodrigues, which displayed a high rate of speciation.[12]

In addition, she observed that some genera pointed to a link with South America, such as the *Buddleia*; the evolutionary migration in this case was from Africa towards the east and west. The genus remained diploid in the entire region of Africa but underwent vigorous speciation and high polyploidy when it reached humid and temperate regions such as the Eastern Himalaya. The same was demonstrated by the *Camellia*, *Magnolia* and *Rhododendron*, which remained diploids in Malaysia but displayed polyploidy in the cold and humid high elevations of Sikkim, Assam and Upper Burma. 'Man has taken over the work of nature and can now produce changes in the genetic make-up of plants by inducing polyploidy through such drugs as colchicine and by X-rays. He will find in the humid forests of South Asia plants that have remained static for 100 million years with which to experiment,' she would remark in conclusion.[13] She believed that conditions in the humid tropics were more uniform than in the temperate arid regions, for which reason the incidence of mutations was negligible.[14]

At the end of the discussion on her paper, she stated that palaeontological evidence had demonstrated the antiquity of humid forest types, making it very important to undertake their study from the evolutionary point of view. She pointed out that very little cytological research had been done on humid tropical flora and the 'little work' she had done revealed that the large trees that composed the upper canopy of the humid forests were diploids—that is, 'they represented primitive types.' Studies on tropical vegetation 'from a genetic and cytological angle' was a serious desideratum, she said, for the 'association of plants in the humid tropics is more than ecological. It is also evolutionary—as the same association found today has existed for millions of years and is responsible for the speciation within the habitat.' Janaki claimed that polyploidy in the humid tropics was rare given the uniform and rigid conditions under which plants grew, for 'natural selection would weed out all forms that cannot adapt to this unchanging environment.'[15]

Conservation of Forests

As the Kandy symposium came to a close, the UNESCO Director General called for a preparatory meeting of experts on the humid tropics, to be held between 22 and 24 March 1956, which included Janaki, the aim of which was 'to consider matters pertaining to research in the fundamental aspects of the natural sciences, of interest in the humid tropical zones; to stimulate the carrying on of such research, to assist it, and in cases where important problems are not at present receiving adequate attention, to initiate or promote the necessary research.' This meeting took place in Kandy again (hereafter referred to as the Kandy meeting) and the papers presented were published by UNESCO in a volume titled *Problems of Humid Tropical Regions* (1958). Janaki's focus on this occasion was on the ecological conservation of forests in the humid tropics; she was perhaps the first Asian scientist to think along these lines. Among the recommendations of the Kandy symposium was the conducting of cytological surveys of the forest flora so that 'the evolutionary tendencies may be brought to light.' Further, alarmed by the fast disappearance of several types of tropical vegetation, it recommended that all governments of tropical countries undertake as a matter of urgency, the formation of natural reserves, and the study of vegetation that was fast depleting or becoming extinct.[16] Both of these were Janaki's suggestions.

At the Kandy meeting, Janaki's focus was on the humid regions of South Asia, primarily the forests of the region, which included India, Pakistan, Burma and Ceylon, and formed part of the Great Indo-Malayan Rain Forest. The growing pressure on the natural resources in this region, and the demand to provide food for the teeming millions, had depleted vast stretches of forests to make way for agriculture. Janaki had after all been a direct witness to the Grow More Food campaign in India in the late 1940s. She spoke of how it had caused 'incalculable damage to forests,' particularly in the Nilgiris and Assam, the consequence of which were annual floods in the plains, including in contemporary times. Reflecting on economic and ecological history, Janaki stated that 'a just balance between land utilized for cash crops and internal consumption' had to be attained, an aspect that had been ignored in the past years, driven by the needs of industry and the export market; she was clearly driving home Sauer's point that the morality of capitalism was yet to distinguish between 'yield' and 'loot.'[17]

Her friend Bor, Janaki noted, as early as 1947, had drawn attention to the want of planned research in tackling problems connected to tropical ecology.[18] Bor's paper had suggested the founding of a research institute in the tropics for 'the application of modern research methods to forestry, agriculture and land use in Colonies and overseas Dominions.' The biggest misuse of land had occurred because it was the administrator who took decisions, rather than the scientist, Bor had rightly noted. To remedy this, Bor recommended that young administrators be given a sound education in ecology 'in its broadest sense, with emphasis on the catastrophic effect which may follow a mistaken policy in regards to land use.'[19] Janaki reiterated that the study of humid forests was a serious desideratum and that, as a matter of urgency, botanists of the world should work on the living flora of the tropical rain forests, which were rapidly undergoing transformation at the hands of man. Her own presentation provided a floristic composition of the humid forests of South Asia, based on E. P. Stebbing's classification of 1922. She pointed to the importance of exploring from the point of view of ecology, the connection between soil and vegetation in tropical forests, and the study of the people of the forests such as the Kadars of the Anaimalais, the

Pannyar Kuruchiars of Wayanad and the Veddhas of Ceylon, all of whom were food gatherers and used wooden digging sticks to uproot wild tubers. Janaki was convinced that these forest dwellers did 'not upset the biological ecology of the forests they occup[ied]. As a food gatherer, man [was] here part of the biotic community of forests.' She stated:

> It is true that he digs for roots, but the damage done in humid luxurious forests is not great. It is true he hunts, but he hunts for his living and not for export. His family is small because he does not reproduce as prolifically as urban dwellers, the time of nursing being prolonged until children are able to digest roots and meat.[20]

Janaki argued that it was only when forest tribes took to cultivation that ecological balance began to be disturbed. It was to the uncontrolled shifting cultivation practices by certain forest tribes in Assam and the Irulas of the Western Ghats that she attributed 'the changeover from the climax' (deterrent to producing a stable community of vegetation or a state of equilibrium); she also held responsible the selective harvesting of certain trees for food and the making of primitive tools—such as the *Caryota urens* for extracting starch and making the digging stick, and *Artocarpus hirsutus* for making dugouts and canoes. Thinking archipelagically, she wondered whether there were similarities in ways of life among tribes across the humid forests of the tropics that had missed the attention of anthropologists.

Any study investigating the ecology of humid forests, Janaki argued, should also undertake a macro-level study of 'primitive man in humid forests and his relationships with the fauna and flora of tropical forests.' According to her, man's role in changing the biotic composition of humid forests occurred through 'two groups of processes.' The first group of processes (called the pioneer effect by Janaki) comprised the changes brought about by man's exploitation of forest trees for timber, firewood or charcoal, activities limited to the periphery of the forests rather than the inner forest region, and without any concern about their availability in the future. The second was the effect stemming from intensive land use by man, such as the establishment of plantations of rubber, tea and coffee in the *shola* forests, or driving animals to graze inside forests as practised in Malabar, activities occurring in the inner forests which led to deforestation and soil erosion. These activities brought about 'a shift in the species composition both in number and quality,' she noted, and more importantly, by breaking up the forest canopy, altered the microflora of the inner forests. Janaki cited the example of pests such as *Lantana*, which had the ability to take over cleared areas in forests and irrevocably change the very profile of a forest.[21]

The traditional practice of setting apart a section of the forest as sacred, as in Assam and Malabar, Janaki noted, contributed enormously to keeping vegetation unaltered. Referring specifically to Malabar, where the sacred

groves are called *kavus*, she continued from her presentation at the Wenner Gren Symposium, she stated: 'They serve as sanctuaries for snakes and
wild life and often become the nucleus of sylvan temples at a later stage.'[22]
Undoubtedly influenced by Sauer, she highlighted at the Kandy meeting the
virtues of self-supporting forms of cultivation, with mixed-cropping, which
she called the *paramba* style, as practised by the indigenous people, especially in the hilly tracts of the southwest coasts of India and Ceylon. This
form of cultivation demanded only a limited forest clearance, with useful
trees in the clearing retained for timber, firewood or as supports for the
growing of pepper vine and other useful climbers; the spaces between trees
were also put to good use, to crops such as ginger, turmeric, yams, bananas
and cassava. The garden vegetables were grown closer to the house and
rice cultivated in the low-lying wetlands. In fact, the Edam *paramba* was
organised in this very manner. A cultivation style practised in the humid
regions from very early times, Janaki found the *paramba* approach closest
to achieving ecological balance in nature, an idea that had struck her as
early as 1932. To her, *parambas* were self-sustaining economic units, 'nearest to nature' and the fact that they had been present in the humid regions
from very early times demonstrated their relevance and significance. The
demand for cash crops such as rubber, pepper, tea, coffee and bananas for
the export market had given rise to the monocropping or plantation style of
cultivation, which soon began to replace the *paramba* style mixed-cropping
cultivation. This shift had led to widespread destruction of useful trees and
consequent soil erosion, on an unprecedented scale (in the Anaimalais, for
instance, where tapioca was cultivated on a plantation basis, and in the Nilgiris, potatoes), altering the very landscape of these regions.[23]

Discussing future plans for the study of humid forests of South Asia,
Janaki stated categorically that it was the duty of every botanist to save all
the plant species in their region from extinction, without exception. She
suggested that the practice of maintaining sacred groves could be undertaken on a scientific basis by creating sanctuaries or protected forests in
different parts of South Asia. Janaki stressed the need to evolve a consistent method for the study of humid forests across the region, much like
Bor had suggested, and more importantly the need to conduct cytological
studies alongside systematic and ecological ones to map the genetic evolution of the floras of humid forests; she was particular that ecological
studies should explore the interconnections between flora, fauna and the
forest-dwelling aboriginals or tribes. The founding of an international centre for chromosome studies, that would facilitate an exchange of slides,
plant materials and research findings, Janaki added, would be crucial for
the study of the floras of humid forests. She also suggested a compilation of
vegetation maps showing the chief kinds of plant associations to be found
in these humid forests, and the use of aerial photography to achieve this
end.[24]

Notes

1 E. K. Janaki Ammal, 'An introduction to the genetical analysis of the humid forest flora of South Asia,' in UNESCO, *Study of Tropical Vegetation*, p. 139.
2 Foreword to *Study of Tropical Vegetation*.
3 Ibid., pp. 11–12.
4 *Indian Forester* 81, no. 5 (May 1955): 296–299.
5 Janaki Ammal, 'An introduction to the genetical analysis of the humid forest flora of South Asia', p. 137.
6 *Indian Forester* 80, no. 12 (December 1954): 759–767.
7 Champion, 'A preliminary survey of the forest types of Indian and Burma.'
8 Janaki Ammal, 'An introduction to the genetical analysis of the humid forest flora of South Asia', p. 137.
9 Turesson, 'The scope and import of genecology'.
10 Janaki Ammal, 'An introduction to the genetical analysis of the humid forest flora of South Asia', p. 138.
11 An identification group as per Hooker and Bentham's classification system based on key morphological characteristics of plants (in this case, a group of plants whose flowers show distinct calyx and corolla) and in contrast to a system based on evolutionary relationships between plants.
12 Janaki Ammal, 'An introduction to the genetical analysis of the humid forest flora of South Asia', 1958, p. 139.
13 Ibid.
14 Ibid., p. 143.
15 Ibid.
16 *Study of Tropical Vegetation*, p. 222.
17 Sauer, 'Theme of plant and animal destruction in economic history', p. 775.
18 Bor, 'Tropical ecology and research'.
19 'Fifth British Empire Forestry Conference, 1947: Review of Papers submitted.' *The Empire Forestry Review* 27, no. 1 (July 1948): 83–128, pp. 120–121.
20 Janaki Ammal, 'Report on the humid regions of South Asia', 1958, p. 49.
21 Ibid., pp. 49–50.
22 Ibid., p. 50
23 Ibid.
24 Ibid., pp. 51–52.

22

LUCKNOW–ALLAHABAD
The Central Botanical Laboratory

I am very 'homesick' for England. . . . I need quiet to work.
Here I am chiefly, an administrator and a writer of Reports.
—E. K. Janaki Ammal (1957)[1]

At Lucknow, space had posed a considerable challenge for the proper functioning of the CBL, particularly the absence of grounds for conducting field experiments; Janaki had to resort to growing plants in pots placed along the verandahs and wherever else possible. The government's search for a permanent abode for the CBL ended when the Old Commissioner's Office building in Allahabad (the birthplace of Jawaharlal Nehru) located within 7-acre grounds ticked all the boxes. Plans would be hastily drawn up for a new building, but nothing would materialise during Janaki's tenure as Director.

The chief activities of the CBL under Janaki's directorship, over the five-year period from the time of its inception in 1954 until 1959, revolved around realising the unfinished Vavilovian project among other aims; she was clearly trying to accomplish something of what Vavilov would have, had he been allowed to travel in India in the 1930s. Her aims were also clearly a reflection of her archipelagic thinking as earlier noted, of her great affinity for the Western Ghats and Assam, her passion for plant breeding and introduction and of her ambition to publish a new chromosome atlas of flowering plants. Her list of planned activities was an ambitious one: creation of new kinds of vegetables, fruits, trees and medicinal plants through polyploidy; survey of Indian waterlilies, these plants being indicators of early land connections; search for new economic plants such as those cultivated by indigenous tribes, like the *Dioscorea* (at Allahabad, she gradually built up a large collection of *Dioscorea* from Malabar, which would be cytologically examined by her assistant R. Sundara Raghavan)[2], *Colocasia* and *Curcuma*; cyto-taxonomic surveys of medicinal plants[3]; compiling a chromosome atlas of Indian flora from published material; studying the genetical composition of different vegetative types such as mangroves; assembling a collection of the useful grasses of India with a view to their improvement;

DOI: 10.4324/9781003267089-22

work on tuberous plants (Janaki would play an important part in introducing the tapioca to Uttar Pradesh); studies on endemic orchids of Assam; survey of *Lantana* species; polyploidy of the medicinal plant *Rauwolfia serpentina* (reserpine, originally isolated from it in 1952, was used in treating high blood pressure and psychotic disorders); work on radioisotopes; and lastly, a survey of the flora of the monazite sands of Kerala.[4]

Let Knowledge Grow

It was in Allahabad, at the Bhola Nath Book Binders in Katra, that Janaki had her growing collection of books re-bound, the spine of each imprinted in gold with the words 'Crescat Scientia Vita Excolatur' (Let knowledge grow, and so let life be enriched). Among the books she acquired in the late 1950s were controversial demographer Sripati Chandrasekhar's *Hungry People and Empty Lands: An Essay on Population Problems and International Tensions* (1954, 1956) and F. E. Zeuner's *The Pleistocene Period* (1959). The former was of immense interest to her, given her deep anxiety about India's over-population (Chandrasekhar however advocated such measures as forced sterilization and easy abortions), while the latter, by considering such aspects as ancient land connections and palaeo-climates, besides changes in the position of oceans, mountains and continents, enriched her cytogeographical and historical framework.

Indian Symposium on Breeding, 1957

The Indian Society of Genetics and Plant Breeding, at its annual general meeting in December 1955, had decided to organise a symposium on 'Genetics and Plant Breeding in South Asia.' It was to be funded by UNESCO and held in New Delhi between 21 and 24 January 1957. A committee was set up to organise the symposium, with the wheat geneticist B. P. Pal as Chairman and M. S. Swaminathan as Secretary, along with members W. J. Ellis (UNESCO, New Delhi), S. M. Sikka, V. M. Chavan, T. R. Mehta and J. S. Patel. The experts invited from outside the country were Åke Gustafsson of the Swedish Institute of Forest Research (referred to as the father of mutation breeding); Otto Frankel, who was among the earliest to warn of genetic erosion (loss of plant biodiversity); and G. L. Stebbins, whose book *Variation and Evolution in Plants* (1950) was considered a keystone in the study of modern evolutionary synthesis. In his inaugural address, Stebbins spoke on the uses of plant breeding to increase the world's food supply. Among those from India presenting papers were K. Ramiah on rice breeding, Pal on wheat, A. Abraham of Travancore on tubers, R. H. Richharia on oil seeds, and N. L. Dutt and R. R. Panje on the sugarcane. The section on cytogenetics saw two presentations: Gustafsson spoke on mutation work in Sweden in relation to plant breeding and Swaminathan on polyploidy and sensitivity to mutagens. Three other sections, on genetics, plant introduction and plant physiology, were represented by such botanists as V. G. Panse, L. S. S. Kumar, R. D. Asana, S. M. Sircar and P. Maheshwari. Otto Frankel presented a paper on the genetical basis of plant introduction, concerning Australia in particular; Frankel advised that a new crop plant should be introduced always ensuring a maximum range of variation.[5]

Surprisingly, Janaki did not attend the symposium. Incidentally, she would be elected Fellow of the Indian National Science Academy that year (1957), the first woman to be conferred this honour, and one that was long overdue.

Indian Taxonomist at Kew

In mid-1956, the Government of India had sought Janaki's advice on the subject of reviving the practice of appointing an Indian taxonomist attached to the Kew Gardens, to work on the collections from the subcontinent. Janaki promptly wrote to her friend, N. L. Bor at Kew, wishing to gather from him, in an unofficial capacity, such details as when the post had been first created, the names of all those who had occupied the position and the remuneration they had received. As far as she knew, K. N. Kaul was the first and the last to occupy the post, for D. Chatterjee who went to Kew after Kaul (in 1948–49), was referred to as 'Indian Visitor to Kew.'[6] Janaki also wished to know from Bor what qualifications were expected of the person if Kew were to welcome him/her. Having gathered that the cost of living had risen in London from when she was last there, she was keen to ensure that the new appointee received sufficient remuneration, commensurate with what other Commonwealth countries were paying their own taxonomists at Kew.[7] We do not however know if she heard from Bor on this, but she would write to him again several months later, but this time, it was for a packet of seeds of the *Ornithogalum virens*, a model organism for the plant cytologist. The chromosomes of this plant stained easily and, being few in number ($2n = 6$) and large, could be viewed clearly under the microscope.[8] Incidentally, at this time, Bor's wife Eleanor was suffering greatly (or may even have already passed) from a 'distressing' ailment; on her death her ashes were scattered in the Azalea garden at Kew, as she had desired.

Homesick for England

True to her nomadic nature, Janaki was restless again. Darlington had rightly observed years earlier that she was not one to stick to a place too long. Time and place to do research was far more important to her than the money and power a high-ranking official position brought with it. She longed to go away to England, desperate to work on a new version of the *Chromosome Atlas*, a publication she hoped would outdo the one by Darlington and Wylie. Janaki wrote to dear friend Col. Stern from Lucknow: 'This is to tell you that I am very "homesick" for England. It is 5 years since I returned to India and officially I am very well-placed but I long to see my "New Chromosome Atlas" published and for that I need quiet to work. Here I am chiefly, an administrator and a writer of Reports.' She hoped the John Innes would accommodate her for two or three years. 'I am very keen to make my Atlas—a cytogeographical one with maps. It will be very different from the II edition by CDD and Wylie—it will have to be!' she declared emphatically. She had thrown a challenge at Darlington and was determined to prove herself to the world. Janaki begged Stern to find her a place to work on the *Atlas*. 'My alternative to England is Michigan U.S.A. but somehow I love England better—after the years I spent there,' she told him. She was also curious to know from him if her tetraploid Magnolias and the tetraploid *Rhododendron yakushimanum* (which had given her immense joy) at Wisley had flowered that year.[9]

Janaki also took this opportunity to enlighten Stern about her home state of Kerala, which had recently 'gone communist . . . there is a fear amongst people that they might lose their land,' she remarked.[10] Malabar (especially Cannanore and Palghat), where she hailed from, was the heartland of communism in Kerala. The government of Kerala, formed under the leadership of E. M. S. Namboodiripad in April 1957, was one of the first democratically elected Communist governments of the world. It made a mark primarily for the introduction of the Land Reform Ordinance and the Education Bill; the Minister for Law and Education (besides power, irrigation, prisons, justice and home) was the leading lawyer V. R. Krishna Iyer, who not only had his early education in Tellicherry but also began his legal career there in 1938. When the government was dismissed by the Centre in 1959, Iyer resumed his legal profession there and became a tenant of Devi Nivas, uphill from Edam, and owned by Janaki's family. Incidentally, Namboodiripad's cabinet of ministers included K. R. Gowri Amma (1919–2021), the first woman from the Ezhava community (which corresponded to the Thiyas of Malabar) to graduate in law, as the Minister for Revenue, Excise and Devaswom, in whose hands lay the primary responsibility of executing the land reforms.

Stern's Model Monograph

A copy of Stern's *Snow-Drops and Snow-Flakes* found Janaki in Allahabad in November 1957. She was exceedingly impressed by the publication: 'Snow-Drops and Snow-Flakes' reached me yesterday and I have not been able to put it down. I even took it to bed with me last night. It is superb and I wish the RHS will take it as a model for other Monographs.' She yearned to produce a work of that nature. 'You are creating or shall I say *have* created a new type of treatise of garden plants—wedding plant geography and Cytology to Taxonomy and Plant History. It is fascinating reading! I thank you for the gift,' she wrote to Stern. Unable to pursue her own researches, and weighed down by the daunting task of erecting a well-equipped cytological labora-tory, while also negotiating 'the red tape associated with getting land and plans etc.,' she was feeling out of sorts. 'I would have been happier working on the many interesting problems relating to plant cytogeography of India. I have a good assistant [S. Raghavan or perhaps K. M. Sebastine][11] but that is not the same,' she remarked to Stern. Janaki was also feeling unwell physi-cally, with her joints becoming increasingly painful and stiff; she wondered if she would 'ever be able to climb a hill again.' Fondly recalling her visit to Stern's Chalk Garden in the company of Darlington a decade before, Janaki wrote: 'I look back with pleasure to the weekend spent at Highdown.'[12]

Darlington's Plan to Visit India

Darlington was at this time planning a visit to India and expecting to be in Lucknow between 26 and 31 December that year (1957), coinciding with the CBL's move to its new home in Allahabad. He was to arrive with wife Gwendolen, but the plan would change and he would travel alone. Inciden-tally, by this time, common friends Haldane and his partner Helen Spurway had been a few months in the country, based at the Indian Statistical Insti-tute, Calcutta (established in 1931).[13] They had arrived in July 1957, at the invitation of P. C. Mahalanobis (1893–1972), founder-director of the ISI and the chief architect of India's Second Five-Year Plan (1956–61).

India's Food Crisis and the Five-Year Plans

In the First Five-Year Plan (1951–56), the government had focused on increasing agricultural productivity and implementing land reforms, but met with limited success; in the Second Five-Year Plan (1956–61), the slant was towards the promotion of large-scale industrialisation so as to put an end to dependence on imports. The intensification of agriculture was crucial to the development of industry, but large landowners' disinclination to invest in better technology led to persistent low-yields and increasing dependency on imported grains. A team from the US Department of Agriculture, organised by the Ford Foundation, produced a report, *India's Food Crisis and Steps to Meet It* (1959), which would play a salient role in repositioning India's agrarian strategy from one based on social reform to that of state-of-the-art technology. The report warned against adopting ideas that promoted cooperative agriculture (as in Socialist countries) and recommended agricul-ture as a means to finance industrialisation. Farmers were persuaded to go in for capital-intensive measures: the use improved seeds, better fertilisers, pesticides, better equipment and machinery, credit and scientific advice, all to ensure higher yields to feed the rising population. The Ford Foundation Report thus led to the birth of the Intensive Agricultural District Programme

(IADP), which would in turn pave the way for the so-called green revolution of the 1960s. The IADP could not claim total success though, for despite increased yields in certain districts, the country still remained dependent on imported foodgrains.[14]

The Baroda Meeting, 1958

On 29 December 1958, Janaki attended the 24th Annual Meeting of the Indian Academy of Sciences at Baroda, where she spoke on cytogeography of some Indian plants. 'The Plant Geographer is concerned not only with the distribution of plants in Space but also in Time,' she stated at the very outset of her presentation: an indication of her definitive turn to deep history, or paleobotany. She went on to explain that it was subsequent to the evolution of flowering plants called angiosperms, which first appeared on earth about 130 million years ago (in the Cretaceous period), that the supercontinent Gondwana had split up, resulting in the Gondwanan medley of plant distribution. Elucidating what we refer to as her archipelagic thinking, Janaki stated that the scattering of related families and genera across South America, Africa, Australia, Arabia and peninsular India necessitated the study of the cytology of certain genera of Indian plants and the occurrence of polyploidy in them alongside similar plants in these other continents. Her research revealed *Magnolia* and *Viburnum* to be diploids in the Northern part of Asia, but polyploidal in the Eastern Himalaya. Among other examples, Janaki cited the case of the diploid *Nymphaea* (the *Nymphaea stellata* of Kerala was closely linked with the *Nymphaea* of Madagascar and East Africa, she observed), which appeared as hexaploids in the Gangetic Plain, Egypt and Europe. The genus *Buddleia* demonstrated something similar: while it appeared as a diploid in Africa, where it perhaps originated according to Janaki, it exhibited high polyploidy in India. As for the Magnolias, whose fossils had been discovered, they had been diploids for a very long period, similar to some colossal trees of the humid tropics of South Asia.[15] The flora of the humid forests, she observed, were therefore similar to the vegetation that thrived on earth in the Cretaceous period, which means that there was something unchanging about them.

On Triploidy

It was from the cytogeographic angle that Janaki also approached the genus *Philadelphus*, comparing the old and the new world species, at the CBL, Allahabad; we may recall that she had worked on the genus at Wisley, in 1950–51, at the suggestion of Gilmour.[16] Her objective at that point was to trace the history of the cultivated *Philadelphus* and report the occurrence of triploids among the offspring of garden hybrids, 'between species widely separated in nature.' Janaki's focus at the present time was in particular on the chromosome morphology and behaviour of the triploid *Philadel-phus* across space and time. The best known among the triploids, 'Belle Etoile' (2n = 39), had as its grandparents the European species *P. coronar-ius* (2n = 26) and the scented Arizonian species *P. microphyllus* (2n = 26). *P. lemoinei* (2n = 26), the hybrid produced from this combination when crossed with the Mexican species *P. coulteri* (2n = 26), resulted in the highly fragrant and purple-tinted *P. purpureo-maculatus* (2n = 26). All *Philadel-phus* species in nature were found to be diploids (2n = 26).

The triploids were in general more fragrant and larger in size compared to the diploids, and were first discovered among offspring of the tri-specific (European, Arizonian and Mexican) hybrid *P. purpureo-maculatus*. Janaki made some cytological discoveries with respect to the origin of triploidy in the *Philadelphus*. She attributed its triploidy to certain anomalous events that occurred during meiosis; the tenacity of the iso-chromosome in the chromosome complex of the genus, which boasted of several species, so far removed from each other, she argued, yielded a useful comparison of the role of the iso-chromosome in such plants as the monotypic genus *Nican-dra*, the subject of her doctoral dissertation.[17] She also found that pollen fertility of a species and its hybrids was related to the behaviour of the iso-chromosome. Incidentally, in 1945, Janaki and Darlington had discovered that in plants with a single iso-chromosome, germination was delayed.[18]

Triploids were not only valuable to the horticulturist and agriculturist but also useful in creating hexaploids. Hexaploids, Janaki pointed out, were to be found in places, where the distribution of diploid species converged, either owing to migration or changes in the physical geography of the land, as exhibited by the *Magnolia*, *Camellia* and *Rhododendron*. The material for the study had been drawn from the *Philadelphus* collections at the RBG, Kew, and thus Janaki had to restrict her study of the evolutionary significance of the iso-chromosome in the genus for 'geographical reasons' (she had to be in Eng-land to further this line of research, but this was not possible immediately) to mapping its behaviour and persistence in successive generations of diploids.[19]

Connections, Similitudes and Border Zones

The Indian Science Congress of 1959 (the forty-sixth session since its incep-tion in 1914) was held in New Delhi in early January. A. L. Mudaliar was General President and the theme of the Congress was the basic sciences. Janaki and J. C. Sengupta represented the BSI at the Congress, which was inaugurated by Jawaharlal Nehru; the welcome address delivered by V. K. R. V. Rao of the University of Delhi. Among the important lectures delivered at the Congress was one on atomic power by Homi J. Bhabha and the spe-cial lectures included one on the origin of maize by American plant geneti-cist Lowell F. Randolph, a close contemporary of Janaki. Janaki took an active and leading role in a discussion on the 'Distribution Pattern of Plants in India,' as her recent researches were all about exploring the chequered evolutionary pathways, the songlines of tropical flora, by uniting plant geography and plant history with cyto-taxonomy, stringing together local singularities, and then federating or synthesising this knowledge to map the global picture, or the mosaic of plant distribution. She began by stating that:

> The present distribution of a species, a genus or a family is deter-mined by many factors. While climatic, edaphic [relating to the physical and chemical properties of the soil] and environmental factors act like a sieve to eliminate those plants which are incapable of surviving the environment, we have also to consider larger and more ancient factors that are responsible for the present day distri-bution of certain genera and families of plants. Thus the study of ancient land connections and palaeo-climates and changes in the position of oceans, mountains and continents have to be considered when we try to interpret the mozaic [sic] of plant distribution.[20]

The distribution pattern of plants in India (typically monsoon flora) depended on the degree of rainfall and the length of the intervening dry spell; higher rainfall and shorter dry spells would take the forest type to the climax of the humid tropics. On the strength of her archipelagic thinking, which was essentially transnational, she argued that mapping plant distri-bution demanded that this pattern be linked with those of the surrounding countries, 'as in many instances the distribution of a genus or even species found in India is continued to one or other of the surrounding countries.' Her focus was on connections and similitudes, and border zones, and in scale, both local and planetary. Janaki alluded to the Gondwana landmass and how it had broken up only after the flowering plants had evolved, a fact revealed by the similarities observed in several genera spread across conti-nents such as Africa, South America, Australia and peninsular India. Janaki observed that a study of the flora of the more recent (geologically speaking) Gangetic Plain, revealed interesting data on the migration of plants from the surrounding regions. Of the methods available for the study of the dis-tribution pattern of plants from the evolutionary perspective, genetic analy-sis was the most rewarding, she stated; the relationships between closely related genera could be mapped by working out the chromosome count, chromosomal behaviour and the morphology of one family:

> When this is done for a number of families then the pattern [the mosaic] of distribution becomes an indicator of not only the move-ment of the plant in the region but of the evolutionary movement of genes that go to make up a particular genus within a family. Such a study is as necessary as the study of the influence of the environ-ment on distribution patterns of the plants in India.[21]

Further, sometimes there could be a co-occurrence of floras composed of annual plants, she noted, such as is seen in the northern regions of India, where the dry and hot periods alternate, with the time between marked by a somewhat cold winter. When 'distribution patterns of these annuals are studied in detail, they are found to link the flora of India with the hotter regions like Africa on one side and the colder regions of Siberia and Western Europe,' Janaki observed, but these remained to be fully investigated, especially from the genetic point view.[22]

Cytogeography of Bamboos

As Director of CBL, Allahabad, Janaki turned her attention to grasses briefly by undertaking a cyto-taxonomic study of the Bambusoideae, in particular to the slender bamboos of Asia and South America and their distribution across the world. The distribution of a plant, according to her, was the consequence of certain causes that had acted upon it over long stretches of time, and across geographies. Bamboos were of much interest to her because they possessed certain characteristics of the Gramineae, were considered 'primitive' and demonstrated a 'slow motion picture of the evolution of grasses as a whole.' Through her study, Janaki aimed to unravel the relationship that existed between the systematic characteristics (attributes or observable features of the plant) and the chromosomal characteristics (such as chromosome number, size and behaviour at meiosis) of the different genera of bamboos. The Indian species of slender bamboos were collected from the Nilgiris, Assam, Darjeeling and the Sibpur Botanic Gardens, while the Chinese, Japanese and some rare Himalayan and South American species she had collected from the Kew and Edinburgh Botanic Garden. The Himalayan slender bamboos and the one from South India and South America were being examined for the first ever time together from the cytological point of view and, very importantly, in comparison with the Chinese and Japanese species– quintessentially, a cytogeographical project. Janaki found them all to have the same chromosome number, of $2n = 48$, indicating that they had remained a cytologically stable group.

More Official Posts

One other official event Janaki attended about this time was the 19th Annual General Meeting of the Indian Society of Genetics and Plant Breeding in New Delhi, on 23 January 1959; she was as usual, the sole woman attendee. New office-bearers were elected at the meeting, with R. H. Richharia as President, A. B. Joshi and P. N. Bhaduri as Vice-Presidents, M. S. Swaminathan as Secretary, N. L. Dhawan as Treasurer, and B. P. Pal as editor of the Society's journal.[23] Among the six Councillors elected at the meeting were Janaki (for the Mid-Eastern zone), P. Maheshwari (Northern zone), S. K. Mukherjee (Eastern zone), A. R. Gopal-Ayengar (Central zone), T. R. Mehta (Western zone) and N. R. Bhat (Southern zone). The Presidential Address delivered at the annual meeting of the Society in January 1960 was, by Richharia (based at the Central Rice Research Institute, Cuttack) on the origins of cultivated rices.[24] The Indian Science Congress was held about the same time in Bombay, but Janaki would not attend.

From 20 November 1958 to 20 November 1960, Janaki also served on the Advisory Committee for the Application of Atomic Energy to Food and Agriculture set up by the Department of Atomic Energy, Government of India. Her friend, Gopal-Ayengar was head of the Medical and Biology Division of the Atomic Energy Establishment at this time.

Lab-Based Indian Women Scientists

The 1950s saw a rising number Indian women making a mark in the laboratory-sciences; examples include, bio-chemists Smita P. Bharani of the University of Bombay, Saraswathy Royan of the Indian Institute of Science, and the experimental biologist Susheela Waravdekar of the Cancer Research Institute, Parel. At least three had earned their doctorates—Rajeswari Chatterji in electrical engineering in 1953 (under William G. Dow) from Janaki's alma mater, the University of Michigan, the developmental biologist Leela Mulherkar of Pune in 1956 from the University of Edinburgh (under C. H. Waddington) and the radiation geneticist Shanta V. Iyengar, also in 1956, from Indiana University (under Hermann Muller).

Speaking of chemists, when Janaki arrived in Calcutta in 1952 as Special Duty Officer of the BSI, a much younger contemporary, Asima Chatterjee (1917–2006), who had earned a doctorate in organic chemistry from the Calcutta University in 1944, had already returned to the city after a brief research stint at the University of Wisconsin and László Zeichmeister, Zurich. In 1954, she joined the University College of Science of the University of Calcutta as Reader in Pure Chemistry. However, it does not appear that Janaki and Chatterjee knew or met each other despite living in the same city, and being women scientists with a common interest in Indian medicinal plants. Chatterjee's chief interest however was in the chemistry of plant products, and her practice was almost entirely lab-based; it involved the development of indigenous drugs from alkaloids, coumarins and terpenoids extracted from Indian medicinal plants such as the *Rauwolfia canescens*, *Alstonia scholaris*, *Swertia chirata* and *Caesalpinia crista* at the laboratory. Janaki's focus on the other hand was on the introduction of polyploidy in medicinal plants with the aim of multiplying their alkaloid or medicinal content (as in the *Rauwolfia*), and her practice was located in the lab-field border zone.

Among women scientists publishing in the *Proceedings of the Indian Academy of Sciences* (Section B) in the 1950s, besides the prolific laboratory

410

scientist Sunita Inderjit Singh, were several botanists such as Rachel John of the University of Lucknow and C. B. Sulochana, L. Yogeswaran, K. Radha, Anna T. Zachariah, K. Bhuvaneswari, P. Gnanam, P. Shanta, T. S. Sarojini and L. Saraswathi Devi (the last two were pursuing their doctorates) of the UBL, Madras, which since 1944 had been under the Directorship of plant virologist, T. S. Sadasivan, a doctorate from the Rothamsted Experimental Station in England; among the women zoologists were Mary Samuel of the Central Marine Fisheries Research Station, Mandapam, and K. G. Raja Bai Naidu of Andhra University, Waltair.

Grüneberg Visits

In 1959, the German-born British geneticist Hans Grüneberg (1907–82) contacted Janaki (perhaps at the behest of Haldane), whom he had seen at several Genetical Society meetings in London (but had never really been introduced to), with regard to a three-month research visit to India he had planned for early October. Grüneberg's work was in the field of animal genetics (he is hailed as the father of mouse genetics) and at one point in the 1950s had chosen the zoology department of the University of Delhi as his research headquarters. His chief objective was the collection of some wild populations of house mice (*Mus musculus*), the skeletal variation of which he had been studying for the past few years; he was well-equipped for the work except for rat-traps, an incubator and a laboratory bench, which he intended to source locally.

Always helpful and generous, Janaki wasted no time in requesting University Professors of Zoology to extend all assistance to Grüneberg during his visit. She was also ready to offer him a place in her laboratory, if he so wished, although it was 'a purely Botanical one.' In addition, she volunteered to reserve a room for him in Allahabad in case he decided to accept the offer, in the same hotel that she had accommodated the Haldanes at when they visited her in June that year. Incidentally, on that occasion, Janaki had shown Haldane and Helen around her laboratory, besides 'several slides of chromosomes of Indian plant species.' She had also driven them to the countryside, 'where Haldane, [much] against the advice of a local Brahmin, jumped into the river Ganges which was supposed to contain the freshwater crocodile, and had a swim.'[25] Since Janaki was to travel to Delhi to attend the Annual Meeting of the Indian Academy of Sciences scheduled for 2–3 October, she let Grüneberg know that she was also more than happy to meet him at the airport;[26] it however turned out that alternate arrangements had been put in place through a colleague's friend and he did not need to bother her.

Grüneberg would visit India again in early 1961, this time to make arrangements for a systematic study of the effects of high radiation in the monazite sands near Quilon in Kerala. The monazite sands were the property of the Travancore Minerals Ltd, a company under the control of the

Atomic Energy Commission's Medical and Biology Division at Apollo Bunder, Bombay, headed by Janaki's friend Gopal-Ayengar; the chief of the Atomic Energy Commission and the TIFR (during the Second Five-Year Plan, the TIFR also built India's first digital computer) was the nuclear physicist Homi J. Bhabha. In fact, the Biology Division (in the 1960s, on Bhabha's invitation, the young Obaid Siddiqui would set it up as a molecular biology unit) was itself at this time organising a study of the high radiation areas of Kerala. Soon, Janaki would embark on experiments on the impact of radiation on plants in the monazite-rich sands of Southern Kerala (which was in any case one of the objectives of the CBL), but more on this later.

Personal Loss

On the family front, there was much unhappiness. In March 1958, her forester brother Kittu passed away, his departure a huge blow to the EK family and a great loss to Janaki personally. No sadness however could stand in the way of the family's love for the outdoors and plants in particular. Elder brother Raghavan, although weighed down by his unending financial woes, sent Janaki a mango seedling, about four months old, because he thought it might interest her on account of its oddity: the stem, leaves and even a small offshoot was almost white in colour, which he attributed to the lime-laden soil or immature seed, but wished to know what she made of it. He had also made a collection of cuttings for her, including some from Mahe, their father's ancestral place, and arranged for date-palm saplings as per her request, through a doctor-friend in South Malabar. 'I hope you are keeping fit and progressing well in your research work,' he wrote to her with much fraternal concern, while also suggesting that she start 'a garden of herbs in SRR [Shoranur]—there is an ideal spot for it near the big tank there.'[27] Janaki would soon make arrangements to buy this property, in Kallipadam, contributing a sum of Rs 2,000 towards it, with the rest paid up by her youngest brother, Varadan. She would call it 'Ashok', perhaps after the *Ashoka-vana* (forest) of *Ramayana*, where Janaki, her namesake (also known as Sita) was confined by Ravana. This piece of land came with a cottage, and both would be registered in the name of Varadan.

On 19 December (1959), which happened to be Darlington's birthday, Varadan passed on; the family would be deeply shaken by the sudden loss of the youngest among the EK siblings. He had left behind six young daughters and an only son, Muthukrishnan; his wife had passed away only a year before. In fact, Varadan had retired to Shoranur just a year earlier, and following his wife's death had begun 'working hard like a Cooly developing the large garden [Ashok],' elder brother Raghavan sadly recalled several years later.[28] Janaki would henceforth play an active role in the education of Varadan's younger daughters, and Ashok would become central to her life.

412

Ashok, the Medicinal Garden

Janaki would begin investing energy and money on Ashok with a view to establishing a medicinal garden and a drug research laboratory there, on the lines of the one in Jammu she had visited in 1959. The property consisted of four plots entirely dedicated to the cultivation of paddy, some garden land and a small cottage. 'I am wondering what is happening at Ashok and how much money is left for finishing the job. I hope it will be completed as soon as the rains are over. I want a *Konayi* [mason] to [work on] it—otherwise it will not look well. Please take an interest in this little cottage I have dedicated to our brother Varad,' Janaki would appeal to Raghavan from Allahabad.[29] Her sister Cousalya at Shoranur was far from well and could not be bothered with these jobs. However, unlike Raghavan who thought it was a 'folly' to invest money in the cottage, as Janaki was doing, Couslaya never stood in her way when it came to developing Ashok.

The fact of the matter was that Raghavan feared the money spent on the cottage would go to waste, if it were left unoccupied for long stretches of time. Janaki's solution to this was to rent it out, then 'it would take care of itself,' she justified. At any cost, she was dead against selling it: 'There will be a time when Varad's children will stay there!! As we are doing at Edam,' she reasoned. She might herself stay there one day, and pay rent to Varad's children, Janaki imagined, or in the event of establishing her own 'Drug Research Institute' there, it could be used to house her staff. In any case, she had no intention of buying any land in Shoranur, she clarified, and if Government were going to do so on her behalf, it would be their choice and not hers.[30] From this it appears that she had already suggested to the Government as regards the establishment of a drug research laboratory somewhere in Malabar or at least in South India, if not in Shoranur. It was a serious desideratum, she felt, for all that existed, was the one in Jammu, devoted entirely to the medicinal plants of the Himalaya.

Notes

1 RBG Kew, Library & Archives: F. C. Stern and Lady Stern correspondence, E. K. Janaki Ammal to F. C. Stern, letter dated 17 July 1957.
2 Sundara Raghavan, 'A chromosome survey of Indian Dioscoreas (communicated by E. K. Janaki Ammal)' and 'A Note on Some South Indian Species of the Genus *Dioscorea*'.
3 Ibid, 'Chromosome Numbers in Indian Medicinal Plants'; 'Chromosome Numbers in Indian Medicinal Plants—II'; and 'Chromosome Numbers in Indian Medicinal Plants—III'; all three papers were forwarded to the journal by E. K. Janaki Ammal.
4 Janaki Ammal, 'Central Botanical Laboratory', 1959.
5 *The Indian Journal of Genetics and Plant Breeding*, special symposium number, 'Genetics and plant breeding in South Asia', 17, no. 2 (1957): 111–414.
6 In 1950, botanist P. Maheshwari and K.A. Chowdhury of the Forest Research Institute had spent some time at the Jodrell Laboratory: Maheshwari on embryological investigations and Chowdhury on wood structure.

7 RBG Kew, Library & Archives: QG 169, Correspondence with Janaki Ammal, 1945–1970, E. K. Janaki Ammal to N. L. Bor, letter dated 14 June 1956.

8 She had grown the plant from seeds (sent to her by Bor) at Wisley in 1949. RBG Kew, Library & Archives: QG169, Correspondence with E. K. Janaki Ammal, 1945–1970, E. K. Janaki Ammal to N. L. Bor, letter dated 13 July 1957.

9 Ibid, F. C. Stern and Lady Stern correspondence, E. K. Janaki Ammal to F. C. Stern, letter dated 17 July 1957.

10 Ibid.

11 In 1955, Janaki would communicate a research note to the Indian Academy of Sciences titled, 'The Immigrant Economic Plants of India', by K. M. Sebastine (BSI).

12 F. C. Stern and Lady Stern correspondence, E. K. Janaki Ammal to F. C. Stern, letter dated 16 November 1957.

13 For the circumstances that let Haldane and Spurway to India and their life and work thereafter see Dronamraju, 'J. B. S. Haldane's Last Years: his life and work in India' and *Popularizing Science: The Life and Work of J. B. S. Haldane*; McOuat, 'J. B. S. Haldane's passage to India: reconfiguring science'; Rao, 'J. B. S. Haldane, an Indian scientist of British origin'.

14 Perkins, *Geopolitics and the Green Revolution*, pp. 176–183.

15 Janaki Ammal, 'Cyto-geography of some Indian plants', 1959.

16 Ibid., 'Chromosomes and the Evolution of Garden Philadelphus', 1951.

17 Ibid., 'Iso-chromosomes and the origin of triploidy in hybrids between old and new world species of Philadelphus', 1958.

18 Darlington and Janaki Ammal, 'Adaptive iso-chromosomes in Nicandra', p. 267.

19 Janaki Ammal, 'Iso-chromosomes', 1958, p. 256.

20 Ibid., 'Distribution patterns of some Indian plants', 1959; and 'The genetic pattern in the distribution of some Indian plants', 1960.

21 Ibid., 'Distribution patterns', 1959, p. 112.

22 Ibid.

23 It was at this meeting that the decision to include papers on animal genetics and breeding in the future issues of the journal was taken.

24 Richharia, 'Origins of cultivated rices'.

25 Dronamraju, *Haldane: The Life and Work of J. B. S. Haldane with Special Reference to India*, pp. 117–118.

26 Wellcome Library: The Hans Grüneberg Papers, PPGRU/37/1, E. K. Janaki Ammal to H. Grüneberg, letter dated 8 September 1959.

27 E. K. Raghavan to E. K. Janaki, letter dated Coimbatore, 8 September 1958, private collection.

28 E. K. Raghavan to E. K. Parvathy, letter dated 10 February 1973, private collection.

29 E. K. Janaki Ammal to E. K. Raghavan, letter dated [Allahabad, 1959], private collection.

30 Ibid.

23

JAMMU & KASHMIR I

A Border Zone of Mixed Flora

There is still a lot of energy in me. Ladak proved it . . . Perhaps
I'll start a little Institute in Malabar 'Hortus Malabaricus' . . .
Before that I must visit England and see you all.

—E. K. Janaki Ammal (1962)[1]

Janaki had already begun to plan a tour of the State of Jammu & Kashmir
in February 1959, hoping to visit the Kashmir Valley in particular, as part of
her research on the genetic distribution of Indian flora. By virtue of its location at the continent's centre, the flora of Kashmir shared features with the
floras of the neighbouring regions and was thus interesting from the point
of view of cytogeography, cytosystematics, history and ecology. This border
zone of mixed flora boasted of diverse habitats, which contained both primitive and evolved genera and species. L. D. Kapoor, the Chief Botanist, Drug
Research Laboratories (DRL), Kashmir, and Dr Col. Ram Nath Chopra,
Research and Technical Director, DRL (of the Indian Medical Service, and
considered a doyen of Indian pharmacology), observed that the Himalayan
affinity was most strongly evident in the forest zone, ranging from 1800 m
to 3000 m above mean sea level, while the Mediterranean and tropical elements were most commonly found in foothill zones.[2]

The idea of establishing the DRL was first mooted at a meeting in Delhi
in January 1940 at which were present Ram Nath Chopra, N. Gopalaswami
Iyengar, the Dewan of the State of Jammu & Kashmir and Sri Ram, a leading industrialist, who had hosted a lunch in honour of the three. Chopra had
a couple of years remaining before retirement from the Tropical School of
Medicine, Calcutta, but had already been invited by the Diwan of Mysore
to establish a drug research laboratory in that state. However, given J &
K's wealth of medicinal flora, Gopalaswami Iyengar believed it was this
state that was in greater need of a DRL; the idea was promptly seconded
by Sri Ram, and Col. Chopra agreed to make this happen, at least in principle. Serious discussions followed at the governmental level and the decision was finally taken to establish one such institution in Jammu, with four

departments under its fold, devoted to: botany, chemistry, pharmacology and the production of drugs from plants endemic to Kashmir.

The DRL began officially functioning from November 1941, despite the limited funds at its disposal. After independence, the CSIR began negotiations with the J & K state for taking over the DRL, and linking it with a chain of laboratories which were being founded at the time. It eventually became a CSIR-controlled institution in December 1957 and was renamed the RRL, Jammu, with a more inclusive programme that aimed to utilise all sorts of natural resources for product manufacturing, including animals and minerals, rather than just plants.[3] The Department of Botany of the DRL had already begun detailed mapping of the flora, and a herbarium was being gradually built up, promising much from the utility point of view.

It was to botanist L. D. Kapoor that Janaki had written to first about her proposed botanical tour of the state.[4] Kapoor replied in a couple of weeks, enclosing a detailed travel itinerary for her. In his opinion, although May was a good month for visiting the Kashmir Valley, the months of July–August were more interesting for botanising at higher altitudes. Offering to show her around the RRL, Jammu, when she visited, he recommended that she also travel to the Gulmarg and Pahalgam regions, for the floristic interest these offered to the botanist. Janaki was to travel either by air or by road for a distance of 200 miles (320 kilometres) to Srinagar, after first obtaining a permit to enter the state. Kapoor suggested that she reside in Srinagar for 3 to 5 days, to visit the Dachhigam Rakh and the Harwan water reservoir to make collections.[5] Dachhigam, 22 km from Srinagar, had been a game reserve since 1910 under Maharaja Pratap Singh, and declared a sanctuary in 1951, while the reservoir was the source of drinking water for the city.

Planning an Active Retirement

In August 1959, when she wrote to R. N. Chopra to enquire about the possibility of a research post for her at the RRL, Janaki had not yet been relieved from the BSI; she was to continue as Director of the CBL at least until October though she was well past the official retirement age of 55. She had by this time already visited Jammu & Kashmir, and felt the RRL suited her perfectly. She had after all some experience in breeding medicinal plants, especially the *Dioscorea deltoidea*. 'I feel this is the best place I should like to continue these researches in India,' she wrote to Chopra. She was also anxious to find employment that would bring her some money, because she had taken on the responsibility of Varadan's youngest daughters. Janaki knew that Chopra's recommendation would go a long way in landing her a job. Additionally, she wrote to him with a request to take up the matter with P. M. Nabar, Director of the newly founded (1959) Central Indian Medicinal Plants Organization (CIMPO); incidentally, Chopra was a permanent member of the CIMPO Council. In her letter to Chopra, Janaki did not forget to enquire about the fate of his 'beautiful gardens' in Srinagar, which had recently encountered some flooding.[6] Chopra's recommendation went in her favour and her anxieties would die down. At least for some time to come.

Officer on Special Duty at the RRL

In 1959, Janaki retired as Director of the CBL, and was appointed Officer on Special Duty by the CSIR (the organisation which controlled the RRLs), to help in organising their newly established Regional Laboratory for the study of medicinal plants, at Jorhat in Assam. She was also put in charge of the Cytogenetic Department of the RRL, Jammu; the plan was to remain in Jammu for one half of the year, and in Jorhat, for the remaining. Her work began with the founding of a pan-Indian medicinal plants garden, containing over a thousand plants. Incidentally, G. S. Puri, who had worked with Birbal Sahni, and had been a fellow participant of Janaki at the Kandy Symposium of 1956, succeeded her to the position of Director, CBL.

She updated Bartlett on the change in her circumstances; that she had retired, but very importantly, that she was still full of energy to carry on with research:

> I am handing over charge to my successor Dr G. S. Puri and will be a free person from tomorrow. In India one is not allowed to work as a Government servant after 55 years. They have now extended the age limit to 60 for scientists and I have reached that great age. I do not feel a bit like retiring and was planning to accept Mr Ralph Peer's invitation to work in California on Camellias. I wanted to send you a copy of his letter. This will now not be necessary—as when I was in Delhi recently—I was asked by the Director General of CIMPO (Council of Indian Medicinal Plants Organization) if I would help them to fit up a Drug Research Lab in Assam. In spite of Nagas and the like I am attached to the great floristic region of India and have said yes. So I am going to be Officer on Special Duty for 2 more years.[7]

Janaki was almost sixty-two, when she wrote this letter to him, and had been hoping 'to wander as a Scientific *Bikshu* through Malaya, Philippines, China, Japan and Hawaii before landing with some Camellias at Los Angeles,' where Peer was located, and finally descending on Ann Arbor, where she would help Bartlett sort out the rich Walter Koelz collection—but this was not to be. 'Perhaps I will yet come,' she told Bartlett, because she had begun dreaming of having her 'New Chromosome Atlas' published from the United States. The past seven years had been hectic for her, but Janaki was pleased that the BSI 'was on its feet once more' and the 'little Laboratory' she had started was 'growing up nicely.' 'I hope workers will be coming forth to make really worthwhile contributions. At present so much time is wasted in official reports,' she would comment in her letter to Bartlett.[8]

The Jammu Mint

Janaki's work at the cytology laboratory of the RRL, Jammu was focused on one of the many research topics that she had set out to investigate at the CBL, namely, the chromosomal study of Indian medicinal plants. If previously, fine morphological characteristics (such as the thickness of leaves, the size of flowers or the presence or absence of hair) were relied on to differentiate between plant species, cytogenetical studies had now taken their place. When it came to medicinal plants however, one could not blindly proceed with genetic determinism of a kind, because the properties of the plant also depended on its habitat. It was thus important to study and compare plant species from different habitats (ecospecies) before selection could be made for cultivation. The active principles of the medicinal plant also varied with seasons, and such variables as temperature and humidity, but very importantly, ploidy levels. The same species could appear in different forms (polymorphism), depending on the ploidy. Polyploidy, arising from the duplication of the chromosome set, was found to enrich the plant's medicinal or economic properties, at least up to a certain level. For instance, the amount of Vitamin C in triploid apples and tetraploid tomatoes was conspicuously greater than in their diploid counterparts, demonstrating a direct correspondence between ploidy and Vitamin C. A knowledge of the chromosome numbers of medicinal plants and the study of genetic variation played an important part in the selection of the variety to be cultivated along commercial lines, and which knowledge Janaki expertly provided.

She strived to induce tetraploidy in the commonly found medicinal plants of the region (usually diploids), through the application of colchicine; this was done to boost the active principles or ingredients of the plants, occurring usually in the form of alkaloids, glycosides, tannins, flavonoids, saponins or essences.[9] One such was the *Mentha arvensis* var. javanica (2n = 72), found widely in Kashmir. Like the menthol-producing mints of Europe, it had a low oil content, with only traces of menthol in its leaves, thus failing to attract the attention of researchers as a useful plant. UNESCO had introduced to the RRL a mint from Japan in 1952, referred to as the Japanese Mint (*Mentha arvensis* var. piperascens, with 2n = 96), which claimed an oil content of 3%, of which 70% was menthol; it was a sterile cross between two varieties (Man-yo and Ban-Bi) found in Japan and began to be extensively cultivated in Jammu & Kashmir. At the cytological laboratory of the RRL, in 1960, Janaki treated the suckers of Japanese Mint with colchicine, to produce a tetraploid (2n = 192) that was richer in oil content (about 4%) than its diploid form and the induced tetraploidy rendered it fertile. From the seeds of this tetraploid, in late 1961, around 200 seedlings were raised, of which 140 were examined for oil; of these, four were found to be richer in oil than even the tetraploid parent. Out of these, one was selected by Janaki for its 5% oil content and gave it the name 'Jammu Mint.' This plant would be vegetatively propagated and studied under field conditions before being released to drug companies (including the American company, Vicks) as a commercially viable menthol-rich mint.[10]

A New Collection of Dioscorea

In early 1961, Janaki received from I. H. Burkill, a copy of his article on the evolution of the *Dioscorea*.[11] She had already read the paper in the *Journal of the Linnean Society*, but was happy to have a copy of it, to add to her growing collection of reprints. Burkill promised to send copies of his other papers for the RRL library. Referring to his *Dioscorea* paper, Janaki remarked that it was a 'grand one!'[12] Her collection of *Dioscorea* had to be left behind at the CBL, Allahabad, when she moved to Jammu upon retirement, and this had made her deeply unhappy, because she had collected most of them painstakingly 'from the Kadars and Paniyars—the aboriginal tribes of Malabar.'[13] She wrote to Burkill: 'As you know I had got together a collection of Indian *Dioscorea* at the Central Botanical Laboratory and given their study to Mr Sundara Raghavan. Since both of us have left Allahabad I am afraid they are not being taken care of. Sundara Raghavan has registered for PhD with a problem on *Dioscorea*, and I am hoping he has taken duplicates with him to Poona where he is now the Systematic Botanist of the Western Circle of the Botanical Survey of India.'[14]

Alongside the Jammu Mint, Janaki worked on inducing tetraploidy in the two Indian species of the *Dioscorea* (belonging to the primitive group Stenophora, with rhizomatous roots, rather than the more evolved group that produced tubers), namely, *Dioscorea deltoidea* of the Western Himalaya (growing in abundance in Jammu & Kashmir) and the closely related *D. prazeri* of the Eastern Himalaya. They were both important sources of diosgenin, used in the preparation of cortisone and other steroidal hormones. In their *Atlas*, Janaki and Darlington had published the chromosome numbers of all the Stenophora species of Asia and Europe, and found them to be diploids (2n = 20). To this list was now added the chromosome number of *D. deltoidea* (2n = 20).[15] While the tuberous *Dioscorea* (within which was included the edible yams) exhibited high ploidy, the Stenophora group was largely diploidal, with the only polyploid reported from Eastern America, and cultivated in the botanic gardens of Europe. In order to increase the active principle in *Dioscorea deltoidea* prior to cultivation, it was important to double the chromosomes by injection of colchicine into the rhizomes; rhizomes of tetraploids were discovered to be sluggish in growth compared to diploids. Two-year-old plants were being propagated at the RRL at this time for studying their diosgenin content.[16] In her letter to Burkill, Janaki observed: '*Dioscorea deltoidea*—which we are interested in here for its cortisone producing diosgenin is after all a diploid 2n = 20. I counted it here—so the diploidy fits very well with the rest of the Stenophora group. We have succeeded in producing a tetraploid here by the use of colchicine treatment.'[17] Janaki would begin a new collection of *Dioscorea* at the RRL.

Terminalia and the Rauwolfia

Janaki worked also on the important medicinal plant genus, *Terminalia*. Over 100 species had been included in the genus, most of which were large trees growing in evergreen and deciduous forests of both hemispheres, yielding commercially valuable tannins. The bark of the *Terminalia arjuna*, for instance, was known to yield a powerful cardio-stimulant, which demanded investigation. The credit for the first ever cytological study on the genus *Terminalia* went to N. W. Simmonds; in 1954, he studied the *T. catappa* (Indian Almond) from the West Indies (a diploid, 2n = 24). As for, Janaki and her research assistant S. N. Sobti, they studied five species of the genus, including varieties, and found its basic chromosome number to be x = 12; this number x = 12 (2n = 24) also turned up for *Quisqualis indica* (*Combretum indicum* or the Rangoon creeper). The cultivated varieties of *T. bellerica* and *T. chebula* growing in the R. N. Chopra Garden of Medicinal Plants (Jammu) were found to be tetraploids (2n = 48) as against the wild specimens (collected from Dehra Dun, the National Botanic Gardens, Lucknow, and Katra in the Sub-Himalayan Range of Jammu), which were diploids (2n = 24). The authors suggested that 'in the cultivated tetraploid races of *Terminalia* there was probably a correlation between ploidy, size of fruit and tannin content' just as they had noticed with *Emblica officinalis* (1958),[18] where there was a definite correlation between size of fruit, higher vitamin C content and increase in chromosome numbers. Janaki and Sobti concluded that it was of utmost importance to the future of the drug industry in the country to conduct chromosome surveys of indigenous medicinal plants in order to discover and commercially exploit high-yielding polyploids, occurring both in the wild and under cultivation.[19]

Among the medicinal plants in which Janaki induced tetraploidy was the *Rauwolfia serpentina*, which had attracted much interest for its sedative properties and effectiveness in the treatment of hypertension. All the wild varieties, collected by Janaki from Rishikesh (perhaps on her way to Banaras with Haldane and Helen Spurway, who had by this time moved to Bhubaneswar, Orissa),[20] Kerala, Bengal and Dehra Dun, and examined at the laboratory (besides the ones reported from Thailand, Pakistan and Darjeeling) were found to be diploids (2n = 22). The drug being in huge demand, Indian states had taken up the cultivation of *Rauwolfia* extensively, but there were marked differences in the alkaloid content of the various forms of the plant in cultivation. In general, the North Indian forms showed a higher alkaloid content compared to the South Indian ones. Janaki induced tetrapoloidy in *Rauwolfia* by treating the seeds, seedlings and stem cuttings with colchicine, and produced plants with 2n = 44, which had far broader and thicker leaves and larger flowers than the diploids; the alkaloid content also doubled, but there was much to be done before it could be commercially exploited.[21]

Between April 1961 and 31 March 1962, Janaki devoted herself (assisted by her researchers) to the genetical studies of medicinal plants such as the camphor-yielding *Ocimum*, the *Mentha* (in collaboration with Sobti and Rao), *Clitoria ternatea* (chiefly conducted by Bezbaruah, at the RRL, Jorhat),[22] *Cymbopogon*,[23] the attar-yielding Rose and the tetraploid garlic.

A Flora of Assam

Janaki was also involved in the making of a flora of Assam (herbarium), with the help of one Hazarika, contributing at least 500 specimens to it. The living plants of Assam were compared with identical species from other parts of India, and in the case of *Solanum khasianum* interesting structural changes had been noted in the chromosomal make-up. In addition, she prepared notes on medicinal plants of Assam, while also investigating several families in detail. By the end of the year, Janaki was ready with a list of economic dicotyledonous plants of Assam for publication, and in collaboration with Hazarika, a preliminary study of the orchids of Assam.

Election to Important Official Posts

In 1959, the Indian Botanical Society had chosen Janaki as President, the first woman to hold this post; it would take three decades before a woman would be elected President of the Society again (the plant cytogeneticist Archana Sharma, 1932–2008). In the following year, Janaki's paper on the cytogeography of some Indian plants and the effect of the Himalayan uplift would be published in the Society's journal[24]; the article was the result of her researches on the humid forests of South Asia begun in the late 1950s and a fair version of the papers delivered at the Annual Meeting of the Indian Academy of Sciences held in Baroda in late December 1958, and the Indian Science Congress a year later. In late 1960, Janaki received the Society's Birbal Sahni Medal (instituted in 1957 by the plant virologist T. S. Sadasivan of the UBL, Madras), awarded annually to an Indian botanist for notable contributions and services to botany in India.

Janaki would also become a life member of the newly founded International Society for Tropical Ecology; the idea for such a Society had been mooted and discussed at the Fourth World Forestry Conference in 1954, the UNESCO's Kandy Symposium in 1956, UNESCO's Symposium on Vegetation of the Humid Tropics held in Bogor, Indonesia, in 1958 and the Pacific Science Congress, also in 1958. It came into being in 1960 at the 47th Indian Science Congress Association at Bombay. Its first President was J. C. Sengupta, who had succeeded Santapau as Chief Botanist of the BSI in 1955. Early in 1961, Janaki would be elected President of the Indian Society of Genetics and Plant Breeding; the Vice-Presidents were A. R. Gopal-Ayengar and A. B. Joshi and the Secretary M. S. Swaminathan. B. P. Pal remained editor of the Society's journal and the Councillors elected included S. M. Sikka, S. Govindaswamy, P. N. Bhaduri, B. S. Kadam, G. P. Argikar and K. Ramiah.

Birthday Wishes from the Himalaya

As always in December, Janaki sent Darlington her birthday wishes, this time from Jammu, where she resided in an old and elegant mansion, very close to the RRL. She would explain to him that 'Jammu-on-Tawi' meant 'The river on which Jammu town stands.' On one occasion, she would send him a postcard, illustrating a house-boat on the Dal Lake in Kashmir. All was not well at this time on the Indian borders with China, with conflicts dangerously brewing over a disputed 2000-mile-long Himalayan border and over India's granting of asylum to the Dalai Lama, post the Tibetan Uprising (1959).[25] 'We are greatly disturbed over the Sino-Indian developments over the frontier question,' she would tell him.[26] Darlington replied sympathetically, and in his letter revealed that he had reverted to his old interest in 'the origin and structure of society, especially Indian society.' He had, after all, taken copious notes on Indian society during his first visit to the country in 1933. Janaki always acted as his India-digest, the go-to person with queries, even those that had nothing to do with biology. This time he wanted information on the 'the criminal castes in India, and about the distribution today of crime amongst different castes.'[27] Janaki would take this up very seriously, and spare no effort in answering his questions fully, besides sending him relevant reading material, which would go into his book, *The Evolution of Man and Society* (1969). Incidentally, it was only recently, in 1956 to be exact, that the actual number of chromosomes in a normal human cell had been confirmed as 46.

Racial Classification of Man

In late 1960, Janaki would send Ruggles Gates and Haldane for their comments, an exegesis on the three races of man by the racist Scottish anthropologist George Robert Gayre (based at one time at the Sagar University, Madhya Pradesh). Gayre and Gates were founding editors of the racist journal, *The Mankind Quarterly*, mouthpiece of the International

Association for the Advancement of Ethnology and Eugenics, begun in 1961.[28] Gates responded that man had a long history, and the classification of man was chiefly 'a matter of semantics.' Citing Darwin, he concluded that 'it was best to call the main types of mankind sub-species . . . The term Caucasicus, Mongolicus, Africanus are simply meant to imply that all mankind is included in our purview.' Gates explained to Janaki that this 'geographic distribution of types goes back to pre-history . . . The Wenner-Gren Anthropological Institute, which is international, uses similar racial symbols. Roughly White, Black and Yellow are generally regarded as the primary races of mankind, to which the Australian aborigines and certain others are sometimes added. One cannot expect unanimity regarding these categories, which are, everyone agrees, the products of evolution and therefore in various stages of separateness.' He went on to claim that all anthropologists believed 'that the Hindu population of India (as distinct from jungle tribes [were] an Eastern extension of the Mediterranean race in Europe.' She was of course deeply concerned about the rising population of India, but this was besides the issue here; what she had taken offence to was the 'making of 3 species of H. Sapiens,' and this she clearly indicated in the form of an annotatation in the margin of Gates' letter to her.)[29]

On the basis of his study of 'the inheritance of hairy ear rims' (with his Indian student, the plant cytogeneticist, P. N. Bhaduri), Gates had claimed that there was a strong relationship between Southern Europe and India; hairy ear rims were frequently found among people from South India and Italy, he noted. In that same letter Gates enquired about her work at the RRL. 'Are you continuing to work in Cytology?' he enquired. He spoke of his visit to India the previous year 'to study jungle tribes' but regretted he 'was never able to visit Kashmir.'[30] In 1959, Gates had travelled to India to study the Kurumbas and Kanikars of the South and the Asurs, Bihors and Muria Gonds of the North (central India in particular).

By the time Janaki's letter enclosing Gayre's paper reached Haldane's Calcutta address, he was (in late 1960) in Sri Lanka to address the Annual Meeting of the Ceylon Association for the Advancement of Science. His reply, sent from 203, Barrackpore Road, Calcutta (location of the Indian Statistical Institute, and his base since his arrival in India in 1957) a couple of weeks later, opened with a humorous depiction of his eventful (and even physically painful) visit to the island country, amusing Janaki no end: 'like Hanuman, I arrived in Lanka by air. As my tail is rudimentary, no one set fire to it. But I did fall down the steps of an ancient bath in Polonnaruwa, and was bumped all over. A week or so later I got back home. As a small minority of bumps continued to be tender, they were X-rayed, disclosing four fractures (2 ribs, and one metatarsus twice). I am therefore partly encased in plaster, like an unfinished idol.' Regarding Gayre's journal, Haldane commented: 'What a ludicrous journal! . . . I take it to be a neo-fascist

423

journal.' He stated categorically that he would have nothing to do with any journal that was linked to Gates. As a matter of fact, he disclosed that they were planning to start 'a real journal' at Monaco, exclusively devoted to quantitative (biological) anthropology. 'Biswas [P. C. Biswas] is no good, Sahni [A. Sahni] makes bad mistakes about reptiles. Guha [B. S. Guha] is a fraud. The only man on the list who has done anything is Sjögren [?],' the highly critical Haldane remarked.[31]

He however had good things to say about Dronamraju Krishna Rao (b. 1937), his doctoral student at the Indian Statistical Institute, who was working on a thesis titled 'Genetic Studies of the Andhra Pradesh Population.' 'Since you saw him [at Calcutta in August 1960], Dronamraju has revolutionised Indian anthropology. He has got good evidence that hairy pinna of the ear is Y-linked, as suspected for 50 years. Its frequency is about 6% in Andhra Pradesh, 20% in West Bengal, 31% in Ceylon. It is a much better character than the blood groups,' Haldane opined. Incidentally, Dronamraju's name was somewhat familiar to Janaki because she had been sent a paper co-authored by him and Helen Spurway on butterflies and their innate preferences for different colour varieties of *Lantana camara* (1960).[32] Haldane was keen on receiving a paper from Janaki, if 'possible genetical rather than taxonomical,' for his *Journal of Genetics*.[33]

Haldane's response to the enclosed Gayre piece was typical of him, imbued with sardonic wit; he even invented a limerick to mock Gayre:

> An editor called Gayre of Gayre
> > Has classified man by their hayre
> > In Mongols its straight,
> > In Negroes crenate,
> > In Caucasians curly and phayre

This, Haldane wrote, was not 'quite right, but it will have to do, even if it makes you [Janaki] a blonde,' a reference to her wavy hair. Insulted by her suggestion that he might know Gayre, Haldane commented: 'I have never heard of it before. I may be in Scotland next year. If so, I will try to discover why it has been permitted to leave its habitat and settle in the once civilized city of Edinburgh . . . I am only surprised that it [*The Mankind Quarterly*] had not got C. D. Darlington on its editorial board. This is the best thing that I have heard about Darlington for some time,' he quipped. That all was not well between the two was obvious, and this certainly had to do with their differing views on the Lysenko affair and on race in general. Haldane insinuated that Darlington was racist, as did several others, some even from within his own camp, following the publication of his first ever book on man, the highly controversial *The Facts of Life* (1953), which reeked of biological determinism and racialism; the truth was however far more complex.[34]

Haldane disclosed to Janaki, by way of observation, that he had a Malayali cook who had curly hair, and a 'Christian from that State [Kerala] called Davis [T. A. Davis], very black, who is trying to propagate coconut palms from cuttings' [something highly improbable]. Both, he remarked, 'look as if they might have some African ancestry.'[35] Davis had been picked by Haldane from the Central Coconut Research Institute in Kayamkulam (Kerala), and brought to the ISI, Calcutta, to work in its Crop Science Unit. It was also about this time that Haldane undertook a fast in protest against the 'discourtesy of the United States Information Service' in Calcutta. He claimed that the USIS had 'forbidden' two young scientists from America and Canada 'to keep a dinner engagement with him to meet some Indian scientists on January 16th.' Haldane had begun the fast on the 17th, and it continued for a few days before it was called off. Janaki however thought his reason for the protest rather 'silly.' For Darlington's information, she enclosed a newspaper-cutting (of *The Statesman*), carrying a photograph of a protesting Haldane, with his left ankle in plaster. 'The photograph will give you some idea of his present Indianized appearance. He comes to scientific meetings in pyjamas and wears "dhoties"—ie loin cloth at home, chews betel leaves while Helen smokes beedies and sips Gin! I could tell you more but I will not,' she joked. She was happy to learn from Darlington that Oliver and Andy (his sons) were in College. 'I hope the girls are equally bright—and good looking too.' She longed to see them all.[36]

Reading Material for Darlington

By June 1961, Janaki had posted a substantial amount of reading material for Darlington on anthropology in India; physical anthropology in particular, and books by D. N. Majumdar. 'You will find blood groups of castes in one,' she wrote to him, referring to Majumdar's *Races and Cultures in India* (1961, 4th edition). 'He has praised Eileen's [their common friend, Eileen Macfarlane] pioneer work on blood groups of tribals,' she noted. Janaki commended Darlington on his articles on marriage, even while she supplemented him with information from her own part of the country: 'In matriarchal [perhaps 'matrilineal' was intended] North Malabar, a man is supposed to marry his father's sister's daughter in preference to his mother's brother's daughter—even though both are permissible.'[37]

Janaki hoped Eileen would return to India and continue with 'her anthropological research . . . She would make a good partner for Prof. J. B. S. who by the way got his Indian citizenship the other day. I wonder if he will also turn Hindu?' Janaki quipped. She had driven Haldane and Helen to Banaras (from Allahabad, in August 1960), and couldn't wait to share with Darlington some juicy bits of news. Haldane, she commented, 'was very respectful in front of the Shiva temple and even more respectful when we came to the Annapurna shrine—the temple dedicated to the wife of Shiva. I like the ladies, he said.' Seeing Janaki stand by in a detached manner, Haldane commented: 'You are not a Hindu. You are a Buddhist, Janaki!' 'Then the two of them and a student of theirs [Dronamraju] waded through the filth of the temple alley in search of a betel shop—to chew pan—I had a difficult time trying to keep the temple bulls from chewing JBS's rose garland the one the priest gave him at the temple of Shiva-Viswanath.'[38] As for Janaki, 'she had a scientific interest' in religion, but 'went into a "trance" about "maya" often.' She was 'intrigued about "Kali" & would talk about her fascination of watching the Bengallees at Calcutta at Kalighat.' Her 'metaphysical thoughts' were however 'beyond me,' remarked a late nephew who knew her intimately.[39]

She longed to see his 'progeny,' Janaki remarked to Darlington; 'I met them in 1956 [actually mid-1955] (on my way) to [from] America. She spoke to him of her impending retirement from the RRL, Jammu, in November 1961, and of her anxiety about the future: 'I do not know what to do after that. I have inherited a family—I lost my youngest brother (not the one [E. K. Madhavan] you met in Krusadai) just little over a year ago—on your birthday [19 December]! His wife died of cancer a year before he died—so I am left with the children—luckily very good-looking girls who will be married early I hope—the youngest is only 12.'[40] One might not perhaps expect a remark of this kind from a world-ranking woman scientist like Janaki, but that was the way she was, full of contradictions and complexities just as anyone else. Protective of her nieces in the extreme, only rarely did she consider higher education for them as an option; all she wanted for them was to be married well and early. Her eugenicist leanings would ensure that they were married to men of good genetic disposition, sometimes to the point of great annoyance to her siblings. Janaki was worried about the trend to marry them off to cousins as had been the practice in the previous generations of the EK family; as a geneticist, she knew this was not good for the 'progeny,' a term she used even in the most ordinary of everyday contexts. 'I am not keen to get more Edavaleth [her father's *tharavad*] blood into ours [Edam],' she would write to brother Raghavan categorically on one occasion.[41] There was no denying that she was a most generous sibling, and a loving aunt to her numerous nieces and nephews, but there were times when they found her much too overbearing for comfort.

A Monsoon Holiday at Edam

In early September 1961, Janaki sat down to reply to one of Darlington's letters (sent to her in July). She had just returned from a long holiday in Malabar, surprisingly, her first since her return to India from America in late 1955. Malabar was now no more part of the Madras Presidency; it was part of the state of Kerala (together with Travancore-Cochin) following the State

426

Reorganization Act of 1956. Kavalam Madhava Panikkar (1895–1963), better known as Sardar K. M. Panikkar, appointed Vice-Chancellor of the Jammu & Kashmir University in 1961, was also a member of the State Reorganization Commission, which recommended reconstitution along linguistic lines. Janaki had on purpose chosen the monsoon season to visit Malabar, 'to see once again the fury of the Arabian Sea,' just as she had as a young girl in Tellicherry. 'It was magnificent,' she remarked to Darlington. Not even the great deluge of 1924 and the widespread havoc it caused, when she was just about to embark on her maiden journey to America, made her fear the Malabar monsoon.

In Malabar, Janaki was hoping to visit Ernad (a region in midland Malabar, comprising Malappuram, Kondotti, Manjeri and Anakkayam; the area was the centre of the Moplah Uprising of 1921), 'where the Ernadis live, to find out for [herself] if fathers married their own daughters ever.' The case where a man sometimes married 'his wife's daughter by a previous husband,' Janaki believed, was the more probable one. Incidentally, her own colleagues at the RRL, Jammu sometimes became anthropological material for her. For instance, there was this person 'from Andhra' who worked with her, and had married his niece (sister's daughter); Janaki noticed that their children were 'puny,' and the son 'something of an idiot' who suffered 'from fits.' She would share this observation with Darlington.[42]

Genetics and Man in India

Janaki's Presidential Address to the Indian Society for Genetics and Plant Breeding late that year (1961), interestingly enough, was titled 'Genetics and Man in India.'[43] 'I wish you were giving this address! I am prepared to read what you have to say—in your name,' she would tell Darlington. She missed him very much; despite everything, he remained her intellectual companion and sounding board for ideas. 'When will you visit India again?' she queried with much affection and remarked: 'I often wish you were not so far away. There are so many things I would like to talk over with you.'[44]

Gates' Impending Visit

That same month, in September 1961, Janaki would hear from Gates, now in his seventy-ninth year, about his impending visit to Jammu & Kashmir with wife Laura Greer. He was to be a guest of the Indian Statistical Institute (ISI), Calcutta; the director P. C. Mahalanobis had arranged for two anthropological assistants to accompany Gates on his excursions through the country. Janaki meanwhile had forwarded to Darlington, Gates' and Haldane's responses to Gayre's paper on the three races of man. She was also keen to know what the chromosome numbers of Gorillas, Chimps and

Orangutans were; his reply was a quick one—48, thanks to Brunetto Chiarelli of Pavia, who had recently completed a count and described them in *Nature*.[45]

Darlington was full of displeasure when he learnt of Gates' arrival in Jammu: 'I am so sorry that you are going to have Gates in India again. So long as he lives he will continue to put genetics in a false light,' he warned Janaki. In a letter to him, post Gates' visit, Janaki commented that 'Mrs Gates was acting as his Stage manager during his visit' and that 'She [was] very American.'[46] Incidentally, this would be Gates's last visit to India (he would be no more in August 1962). Darlington was at this time reading the controversial American physical anthropologist Carleton S. Coon's *Faces of Asia* (1958), the result of Coon's travels in India for the US Air Force. 'He is the only Anthropologist I know who writes with a genetic point of view,' he would remark to Janaki. Darlington was working on a 'history of society' but the task of revising his earlier books, *Chromosome Botany* and the *Elements of Genetics* had come in the way.[47] Janaki was tempted to send him more books on anthropology, a subject she had wanted to pursue herself, after retiring from the BSI; 'There are interesting aboriginals in Malabar,' but because Gates was planning to write about Todas and Kotas, she had decided not 'to join his camp.' Her old dislike for Gates (he was far from popular for his conservative scientific views on racial origins and differences, and his overt anti-Semitism) had surfaced once again; the trouble he had caused her at Coimbatore in the 1930s was still fresh in her mind.[48]

Planning a Holiday in Kashmir

Ahalya (one of Varadan's daughters) and Shantha (Lekshmi's eldest daughter) had joined Janaki at Jammu for a holiday. A nephew was also expected to arrive soon. An ever-doting aunt, even if somewhat of a disciplinarian, Janaki had planned to take them to Kashmir later that year (1961); the RRL was in the process of establishing a 'Drug Farm for Temperate Himalayan Medicinal Plants' in the valley. She was at this time writing up a paper with junior researcher Sobti, on the chromosome relationship in the several species of the medicinal plant *Calendula*. She had discovered that the evolution of the *Calendula* species was similar to that of the *Viburnum* (investigated by her at Wisley in the early 1950s) in that all showed different basic chromosome numbers, the result of hybridisation and back-crossing (crossing of a hybrid with one of its parents or an individual genetically similar to the parent). Janaki also discovered that the tetraploid *Calendula* ($2n = 44$), like the *Narcissus*, provided the best garden form for cultivation. Thus, it was from Jammu that for the first time ever a new basic number $x = 11$ for the genus *Calendula* was reported, in the species, *C. persica* ($2n = 44$).[49]

High Altitude Agriculture

In April 1962, an attempt was made to establish a small agricultural research unit at Leh (Ladakh) in the state of Jammu & Kashmir under the direction of the agricultural scientist Boshi Sen of the Vivekananda Laboratory, Almora, who was close to Nehru; he had been awarded the Padma Bhushan in 1957. Nehru had suggested growing food crops on an experimental basis in Ladakh to feed the troops deployed in the high-altitude regions of India, and personally requested Sen to carry out research, until a competent person could be sent up for the purpose. Sen was asked to meet the Deputy Commissioner of the District of Ladakh to discuss the matter, and set up a small research centre at Leh under his direction; he was provided with Rs 10,000 by the ICAR. This marked the beginning of high-altitude agricultural research in India. In 1962, the institution would be taken over by the DRDO.

Preliminary work was begun by Sen and botanist M. C. Joshi at the Murtse Veterinary Farm, after much delay on the part of the Kashmir Government to provide formal sanction for the project. A small unit for making meteorological observations was also set up at the experimental fields of the Farm. On account of the scanty rainfall in the region, irrigation was crucial for the growing of crops, so a well was sunk, which provided plentiful water. A preparatory experiment involving eight introduced strains of wheat and one local strain was begun, to examine the possibilities of autumn sowing. Thanks to Col. K. A. Raja, Deputy Commissioner for the District (1959–60), some sowing had been attempted in the winter of 1960, but the experiment proved a failure on account of the damage caused to the crops by snow and birds. Birds also destroyed most of the cereal crops such as wheat, barley, rye and milo (*Sorghum bicolor*), despite using indigenous methods for driving them away. Sen introduced oats and rye to the region in 1961. Experiments were also made in growing vegetables such as peas, beans, tomato, okra, capsicum, cabbage, cauliflower, spinach, and Brussels sprouts and broccoli (with seeds obtained from Britain), root crops such as onion, potato, sweet potato and carrot, oil seeds such as mustard and sunflower, grasses such as the Giant Star grass (*Cynodon plectostachys*) and Love grass (*Eragrostis curvula*), legumes and forage crops.[50]

It was at the behest of the Defence Minister V. K. Krishna Menon that Janaki's service was enlisted to examine the possibilities of agriculture in Ladakh and on how the region could be made self-sufficient as far as food was concerned. A couple of months prior to her visit to Ladakh, and exactly when Boshi Sen began his preliminary experiments at the Murtse Farm, Janaki contacted the Director of the Kew Gardens for suggestions 'as to useful plants for introduction into Ladakh.' The chief crop in the region was barley, of which a naked variety grew at high altitudes. In general, the crops that grew at places like Nepal such as the ones described by the

Japanese expedition of 1952/53 in their book *Land and Crops of Nepal Himalaya* (ed. Hitoshi Kihara) were thought suitable for Leh as well, given their comparable altitudes, but only after due consideration of the terrain, the presence of weeds, the methods of cultivation and availability of water. Kew suggested that some Andean vegetables could be tried like *Chenopodium quinoa* and certain *Coleus* species, which produced edible tubers at high altitude, as well as cold-resistant potatoes such as those being bred at Colombia and the Institute of Arctic Agriculture in Russia, Norway and at the Wageningen in the Netherlands.[51]

Secret Mission to Ladakh

On the morning of 19 June 1962, at the break of dawn, Janaki set out for Ladakh by road from Srinagar on a scoping mission accompanied by Col. K. A. Raja (now Deputy Secretary, Ministry of Defence), Brigadier B. S. Bajwa, Director of Military Farms, Dr V. Ranganathan, Deputy Chief Secretary and Dr. Sheema, Chief Scientific Officer to the Western Command. The group reached Sonamarg (at 2800 m at 9.30 am, some 80 km northeast of Srinagar, where they spotted several medicinal plants, of which *Hyoscyamus* (henbanes of the nightshade family, Solanaceae) and *Verbascum* (mulleins of the figwort family, Scrophulariaceae) were common. The meadows at Baltal, 15 km north of Sonamarg on the Indus River at the base of Zoji La (pass), were full of flowers and included medicinal plants like the *Podophyllum* and *Rheum emodi*, or the red-veined rhubarb. As they moved closer to Kargil, *Physochlaina* (a small genus of the nightshade family) began to appear in large patches, which in Janaki's opinion was a potential source of revenue, they being rich in tropane alkaloids. They also caught sight of *Artemisia* and *Ephedra* aplenty; a source of the medication, ephedrine, *Ephedra* was at present being used just as fuel by the locals. Every village they passed by flaunted the double-yellow rose (*Rosa foetida*), and the double-pink form, with similar leaves. She found *Rosa webbiana*, both the pink and dark red varieties, to be quite tolerant of the arid conditions, and in her view, to be encouraged as a soil-binder. At 3962 m they ran into two species of the *Ferula*, of which *F. narthex* was the asafoetida-yielding one, again a potential plant for cultivation along commercial lines. It would be nightfall by the time the team arrived at Kargil. Resuming their journey early next morning, they located a plateau near the cantonment by the side of the Indus as an ideal spot for an experimental farm. The whole of that day went by in traversing the 140 miles between Kargil and Leh, the landscape having turned 'even more rugged and the mountains treeless.' They were able to reach Leh only by night, having encountered terrifying avalanches en route, but there is no doubt Janaki was thrilled by all the adventure.

She would later have the opportunity to inspect the work done at the Murtse Experimental farm by Boshi Sen and Joshi. 'Mr [M. C.] Joshi has done very good pioneering work on selection of suitable varieties for Ladakh from 240 types of seed of which the most suitable forms are now being propagated at Ranvirpura Farm—16 miles from Murtse, for field trials on a large scale,' she noted. The group also visited Ranvirpura, the soil of which was found to be poor and alkaline. Janaki discovered that the fertile top soil had been taken away for making brick, leaving the unfertile subsoil for cultivation. She recommended that the soil be improved through a generous application of green manure. The building in the Farm was suitable for a botanical laboratory, but if this was to be accessed, a transport vehicle was essential. The future programme of the Farm, she advised, should include hybridisation experiments involving the local barley and wheat, of which suitable varieties could be chosen as parents, and also between wheat and rye as was being done in Russia.

The cytogenetic experiments could be taken up, Janaki suggested, in collaboration with the University of Kashmir and the RRL, Jammu. Ladakh was exceedingly rich in medicinal plants that would yield easily to commercial exploitation, of this she was certain; she mentioned in particular of the *Podophyllum, Artemisia, Ephedra, Physochlaina praealta, Hyoscyamus niger, Saussurea lappa, Ferula narthex* and *Aconite* among several others. Among the positive aspects of the Farm, Janaki observed, were a good supply of water and large patches of marshy land, which could facilitate research on aquatic plants, fish culture and soil conservation.[52]

A few weeks after her return, Janaki wrote to Darlington about her exhilarating journey: 'I am back from a most interesting and breathtaking trip to Ladak—taken at the request of our Defence Minister to advise about making Ladak self-sufficient.' She had brought back a large collection of plants and seeds, which had been handed over to B. P. Pal and his staff at the IARI. 'They grow a naked barley called "grim",' she enlightened Darlington. The 65-year-old Janaki had ascended up to almost 16000 ft (Leh was itself about 13,000 ft), but the altitude, she observed did not affect her 'one bit.' In fact, she was eager to return to Ladakh to conduct a detailed study, despite the inhospitable weather conditions there. 'This time I was rushing about in jeeps with Brigadiers & Generals (Secret Mission!) and had to pass by plants I wanted to collect,' she remarked to Darlington. Between botanising and inspection, her team had also visited monasteries 'which are generally perched up on mountain sides.' The Ladakh trip was a culmination of Janaki's work at Jammu and she was undecided about what next to do: 'I will be too old for employment by CSIR! . . . There is still a lot of energy in me. Ladak proved it . . . Perhaps I'll start a little Institute in Malabar 'Hortus Malabaricus.' Before that I must visit England and see you all,' she would tell him.[53]

Figure 23.1 A Telling Portrait of Modern Indian Science. Fellows of the National
Institute of Sciences (later called INSA) with Nehru, at its Silver Jubilee Celebrations in New Delhi on 30 December 1960. Janaki, the only
woman in the group can be seen, relegated to the margin.

Source: Courtesy of the late mycologist, Dr C. V. Subramanian.

Figure 23.2 Janaki at her office at the Regional Research Laboratory, Jammu, in the
early 1960s. S. N. Sobti behind her, and C. K. Atal to her right.

Source: Courtesy of the RRL, Jammu.

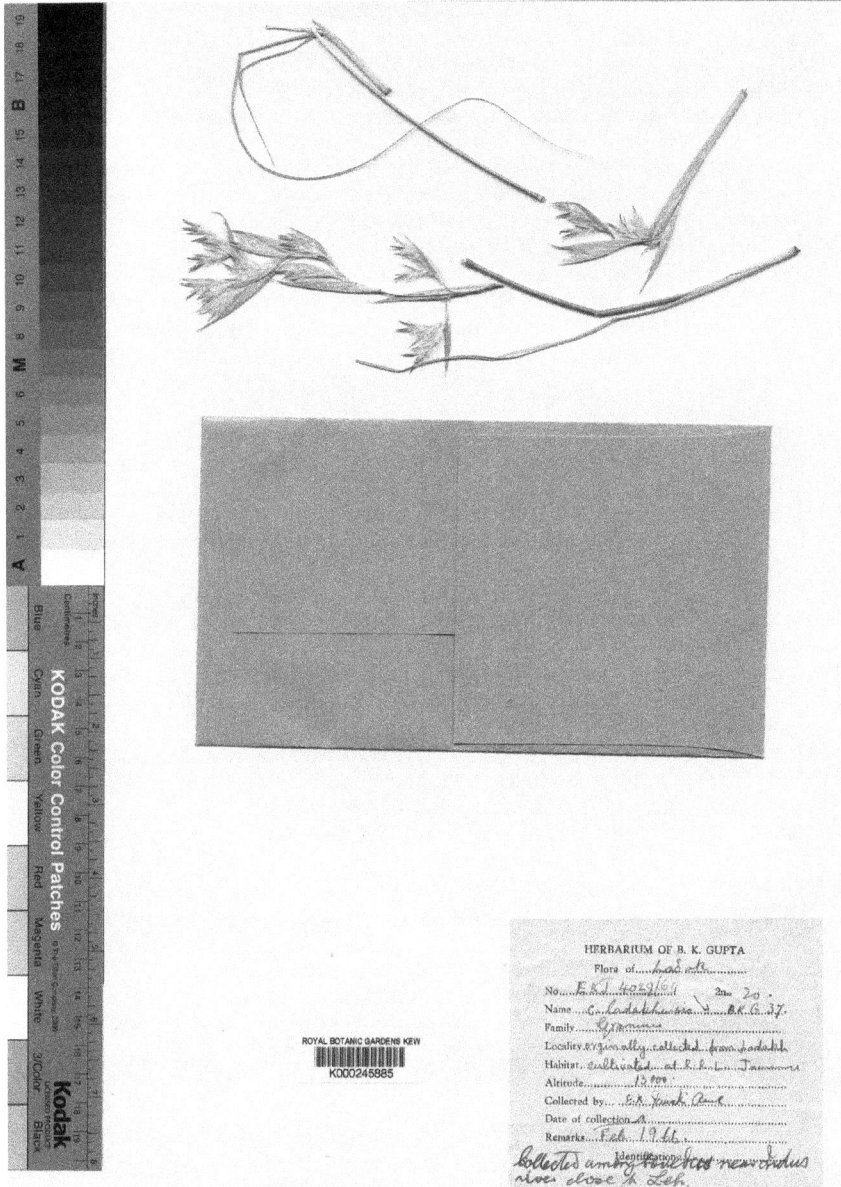

Figure 23.3 'Flora of Ladak.' *Cymbopogon ladakhensis*, originally collected by Janaki from Ladakh in 1962 (at 13000 ft), and grown at the RRL, Jammu. 'Herbarium of B. K. Gupta.' February 1966.

Source: © copyright of the Board of Trustees of the Royal Botanic Gardens, Kew. http://specimens.kew.org/herbarium/K000245885.

Notes

1 DP: C 109 (J 114), E. K. Janaki Ammal to C. D. Darlington, letter dated 13 August 1962.
2 Chopra and Kapoor, 'Some botanical aspects of Kashmir', p. 116.
3 'History of regional research laboratory, Jammu', *Souvenir of the XIII Annual Conference of the Indian Pharmacological Society*, in memory of Col. Sir Ram Nath Chopra, RRL, Jammu, 30 September–2 October 1980, pp. 11–12.
4 Archives of the RRL, Jammu & Kashmir (ARRLJK): E. K. Janaki Ammal to L. D. Kapoor, letter dated 22 February 1959.
5 ARRLJK: L. D. Kapoor to E. K. Janaki Ammal, letter dated 13 March 1959.
6 Ibid, E. K. Janaki Ammal to R. N. Chopra, letter dated 4 August 1959.
7 BHL, UoM: University Herbarium Records (University of Michigan) H. H. Bartlett series, Box 11, E. K. Janaki Ammal to H. H. Bartlett, letter dated 15 October 1959.
8 Ibid.
9 The term 'active principle' refers to '*the intrinsic chemical substance which induces pharmacological activity*'. Capasso, Gaginella, Grandolini and Izzo, 'Active principles.'
10 Janaki Ammal and Sobti, 'The Origin of the Jammu Mint', 1962.
11 Burkill, 'The organography and evolution of Dioscoreaceae, the family of the Yams'.
12 RBG Kew, Library & Archives: I. H. Burkill—Miscellaneous Letters, folio 2, E. K. Janaki Ammal to I. H. Burkill, letter dated 30 April 1961.
13 Ibid.
14 Ibid., folio 1, E. K. Janaki Ammal to I. H. Burkill, letter dated 17 March 1961.
15 Sundara Raghavan had recorded 2n = 40 in his paper, 'A chromosome survey of Indian Dioscoreas'.
16 Janaki Ammal and Singh, 'Induced Tetraploidy in Dioscorea deltoidea', 1962.
17 RBG Kew, Library & Archives: I. H. Burkill—Miscellaneous Letters, folio 1, E. K. Janaki Ammal to I. H. Burkill, letter dated 17 March 1961.
18 Janaki Ammal and Sundara Raghavan, 'Polyploidy and vitamin C in *Emblica officinalis*', 1958.
19 Janaki Ammal and Sobti, 'Polyploidy in the Genus *Terminalia*', 1962.
20 Haldane had established a Genetics and Biometry Laboratory at Bhubaneswar in 1962.
21 Janaki Ammal, 'Tetraploidy in Rauwolfia serpentina', 1962.
22 Janaki also induced tetraploidy in *Catharanthus roseus* (the rose periwinkle), used medicinally to treat diabetic carbuncles; see Janaki Ammal and Bezbaruah, 'Induced Tetraploidy in Catharanthus roseus', 1963.
23 Janaki Ammal and Gupta, 'Oil content in relation to polyploidy in Cymbopogon', 1966.
24 Janaki Ammal, 'The Effect of the Himalayan Uplift on the Genetic Composition of the Flora of Asia, 1960.
25 Gopal, *Jawaharlal Nehru: A Biography*, pp. 418–434.
26 DP: C 109 (J 114), E. K. Janaki Ammal to C. D. Darlington, letter dated 19 December 1959.
27 Ibid., C. D. Darlington to E. K. Janaki Ammal, letter dated 18 January 1960.
28 For more on the journal, and the view on race in the post-war world, see Schaffer, ' "Scientific racism again?": Reginald Gates, "The Mankind Quarterly" and the question of "race" in science after the second world war'.
29 DP: C 109 (J 114), R. R. Gates to E. K. Janaki Ammal, letter dated 2 January 1961.

30 Ibid.
31 For a review of the history of physical anthropology in India, see Reddy and Malhotra, 'Physical Anthropology as a Science: A Commentary.'
32 DP: C 109 (J 114), J. B. S. Haldane to E. K. Janaki Ammal, undated letter (January 1961).
33 Ibid.
34 For a discussion, see Harman, *The Man who Invented the Chromosome*, pp. 235–252.
35 DP: C 109 (J 114), J. B. S. Haldane to E. K. Janaki Ammal, undated letter (January 1961).
36 Ibid., E. K. Janaki Ammal to C. D. Darlington, letter dated 19 February 1961.
37 DP: C 109 (J 114), E. K. Janaki Ammal to C. D. Darlington, letter dated 3 June 1961.
38 Ibid.
39 Hari Krishnan, 'Nachee', private collection, July 1993.
40 DP: C 109 (J 114), E. K. Janaki Ammal to C. D. Darlington, letter dated 3 June 1961.
41 E. K. Janaki Ammal to E. K. Raghavan, letter dated 12 August 1960, private collection.
42 DP: C 109 (J 114), E. K. Janaki Ammal to C. D. Darlington, letter dated 11 September 1961
43 A revised and expanded version of Darlington's provocative *The Facts of Life* (1953) would be published as *Genetics and Man* in 1964.
44 DP: C 109 (J 114), E. K. Janaki Ammal to C. D. Darlington, letter dated 11 September 1961.
45 Chiarelli, 'The chromosomes of Orang-Utan (*Pongo pygmaeus*)'.
46 DP: C 109 (J 114), E. K. Janaki Ammal to C. D. Darlington, letter dated 5 October 1961.
47 Ibid., C. D. Darlington to E. K. Janaki Ammal, undated (late October 1961).
48 Ibid., E. K. Janaki Ammal to C. D. Darlington, letter dated 5 October 1961.
49 Janaki Ammal and Sobti, 'Chromosome Relationship in Calendula Species', 1962.
50 Sen, 'Report for the Establishment of Small Agricultural Research Unit at Leh (Jammu & Kashmir) 1961–62', private collection.
51 RBG Kew, Library & Archives: Director George Taylor's Memorandum to Russell, dated 18 April/27 April 1962.
52 Janaki Ammal, 'A Note on Tour to Ladakh', private collection.
53 DP: C 109 (J 114), E. K. Janaki Ammal to C. D. Darlington, letter dated 13 August 1962.

24

JAMMU & KASHMIR II

High Altitude Flora, Polyploidy and Variation

I need help. I have still a lot to do . . . some of the distribution
maps are very exciting—especially of the Himalayan plants.
—E. K. Janaki Ammal (1969)[1]

Anxious about her future post-retirement from the RRL, Janaki once again
approached R. N. Chopra, this time, to explore the possibility of holding
an honorary professorship at J & K University. Chopra had been working with Janaki to commission the land lying to the north of the RRL for
yet another garden of medicinal plants. He had also written to Dr Kapoor
[L. D. Kapoor] with regard to this.[2] Chopra wrote to the Prime Minister of the State, Bakshi Ghulam Mohammad and also spoke on the phone
with Sardar K. M. Panikkar, the Vice-Chancellor of the University about
Janaki. A journalist, scholar and writer, Panikkar had served as Ambassador to China until 1952, during which time he had famously established
strong ties with Chiang Kai-Shek, witnessed the high point of the Revolution (which began in 1946, led by Mao Zedong, Chairman of the Communist Party of China) and the Communist ascendance of October 1949; his
Chinese experience during the revolution and after, incidentally, formed the
content of his *Two Chinas* published in 1955. A couple of years previous
to that, Panikkar had published his magnum opus *Asia and Western Dominance* (1953). At the end of his assignment in China, he would serve as
ambassador to Egypt (1952–53) and later France (1956–59) but had been
forced to return to India having suffered a severe stroke. Within months of
his recovery, in 1961, he was appointed Vice-Chancellor of the Jammu &
Kashmir University.[3]

Chopra had also approached Krishna Menon about Janaki, but was
yet to hear from him; in addition, he would contact Ghulam Mohammed
Sadiq, the Education Minister about her appointment. Panikkar was hopeful, Chopra informed Janaki, but Krishna Menon remained the key person
to decide on the matter. In the meantime, the Government had decided to

DOI: 10.4324/9781003267089-24

transfer the Botanical Research Laboratory of Kashmir to the University; if this were to happen, Chopra assured Janaki, she would have no dearth of employment opportunities: 'I am very hopeful that something satisfactory will happen and you will be able to carry on work in the RRL, Jammu. It was also Chopra who had suggested that she run up to Delhi to meet Krishna Menon personally, so as to expedite her appointment.[4] Janaki did accordingly, and it helped. At Delhi, her niece and husband, the diplomat K. V. Padmanabhan (posted previously at Paris, in 1949), would host her at their Lutyen's residence, 7B, Wellesley Road (today the Dr Zakir Hussain Marg). Chopra would soon reach her the good news: that her appointment as honorary professor at Kashmir University had been accepted. However, the National Research Professorship (also called the Emeritus Professorship, awarded to those who had made exceptional contributions to their fields), he warned, would have to wait.[5]

Doctoral Supervisor

By the end of 1962, as soon as she retired as a CSIR Scientist, Janaki accepted the honorary professorship of Botany of the University of Jammu & Kashmir, which entailed the supervision of research work of PhD students and lecturing to postgraduate students, but would continue to work from RRL, Jammu, where she was in charge of the cytogenetic division. Janaki's research focus would remain the plants growing at high altitudes. Janaki was also often invited by Indian Universities to act as an external invigilator on doctoral theses in botany. In November 1962, for example, she would examine the thesis of the future ecologist, V. M. Meher-Homji of the University of Bombay, with F. R. Bharucha as the internal referee. She would conduct an extension course in Taxonomy at the Birbal Sahni Institute of Palaeobotany in Lucknow.

On 15 April 1964, Janaki would eventually be appointed Emeritus Scientist at the University of Jammu & Kashmir. She would hold this position until 1969 and work on the improvement of medicinal and aromatic plants such as the *Cymbopogon, Mentha, Datura*[6] and *Solanum* using colchicine and radiation (X and Gamma rays).[7] She would also compare the performance of select medicinal plants from North and South India in the field and work on the breeding and improvement of the *Pyrethrum* (Chrysanthemum). Importantly, she now had five research students (including two women students) working on doctoral projects under her guidance; the research topics included the genus *Cymbopogon* with a view to improve oil content,[8] inheritance studies in the genus *Capsicum*, cytogenetic studies in Indian Roses, improvement by selection, hybridisation and mutation of the *Rauwolfia*[9] and inheritance studies in *Solanum*, which yielded the highly toxic glycoalkaloid, solasodine.[10]

Her students found her formidable. They addressed her as 'Madam Ammal.' She was always watchful of what they did when they were out and about; 'one wouldn't walk into her office chamber without a knock or formal permission,' one of her students recollected. One had to judge her mood, before deciding to have a conversation with her. She kept a framed photograph of Nehru on her working table, was always full of energy, and spoke to her students passionately about Darlington's researches, and sometimes on the contributions of La Cour. She would also discuss her ideas with them on cytogeography, alloploidy and the making of inter-specific crosses, and the finer aspects of the use of colchicine, in which she was particularly skilful. One student remembered that she draped herself like a 'sanyasi' in pleasantly scented saffron-coloured silk saris, tied her long hair into a loose knot and wore 'Gandhi' style slippers on her feet. She appeared much taller than she actually was and walked with a distinct slant to the left, he vividly reminisced, the aftermath of an injury sustained from a fall. Janaki was always fastidious about maintaining field notebooks and labelling plants, and insisted that her students emulate this. One of her major interests in Jammu was plant introduction; she introduced to Jammu different species of the *Cymbopogon* collected from Malabar (and the Pykara Falls in the Nilgiris) besides, it is said, the tapioca and the curry leaf plant (*Murraya koenigii*). She also kept a large number of cats, large and small, which freely roamed about her house, often making themselves comfortable in her drawing room, and refusing to budge even when there were guests. One day, she insisted that a particular student take home a kitten for rearing. The kitten stayed happily with him for a week, but all of a sudden, Janaki demanded that it be returned to her; when she was reunited with the kitten, she let out such a huge sigh of relief, that it left the student completely baffled at her behaviour.[11] She was no doubt a complex, and whimsical person, even somewhat eccentric at times, but almost everyone who knew her intimately were touched by her great generosity and concern for their welfare.

There were days when Janaki invited her students to the Ram Nath Chopra Garden to de-weed, till, and ready the plant beds for trials. Sobti, now a member of staff, was assigned the task of monitoring the activities. If anything went wrong or not quite how she had planned it, the gentle Sobti would have to face the brunt of her fury. It is said that Janaki disliked the opposite sexes intermingling; she would do everything possible to segregate them while they worked. She had given up eating rice by this time, replacing it with wheat. Sometimes she would fry *jalebis*, each shaped like an alphabet in the English language, to the amusement of her many guests. At other times, they would be served deliciously cool watermelon as dessert, after giving the fruit a Janaki-style treatment; this involved immersing the fruits for several hours in the icy cold stream (a tributary of the river Tawi ran across the property), before being pulled out, wedged and relished.

Darlington Plans a Visit to India

In June 1963, A. C. Joshi, Vice-Chancellor of Punjab University, Chandigarh, invited Darlington requesting to India to spend two to three months (in the winter of 1964), as a Visiting Professor at the University. 'We can organize a course of lectures to which we can invite workers in genetics from other Universities and institutes in the country,' wrote Joshi. Having heard that Darlington was working on human genetics, he offered to help him further that line of research: 'India, with its castes, may provide a very good opportunity for such studies. We can expect fruitful results. After some time it is likely that there would be a good deal of inter-marriages among the different castes, the caste system itself may break down under the impact of industrialisation and the kind of material which is available to-day may disappear.' Darlington was keen to accept the invitation and immediately went about preparing an itinerary; his idea was to spend two months at Chandigarh, and then travel to Jammu to be with Janaki for a week. There was also a plan to visit Gopal-Ayengar at BARC and S. P. Ray Chaudhuri at the Department of Zoology of Banaras Hindu University, before going to Agra and finally Delhi, to board the plane back to England but not before meeting Janaki once again, as also Panchanan Maheshwari of the Delhi University; Maheshwari incidentally would be elected Fellow of the Royal Society in 1965. Darlington in the meantime ran into K. N. Kaul at Oxford, who also assured him that he would find the trip very useful, and even offered to accompany him on excursions to see villages and 'possibly hill tribes.'[12] However, Darlington's trip did not materialise, and it left Janaki deeply disappointed.

A Supportive Family

To Janaki her family meant the world; she loved to hear from them. She was a good communicator herself, never failing to keep in touch with her siblings despite her pressing professional commitments. Sister Cousalya from Shoranur, a keen rose gardener, would write to her as often as she could. She was helping Janaki with the maintenance of Ashok, even though she was not keeping too well, making sure its bills were paid on time for instance. She would also give Janaki all the news from her part of the world: 'We are still having rain every day, and the garden is over growing. Harvest is in full swing and people come around with paddy for sale,' she would write on one occasion.[13]

In February 1962, when Janaki was in Jammu, younger brother Madhavan, sent her a small consignment of wild cotton besides several other seeds from Valiyakara, the northern islet of Suheli Par, a coral atoll in the Laccadive Islands; it was a region strikingly rich in marine fauna, much like the Krusadai Island. Madhavan had been hired by his close friend

439

Murkoth Ramunni of Tellicherry (a former RAF pilot and youngest son of Murkoth Kumaran, the well-known Thiya literary figure), Administrator of the Islands, to train the islanders to use mechanised boats and improved fishing gear. The 63-year-old Madhavan was hoping to be in Minicoy by the middle of March and was studying the chart of the island drawn up by one Lt. K. C. Divakaran, ahead of his visit there. 'Valiyakara has yet to be surveyed for its configuration and area. Suheli Par is the South-Western-most atoll with a Barrier Reef enclosing a lagoon,' he enlightened his sister, in a letter sent from Edam (he had come to Tellicherry to exercise his vote in the third Lok Sabha elections), which was accompanied by a sketch of the atoll and a graphic description of this remote location: 'Valiyakara was previously completely overrun with jungle with scattered coconut trees. Water is brackish with lime content. Banyan trees were in existence but no screw pine. Now the coconuts have been well planted and a few temporary huts built, During the four months of the monsoon nobody stays. The rains are comparatively heavier than the other Islands except Minicoy where it is 100 inches. There is a Mosque of rude construction and the tomb of a Tangal much venerated . . . Many miracles are ascribed to him including protection from storms.' Very fond of Janaki, and always eager to send her a botanical collection, Madhavan wrote: 'I wish I could visit Cheriyakara [on the south-east of Suheli Par] to make a collection for you. It would be grand if we could go together to the Islands. Landing from the "Sea Fox" into the smaller boats that would cross into the Lagoons is a problem. We are perfecting this operation. The usual gangway ladder gets smashed when the waves raise the boat below.'[14]

Madhavan's letters to Janaki were invariably about his work or the collections he was making for her. In his last one however, he revealed to her that he was growing 'a Kon-Tiki type of beard—saves the trouble of shaving during cruises,' he joked. He was obviously inspired by his younger contemporary, the Norwegian adventurer and ethnographer, Thor Heyerdahl, who had led the Kon-Tiki expedition of 1947. Madhavan knew that he needed to gain the islanders' confidence before convincing them to use mechanised boats and improved fishing methods. Madhavan's adventurous plan was thus 'to stay on one of these Islands during the four months of the monsoon completely cut off from the mainland. This might be Minicoy or Ameni [Amini] or better still Chetlat.' Such an extreme situation, he believed, would be the ideal time to teach the islanders on how to make better boats and fishing gear.[15]

Janaki's nieces and nephews were most willing to be of service to her as plant collectors. Being interested in high-altitude flora at this time, Janaki had requested her nephew in the army, Major Ravi Krishna (late brother Kittu's younger son) to send her a collection from Leh or Kargil. He would write to her from an undisclosed location that he was 'trying to get the plants,' but was finding it a challenge to reach them to her. In response, Janaki suggested that he despatch the plants to her young friend, the botanist

Triloki Nath Khoshoo (he was Chairman of the Botany Department of the Jammu & Kashmir University in Srinagar at this time, but on the verge of moving to Lucknow to work with K. N. Kaul at the National Botanic Gardens), who would take care of them for her. Despite the army planes not flying in that direction, Ravi was determined to reach the plants to her.[16]

Friends Pass On

The year 1964 saw the demise of two major international figures, both closely associated with Janaki's life: Jawaharlal Nehru on 27 May (after eighteen years as Prime Minister) at Delhi and Haldane on 1 December, at Bhubaneswar. Nehru's health had begun to fail in the aftermath of the Sino-Indian War, and he had spent several months in Kashmir (in 1963), recuperating from his illness. With his death, a major leadership crisis would ensue in the Congress Party and eventually, Lal Bahadur Shastri would succeed him. As for Haldane, who had died of cancer, he had willed his body to the Rangaraya Medical College, Kakinada, for medical research and teaching. That same year, Darlington would suffer a massive heart-attack.

Reflections on Polyploidy and Variation

An international symposium organised by the Indian Society of Genetics and Plant Breeding on 'The Impact of Mendelism' was held at the IARI, Pusa (Delhi), in 1966, and the Society's journal under the editorship of B. P. Pal (since 1941) brought out a special symposium number that year. With 500 members, the Society's President at this time was Swaminathan. Pal delivered the welcome address, presided over by the Minister of Agriculture, Shah Nawaz Khan. Several important geneticists visited the country. Arne Müntzing (Institute of Genetics, University of Lund) spoke on 'The origin and evolution of cultivated plants,' followed by H. Kihara on 'Factors Affecting the Evolution of Common Plants' and Swaminathan on 'The Origin of Macro from Micro-mutations and Factors Governing the Direction of Micro-mutational changes.' That biochemical and molecular genetics had come to stay was reflected in the papers delivered at the symposium; two Indian women geneticists presented papers under this section, Rebecca Sinha of the Banaras Hindu University and one Miss K. Pandharipande. Janaki's paper, 'Hexaploidy in the Evolution of Some Flowering Plants,' was presented as part of the session on Cytogenetics, which had on its panel D. N. De, H. K. Jain, A. Abraham, C. A. Ninan, P. Gopinath, T. N. Khoshoo, S. V. S. Shastry, G. B. Deodikar and L. S. S. Kumar. As she had not submitted a fair version of her paper by the time the special number went to press, it did not appear in the publication.

In her symposium paper, Janaki had argued that hexaploidy, 'a recent or post-ice age phenomenon,' set off new evolutionary cycles in flowering plants. On the other hand, the American evolutionary biologist Stebbins had suggested that polyploidy acted as a deterrent to evolutionary progress, albeit providing a wider setting for hybridisation. While Darlington had speculated that roughly 50% of angiosperm species were polyploid, Stebbins estimated it to be much lower, at 30–35%. Stebbins referred to polyploids as 'dead-ends' and gave them little role in diversification; his view was that polyploids only contributed to 'evolutionary noise.'[17] Evolutionary action was at the level of diploids and not polyploids as far as Stebbins (as well as some of his contemporaries) was concerned, and although polyploidy had resulted in much variation, they were on old themes, rather than significant new departures; a new polyploid species was thought to have originated through a single polyplodisation event and therefore exhibited a high level of uniformity. Alternatively, an allopolyploid would exhibit only homologous or non-segregating variation. Stebbins believed that chromosome doubling in most cases only had a decelerating impact on evolutionary change.[18] His student, the fern systematist Warren H. Wagner Jr (based at the University of Michigan since 1951), who at one time edited the *Indian Journal of Pteridology*, held a similar view. Wagner also claimed that polyploidy never played a major role in the evolution of plants, and that the study of polyploidy had led researchers to be 'carried away with side branches and blind alleys that go nowhere.'[19] However, contradictory to his view that polyploids had limited long-term evolutionary potential, Stebbins discovered several cases of ancient polyploidy in flowering plants and other seed plants.[20] Consequent to this he was willing to concede that at some point in history important families such as the Magnoliaceae and Ericaceae had undergone polyploidy, that is, if flowering plants with x = 12 or higher were ancient polyploids.[21] We know now that polyploids have multiple origins, and contra Stebbins their genomes are highly dynamic and anything but uniform.

At Wisley, based on her work on the *Magnolia*, *Camellia* and *Rhododendron*, Janaki had identified hexaploids as an interesting group of plants, found in locations where the distribution of diploid species had converged as a result of migration or due to changes in the actual physical geography of the land. While hexaploidy expressed itself as gigantism in the culms of Bambusae and Agavaceae, and of flower size in both *Magnolia* and *Camellia*, in the *Rhododendron* it had exactly the opposite effect: it made the plant smaller, associated with which were physiological characters, making it perfectly suitable for alpine distribution (i.e. to grow at a higher altitude, above the tree line, and in harsh conditions such as low temperatures, aridity, intense ultraviolet radiation and a short growing season). Although, some hexaploids resembled the diploid in external morphology (she cited the more recently evolved cultivated species such as the *Narcissus* and *Dianthus*), their chemical composition could be completely different, akin to what she had discovered happening in the medicinal plant, *Acorus calamus*

(Sweet Flag). In the discussion that followed, Janaki cited the example of the hexaploidy, *Narcissus poeticus*, to substantiate her point about hexaploid daffodils being much like their diploid counterparts in resemblance.

Janaki also responded to the papers by S. K. Mukherjee and P. K. Majumdar on 'Genetics and Breeding of Tropical and Subtropical Fruits,' by sharing her experience in breeding and introducing polyploid tropical fruits. She revealed that the tetraploid guava made by her (perhaps at Wisley) was sent to India and grown at the Agri-Horticultural Society's Gardens in Madras (she had sent the plant to Africa as well), and as for her interspecific Citrus hybrid (a cross between Citron and 'Malta Lemon') and its F2 hybrids, they were being grown in Jammu.[22] To Mukherjee's question on how one might explain evolution in Anacardiaceae (especially in *Mangifera*, where all species show 2n = 40 and n = 20 and are distributed in the tropics, particularly in the Malaysian region), when polyploidy is the norm for evolution, Janaki had a quick answer, despite it being a contentious issue. Many Malaysian plants had remained diploid, she stated, because of the uniformity of environment found in the humid tropics, a view point she had put forward at the UNESCO symposium at Kandy in the late 1950s; the more stressful an environment, greater the need for adaptation through allopolyploidy, a phenomenon associated with interspecific hybridisation. In fact, in early 1959 (based at the ISI, Calcutta), Haldane had remarked that alloploidy was 'the most important amendment to Darwin and Wallace's [Alfred Russel Wallace, who independently put forth the theory of evolution by natural selection] account of evolution.'[23]

Dobzhansky had argued that tropical environments were more constant than temperate ones in a geological sense, a point of view Janaki also subscribed to; the temperate zones, owing to the Pleistocene glaciation, suffered drastic climate and biotic changes, as a result of which the present flora and fauna of these places were almost wholly newcomers. In other words, there were far more evolutionary opportunities such as polyploidy in temperate zones.[24] And yet, the tropics demonstrated greater biodiversity, and polyploidy was recognised as a major factor in the accumulation of plant diversity. All this was rather confounding not only then but also today. Dobzhansky speculated that adaptive changes which kept living organisms acclimatised to their changing environments occurred in both temperate and tropical zones, and that the 'constant evolutionary turmoil' prevented evolutionary stagnation in both zones, equally.[25]

While Janaki's hypothesis was that conditions in the humid tropics were far more uniform than in temperate arid regions, and consequently the incidence of mutations was negligible (that is the presence of diploids was high), the British cytogeneticist, Irene Manton (1904–88, of the University of Leeds, worked on ferns and algae, elected FRS in 1961) held the exact opposite view: Manton argued that speciation was faster, with a greater incidence of polyploidy, in the tropics than in the temperate zones.[26] Manton's view was based on her study of the cytology and taxonomy of the pteridophyta flora (gymnosperms) of Ceylon in the 1950s, while Janaki's was chiefly based on studies of flowering plants (angiosperms). Manton had found a far greater number of polyploids among Ceylon pteridophyta than in the northern temperate latitudes, based on which she debunked the understanding that polyploidy was an adaptation to cold, and connected to the Ice Age as suggested by geneticists like Dobzhansky. The results from her survey of Ceylon ferns led her to conclude that evolution proceeded faster in the tropics than in temperate latitudes.[27]

Janaki's inferences, which were based on the strength of her extensive field and cytogenetic explorations, and revolved around her study of flowering plants (angiosperms) in the forests of the humid tropic and temperate zones, appear to have been far more nuanced than Manton's. In fact, it is difficult to name anyone in the world at this time, who had studied flowering plants as extensively as Janaki had done, cytogenetically and in both the tropical and temperate regions. She was certain that all angiosperms had undergone at least one ancient polyploidal event and that polyploidy was one of the most important drivers of diversity among flowering plants. Floristic comparisons suggested that the occurrence of polyploidy, a speciation mechanism, varied according to latitude, altitude and habitat (the geographical distribution of polyploids), but adopting a genecological approach, Janaki also demonstrated that in the associations existing in humid forests there was a clearcut evolutionary hierarchy in the families found in the different registers of the flora in a given location. The highest canopy comprised the 'primitive' genera such as the Magnoliaceae, the middle register of woody climbers, occupying an intermediary position on the evolutionary scale, and the bottom-most of the most evolved of the Gamopetalae (as per the classification of angiosperms provided by Bentham and Hooker; technically, plants that produced flowers with distinct calyx and corolla) such as the *Strobilanthes*.

The crux of the matter was that latitudinal analyses of polyploidy by scientists were biased towards floras of Europe and North America, with very little work done on the tropics or the temperate zones of the Southern Hemisphere. According to Janaki, at one point in time, humid forests were great centres of speciation by hybridisation (by autopolyploidy and allopolyploidy), but which had (over 100 million years, she would suggest) grown static, with a population now chiefly made of secondary diploids. She pointed to the need for more extensive studies of the genetic composition of the flora of the humid forests, to obtain a clearer picture of evolution in the tropics and the planet in general, a synthesising project that continues into the present time. By this she meant, conducting highly detailed studies of the evolutionary pathways (plant phylogeny or the evolutionary chronicle of a plant's life on earth, both extant and as fossils, and the origin and extent of polyploidy), the geographical distribution of cytotype diversity (the extent of cytotypic variation, that is whether as diploids, tetraploids and so on— within and among species), and life-history characteristics (traits such as time taken to germinate, growth pattern, number of offspring, longevity and so on, which determine the organism's fitness to adapt to a changing environment), all of which were crucial in mapping biodiversity. This knowledge could then be used to conserve unique ecosystems.

Personal and Professional Concerns

In June 1966, the southwest monsoon had at last arrived in Malabar, and the weather had cooled down somewhat. Shoranur was suffering from an acute scarcity of water, Raghavan reported. They were having to resort to the 'Engineer's well' near Coustubham, Cousalya's residence. He was still hoping to sell Sugvir rather than rent it out: 'when I get offers of high price for it, I am tempted to sell it, even though I love it and the many plants and trees, my good friends they are.'[28] Parvathi at Edam had recently received a remittance from Janaki, and would send a reply to thank her; she also gave her some local news: 'Schools have all re-opened and there is the usual scramble for admission and rush to buy books and new dresses and umbrellas. The devaluation of the rupee is likely to increase the price of everything!' Parvathi incidentally had no children of her own. In her letter, she also shared news of Madhavan, 'busy in his own way, making boats and sails,' on the verandah of his house at 38, Periapalli Street, Mylapore in Madras. The family had been informed of Janaki's move from Jammu to Srinagar and her appointment at the University as Emeritus Scientist. 'I suppose that is a better place but a bit far away and much colder in winter . . . Hope you are taking care of your health and not over-exerting yourself,' wrote, a concerned Parvathi.[29] Only a few months later, Janaki would be laid up with a bad bout of bronchitis; her family grew anxious, and suggested she take a break, but she would not relent. She now had a young personal assistant named Vishnu; she badgered him into completing his matriculation. Younger brother Madhavan would join her on a holiday sometime in May (1967); this would be their last meeting as he would sadly pass away on 21 October that same year.

Edam was now like an albatross around their necks. Janaki was the only one in any position to help, at least financially, but she was far away in Kashmir. Raghavan's journal described the tragic state of affairs at Edam, a scene that will not fail to resonate with us:

Pappu [brother Padmanabhan, who remained unmarried, and had worked in a Government department] can still be very useful with his experience in Revenue and Magisterial work to the family . . . we should try and keep him cheerful . . . what he needs is a change of place now . . . a car drive around the town and fine scenery. He loves to talk to poor beggars and school children as he did at Shoranur. This life alone in a small room in Edam—not even a chance to gaze at the blue sky is disgusting to him. However much I like

445

to be of service to Edam I am now useless. . . . I can't hear properly or recognize my friends. . . . Swallowing lot of costly pills daily keeps me on my legs. . . . Hopeless for me to bear the multifarious responsibilities of a big house like Edam, its ever tense atmosphere, servants' problems—the staircases, big door steps . . . on top of all the Dharmadom house [Varada Mandiram was lost, as a result of a legal wrangle over its possession] & Devi Nivas debacle! Do leave me alone—Now it is time for me to seek peace of mind in a small cottage in a locality where I can gaze at some river or sea scenery & indulge in some gardening, my best TONIC in old age![30]

The year proved to be one of immense loss to Janaki. Besides younger brother Madhavan, she lost her good friend and confidante, the English horticulturist, Col. F. C. Stern. An exact contemporary of Raghavan, who loved plants equally if not more, Stern was 83 when he died at Worthing, Sussex, on 10 July (1967).

Friends Fading Away

In November that year, Janaki sent Darlington, now wholly immersed in the study of man from the point of view of genetics, a paper on a case of trisomy in a dizygotic twin by her young nephew E. K. Karunakaran (Devayani's eldest son), a paediatrician who had a diploma in child health from the Royal College of Physicians of London.[31] 'I am sure you will be interested in this case,' she wrote to him. It was during the 1950s and 1960s that several of the common chromosomal anomalies such as trisomy 21 in Down's syndrome were reported. The government had planned to employ cytologists in every large hospital in Delhi and 'Dr Swaminathan [plant geneticist M. S. Swaminathan] of IARI is helping with the project,' she enlightened Darlington. She was now Emeritus Scientist and Honorary Professor of Botany at the University of Jammu & Kashmir, she told him. 'Getting old and feeble,' she added as a post-script.[32] She was also becoming forgetful, so much so that she did not convey the news of brother Madhavan's death to him; Darlington had after all met him at Krusadai in the early 1930s.

Janaki reached the end of her term (of five years) as Emeritus Scientist in May 1969. She had had no news from England for quite some time now. Not even Pio Koller, who often wrote to her, was in touch. 'Six out of my seven students have submitted their thesis for PhD. The last one is working on roses of India and is far behind. (She got married and produced a baby instead),' commented Janaki disapprovingly in a letter to Darlington. Meanwhile, Janaki had begun compiling a book on the cytogeography of flowering plants and wished 'so much that it could be a joint publication' with him. She needed that extra push to accomplish something like a book—'someone to discipline her,' she would say in the 1930s—and only Darlington could do that for her. Although she often claimed she was nearing completion of her *Atlas*, in truth, she was far from it. 'I need help. I have still a lot to do ... some of the distribution maps are very exciting—especially of the Himalayan plants,' she would confide to him. Meanwhile, the Nepal Government had sought her assistance with the breeding of medicinal plants. 'I hope they will not change their mind when they know I am a very old lady!' she would comment.[33] Although, she felt it was time for a 'final retirement,' she felt 'quite energetic still,' and wished very much to visit Sikkim, a naturalist's paradise made famous by Joseph Hooker.[34]

Cinderella of the Sugarcane Station

By far, the biggest piece of good news to reach Janaki during this period was from Coimbatore, informing her that her *Saccharum-Zea* cross (created in 1936) had flowered after thirty long years! Janaki had referred to her cross as 'the Cinderella of the Sugarcane Station,' because Venkatraman and Gates had suspected the genuineness of her cross, by far the widest intergeneric cross involving the sugarcane; to make matters worse for Janaki, the cross had not flowered even after four years, as a result of which she had to limit her study to the somatic chromosomes. Now, after almost three decades, her hybrid had flowered, much like in a fairy tale.

Geneticist P. A. Kandaswami of the Sugarcane Breeding Station, made the best use of the situation, and studied the hybrid's chromosome behaviour at meiosis, but Janaki found the work inadequate for her purposes. His research had shown that 40 chromosomes of *Saccharum* formed 20 bivalents, while the *Zea* chromosomes remained univalent (with unpaired chromosomes), confirming the somatic number of 52, reported by Janaki in 1941.[35] Janaki remained convinced that if she and Darlington had put their heads together, they would have done a far better job at investigating the cross. She sent Darlington a reprint of Kandaswami's paper and in a post-script reminded him of what Gates had said about her cross in 1938; 'the Blighter!' Janaki swore in contempt, the incident still fresh in her mind despite the decades that had passed by.[36] Janaki could not wait to see her hybrid in flower, and despatched a letter to J. Thuljarama Rao, Director of the Breeding Station. 'It would be a pleasure and privilege to have you visit us any time convenient to you,' Rao replied.[37]

Janaki would enclose Rao's letter for Darlington to read, because in it he had disclosed for the first time, that SG 63–32, one of the seedlings she had produced by the pollination of POJ 2725, a Java hybrid cane, with *Imperata cylindrica* (a weed) in the late 1930s, had served (and was continuing to) as the best parent ever all these years. It had attained the status of a 'Stud Bull.'[38] Over the past two decades, it had been repeatedly selected as pollen parent for at least 23 Coimbatore canes (Co. canes). Although it could not be released as a commercial cane on account of its low sucrose content/yield, Janaki's *Saccharum-Imperata* hybrid had proved itself as an extremely desirable sire thanks to its 'very good quality,' fecundity and robustness.[39] Two of the 23 Co. canes it had brought forth into the world, Co. 997 and Co. 1007, had even lent themselves to commercial cultivation. Co. 997 was in fact the most popular commercial cane in the state of Andhra Pradesh, owing to its rich sucrose content; at an early age it recorded a sucrose content of 17%, and also gave a record yield of 130 tons of cane per acre in a cane competition plot. As for Co. 1007, it had captured a market in Uttar Pradesh.

'Don't you think I deserve an FRS?'

That the Station had been using Janaki's intergeneric hybrid as a parent to much benefit and for so long made her immensely happy, but it came also as a rude shock to her that the fact had been hidden from her all this time. It was clear that her work had been taken for granted: 'I get no credit for all I have been doing—not even a mention. That is India! Don't you think I deserve an FRS,' she would remark angrily. She was deeply unhappy that her work had gone unrecognised in her country, but the sad truth was that it was not very different elsewhere either. Given the great significance of her contributions to the field of evolutionary biology—plant cytogenetics, breeding and cytogeography—a good part of which was accomplished on British soil during the war, and its aftermath, she indeed deserved an FRS (Fellow of the Royal Society); had this happened, Janaki would have been the first ever 'non-white' or coloured woman scientist to have been so recognised by that great scientific institution, and deservedly so.[40] This was not to be. As a matter of fact, Irene Manton, Janaki's younger contemporary, who she occasionally ran into at the Jodrell and met at international scientific gatherings, had already been made FRS in 1961.[41] As for Janaki's Indian male contemporaries in the field of plant genetics, Panchanan Maheshwari was conferred an FRS in 1965, and B. P. Pal and the much younger Swaminathan in 1972 and 1973, respectively.

Darlington was thrilled to hear that her intergeneric cross had flowered after so very long. 'The news of your *Saccharum-Zea* hybrid is most exciting. Would it be possible for you to bring it to England, perhaps in connection with our next Chromosome Conference?' he queried. The Conference was scheduled for September 1970 in Oxford, but Janaki would not attend. Darlington was also keen to know about her project on the cytogeography of flowering plants, for which she had been accumulating material. 'Perhaps later you could send me a note of your idea of the contents and a list of the maps and I would try to make some suggestions,' he told her patronisingly.[42] This would not happen and the book project would never take off.

Retires as Emeritus Scientist

In 1969, Janaki was on the verge of leaving Jammu & Kashmir, having exceeded the age for an Emeritus Scientist; she was now 72. 'I don't know what to do with myself. I'll go & work with the lepers I saw in Rishikesh,' wrote a despondent Janaki to Darlington.[43] In less than a month, she would write to him again from Srinagar where she had gone to take one 'last look at beautiful Kashmir before retiring. . . . It is as beautiful as ever. I am sitting in the old Residency Garden [in Srinagar] with its wonderful collection of old trees and grand old trees of *Magnolia grandiflora*. I do not know what is before me when I leave Jammu and Kashmir.' A. R. Gopal-Ayengar, her friend, had invited her to the Tata Institute of Fundamental Research in the capacity of adviser, where he was the Director of the Biological Unit. 'What advice can I give in the latest trends in Biology, which is all bio-chemistry [molecular biology]? So I am diffident but if it comes my way I shall accept it,' she told him.[44] From the late 1950s, molecular genetics had reigned supreme; a genome revolution had taken over the world of biological science. It was now all about genes; chromosomes were not glam-orous anymore. Darlington had himself turned reclusive, no more at the centre of things in a world dominated by molecular biology, psychometrics and human genetics, and where evolutionary biology had become a profes-sional field, with its own set of sophisticated tools.[45] If nothing else, Janaki was hopeful that if she went to Bombay, she could cytologically analyse her *Saccharum-Zea* hybrid in detail.[46]

All but one of Janaki's students at the Jammu & Kashmir University had submitted their doctoral dissertations. 'I feel it was a waste of time guiding some of them,' she would remark critically. 'They were more inter-ested in the degree than the work—all except one—B. L. Kaul, who has been working on the effect of chemicals on chromosomes,' almost echoing Darlington's words from three decades ago. Kaul who had won a schol-arship to the Max Planck Institute at Cologne had just returned having completed his postdoctoral project. Her personal assessment of the quality of science (agricultural and biological science chiefly) in India was equally damning: 'Except for Swaminathan and his School in Delhi—Indian Sci-ence is a mere copying of what is being done elsewhere and I am very disappointed at the work in the Universities. Students are more interested in politics than research.' She was nearing the completion of her 'Chro-mosome Dictionary of Indian Medicinal Plants,' she claimed, which she hoped students would find useful. It was an excellent idea, and she had worked out several chromosome counts of medicinal plants already, but it was far from complete. Like her other book projects, the cytogeography of flowering plants and the chromosome atlas of tropical plants, this project too would be shelved. In her letter to Darlington, she enquired after his children; Oliver Darlington was studying at her alma mater, the Univer-sity of Michigan at this time. She ended her letter wistfully, and as usual alluded to her end: 'It is getting time for me to depart from this world but I would love to have a last look at Wisley & Kew and London and you before I depart.'[47]

449

Notes

1 DP: C.109 (J 115), E. K. Janaki Ammal to C. D. Darlington, letter dated 21 May 1969.
2 ARRLJK: R. N. Chopra to E. K. Janaki Ammal, letter dated 3 October 1962.
3 After a stint as Vice-Chancellor of the Mysore University, Panikkar passed away in 1963.
4 Ibid., letter dated 6 October 1962.
5 R. N. Chopra to E. K. Janaki Ammal, letter dated 12 October 1962, private collection.
6 Janaki Ammal and Zutshi, 'Tetraploidy and alkaloid content in Datura metel', 1970.
7 See for example, Janaki Ammal, Sobti and Handa, 'The interrelationship between polyploidy, altitude and chemical composition in Acorus calamus', 1964; Janaki Ammal, Singh and Dhar, 'On the occurrence of natural population of Rauwolfia serpentina in Jammu', 1966; Janaki Ammal and Kaul, 'Cytomorphological studies in autotetraploid Asparagus officinalis', 1967.
8 Janaki Ammal and Gupta, 'Oil content in relation to polyploidy in Cymbopogon', 1966.
9 Janaki Ammal and Dhar, 'Heredity versus Environment in the production of alkaloids in Rauwolfia serpentina', 1967.
10 Zutshi, 'The Transference of Solasodine from Solanum incanum to S. melongena', 'Variation in the Morphology and Solasodine Content of Two Races of Solanum incanum' and 'Interspecific hybrids in Solanum'.
11 Dr B. L. Bradu's personal note on E. K. Janaki Ammal, dated 13 July 2007, given to the author during her visit to the RRL, Jammu. Bradu worked on the Ocimum gratissimum for his doctoral dissertation submitted to the University of Jammu & Kashmir (in 1991).
12 DP: H 115, A. C. Joshi to C. D. Darlington, letter dated 15 June 1963, and Darlington to Miss Coole (Specialist, Tours Department, The British Council, London), letter dated 1 August 1964.
13 E. K. Cousalya to E. K. Janaki, letter dated 9 October 1962, private collection.
14 E. K. Madhavan to E. K. Janaki Ammal, letter dated 24 February 1962, private collection.
15 Ibid.
16 Major Ravi Krishna to E. K. Janaki Ammal, letter dated 17 February 1964, private collection.
17 Stebbins, Variation and Evolution in Plants, pp. 358–359, 366.
18 Ibid., pp. 147–148.
19 Wagner, 'Biosystematics and evolutionary noise'.
20 Stebbins, Variation and Evolution, p. 365.
21 Ibid., Chromosome Evolution in Higher Plants, p. 124.
22 The Indian Journal of Genetics and Plant Breeding, Vol. 26A, Special Symposium Number, 1966, p. 166.
23 Haldane, 'The theory of natural selection today'.
24 Dobzhansky, 'Evolution in the tropics'.
25 Ibid., p. 219.
26 The Indian Journal of Genetics and Plant Breeding, Vol. 26A, Special Symposium Number, 1966, p. 273; Manton was particularly credited for her sharp observations, of botanical specimens, using the electron microscope from the late 1940s. She received the Linnean Society's medal in 1969, sharing it with the British ichthyologist Ethelwynn Trewavas.

27 Manton, 'The cytological evolution of the fern flora of Ceylon', p. 185.
28 E. K. Raghavan to his youngest daughter, letter from mid-1966, private collection.
29 E. K. Parvathi to E. K. Janaki, letter dated 9 June 1966, private collection.
30 E. K. Raghavan's personal notes, private collection.
31 Karunakaran and Pai, 'E Trisomy syndrome in a dyzygotic twin.'
32 DP: C.109 (J 115), E. K. Janaki Ammal to C. D. Darlington, letter dated 29 November 1967.
33 Ibid., letter dated 21 May 1969.
34 Ibid.
35 Janaki Ammal, 'Intergeneric hybrids of Saccharum', 1941.
36 DP: C 109 (J 115), E. K. Janaki Ammal to C. D. Darlington, letter dated 21 May 1969.
37 Ibid., J. Thuljarama Rao to E. K. Janaki Ammal, letter dated 29 May 1969.
38 DP: C 109 (J 115), E. K. Janaki Ammal to C. D. Darlington, letter dated 11 June 1969.
39 Ibid., J. Thuljarama Rao to E. K. Janaki Ammal, letter dated 29 May 1969.
40 It would take seventy-five years for the Royal Society to elect an Indian woman (or of Indian extract) to the Royal Society as Fellow (the Royal Society began electing women as Fellows in 1945); the first to be so recognised was Pratibha Gai in 2015, followed by the microbiologist Lalita Ramakrishnan in 2018, and the clinical scientist Gagandeep Kang in 2019.
41 Irene Manton's sister, the zoologist Sidnie Manton, was elected FRS in 1948.
42 DP: C 109 (J 115), C. D. Darlington to E. K. Janaki Ammal, letter dated 29 May 1969.
43 Ibid., E. K. Janaki Ammal to C. D. Darlington, letter dated 11 June 1969.
44 Ibid., letter dated 8 July 1969.
45 Harman, *The Man who Invented the Chromosome*, p. 256.
46 DP: C 109 (J 115), E. K. Janaki Ammal to C. D. Darlington, letter dated 8 July 1969.
47 Ibid.

25

TROMBAY

A Radiation Interlude

Janaki accepted A. R. Gopal-Ayengar's invitation to join the Biology Unit of the Bhabha Atomic Research Centre, Trombay in the capacity of adviser; she would remain there a little over a year, until late October 1970.

Anekkal Ramaswamiengar Gopal-Ayengar (1909–92) had graduated in Biology from Mysore University in 1929 and completed his MSc in 1933, after which he joined his alma mater as lecturer in cytology. In 1938, he won a Vincent Massey Fellowship to go to the University of Toronto, from where he obtained MSc and PhD degrees. After teaching and researching in Canada for a few years, Gopal-Ayengar moved to the United States, where at the Barnard Free Skin and Cancer Hospital, Washington University, St. Louis, he was Kettering Research Fellow. He was one of the first to isolate DNA from mouse chromosomes, the details of which would be published in *Cancer Research*.[1] Gopal-Ayengar returned to India in 1947, to embark on a career in industrial research. He first took up the position of Chief Research Cytologist at the Tata Memorial Hospital, Bombay, where he would continue to be until 1951. A year later, he was appointed Assistant Director of the Biological Division of the Department of Atomic Energy, Government of India (Trombay) and a decade later, became its Director. Following physicist Homi Bhabha's death in 1966 in a plane crash, the Atomic Energy Establishment was renamed the Bhabha Atomic Research Centre (BARC). An atheist, and a connoisseur of Carnatic music, one of his biological interests was the study of the mutagenic effects of radiation on plants, for which he began the Gamma Garden in BARC.[2] Gopal-Ayengar and his associates, K. P. George, K. K. Nayar, K. Sundaram and K. B. Mistry, worked on inducing mutations in plants using gamma rays between 1969 and 1972, and published on radiosensitivity, understood in terms of meiotic abnormalities in pollen mother cells, an interest shared by Janaki and the mouse geneticist Grüneberg. The BARC group worked on plants growing in the monazite-bearing high-radiation areas of the Kerala coast and Manavalakurichi in Tamil Nadu. Janaki established close personal and professional ties with this group.

Janaki was attracted to Trombay because it offered her facilities for the study of her *Saccharum-Zea* hybrid and also the *Solanum khasianum* material she had brought with her from Jammu. Janaki had hurried to Coimbatore

DOI: 10.4324/9781003267089-25

to see her *Saccharum-Zea* hybrid in flower and had brought back with her a plant in the hope 'to do something to make it flower here,' but the experiment would not meet with success, which meant that she would have to go back to Coimbatore to study its behaviour at meiosis. 'I shall send you my findings,' she promised Darlington, and described her Trombay spell as a 'Strong Interlude' in her life. With the *Solanum khasianum*, however, she was lucky; in collaboration with Bharati Bhatt, a biologist from BARC, she produced tetraploids and 'some interesting mutants' by subjecting the material to gamma rays and colchicine. This was the first time tetraploidy had been successfully induced in *Solanum khasianum* by anyone anywhere in the world. 'It has been a very important source for Solasodine, used in the synthesis of Cortisone,' she explained to Darlington. Simultaneously with the profusely illustrated longer version sent to the *Proceedings of the Indian Academy of Sciences* (Biology Section), Janaki would send a short note on this to the *Current Science*.[3]

At BARC, Janaki also collaborated with V. Abraham of the Nuclear Agriculture Division, to publish a paper on the cytology of the genus *Apluda*, a monotypic genus (with only one species in the genus like the Nicandra) belonging to the tribe Andropogoneae, to which sugarcane belonged. Hooker had described the single species *Apluda varia* as 'unstable and polymorphic,' and in 1960, Janaki's friend Bor had lumped the two subspecies and their varieties, under *Apluda mutica* Linn. She was certain the species deserved a thorough cytotaxonomical investigation. The *Apluda* that Janaki found on Trombay Hill was a slender variable grass, used as fodder for horses. She examined twelve plants selected at random and studied their behaviour at meiosis in the pollen mother cells and discovered one or two supernumerary chromosomes, besides ten bivalents; these were smaller in size than the normal chromosome complex, with which they did not pair.

In their *Chromosome Atlas*, Darlington and Janaki had warned that it was not sufficient to just confirm the chromosome number of a plant, but important to look for the genetically inert supernumeraries or B chromosomes, which when present in large numbers could cause infertility and loss of vigour in the species. In fact, Janaki had been observing and studying the significance of these B chromosomes as early as the 1930s (in the Para-Sorghum for instance). In the short note on *Apluda* published in 1970, Janaki and Abraham stated that the 'origin, distribution and evolutionary significance of these supernumerary chromosomes' could explain the polymorphism that Hooker had observed in this widely distributed genus. An extensive survey of the genus in other parts of India, she believed, would throw up evidence on polyploidy in the species.[4]

A Defeated Man

Darlington had published his epic *The Evolution of Man and Society* (1969) by this time; this was yet another controversial volume from him.[5] In the preface, he warned that the book was intended to raise more questions than provide answers; it was a world history written in terms of gene-flows (or gene-flux), and located man and his plants as part of the same evolutionary system. In his book, Darlington acknowledged Vavilov and his colleagues for their help and encouragement in Russia and Transcaucasia in 1929 and 1934, Hitoshi Kihara and F. A. Lilienfield for introducing him to Japan in 1933, 'Dr E. K. Janaki Ammal who guided [his] steps in India (in Travancore and in Orissa) in 1933 and 1937,' and Eileen Erlanson who introduced him to many American problems between 1932 and 1967. He also acknowledged his father (who had brought to his notice problems of dialect, place names and the poet Lucretius), besides J. H. Hutton, the anthropologist and administrator of the Indian Civil Service, I. H. Burkill, the botanist and *Dioscorea* expert, who was also well known to Janaki.[6] Janaki thought it was a 'great book'[7]; the visiting Pio Koller had reached her a copy at Trombay. Koller had recently retired as Professor of Cytogenetics at the Chester Beatty Research Institute, London, which used to be the research wing of the Royal Marsden Hospital and a constituent college of the University of London. Darlington's large tome would keep her occupied for quite some time to come; perhaps it was Koller who photographed the aging Janaki seated amidst the rocks on the beach at Trombay, with the volume near her.

In an earlier letter to her, Darlington had come down heavily on Haldane: 'You were unnecessarily hard on Haldane!' she would berate him in her reply,[8] but he was relentless. In yet another letter, he continued his attack on Haldane: 'Apropos of Haldane. No nation could survive a large number of such men. His betrayal of Russian Genetics was on a par with his treatment of Lawrence [W. J. C. Lawrence] at the J. I. H. I.: both nasty jobs.' He remarked that he would rather listen to her telling him a thing or two about his book and how it 'could be improved or corrected,' rather than discuss Haldane.[9] Within a short time of this publication, on 9 January 1970, Darlington's second son Andy (Andrew Jeremy), who had completed a doctorate in plant genetics and some postdoctoral research in a Stanford University Laboratory in Palo Alto, took his life. It is believed that he had inherited mental instability from both sides of his family, and had suffered long spells of depression starting from his late teens. Andy had begun to detest his father for his reactionary views and intimidating style, so much so that he evinced little interest in his father's magnum opus.[10] The suicide of his favourite and brilliant son would crush Darlington; a defeated man every way one could imagine, personally and professionally, he would turn to gardening for solace and sanity.

Janaki was deeply distraught when she received news of Andy's death, and could not get herself to write to Darlington for quite some time. 'I should have written long ago before this but after hearing about Andy's death I found it difficult to write. I still cannot understand why it should have happened. He was so loveable. I think of your children as I left them years ago. I had a glimpse of them as I passed through London on my way back to India from Oakridge in 1958 that's all. They looked well & happy then,' she wrote overcome by deep sadness.[11] It was actually in 1955 that she had seen them last, on her return journey from America; age was obviously catching up with her.

Venkatramanism

Janaki would write again to Darlington after a week. The only matter of this letter was her intergeneric hybrid (SG 63–32), the Station's 'Stud Bull'—the cross between the Java cane (POJ 2725) and the *Imperata*, which she had confirmed to be a genuine hybrid, way back in 1941. She enclosed for his perusal the parentage of the best canes of Coimbatore, 'all of which' she noted had 'the 'blood' of [her] hybrid.' However, in the parentage assigned to the new canes, SG 63–32 had been described as a 'Parthenogenetic derivative! of POJ 2725' (which meant that it was merely a product of asexual reproduction), whereas in truth it was the result of sexual reproduction, with Janaki's cross as the male parent. This important fact had not only been cunningly kept away from her but also erased from history. 'This is a Venkatramanism!' she would remark in utter disgust. Not one to let this pass, and deservedly so, Janaki took it up with her friend B. P. Pal, Director General of the IARI, but it does not appear anything came of this.[12] Meanwhile, Janaki had been offered a place at the

Figure 25.1 Janaki on the Trombay beach, with Darlington's book, *The Evolution of Man and Society* (1969), beside her. October 1970.

Source: Photo probably by Pio Koller. Courtesy of the EK family.

Centre for Advanced Study in Botany (CAS), University of Madras at Chepauk, under the directorship of a far younger contemporary, the plant virologist T. S. Sadasivan.

Notes

1 Gopal-Ayengar and Cowdry, 'Deoxyribose Nucleic Acid from Isolated Chromosome Threads in Experimental Epidermal Methylcholanthrene Carcinogenesis in Mice'.
2 'Anekal Ramaswamiengar Gopal-Ayengar (1 January 1909–09 September 1992)'.
3 Janaki Ammal and Bhatt, 'Induced Tetraploidy in Solanum khasianum Clarke', 1971; 'Observations on the Glyco-Alkaloid Content of Diploid and Tetraploid Solanum khasianum Clarke', 1971.
4 Janaki Ammal and Abraham, 'On the Occurrence of Supernumerary Chromosomes in the Genus Apluda', July 1970.
5 See Harman, *The Man Who Invented the Chromosome*, pp. 228–234 and 248–250.
6 Darlington, 'Acknowledgements' in *The Evolution of Man and Society*.
7 DP: C 109 (J 115), E. K. Janaki Ammal to C. D. Darlington, letter dated 20 October 1970.
8 Ibid.
9 Ibid., C. D. Darlington to E.K. Janaki Ammal, letter dated 26 November 1970.
10 Harman, *The Man Who Invented the Chromosome*, pp. 253–254.
11 DP: C 109 (J 115), E. K. Janaki Ammal to C. D. Darlington, letter dated 20 October 1970.
12 Ibid., letter dated 29 October 1970.

26

MADRAS III

The Madras Mint, Solanum and Other Stories

My little lab, the Botany Field Research Station is ten miles
out of Madras. It is my hermitage—beautiful with rice fields
and coconut trees. I have now added sugarcanes and bananas
besides medicinal plants.

—E. K. Janaki Ammal (1971)[1]

Graduating from Presidency College, Madras in 1934, Toppur Seethapathy
Sadasivan (1913–2001) went to Lucknow to secure a Master's degree in
Botany under Birbal Sahni. Sadasivan undertook doctoral research on a government scholarship at the Rothamsted Experimental Station in Harpenden
in plant virology under the guidance of F. C. Bawden and S. D. Garret,
and completed his dissertation in 1940. A little later, he returned to India
and began his career as a microbiologist at the Punjab Agricultural College,
Lyallpur. When he was offered a Reader's post at the University Botanical
Laboratory (UBL) of the Madras University, he seized the opportunity, and
eventually became its Director in 1944, succeeding M. O. P. Aiyengar.

At the UBL, Sadasivan successfully developed a prolific research group,
several of whom were women, to undertake studies on soil-borne diseases
which attacked cash crops such as cotton, pea and rice. Algology, which had
been Aiyengar's focus, was given equal prominence in the research agenda
of the UBL under Sadasivan. Several of the concepts he developed through
the study of plant pathogens, such as 'saprophytic ability' and 'rhizosphere
effect,' were put to use by mycologist C. V. Subramanian (who would go on
to win the Shanti Swarup Bhatnagar Prize) in his classification of Hypho-
mycetes. Considering the substantial and pioneering work being done at the
UBL under Sadasivan's leadership, in 1964, the University Grants Commission recognised it as a Centre for Advanced Study.

Janaki joined the CAS in November 1971, at the invitation of Sadasivan.
Her return to Madras to the metamorphosed UBL, her alma mater, was like
homecoming after a prolonged scientific pilgrimage. Interestingly, it also
signalled a return to her first love, the *Saccharum*.

DOI: 10.4324/9781003267089-26   457

Revisiting the Sugarcane-Maize Hybrid

Janaki took up residence at the Field Research Laboratory, Maduravoyal, on the outskirts of Madras; she had nieces and nephews living in the city, but she wanted to be independent. The first task at hand was a quick trip to Ernakulam to attend the wedding of a niece (Varadan's youngest daughter, who had resided with her in Jammu and studied at the university there). Janaki's elder brother Raghavan was unable to attend, so she had an important role to play at the marriage. At this time, the aging Raghavan and his wife were residing in Madras, at Nathan's Colony, Chetput, with their youngest daughter. 'It is a great consolation,' Raghavan would write to younger sister Parvathi at Edam, that 'Jani [Janaki] decided to attend it [the marriage], even though she had not enough funds. Where there is a will, there is a way.' This was the first time a member of the EK family was marrying a Southerner (from Travancore-Cochin, rather than from Malabar). 'May it prove beneficial in every way. Chitra is an . . . educated girl and is sure to be happy. Varadan will bless her from [the] heavens!' Raghavan would remark with a tinge of sadness.

Janaki had brought with her from BARC, *Solanum khasianum* (diploid and the gamma-rayed tetraploid), *Mentha citrate* and *Mentha arvensis*, the *Saccharum* x *Zea* hybrid and *Saccharum* x *Imperata* F2 seedlings. She had just settled down to work in Madras after the Ernakulam visit, when she received news from Coimbatore that her *Saccharum* x *Zea* hybrid, 'perhaps the widest plant-cross ever made,' was about to flower (again), this time as the result of treating it with gibberellin. Though the hybrid had not flowered at BARC, she had requested Gopal-Ayengar to persist and 'grow it in the gamma field there.'[2] She rushed to Coimbatore. In a letter posted from the Sugarcane Breeding Station to Darlington, she included a few observations on the chromosome behaviour of the hybrid. At Coimbatore, perhaps it was at the Agricultural College that she had chanced upon a beautiful collection of wild rice from Africa, many of which were tetraploid 'and looking like little bamboo plants!' 'I wish you would visit once more,' she pleaded with him.[3] She had not been able to attend the Chromosome Conference at Oxford, which Darlington would report 'was a success.' The next one was to be in Jerusalem, where he hoped they would be able to reunite.[4]

Janaki wrote again to him on the very next day; she was back in Madras having travelled by an overnight train. She enclosed a short paper on her hybrid's somatic chromosomes, 'which showed a variation in the number of B chromosomes' (or the supernumerary chromosomes, which are not essential for the life of the species but sometimes had a role to play in fertility and hybrid vigour) of the maize parent (*Zea mays*) in the hybrid. In 1938, the hybrid had the expected 40 + 10 chromosomes and two B chromosomes, but as a result of vegetative propagation for over thirty years, Janaki discovered that the number of B chromosomes had increased to six or more. As an acknowledgement of their help with the study, she included as co-authors, the names of D. Jegathesan and T. V. Sreenivasan of the Breeding Institute.[5] She wanted Darlington to review her paper, and recommend changes or make suggestions.[6] This time again, Janaki was unable to complete her studies on the pollen mother cells because they were not yet in the right stage of development. She hoped to accomplish this during her next visit. Speaking of which, the Sugarcane Breeding Institute had been subsumed into the ICAR in 1969 and seven years earlier (in 1962, the Golden Jubilee year), a new research centre had been opened in Kannur to house the world collection of sugarcane germplasm.

Janaki began to grow anxious about the paper she had sent him for review. And could not resist writing again after a couple of weeks: 'I hope you received the paper on somatic chromosomes of *Saccharum* x *Zea* hybrid. You can do what you like with it—if there is anything interesting in it, as I think there is. I did not get enough good pollen mother cells to study meiosis in detail . . . I will report in detail when I get more material . . . This brings you my best wishes for your birthday—and all the best for the New Year. I hope we will see you in India as you pass by to New Zealand!'. To his suggestion of meeting at Jerusalem at the upcoming Chromosome Conference, Janaki commented cryptically, 'I hope there will be a Jerusalem to meet.' She had only read parts of his 'mighty volume,' but promised to tell him what she thought of it once she was through.[7] When she did not hear from him for two months, Janaki wrote again to enquire about her paper. 'I wonder what you did with it. If you think it is not worthy of publication kindly return the same to me,' she demanded, somewhat peeved by what she thought was his indifference.

She would soon have some good news from BARC: her plant had produced an inflorescence upon irradiation with gamma rays! However, they had delayed reporting this for some reason. She had been trying to induce polyploidy in the hybrid by treating it with colchicine at Maduravoyal but had not met with success; she had also instructed her contacts at Coimbatore to induce tetraploidy in it 'somehow.'[8]

Meanwhile, Janaki heard from Darlington, who suggested a thorough revision of her paper on *Saccharum* x *Zea*. Wasting no time, Janaki began rewriting her paper, incorporating his suggestions, and sent it back to him. On this occasion, she enquired if he had had the chance to see A. Fedorov's *Chromosome Numbers of Flowering Plants*, which the Academy of Sciences of the USSR (the V. L. Komorov Botanical Institute) had recently published (1969). She had not seen it herself it appears, but was keen to know, if he had any plans to bring out another edition of the *Chromosome Atlas*. Janaki then mentioned her own plan of producing 'a Chromosome Catalogue of Medicinal Plants of India.'[9] From the time she returned to India, Janaki had been claiming to be working on one or the other book, including her own version of the *Chromosome Atlas* (of tropical flowering plants), as well as a

Dictionary of Medicinal Plants (perhaps this was the same as the 'Chromo-some Catalogue of Medicinal Plants'), but none of these would materialise.

An inexplicable diffidence weighed her down when it came to consolidat-ing her researches in the form of a book, which comes as a surprise given her supreme self-confidence otherwise. A feeling of loss and frustration had begun eating into her core, ever since her return to the country, however well-placed she was professionally. She was disgruntled with her research career and had begun to think of her life as a failure. Darlington's verdict remained the ultimate word for her, his approval mattered most, even when she admonished herself for this excessive reliance on him. She had so much to prove to Darlington, and even more to herself, but felt time was not on her side. As soon as Janaki's revised paper had reached him, Darlington sent it away to the editor of *Heredity*, accompanied by a somewhat imperi-ous note: 'I enclose for *Heredity* a paper which I have been fostering for 20 years. It has all the defects of India but still I believe it ought to see the light since it describes the most-remarkable plant hybrid ever produced.'[10] Janaki's paper would eventually appear as a note and comment rather than an article in *Heredity*: 'You will see when the proofs arrive that a number of changes have been made in the text and that the figures of the glumes have been omitted,' the editor explained to Janaki.[11]

In early December 1971, Janaki would admit that she found Darlington's comment (on her *Saccharum* x *Zea* paper) that 'the abnormalities observed at anaphase are presumably in the A chromosome' most apposite. She was studying meiosis in the hybrid and was 'seeing baffling configurations.' 'I wish I could show you the slides. I work all alone here—occasionally I visit the Sugarcane Breeding Station at Coimbatore or one of the workers there come to me,' she told him. Incidentally, Janaki would introduce to the Maduravoyal Field Station, a banana (*Musa*) from Assam, the pulp of which was suitable for use as baby food. She would send photographs of the *Musa* to both J. Joseph of the BSI, Coimbatore and N. W. Simmonds of the Scottish Plant Breeding Station, seeking help in identifying it. One of her aims was to start a germplasm collection of wild Indian species of the *Musa*, with special reference to South Indian varieties.

In her letter to Darlington, Janaki also spoke of the ongoing third Indo-Pak war (3 December–16 December 1971), and all of sudden it took on a nationalistic tone: 'We are going through our third war with Pakistan. The enemy is trying to walk into Jammu—so as to cut off Kashmir—I hope this time we will not stop before we take back territory occupied by Pakistan in Kashmir.' She now had good things to say about Indira Gandhi's abilities: 'You are right, Indira is stronger than her father—I do not know why the west is on the side of Pakistan after all the atrocities done in E. Pakistan. It is now declared Bhangla Desh [Bangladesh]!' In the same letter, she wished Darlington well with his impending tour to Australia and conveyed her 'salaams' to common friends H. N. Barber, Otto Frankel and Ann Wylie.[12]

Janaki was evidently unaware of Barber's untimely death in Sydney (from a rare form of cancer) a few months earlier, on 16 April 1971, at the age of 56. Meanwhile, Janaki had run into Pio Koller in Madras (she had met him in Bombay some time ago), in the company of Ann Wylie (the co-author of the new *Chromosome Atlas*), who had trained under her at Wisley.[13] Wylie had been teaching at the University of Otago since 1961.

Crafting the Madras Mint

We may recall that Janaki had produced a colchiploid (a colchicine-induced polyploid) of the Japanese Mint at RRL, Jammu, in 1960, which she called the Jammu Mint. Owing to its thick stem in comparison to the Japanese one, it was found unsuitable for commercial cultivation. To remedy this problem, Janaki backcrossed the Jammu Mint with the Japanese Mint, thereby producing a triploid form of Japanese Mint. In 1970, it flowered for the first time at the Field Research Laboratory, Maduravoyal. Named Madras Mint, it began to be propagated at Madras and Coimbatore to determine how its oil content varied across habitats, a necessary piece of information to ascertain its viability as a commercial crop.[14]

Concerns About Country and Family

'India is going from bad to worse in every way. Calcutta is become a slaughter house! In Malabar life is very much in danger—so also property,' Janaki remarked in a letter to Darlington in early 1971.[15] In West Bengal, the CPI (Marxist) had formed a United Front government in coalition with the Congress in 1967. That year, a peasant uprising, the May 1967 rebellion, organised by a small breakaway faction of the CPI(M), headed by Charu Majumdar and Kanu Sanyal, had broken out of the twin villages of Naxalbari and Phasidewa in Darjeeling district, inhabited by landless and largely illiterate Santal peasants and tea-garden workers. In November, the United Front government was dismissed by the centre and the Congress formed a minority government, but this would again be short-lived. Bengal was brought under President's Rule. Fresh elections were held in 1969, and CPI(M) emerged as the largest party in the Legislative Assembly, but in March 1970, within a few months of forming the government, Chief Minister Ajoy Mukherjee was forced to resign and the state was once again put under President's Rule. In the years 1970–72, the police carried out an 'annihilation campaign' against the Naxalites, leading to extreme violence and bloodshed in Calcutta, after which elections were held once again. Naxalbari would come to be identified with one of the most violent peasant agitations in modern India.

Kerala witnessed something similar. Here too, the Communist Party of India remained allied with the Congress throughout 1970–77, forming a coalition government called the *Aikya Munnani* (United Front), with the CPI leader C. Achutha Menon as Chief Minister; the first Achutha Menon ministry in 1969–70 had lasted less than a year. President's Rule was imposed between August and October 1970, after which fresh elections were held to elect the fourth Kerala Legislative Assembly. The United Front won a majority and Achutha Menon was again sworn in as Chief Minister. During his seven-year tenure, Menon began development projects (he propelled forward the so-called Kerala model of development characterised by its high literacy rates, widespread access to health care and education, and a politically conscious population, all despite low or moderate incomes). Very importantly, he founded several institutions of excellence such as the Sri Chithira Tirunal Institute of Medical Sciences, the Centre for Development Studies, the Centre for Earth Science Studies, Keltron, Kerala Agricultural University, Cochin University of Science and Technology, the Forest Research Institute at Peechi and so on, attracting such eminent people as M. S. Valiathan, K. N. Raj, K. P. P. Nambiar and Laurie Baker to the state. However, what has never been acknowledged is Janaki's role in drafting a science policy for Kerala during this time; it was to Menon's credit that Janaki was brought in as an expert consultant, the only woman on the committee formed to draft the policy. It was also under his leadership that the Kerala Land Reforms Act of 1963 was amended in 1969, and the *jenmi* system abolished, with the full right of ownership conferred on the tillers of the soil. This development adversely affected landed families, including Janaki's, even if it had already been reduced to just a pale shadow of what it was during her father's time; perhaps it was during this period that the family lost their hold on Varada Mandiram in Dharmadom.

In 1971, Janaki's sister Cousalya passed away, survived by her husband, the litterateur Malabar K. Sukumaran, and daughters Tara and Leela. With Cousalya's departure, Shoranur had become a desolate place for Janaki. Brother Raghavan and wife had already moved out of Shoranur to live with his daughters (he had not yet sold Sugvir though), and brothers Vasudevan and Varadan were long gone. Janaki would however visit Shoranur often to look up Ashok, which had begun to develop as a full-fledged medicinal/ethnobotanical garden. It was about this time that she had suggested to brother Raghavan the formation of a 'club' to govern Edam; she had attempted such a thing earlier too but no consensus had been arrived. Like before, Raghavan believed the idea to be a good one, but the fact that members were scattered far and wide was proving a deterrent. Nephew E. A. Ram Mohan (the late Vasudevan's son) was to be President; Savithri, Leela, and the other nieces were to represent the various branches of the 'EK sons and daughters.' The proposal was generally approved, and Janaki kept informed.

'Poor Jani is doing her best for Edam & every one. Only she & Sumi [Sumithra] have their own ways. Jani spends money, but Sumi tries to make money from everyone. So . . . clever & also efficient she is. She is now looking a wreck—swollen face and leg,' 87-year-old Raghavan would write to his youngest daughter, frequently apologising for his illegible writing: 'Your old father's head and hand are not steady now. So read patiently.' As Edam's eldest surviving member, he was determined to see that 'everyone of EK family . . . co-operate[d] to keep Edam intact in dignity—even if so many members are in distant places—. . . honest souls our father & mother—even though they did not believe in any religion.' He very much wanted to visit his father's tomb at Edavaleth House, Mahe, but the Mahe bridge was under repair, and it was a tedious journey to make for one at his age. Their mother was buried on the grounds of Varadamandiram in Dharmadom, but the tenant had become a bother; the family wondered what best they could do to get it back, but this was not to be.[16]

Conservation of Genetic Resources

Otto Herzberg Frankel (1900–98), the Austrian-born Australian developmental geneticist, as earlier noted, was one of the first to warn of the dangers of the loss of plant biodiversity (the term biodiversity had not yet come into use), a concern that Janaki herself shared very deeply. It was not only a concern for biodiversity that drove them, but for genetic diversity, the need to enrich the gene pool of both domesticated species and wild and primitive cultivars. Janaki was an early pioneer of genetic conservation practices. We may recall that even as early as the 1930s, she had precociously sensed the threat to genetic diversity in the breeding practices of the Sugarcane Breeding Station at Coimbatore under the direction of T. S. Venkatraman. In mid-1950s, Janaki had openly articulated her anxiety over the loss of genetic diversity, on at least three scientific platforms. She considered the centres of origin (or diversity) of crop plants as extremely important to the Vavilovian plant breeder (she described herself as such, on one occasion), in introducing new disease-resistant and early maturing varieties, and thus deserving of conservation. Calling them natural germplasm banks, Janaki stated that the centres were to be conserved in situ by all means, for their potential value to food security. Frankel, we might recall, had visited India in January 1957 to attend a symposium on genetics and plant breeding organised by the Indian Society of Genetics and Plant Breeding, which however Janaki had not attended.

It was Frankel who had coined the term, 'genetic resources' in 1967, along with the Ulster-born woman plant geneticist, Erna Bennett (1925–2012), who worked with the United Nation's Food and Agricultural Organization. The term was coined to explain that genes themselves were a resource, and one that was depleting at a fast pace, with modern pedigreed seeds (the elite crop varieties) created by corporate laboratories replacing traditional peasant seeds, witnessed conspicuously during the years of the so-called green revolution. A few years later, Frankel would present his classic paper, 'Genetic

Conservation, our evolutionary responsibility' at the International Genetics Congress held at the University of California, Berkeley (1973) and in 1981 his book titled *Conservation and Evolution* (with Michael E. Soulé) would be published. The book argued that the world was dependent on genetic resources such as the wild species of a plant and on a wide range of different species (interspecific and intraspecific) to improve crops (as also livestock, forestry and fisheries) by making them disease-resistant and high-yielding. Thus, the protection and sustained use of genetic resources was crucial to increasing food security, improving nutrition and livelihoods in the world.

INSA Project on Ethnobotany

In December 1971, the Indian National Science Academy (INSA) sanctioned and funded a project, 'Ethnobotanical Studies of South Indian Tribes' to be based at the Field Research Laboratory in Maduravoyal, with Janaki as Principal Investigator; funds would be released in February 1972. Field stations were sites that combined laboratory and field practices elegantly, and were usually located in the suburbs, on the outskirts for obvious reasons. Today however, the Maduravoyal Field Station, located on the Chennai-Bangalore highway, has witnessed pressures on land and high levels of air pollution, a situation far from ideal for botanical or agricultural research.

Janaki in the 1950s, we know, had included ethnobotanical research (the study of the dynamic relationships between peoples, plants and environments, from the distant past to the immediate present) among the list of objectives of the CBL but that aim had remained largely unaccomplished. She hoped, she could realise it now through the INSA project. The INSA project's goals were four-pronged: to obtain full and first-hand information regarding the botany of plants collected or grown by primitive tribes of Kerala and Tamil Nadu who lived in the hilly tracts bordering the Western Ghats of these two states; to discover and evaluate the culinary or medicinal properties of the plants used by tribes; to bring together a living collection of such plants as may prove useful for introduction in agriculture, horticulture and medicine; and finally, to initiate cytological and genetical studies on selected plants for enhancing medical and biological resources. The ethnobotanical project was really only a modest scheme with two research fellows and a field-man, and a contingency grant of Rs 2,500. Incidentally, Janaki was at this time receiving no salary but only an allowance, as principal investigator. The group, including 75-year-old Janaki, suffering from frequent bouts of pain caused by her fibrositic condition and history of fractures, but nevertheless untiring, explored the forests of Tamil Nadu and Kerala to gather information from tribals about plants foraged or grown by them for food and medicine. Plants and seeds brought back from field trips were grown in the 'ethnobotanical garden' at Maduravoyal, to be later subjected to cytogenetic and chemical analysis.

The first field trip was undertaken over the week between 22 and 31 March 1972. On the 23rd, Janaki and senior research fellow T. V. Vishwanathan, accompanied by the field-man A. Sundararajan arrived in Coimbatore by an overnight train from Madras. Incidentally, Vishwanathan had just registered (or was about to register) for a PhD in Botany under K. S. Manilal at the Calicut University. At Coimbatore, they met with J. Joseph, head of the BSI, Southern Circle, and spent some hours studying the collection of the 'Flora of Nilgiris' in the herbarium; incidentally, Joseph had replaced K. M. Sebastine, who it appears had passed away a few years ago.[17] On the following day, the team set out nice and early for Ootacamund. En route, half-way between Mettupalayam and Coonoor, they stopped by at the 14-acre Burliar State Horticulture Farm (established in 1871 by at the British), with its large collection of fruit trees, where they were introduced to three Irula men who worked as casual labourers on the farm. After a brief interview with them about plants that they grew and their uses, the team proceeded to Ooty, where they first visited the Botanic Gardens. With the help of the Curator of the Gardens, Janaki and her research assistants interacted with a few Toda men and also saw their distinct-looking huts on the upper side of the Gardens. Their next halt was the Indo-German Nilgiris Development Project, where they met with the Director A. R. Bhaskaran, to enquire about the possibilities of meeting Badagas. After what was a purely exploratory tour, they returned to Coimbatore for the night.

Early next morning, Janaki, Vishwanathan and Sundararajan left for Avalanche (some 28 km from Ooty), with the intention of collecting in the Forest Reserve there; the height of the place was 2,125 m above mean sea level, and the soil loamy. Almost twenty specimens were collected from Lakkidi in Avalanche; they also collected some orchids from the *sholas* (grassland complex). At the end of the day, at the Forest Rest House in Avalanche, the collected specimens were 'poisoned' using mercuric chloride and methyl alcohol, placed between blotting papers and pressed; this process would be repeated each evening during the collecting period. More collecting followed on the next day, at Mulli, located at a height of 1,925 m, with black humus-rich soil, and home to a few Toda families, some of whom they met with. On the penultimate day, collecting was done at the trout hatchery area in the *sholas* of Avalanche but not much of interest was to be had at this location. The Avalanche Power House region was chosen for the final day's excursion, from where fourteen specimens were collected. On the 30th they returned to Coimbatore, but would continue their journey further north to Tellicherry, to make a short call on Edam, where sister Sumithra ruled the roost. They boarded a train to Madras a day later.[18]

Repeated visits to the tribal areas were necessary to instil confidence in the people, as at first they would be reluctant to divulge information, especially about those plants used in medicine. Of the three years devoted to the project, the last two were more intensive than extensive: a small group from a particular tribe would become their focus, rather than a large population. The final phase was also devoted to intensive cytogenetic and cytosystematic investigations on the plants collected, which involved tracing their relationship to cultivated plants and reflecting on their possible uses in agriculture, horticulture and medicine.

Communicates with Darlington

Upon return to Madras, Janaki's thoughts briefly lingered on Darlington; she wondered if he had returned from his tour to Australia and New Zealand, and whether he had met with John Inneans Barber and Ann Wylie. Only a few days earlier (in August 1972), she had received offprints of her paper on the *Saccharum* x *Zea* hybrid. 'I have been receiving requests from all over the world! I have to thank you for editing it so well,' she wrote to Darlington. She had now completed meiotic studies on the hybrid as well. 'I would like you to read the paper and improve it,' she wrote to him. She also discussed her INSA project with him: 'Some of them have "secret" medicines and herbs for family planning etc.' Janaki and her research assistant had collected a number of medicinal plants from the Paniyars and Kadars, the 'food hunters of Malabar.' 'My next visit will be to the Todas of Nilgiris,' she informed him. Janaki had been reading W. H. R. Rivers' book on the Todas, in preparation for the visit. 'If you can get it, please read it. They are a unique people,' she remarked.[19] She had been awaiting the return from Russia of Jegathesan of the Sugarcane Breeding Institute to finalise their paper on meiotic studies of the *Saccharum* x *Zea* hybrid and send it to the Chromosome Conference at Jerusalem (1972). She would eventually however drop the idea, because it would be a mad race against time.[20]

Drafting a Science Policy for Kerala I

On 30 September 1972 in Trivandrum, Janaki chaired a meeting of the State Committee on Science and Technology, Government of Kerala (headed by C. Achutha Menon, its visionary Chief Minister). This was the first ever time she had been invited to an official meeting by her home state. Members of the Committee included K. K. Nair, Chief Conservator of Forests, M. C. Jacob, a retired IFS officer (with experience in Assam), P. N. Nair, Forest Utilization Officer, C. Chandrasekharan and K. J. Joseph, both Deputy Conservators of Forests, and K. P. S. Menon, Under Secretary, Planning and Economic Affairs. The purpose of the meeting was to discuss an 'approach paper' for the Science and Technology Component of the Fifth Five Year Plan. As Chairperson, Janaki opened the discussion with thoughts on how to tackle the issue of unemployment among the well-educated (no one could understand this better than her after all): 'Kerala has a wealth of a large number of highly qualified graduates, Post Graduates etc., who are not able to get any employment or research facilities in their fields. The Universities are the main organizations where they can hope for an employment at present. If technical departments like Forest Departments can engage some of these educated men [Janaki forgot to include women!], for works like Flora revision, investigations regarding plant diseases etc. this will, besides helping the department solve its problem, help the educated young men also to make use of their talents.' In addition, she emphasised the urgent need to undertake compilation of a flora of the state under the guidance of the BSI. As for the flora of Kerala, the Committee concluded there was insufficient staff to undertake the task, and instead suggested that at least twelve doctoral studentships be offered on the subject.

To M. C. Jacob's suggestion that it was essential to tap into the great wealth of medicinal plants that existed in the state by increasing the acreage under their cultivation, there was consensus that it should not be at the cost of forest lands. Janaki offered that it was worthwhile to 'try some special varieties of *Cymbopogon* and *Rauwolfia* as done in Nepal.' Members of the Committee unanimously agreed that a full-fledged research institution for the study of forests (teak in particular) was crucial for the state and should be established at the earliest (the Kerala Forest Research Institute, Peechi would be set up in 1975); the study of various aspects of wildlife and its management was also a matter of urgent concern, they opined. Immense anxiety was expressed over the 'alarming rate at which the forest resources in Kerala [were] being destroyed.'[21]

A Damning Hydroelectric Project

Meanwhile, in 1970, the Kerala State Electricity Board (KSEB), had proposed the building of a dam across the Kunthipuzha in the Silent Valley region of the Western Ghats near Mannarkkad, Palghat district, for a hydroelectric project. The Kunthipuzha was a tributary of the Thuthapuzha, itself a tributary of the Bharathapuzha. The KSEB claimed that the project would generate the much-needed power the state required and would irrigate about 100 km in the Malappuram and Palghat districts besides providing employment to several thousand people during the construction period. As early as 1929, the British had discovered the place to be perfect for producing hydel power but its technical feasibility was explored only in 1958. Many however feared that this would drown almost 8.3 square km of humid evergreen forest, largely untouched by man.

Among the first scientists to visit the Valley for purposes of research was the American (turned Indian citizen) herpetologist Romulus Whitaker in 1969, and the American primatologist, Steven Green, who had studied the Arashiyama Japanese macaques in Kyoto before arriving in India in 1970. Whitaker would go into the Valley once or twice a year, looking for intact forest (the presence of elephants to him was a sign of a well-preserved

467

forest), reptiles and amphibia; everything there, he remembers, was maximal. William Logan, the Malabar Collector and friend of Janaki's father, in his *Malabar Manual* (1887) had described the region as 'an enormous tract of mountainous forest and grassland,' from where timber could not be extracted, it being so dense and impermeable.[22] Although, the Silent Valley area itself measured only about 9,000 hectares, it was fortified by the Nilgiri and Nilambur forests to the north and the Attapadi forests on the east, making a total of 40,000 hectares of virgin rainforest. In February 1973, the Planning Commission gave formal sanction to the project at an estimated cost of Rs 25 crore. The shortage of funds stalled the project, but the story had just begun.

Lecture on Plants and Man

On 16 November 1972, Janaki delivered the Second Silver Jubilee Lecture on 'Plants and Man' at the Birbal Sahni Institute of Palaeobotany, Lucknow, at the invitation of Savitri Sahni. An excellent and succinctly worded lecture, it provided an overview of all aspects of her research until then, in a manner intelligible to lay listeners. She began her presentation in a slight vein of pessimism, speaking about man's destructive and aggressive tendencies, the evidence of which was all around: 'Mankind is now passing through an era of preoccupation with many disturbing and destructive thoughts. The Atom bomb and the still more destructive Hydrogen bomb, population explosion, pollution of air and water, mass killings of fellowmen, all these are engaging man's attention more and more every day . . . In the midst of this to talk on plants and man may seem trivial or even puerile.' Janaki cited the old saying, 'All flesh is grass,' to refer to the period after the angiosperms appeared in the early Cretaceous times, a reference to the close relationship that existed between the plants and animals of the different geological epochs.

The unexpected emergence of flowering plants however still remained a mystery, 'an abominable mystery,' she would remark, quoting Darwin. Speaking about the interconnectedness of nature, Janaki explained that angiosperms not only triggered the evolution of mammals but also of insects, for the alluring mechanisms adopted for cross-pollination would have played an important part in bringing about 'the rapid differentiation of flowering plants into families, genera and species by hybridization.'[23] She cited examples from her own research, of primitive families of plants such as the Magnoliaceae, 'in which living species [were] exact representations of fossils of earlier ages.' The Ice Ages, Janaki stated, were 'great testing times' for plants and animals, and that with each Ice Age, something novel and singular had appeared; if the origin of seed plants (or spermatophytes, under which are grouped gymnosperms and angiosperms) occurred during the Permo-Carboniferous Ice Age (a period of significant glaciation), the last Ice Age saw the arrival of man. 'He rose from a line of descent shared by the great apes of which only four are living in the present day . . . So close are we to the apes, that we share with them the same blood groups, and our chromosome number 46 is very close to 48 in the apes,' she observed.

She also referred to the practice of shifting cultivation and how this form of 'primitive agriculture' had led to the destruction of forests in Assam and Kerala when uncontrolled, a point she had first made in the mid-1950s in her symposium paper on subsistence agriculture. It was possible to reconstruct the story of domestication of wild plants by man, Janaki explained; based on archaeological, cytological and genetical studies, the centre of origin (centre of diversity) of most cultivated plants could be established, she stated, citing Vavilov, while also drawing attention to the importance of the conservation of genetic resources. 'The excellent work of Vavilov is probably well known to all. These centres are of importance to the present plant breeder for introducing new characters into his cultivated crops. They are natural "germplasm banks" to be preserved and to be resorted to for such genes that will give resistance to diseases or earliness to the cultivated crops,' Janaki remarked. She also brought in the example of the sugarcane, and explained how tracing its ancestry was 'both fascinating and time consuming,' as every one of the varieties selected could have its own unique genealogy.[24]

Having discussed man's exploitation of plants, both wild and domesticated, she turned to their place in the field of medicine and referred to 'the science of Ayurveda.' Janaki spoke about the herbals, which marked the beginnings of systematic botany; in fact, she said, 'the art of medicine arose in India and spread into Greece through the Arab.' Just as she had in the introduction to her *Memorandum on the Reorganization of the Botanical Survey of India*, Janaki alluded to the story of Jivaka (who later became the physician of King Bimbisara, and the Buddha), a student of medicine at one of the oldest Buddhist Universities, perhaps Taxila. Jivaka was asked by his teacher Atreya to 'collect, identify and mention the medicinal properties of plants from one Yogana on every side of the University.' Janaki playfully commented, 'I wonder if any of our students would meet the requirements of Taxila today, for a Yogana is 9 miles.'

As a culmination of the lecture, she summed up man's changing relationship with plants thus: beginning with a utilitarian interest in plants, turning to their systematic study, followed by the study of plant anatomy, its physiology, ecology and pathology. In the twentieth century it had also become possible, with the rediscovery of Mendel's work, to breed plants and study its cytology and genetics. Having learnt the mechanism by which plants passed on characteristics genetically from one generation to another, man was able to manipulate genes through artificial means, to make plants to order. 'The boundary of Human knowledge has been extended by the study of plants. To apply this knowledge for the benefit of humanity is the vital part plants have now to play,' she concluded.[25]

The Diamond Jubilee of the Sugarcane Institute

That same month, after her Lucknow lecture, Janaki attended the symposium in connection with the Diamond Jubilee celebrations of the Sugarcane Breeding Institute, Coimbatore. She had accepted the invitation to present a paper and chair a session; she took the opportunity to mention to the Director of the Institute, S. S. Shah, that she had established an ethnobotanical garden at Shoranur, and that she had 'some rice fields attached to the land, which could be used to try out strains of sugarcane suitable for Kerala.' She added that the fields were at present fallow, and could be utilised for agricultural experiments of some kind.[26] At the symposium, Janaki presented a paper on the use of modern cytogenetics in cane breeding, as did her co-author of the *Saccharum* x *Zea* paper, Jegathesan, who had just returned from Russia.

More Ethnobotanical Field Surveys

Towards the end of the year (1–12 December 1972), Janaki went on her second field trip, with senior research fellow Vishwanathan and the junior research fellow, Durairaj Rajiah, this time, to regions within Malabar such as Muzhakkunnu, Iritty, Mattanur, the Sugarcane substation at Cannanore, and the Indo-Russian Farm at Aralam, where they met with members of indigenous tribes, particularly Paniyans and Kurichiyans. Edam would be their base camp, from where they would set out every morning after breakfast and after a long working day, return for the night. Sumithra, who had been a gracious hostess when they visited earlier in the year, was now sadly no more. Her elder sister Parvathi was now managing Edam's affairs, although she was herself unwell, and so was Janaki's elder brother, Padmanabhan who spent the major part of the day reading or resting, confined to his bed in the outhouse. His love for nature was such that a couple of mynahs and squirrels regularly visited to feed off from his hands. In less than two years, Padmanabhan would also be no more, leaving Parvathi alone to maintain the sinking ship that was Edam. Janaki would however visit often and do everything possible to somehow keep it afloat.

Ashok, as a Sub-Station

From Tellicherry, Janaki and her companions travelled to Alwaye by train to meet Janaki's nephew V. Muthukrishnan (the late Varadan's son), who worked for the Premier Tyres Ltd in Kalamassery. Janaki had made this journey to convince her nephew to lease out Ashok, the 2-acre property and house at Kallipadam, Shoranur, that he had inherited from his father. Her plan was to use it as a Substation for the Project on Ethnobotany in South India, for the duration of one year (1 January 1972–73), at a rent of Rs

100 per month.[27] As soon as she received the lease agreement, duly signed by her nephew, Janaki shot off a letter to the pteridologist B. K. Nayar (1927–2012), Executive Secretary of INSA, to let him know that from now on the 'work is to radiate from three Centres:' Shoranur in Central Kerala, Coimbatore in Tamil Nadu, and Madras, which would be the headquarters. She assured Nayar that the 'land in Shoranur had now been acquired on lease from the owner Mr. V. M. Krishnan.' A sum of Rs 20,179 had earlier been sanctioned by INSA for the project, for the period, 1972–73; to this, an extra sum of Rs 2,000 was added in the current year (1973–74), over and above the contingency and travel allowance. Of the total amount, Rs 1,000 was paid in advance for the development of the 'Shoranur Substation.'[28]

It is astonishing that a woman scientist of her stature, growing frailer by the day, was travelling so tirelessly and in modes of transport far from comfortable, and worse still, spending precious research time to list out the myriad expenses incurred on the project over time, to be despatched to authorities such as the University and the INSA. The scientist-administrators at these institutions, who had to sanction her bills, were for the most part far junior to her, and with much less international exposure and intellectual output to their credit. Perhaps living a spartan existence came easily to her, given the life endured during the war years in Britain, but it was also a reflection of her monastic outlook, and yearning to be a pilgrim of science, a scientist-*bhikshu*. She had chosen such a life for herself.

A Summing Up

Janaki and her research assistants had interacted to some extent by this time with the Todas of the Nilgiris, Irulas of Coimbatore, Kaniyakarans of Nagercoil, Paniyans of North and South Kerala, Cherumas of Central Kerala, Kurichians of North Kerala and the Malayans of Cannanore. 'We have in the Todas of the Nilgiris the starting point for the evolution of New cultivated plants,' Janaki observed; the plants used by the Todas (temperate) differed from the those in the plains as 'Toda land occupies a high elevation with a flora related to the Himalaya.' Among the fruits they used was the Nilgiri species of the *Rubus*, 'the wild raspberry,' a diploid, thus providing evolutionary opportunities to the cytogeneticist, for the creation of culti-vated polyploids. From the Kurichiyans of Nedumpoyil they learnt about their shifting cultivation practice, and collected bananas, wild Curcumas and *Zingiber* from their habitat; perhaps these were 'the ancestral forms of the cultivated ones in Kerala,' she observed. *Mrithasanjivani* of the *Rama-yana* (the Hindu god Hanuman, using his powers of flight, transported an entire mountain on which medicinal herbs grew, to the battlefield where Lakshmana lay unconscious having been hit by an arrow; what better tex-tual and visual representation of 'man and his transported landscape,' than this from mythology!)'was the most spectacular' plant that they had the opportunity to see, thanks to the Cherumas (and Kadars) of South Kerala (chiefly confined to Pattambi, Shoranur and Palghat district), which when grown and after flowering was identified as an *Asclepias*, the *Holostemma annulare* (now considered to be part of the family Apocynaceae).[29] Several plants have since become contenders for the title of *Mrithasanjivani*, including the *Selaginella bryopteris*, and *Dendrobium plicatile*.[30]

Janaki was also interested in the genus *Sarcostemma* of the Dogbane family, generally known as the climbing milkweed, the soft-stems being filled with milky white latex, which was thought to be poisonous; she hinted that if used in minute doses, it could serve as a cardiac stimulant. *Alpinia galanga* (the lesser and greater *galangal*) she was of the opinion had several good medicinal properties, but only if their 'friend Pai' (B. R. Pai, Professor of Organic Chemistry, Presidency College, Madras) helped them to know more about the alkaloid it contained. She would occasionally send samples for chemical analysis also to botanist L. D. Kapoor of the National Botanic Gardens, Lucknow (previously he was based at the Drug Research Laboratories, Kashmir).

It was Durairaj Rajiah, the junior research fellow, who studied the Kanikarans of Nagercoil (collecting 14 edible and medicinal plants from them), besides the Kuravas of Chinglepet District in Tamil Nadu. In the first-half of 1973, plants collected from the Cherumas and Kadars were planted at Ashok in Shoranur: 'Thus a beginning was made for the formation of an Ethnobotanical Garden of Kerala,' Janaki would remark in her project report from the period.[31] In all, about 600 specimens were identified. Living plants were also grown at the Maduravoyal Ethnobotanical Garden and later subjected to cytological and chemical analysis. To solve plant identification problems Janaki would send Durairaj to her friend B. G. L. Swamy or to E. Govindarajalu at the Department of Botany, Presidency College. In August 1973 Durairaj resigned from the project as he had been offered a teaching job elsewhere;[32] Janaki and Vishwanathan would be forced to carry out the project as a two-some.

Cytogenetics of the Solanum

At the end of their exploratory field trips and chromosomal analysis, Janaki was able to trace the ancestry of some of the lesser-known cultivated plants and to discover from the tribal people new uses of wild plants, including *Solanum, Curcuma, Yams, Dioscorea* and *Zingiber*.

Cytogenetic studies were restricted to the genus *Solanum*, which yielded glycoalkaloids used for the production of steroid hormones. *Solanum xanthocarpum* was discovered to be used as a contraceptive (Janaki used the term 'family planning' rather than 'contraceptive' in her project report) by the tribals. While, the Jammu variety (based on research by her doctoral student at Jammu, Usha Zutshi) was reported solasodine content to be

1.85%, the one collected by Janaki and her assistants from Kondotty in Malappuram yielded 2.2%. They had collected four races of the *Solanum*, of which the white one was derived from a single fruit given to them by Valli Moopan, the Paniya headman of Iritty, Cannanore, who said that it was used as an oral contraceptive; the Paniyas also used *Calatropis gigantia* as a contraceptive, Janaki recorded. She described the Paniyas as 'an aboriginal Tribe who lived on roots collected from the forest.' They were, she noted, 'famous for tiger hunting by spears as some of the African Tribes with whom they show some affinity,' a matter of interest to the human geneticist in her. The seeds of the 'Paniya' variety of *Solanum incanum* were grown in Maduravoyal, and a uniform population of white fruits obtained. This was the only variety to produce white fruits, as all the rest bore green ones. Janaki discovered that it was also so easy to cultivate, and with its high yield, a much better choice as a source of glycoalkaloid than the *Solanum khasianum*.[33] Janaki and Vishwanathan jointly and individually published papers on their research on *Solanum*.[34]

Incidentally, Janaki had by this time successfully produced a thornless variety of *Solanum khasianum*; several agencies, chiefly research institutions (specialising in genetics and plant breeding, or chemistry) and pharmaceutical companies like the Jawaharlal Institute of Postgraduate Medical Education and Research (JIPMER), Pondicherry, the College of Agriculture, Calcutta and the West Bengal Pharmaceutical and Phytochemical Development Corporation, would write to her for specimens of the thornless *Solanum*. By 1976, however Janaki had stopped growing *Solanum khasianum* in Maduravoyal, and had given away all her material to Bharati Bhatt of BARC, with whom she had co-authored a paper on the *Solanum* a few years before.

Concern About Termination of INSA Project

'I may add that this is the first time that the ethnobotanical aspect of Indian plants is being linked with genetic and chromosome studies,' Janaki wrote to D. S. Kothari, Chairman of the University Grants Commission (1961–73), in September 1973; incidentally, Kothari had been honored with the Padma Vibhushan that year. This work, Janaki noted, would prove useful in all future programmes devoted to breeding and improvement of cultivated plants. The letter to Kothari was prompted by a notification from INSA, of which she was a Fellow, that the project she was heading would be terminated on 28 February 1974. It came as a complete shock to her that this significant project to which she had given so much was being made a victim of INSA's 'general policy of economy.' Beside herself, she wrote to Kothari, 'When the rest of the world is clamouring for preservation and utilisation of wild plants, we in India are doing very little in this direction. We are in fact destroying our native flora very fast, by clearing our forests for cultivation.

I hope the Indian National Science Academy will reconsider its decision to terminate this unique project and instead extend it both in time and space to cover still wider fields of investigation in Ethnobotany.' Janaki copied the letter to T. S. Sadasivan and to M. S. Swaminathan, Director General of the ICAR.[35]

Swaminathan's reply reached her in the shortest possible time, fully agreeing with her that the project should not be interrupted 'at the most interesting phase.' He assured her that even if INSA decided not to lend support to the project, he would request the ICAR to do so. 'As you know we are deeply interested in the collection and utilization of all primitive cultivars,' he added.[36] Wasting no time, Janaki wrote up a fresh research proposal, but along the same lines as the original INSA project, and sent it to the ICAR for consideration. Meanwhile however, she heard from INSA that they had decided to extend her project for one more year (until February 1975), but with the usual disclaimer, 'subject to availability of funds.'[37] Incidentally, the ICAR was interested at this time in her Jammu collection of Cymbopogons, which it wished to have for their Lemon Grass Research Station at Odakkali in Cochin (established in 1951).

Sadasivan's Successor

Meanwhile in 1973, T. S. Sadasivan retired as Director of CAS and was succeeded by the mycologist C. V. Subramanian (1924–2016). Subramanian had first met Janaki at the Seventh International Botanical Congress in Stockholm in July 1950, and again within a few months in London at the India House, at a reception organised by the Indian High Commission in honour of Jawaharlal Nehru who was in the city to attend the Commonwealth Prime Minister's Conference. Janaki, who was attached to the Wisley Garden at this time, was quick to recognise Subramanian at the reception, and kindly offered to introduce him to Nehru. After her return to India, he would often run into her at scientific meetings, as he did at the Silver Jubilee celebrations of INSA in Delhi on 30 December 1960. Now being part of the same Department, Subramanian had ample opportunity to get to know her. He was deeply inspired by her monk-like devotion to work irrespective of her advancing years, her indomitable spirit in the face of challenges, and her outspokenness, spartan life and unassuming nature despite her stature as a world-ranking plant scientist. Perhaps, it was this background that made him supremely eligible to write a biographical note on her.

Drafting a Science Policy for Kerala II

Janaki would attend the meeting of the State Committee on Science and Technology, Government of Kerala, on 13 June 1973, at the New

Conference Hall in the Secretariat, Trivandrum, the agenda of which was the finalisation of the 'Approach Paper' for Science and Technology, to be adopted by the Kerala state. She travelled by air to Trivandrum, where she was met by a grand-niece who lived in the city. She appears to have attended the subsequent meetings too, and in early 1974 she wrote to K. K. Nair, Special Chief Conservator of Forests, Kerala, that she was eager to know 'how far the recommendation of the committee on Forestry were acceptable to the Science and Technology Committee' and if there was at all going to be a Forest Research Institute in the Fifth Plan. She enclosed a list of important timber trees in Kerala, whose seeds were required for germination in Madras.[38] Janaki also approached him for suggestions 'regarding the procedure to be adopted for the ethnobotanical study of the Tribals living in [the] State forests.' Not only did K. K. Nair not have much to offer, but he even put a spoke in the wheel: 'As you know tribals are living scattered all over the Western Ghats, in the interior of the forest areas. If a detailed study is to be made, it might be necessary to have a Botanist with one or two assistants and a vehicle to work on this project covering district by district and the Western Ghats from one end of Kerala State to the other end . . . As you know, the Members of the Committee [of Science and Technology] are already having their own work and it might not be possible for them to devote much time for visiting even a few tribal centres.'[39]

Battling On

The year 1974 was a bad one for Janaki. Early that year, she fractured her left elbow again (the one that had been operated upon in Edinburgh in 1939), and also 'the socket of her right knee.' 'Survived both. I can walk only on level ground now,' she wrote to Darlington in May. She had just begun to walk independently when B. L. Burtt of the Royal Botanic Garden, Edinburgh, arrived in Madras. She was delighted to have his company. 'Your visit to Madras was a memorable one and a source of joy to me. I have been so cut off from my friends and colleagues in U. K. You brought back memories of Kew, Wisley & Edinburgh,' she wrote to him after he had departed. She was unhappy, she told him, being unable to revise her work on the cytogeography of Rhododendrons, one of her favourite genera. 'The nomenclature must have changed very much since I worked on the genus at Wisley,' she remarked wistfully. She had not been able to send Santapau a paper on the Indian species of *Rhododendron*, for the same reason. Janaki wondered if Burtt could assist her in some way, for she had no means to travel to England herself. In fact, at one point, she had contemplated approaching the British Council for funds but her age (now almost 77) was not in her favour. She was 'still energetic,' she told Burtt, having just returned from a ten-day collecting trip to Malabar, 'chiefly medicinal plants and also grasses which come up during the monsoon'; she now had a young Research Fellow (Z. Abraham) she was training in the cytology of grasses.[40] Burtt in his reply assured her of all assistance in getting her work on the *Rhododendron* up to date: 'It is such a pity that the results of so much skilled labour are not available to those many folk now interested in the genus and family.'[41] He had forwarded Janaki's letter to the Regius Keeper of the Garden, Douglas M. Henderson, who was quick to reply that he would be happy to help, and see her work to completion. There was also the possibility, Henderson said, that her paper could be included in the 'publication notes from the Royal Botanic Garden, Edinburgh in which much research on Rhododendrons [had] already been published.'[42]

Connecting with Calicut University

During her recent collecting tour in Malabar, Janaki had visited the Calicut University, where at the botany department she met the young professor Kattungal Subramaniam Manilal (b. 1938); his wife was her grand-niece (eldest sister Lekshmi's granddaughter). At the university she had also seen V. V. Sivarajan's doctoral thesis on the flora of Calicut, completed under the guidance of Manilal. 'It is quite good,' she would comment to Burtt, who was the external invigilator; she had found the thesis very bulky 'due to spacing of typing, but otherwise, good.'[43] Burtt though was unhappy with it: 'Do please influence to prevent PhD students being put on to write a local flora of this type. It tests their ability for hard grinding work, but it gives the examiner very little chance to assess intelligence . . . When the whole flora is enumerated there is no space in the thesis, nor time in the actual work I fear, for any of the more interesting problems that demand thought. I have protested about this so often, and I really think I have made some little progress in propagating this point of view. But Indian Universities are so numerous!'[44] Sivarajan would nevertheless be awarded his doctorate in June 1975.

Janaki also mentioned to Burtt that she hoped to visit the Gersoppa Falls (also called Jog Falls) on the Sharavathi River between Bombay and Mysore, to collect *Hubbardia heptaneuron* 'Bor,' a monospecific genus endemic to the state of Karnataka and named after her friends C. E. Hubbard and N. L. Bor of the Kew Gardens. It was two years since Bor had died, his ashes scattered in the Azalea garden at Kew, like those of his wife before him. Janaki recalled that she had promised Bor she would collect the plant, but 'now he [was] no more.' 'What a wonderful personality he was,' she would remark to Burtt, missing Bor very much. She requested Burtt to let Hubbard know that she was going to complete her work on bamboos 'some day.' Moreover, the Calicut University had requested her help in starting a bamboo garden.[45] Janaki had introduced the *Melocanna bambusoides*, a clumping bamboo, to Kerala; it was thriving in the Edam homestead, as were the *Ochlandra*.

A Modern and Anti-colonial Hortus Malabaricus

Among her future plans was a journey 'one of these days' to Dehra Dun, to get seeds of bamboo 'if they are in the flowering season, otherwise new shoots.' It was to Burtt that Janaki disclosed that she had been trying to lure a 'retired botanist of Malabar' to work on 'a modern version of *Hortus Malabaricus.*' 'Ambitious—am I not,' she exclaimed.[46] Her interest in the Dutch compendium arose chiefly out of her investigations in ethnobotany, on the primitive cultivars of food, and the economic and medicinal plants used by the aboriginal tribes of South India. She not only wanted to rework the compendium in chromosomal terms, but also to decolonise it, supplementing it with new plants and new knowledge collected directly from the forest dwellers—in her worldview, they were the primary interlocutors of indigenous medical knowledge. Coincidentally, at this time, K. S. Manilal would apply for a UGC grant to work on a translation and annotation of the *Hortus Malabaricus* (from the Latin to English). When he divulged his intention to Burtt, the latter promised to help him as long as he re-collected the Zingiberaceae of the Hortus for the Edinburgh Garden.[47]

Drafting a Science Policy for Kerala III

At the meeting of the Kerala Government's Committee on Science and Technology held on 7 July 1975 at the Secretariat in Trivandrum, Janaki announced a plan for conducting a detailed study of ethnobotany based at the Calicut University in collaboration with the botanist B. K. Nayar, Professor and Head of the Department. Their project was titled: 'Ethnobotanical studies on the Tribals of Malabar, with special reference to medicinal plants, folk medicines and health practices among the tribals.' Janaki and Nayar had proposed the names of T. V. Vishwanathan and 'Mrs Nayar' as associates. N. K. Panikkar, Chairman of the Committee, however replied that he would only be able to consider it for the next Five-Year Plan period. Nayar incidentally would go on to play an important (albeit negative) role in the Silent Valley controversy, but more on this, later. At the meeting, Janaki also reiterated the need to undertake a study of the medicinal plants of Kerala. Incidentally, that same year, she would be elected Fellow of the Indian Society for Genetics and Plant Breeding.

UGC Project on Primitive Economic Plants

By September 1975, the tenure of the INSA project, 'Ethnobotanical Studies of South Indian Aboriginal Tribes' had come to an end. Janaki cleverly redrafted the INSA project proposal and sent it to the UGC for funding, with a slightly altered title; it was now called, 'Ethnobotanical Survey of South

Indian Economic Plants with special reference to collection and utilization of primitive cultivars.' The UGC sanctioned the project by the end of that year. This was a difficult time for India. In 1975, a state of Emergency had been declared by Prime Minister Indira Gandhi across the country, claiming there were imminent threats to the State, internal and external. It gave Mrs Gandhi the power to rule by decree, which saw elections suspended and civil liberties repressed. The Emergency was in effect between June that year and March 1977, a period marked by an unprecedented scale of human rights violations, including an aggressive mass-sterilization campaign led by Sanjay Gandhi, son of the Prime Minister.

Janaki set out on a collecting tour in mid-November 1975, of primitive cultivars of medicinal and economic plants, to Malappuram, Calicut and the Wayanad districts; William Jebhadas and Z. Abraham, the two research fellows working on the new UGC-funded ethnobotanical project, accompanied her. A month later, another collecting excursion would be undertaken in the same region. It had now become usual practice to alight at the Calicut Railway Station and travel to the university to look up the plants they had planted on the campus. Manilal had by this time embarked on his own UGC-funded *Hortus Malabaricus* project. From Calicut, Janaki and her research assistants proceeded by road to Cannanore, to make a collection of wild pepper growing on the Ghat Road, and at Iritty and Nedumpoyil. It was in the last-mentioned place that, way back in the 1930s, Janaki had noticed for the first time, a fine example of the *paramba*-style/subsistence cultivation involving controlled slash and burn and polycropping. Exhausted from all the collecting, they would head towards Tellicherry, where at Edam they would unwind and rest, before sorting out the collection and pressing the plants. Janaki would spend a good part of the night conversing with Parvathi and catching up on family news. Collecting over the next couple of days would happen with Edam as base; they would leave early in the morning for Wayanad and Sultan Battery, and return in the evening after a long day's collection.

Updating Koller and Darlington

She enlightened Koller on her new project: 'Our new project funded by the University Grants Commission (UGC), is on Primitive Cultivars of our Economic Plants. This will give me a chance to go back to my favourite genus, *Saccharum*. I am of the opinion that it arose in India and crossed over the E. Indies and New Guinea, when there was a land connection between India and the Islands of Java, Borneo etc. American botanists think the other way [originated in the East Indies and New Guinea and crossed over to India]!' Janaki revealed that she longed to go to Assam, her 'favourite region, where so many related genera exist[ed].' She also told Koller about her student Vishwanathan's work on the *Solanum*: on how the improved variety was

478

being cultivated 'commercially for the production of sex hormone for family planning, so badly needed in India.' Janaki also gave him news of their common friend, Eileen (MacFarlane), who had just 'lost her husband, Mr Mac-farlane.' Eileen had worked with Koller for two years at the Royal Cancer Hospital, on the cytology of irradiated tumours and human blood groups. In a recent letter to her, Koller had enclosed a picture of his daughter Christa (who perhaps was not yet born when Janaki was in England), to which Janaki responded warmly: 'Why isn't she married? She is so pretty! And talented.'[48]

Janaki had also written to Darlington, in late October 1975, enquiring whether he had received her unregistered letter, in which she had sent her paper on the *Saccharum* x *Zea* hybrid (meiotic study). She reminded him that she had 'switched to Ethnobotany in search of plants used by our tribals both as food & medicine' and that their *Solanum* research had brought them 'some publicity.' She would tell him more; on how she had turned all her attention to ethnobotanical research, despite it being unfashionable in a world dictated by molecular biology, and about the urgent need to preserve primitive cultivars:

> We now have high alkaloid containing tetraploids from the ones collected from the Paniya tribe—It is going into large scale cultivation—A student of mine—Vishwanathan crossed *Solanum incanum* with *Solanum indicum*. It gave quite high yield of Solsadine [Solasodine]. Heterosis? I feel these two species have gone into the making of the cultivated brinjal, *S. melongena*. Everyone is chasing RNA and DNA these days when there is so much waiting to be discovered even without them. I am considered out of date, but that does not worry me one bit! At the rate our forests are being cut down, our native flora will disappear soon. I have now taken up a new project on the collection and preservation of primitive cultivars of economic plants. Not enough money is available for this![49]

Janaki longed to see Darlington's children, and his grandchildren: 'I wonder if that will be possible,' she wrote to him. She was also wanting to revise her paper 'on the cytology of the Himalayan Rhododendron in relation to the new names given to the species,' about which she had spoken to authorities at the Royal Botanic Garden, Edinburgh. 'This would entail travelling to Edinburgh, where at the Botanic Garden she had deposited her herbarium species.'[50]

Penniless Again

A meeting was to be held in Lucknow in late 1975, of senior scientists of the CSIR plant-based laboratories, for a stocktaking of work done and those in progress and of future plans in general on glyco-alkaloid-containing *Solanum* and *Dioscorea*. Sadasivan contacted Janaki for her side of the story: 'Published and unpublished data (in so far as it's safe to release) could be referred to in the note,' he hinted. Assuring her that utmost caution would be taken in presenting the highlights at the meeting, he enquired, 'Are we riding together by the evening IC 440 to Delhi? . . . It will be a pleasure if you could make it.'[51]

All that Janaki had been receiving by way of money, since 1972, was the allowance from INSA, for her role as principal investigator on the ethnobotanical project. When INSA stopped funding the project in early 1975, Janaki was in dire straits, and was forced to present the issue to the well-connected Sadasivan, who was more than willing to find a way out. There were people who suggested that she write to Prime Minister Indira Gandhi directly, for after all Janaki had known her father so well, but she refused saying she would not ask anyone for favours, especially Indira, whom she had met as a very young girl several times in Delhi at the Teen Murti Bhavan. Given Sadasivan's immense respect for Janaki and his warmth and generosity, he influenced the Executive Council of CIMPO (a CSIR institution established in 1959, and later renamed the Central Institute of Medicinal and Aromatic Plants/CIMAP) to consider her request for a Fellowship of some sort. The Council of which he was a member had been disbanded at this time and a new one elected in its place, but not one to give up easily, he approached the new Chairman, who luckily for Janaki, readily consented. Sadasivan could not wait to give her the news. In a letter posted from Lucknow, to which place he had travelled for the CSIR meeting, he clarified to Janaki that the 'nature of the stipend would not be termed a Fellowship but will be as Specialist Consultant with special reference to steroid/alkaloid plant improvement programme and establishment of a germplasm collection at CIMPO, Bangalore. . . . Let them call it by any name. As long as they give you a good allowance. That is all that matters as far as I am concerned. Rest assured I shall pursue the matter to a logical end.'[52] His thoughtful intervention made life a little easier for Janaki, though the situation could certainly have been far better for someone of her eminence.

Such issues however rarely ever came in the way of her project; work would go on, even if she had to borrow from her assistants to tide over a sudden research-related expense. She would be fastidious in returning the money too, and always with a big thanks in the form of a fruit or something nice. She just found it very 'fascinating to meet tribal people.' 'I have always been interested in anthropology. This is a phase that needs investigation. Tribes are getting civilized,' she would write to Koller, conveying her sense of urgency in the matter.[53]

The Jammu Symposium

Indefatigable at seventy-nine years, Janaki travelled to RRL, Jammu in April 1976, to attend the Symposium on Medicinal and Aromatic Plants. She would be greeted with much warmth by her former colleagues and doctoral students, most of who she had not seen for several years. Janaki's paper at the symposium was on 'India's wealth in medicinal and aromatic plants: Its exploitation and improvement.' She was a good public speaker, always taking the effort to entertain and not just enlighten, and often drew examples from India's cultural heritage. Her dramatic opening sentence read: 'The World's first symposium on Medicinal plants in relation to diseases was held on the sacred slopes of the Himalaya during the seventeenth century BC, which is also the age of the compilation of *Charaka Samhita*. A detailed account of this symposium and its participants is given in the very first chapter of the *Samhita*.' It mentioned 'the names of 54 participants [including] two women, Maitreyi and Gargi,' she noted. She then alluded to the first medicinal and aromatic plants symposium organised by the RRL, Jammu in 1961, which had been inaugurated by Karan Singh (the Minister of Health), and presided over by 'Sir Ramnath Chopra, the Founder of the Drug Research Laboratory and the Father of medicinal plant organizations in India.' By doing this, she had ingeniously traced Chopra's lineage to Charaka, despite the 3,500 years between them, and hers to those women scholars of yore. This was not only a tribute to the two medical men, Charaka and Chopra, but also to the Indian women philosophers, Maitreyi and Gargi.

Once again, Janaki would state that the chief problem before the world was 'population explosion,' and that it was important to find ways of reducing the high rate of increase in the population: 'it is time to pause and meditate as did the ancient sages, for a solution cannot be delayed,' she warned. The genus *Solanum*, she told her eager listeners, had begun to show positive results in providing cheap and reliable methods of birth control. She spoke of how, with her collaborators in Madras, she had taken up the study of the five races of *S. incanum* bearing white fruits, used by the Paniyas of Iritty in North Malabar as an oral contraceptive. A survey of *Solanum* species had been conducted at the RRL, Jammu, she noted, way back in 1964, and *S. khasianum* had been found to contain the highest percentage of glyco-alkaloids. This attempt, she stated, was the first and most crucial step in the genetic improvement of medicinal plants in India. It was important to unite chemical analysis and cytogenetic investigations, so that the plant's past history could be uncovered, she stated. 'Today, many lines of improvement are open to the medicinal plant breeders,' she noted. As always, using the pronoun 'he' most unconsciously, she explained: 'If he finds a plant is a diploid, he can make a polyploid of it—by treatment with chemicals. He can induce mutation by chemicals. He can treat it with X- or gamma-rays or radioisotopes or he can induce enzymes by tissue culture. Hybridization with a type of higher potency is an aid and even interspecific and intergeneric hybrids can be made.'

Always on the lookout for similitudes and/or connections, Janaki spoke of how there was a great resemblance between the forests of Kerala and the west coast of India, with those of Assam; Assam was 'the most important centre of origin of most of our cultivated plants and also a region of active speciation as shown by genetic analysis of its plants.' Further, most of the medicinal plants came from the Himalayan region. She invoked the *Charaka Samhita* once again, to describe how the forest-dwelling sages had discovered the use of medicinal plants through their contact with tribals or forest dwellers, who at one point were the original inhabitants of large stretches of land. 'Today due to indiscriminate destruction of forests, their habitat has been much reduced. They are also changing their way of life due to the easy communication available these days. In other words, they are getting civilized and that is as it should be in a democratic country such as ours,' she would remark but warned that there was an urgent need to record the knowledge that had come down through generations (of the tribal people), before it was all lost.

'Since the genetic variation in a flora is markedly different from an adjacent one, it is very important that geographic races of all our medicinal and economic plants are preserved as "germplasm banks" for later use,' she added. Speaking of the genus *Cymbopogon*, the most important of the aromatic grasses, Janaki cited her friend Bor's book on Indian grasses (*The Grasses of Burma, Ceylon, India and Pakistan*, 1960) in which he had described the 'mosaic of characters' of the family Gramineae; we may recall, that she had used Bor's phrase also in her presentation on 'Distribution Pattern of Plants in India' at the Indian Science Congress meeting, Delhi (1959). In passing, she mentioned with pride the two hexaploid varieties of *Cymbopogon*, which she had introduced from Assam and which had been selected for cultivation along commercial lines. 'I feel that there is more to be collected—and that our collections should go beyond India,' she would comment. There was also the need to discover primitive cultivars, which could be used in breeding disease-resistant varieties of these aromatic and medicinal plants (her ongoing project).[54]

Connecting with the Kerala University

About this time, Janaki would contact cytogeneticist C. A. Ninan, Professor and Head of the Department of Botany, of the University of Kerala, with a request to nominate her as a guide to doctoral students: 'I am eager to see my two Research Fellows Mr. William Jebhadas and Mr. Abraham registered for their PhD degrees at the University of Kerala.'[55] It was his turn to do her a favour; Janaki had only recently nominated Ninan for a Fellowship of the Indian Academy of Sciences. Ninan was of course deserving of this, and had just published a book titled *An Inventory of Germplasm of Plants of Economic Importance in South India* (1976), jointly with A. Abraham and four others of the University of Kerala. Janaki would have to wait long however to hear from the University.

National Programme for Ethnobiological Research

On 21 September 1976, under the Chairmanship of M. S. Swaminathan, Director General of the ICAR (parent organisation of the IARI), a meeting was convened in New Delhi of the Inter-Organizational Panel for Food and Agriculture. A joint team of experts was formed to assess the current status of the ethnobotanical survey of the tribal regions of India. The group, which included Janaki, was asked to submit a report 'as to how the biological resources found in these communes could be conserved and utilised for socio-economic improvement of the tribals on one hand and the country in general on the other.' The working group met on 16 April and 13 July 1977 to work out a national plan for ethnobiological studies in India. Janaki was the only woman scientist in the group and the seniormost, and it was she who set the tone of the discussion; the group members included T. N. Khoshoo, C. Kempana, K. L. Mehra, R. Nagersenkar, K. K. Tiwari, S. K. Jain, R. N. Bhargava, B. K. Roy Burman and J. K. Maheshwari.

Ethnobiology was defined by the working group 'as the study of the past and present inter-relations of primitive human societies and the surrounding flora and fauna.' They concluded that there was an urgent need for setting up a coordinated programme on ethnobiology, 'since the natural biological resources in the tribal and other backward areas are becoming scarcer as a result of their indiscriminate and unplanned management.' The group observed that tribal cultures were rapidly undergoing transformation owing to urbanisation, industrialisation and 'change in subsistence economic pattern,' and that 'material traits which are found today may become extinct in the near future . . . only to be seen in museums as relics of past cultures.' It suggested a three-pronged approach: preparation of a bibliography on ethnobotany and ethnozoology; the identification of centres for field work and ethnobiological surveys of plants and animals; and a 'follow-up programme by way of documentation, conservation, phytochemical screening and pharmacological assessment.' The BSI and Zoological Survey of India in Calcutta, the National Botanical Research Institute, the Central Drug Research Institute and the Birbal Sahni Institute of Palaeobotany in Lucknow were among the research institutes and organisations proposed for the implementation of the project.[56]

Figure 26.1 Paniya head-man, Aandi moopan (Valli moopan) and his hut. 1970s.
Source: Photo probably by Janaki. Private Collection.

Notes

1 DP: C 109 (J 115), E. K. Janaki Ammal to C. D. Darlington, letter dated 7 December 1971.
2 Ibid., letter dated 27 November 1970.
3 Ibid., letter dated 18 November 1970.
4 Ibid., C. D. Darlington to E. K. Janaki Ammal, letter dated 26 November 1970.
5 Janaki Ammal, Jegathesan and Sreenivasan, 'Further Studies in Saccharum x Zea Hybrid I. Mitotic Studies', 1972. Janaki would conduct meiotic studies as well, and write a paper with Sreenivasan and Jegathesan as co-authors, but it does not appear to have been published for some reason. She discovered the hybrid to be sterile.
6 DP: C 109 (J 115), E. K. Janaki Ammal to C. D. Darlington, letter dated 27 November 1970.
7 Ibid., letter dated 16 December 1970.
8 DP: C 109 (J 115), E. K. Janaki Ammal to C. D. Darlington, letter dated 27 February 1971.
9 Ibid., letter dated 28 June 1971.
10 Ibid., C. D. Darlington to J. L. Jinks, Department of Genetics, University of Birmingham, letter dated 30 July 1971.
11 Ibid., J. L. Jinks to E. K. Janaki Ammal, letter dated 3 August 1971.
12 DP: C 109 (J 115), E. K. Janaki Ammal to C. D. Darlington, letter dated 7 December 1971.
13 Ibid., letter dated 27 February 1971.

14 Janaki Ammal and Sreenivasan, 'Observations on the Cytology of the Madras Mint', 1971.

15 DP: C 109 (J 115), E. K. Janaki Ammal to C. D. Darlington, letter dated 27 February 1971.

16 E. K. Raghavan to his youngest daughter, letter dated 24–28 August 1970, private collection.

17 At one time, Sebastine had worked under Janaki, at the Central Botanical Laboratory, Calcutta. In July 1975, she would send J. Joseph, at his request, a 'beautiful photo' of Sebastine.

18 T. V. Viswanathan, 'Tour Report, INSA Scheme on Ethnobotanical Studies on Aboriginal Tribes', conducted between 22 and 31 March 1972, private collection.

19 DP: C 109 (J 115), E. K. Janaki Ammal to C. D. Darlington, letter dated 11 August 1972.

20 Ibid., letter dated 1 September 1972.

21 'Minutes of the Meeting of the Committee on Science and Technology held on 30.9.1972', private collection

22 D'Monte, 'Storm over Silent Valley', p. 31.

23 Janaki Ammal, 'Plants and Man' 1974.

24 Ibid., pp. 4–5.

25 Ibid., pp. 5–6.

26 E. K. Janaki Ammal to S. S. Shah, letter dated September 1972, private collection.

27 V. M. Krishnan to E. K. Janaki Ammal, letter dated 16 January 1973, private collection.

28 Copy of letter from E. K. Janaki Ammal to B. K. Nayar in private collection.

29 E. K. Janaki Ammal, 'Significant results obtained during 1972', unpublished, private collection

30 See Ganeshiah et al., 'In search of Sanjeevani'.

31 'Report of Work Done during 1.1.1973 to 30.6.1973', unpublished, private collection.

32 Durairaj reminisced in a conversation with the author that Janaki was very fond of Robusta bananas, which he would bring her occasionally. He also remembers learning from her the elegant way to eat a ripe papaya fruit: by cutting out a neat big wedge with the rind intact, clearing the seeds, and then scooping out the soft flesh with a spoon! He also recalled how Janaki would occasionally fetch from her cupboard at the Maduravoyal Field Station, a cardboard box containing Nehru's letters to her, and read one or two out to him; he vividly remembers seeing Nehru's handwriting, his signature 'JN', and his 'My dear Janaki'. These letters are unfortunately not traceable; the family thinks they were destroyed along with most of her letters and papers, in 1984!

33 Janaki Ammal and Vishwanathan, 'A High Alkaloid containing Race of Solanum incanum Linn. collected from the Paniyas of Kerala', 1974; Vishwanathan, 'A new source of glyco-alkaloid Solanum trilobatum Linn., and its tetraploid derivative'.

34 Vishwanathan was awarded a PhD degree in August 1976 by the University of Calicut for his dissertation 'Cytological Studies of some medicinal plants of the genus Solanum utilized by the Tribals of Kerala and Tamilnadu', under E. K. Janaki Ammal. His batch-mate at the University, K. C. Alexander, completed his PhD on 'Smut Diseases of Sugarcane', under K. Ramanathan, and found employment as radiobiologist at the Sugarcane Breeding Institute, Coimbatore. He was in charge of using gamma rays to induce mutations and had several opportunities to interact with Janaki.

35 E.K. Janaki Ammal to D. S. Kothari, letter dated 17 September 1973, private collection.
36 M. S. Swaminathan to E. K. Janaki Ammal, letter dated 24 September 1973, private collection.
37 B. K. Nayar, Executive Secretary, INSA to E. K. Janaki Ammal, letter dated 27 November 1973, private collection.
38 E. K. Janaki Ammal to K. K. Nair, Special Chief Conservator of Forests, Government of Kerala, letter dated 9 January 1974, private collection.
39 K. K. Nair to E. K. Janaki Ammal, letter dated 17 May 1974, private collection.
40 Royal Botanic Garden, Edinburgh, Library & Archives: E. K. Janaki Ammal to B. L. Burtt, letter dated 12 October 1974.
41 Ibid., B. L. Burtt to E. K. Janaki Ammal, letter dated 22 October 1974.
42 Ibid., D. M. Henderson to E. K. Janaki Ammal, letter dated 18 October 1974.
43 Ibid., E. K. Janaki Ammal to B. L. Burtt, letter dated 12 October 1974.
44 Ibid., B. L. Burtt to E. K. Janaki Ammal, letter dated 22 October 1974.
45 Ibid., E. K. Janaki Ammal to B. L. Burtt, letter dated 12 October 1974.
46 Ibid.
47 Ibid., B. L. Burtt to E. K. Janaki Ammal, letter dated 22 October 1974.
48 DP: Pio Koller correspondence (J. 122–J. 139), E. K. Janaki Ammal to Pio Koller, letter dated 16 January 1976.
49 DP: C 109 (J 116), E. K. Janaki Ammal to C. D. Darlington, letter dated 20 October 1975.
50 Ibid.
51 T. S. Sadasivan to E.K. Janaki Ammal, letter dated 12 August 1975, private collection.
52 Ibid., letter dated 11 November 1975, private collection.
53 DP: Pio Koller correspondence (J. 122–J. 139), E. K. Janaki Ammal to Pio Koller, letter dated 16 January 1976.
54 Janaki Ammal, 'India's wealth in medicinal and aromatic plants: Its exploitation and improvement', 1977.
55 E. K. Janaki Ammal to C. A. Ninan, letter dated 31 July 1976, private collection.
56 T. N. Khoshoo, 'Preamble, All India Coordinated Research Project on Ethnobiology', a draft, private collection.

27

MADRAS IV

Forest Tracts and a Protest Movement

I have made up my mind to make a chromosome survey of the
Forest Trees of the Silent Valley, which is about to be made
into a lake by letting in the waters of the River Kunthi.
                              —E. K. Janaki Ammal (August 1977)[1]

The 1970s were the busiest for Janaki in terms of travel, as a result of which
her publications from this decade are relatively few. She was very glad that
her work often took her to Edam, at least twice a year, if not more.

In July 1976, A. Mahadevan had taken over from C. V. Subramanian as
Director of CAS, Madras; Subramanian had been awarded a Jawaharlal
Nehru Fellowship, which would keep him away from the Centre for four
years. In early September that year, Janaki travelled by air to Trivandrum,
from where she would embark on a botanical tour of Nagercoil, Cape
Comorin, Manavalakurichi, Vettoornimadam, Kulasekharam and Cheppa-
lathodu over the period of a week, before flying back to Madras with her
collection. The University of Madras did not take well to her travel by air;
in a letter to her, the Registrar stated that 'there was no provision in the
rules permitting air travel during tour.'[2] Janaki was forced to explain that
the tour 'to the extreme south of Tamilnadu' had been undertaken for three
chief reasons: to visit the Kerala University Botanical Garden at Trivandrum
'for which prior appointment was fixed with Prof. C. A. Ninan, Head of the
Dept. of Botany for the 4th Sep.'76 as I was unable to go earlier'; to visit the
Kanikkars of Kanyakumari District 'to have an on the spot Ethnobotanical Survey of Economic Plants and their Primitive cultivars for which prior
arrangements were made with the Forest Department of Tamilnadu'; and
for the purpose of making a special collection of littoral plants chiefly *Vinca
rosea* growing on the monazite sands of the Cape region including Manavalakurichi. She had requested special permission to travel by air in order 'to
complete the programme in time midst the hazardous conditions encountered in the forest tracts and the Monazite sands,' and to have 'the least
delay in getting the plants collected back to Madras from the radioactive

DOI: 10.4324/9781003267089-27   487

sands,' given the importance of this work dealing with radioactive materials, and in the interests of public safety.[3]

To Darlington, from the Cape

Janaki loved the sea and would take every opportunity to sit on the sands and gaze at the waves endlessly. At Cape Comorin, she took some time away from her research assistants to write to Darlington: 'I am writing from the extreme South—and looking out into the sea, my thoughts went far away into the past. I came here to collect *Vinca rosea* from the Monazite sands. Years ago [the early 1930s, when she was teaching at the Maharaja's College of Science, Trivandrum] I found a mutant with pubescent leaves [with a hairy surface]—I have found it again! *Vinca* is become an important plant and has alkaloids used as a remedy for blood cancer. I had done some work in *Vinca* and even produced a tetraploid—I may do that again.' She hadn't heard from him that whole year; the last he had written was in 1975. In her letter, she told him that she was soon going to meet a tribe called Kanikars, 'who dwell in forests round about these parts.' Her present project, she explained, was about locating primitive cultivars of some native economic plants. Janaki also let him know that she had been made a 'consultant' to CIMPO, which was part of the CSIR, without however telling him that it was Sadasivan's intervention that had made this possible. She had not given up on her *Saccharum* x *Zea* paper; 'I hope you will make it worthy of publication in *Heredity*,' she appealed.[4]

Darlington Writes Back

In a couple of months, Darlington would write back to say that he did not think her paper was publication-worthy, just yet. 'The verdict on the maize paper did not surprise me. It was for the same reason, that I hesitated to send it for publication. The hybrid flowers only once, and that when I was in Jammu and the Coimbatore people messed the material—"took charge of it" so to say,' Janaki complained. She was hoping to succeed in making it flower at Maduravoyal, where she would have better control over things.[5]

Darlington was at this time working on a new book, *The Little Universe of Man* (1978); she was looking forward to reading it. 'Why Little Universe? Is it not expanding?' she joked. In his letter it appears that Darlington had attacked Indira Gandhi for her obduracy and draconian policies (in the place of praise that he had for her earlier), and very rightly so, but Janaki stuck to her position that Indira Gandhi was doing good, even if forcefully: 'India is progressing in its own way. I think the British Press is inclined to be too critical. I wish you would visit India and see for yourself what is happening here—No more Rajas. Equality of status for men and women—Caste disappearing. All those are healthy signs. There had to be some cleaning up of lingering "ailments," and Indira came forward to do it. She is stronger than her father!' she would comment. Surprisingly, her attitude towards Mahatma Gandhi, who had made such an impact on her in her early years, had changed; she now criticised him for his Luddite approach to life: 'As for Gandhi—he would like to see us going about on a bullock cart!' She seemed oblivious to the fact that she and her team were still relying on the bullock cart at times to reach the interiors of forests in search of primitive cultivars.

She remarked to Darlington that tribals were now 'dignified and [did] not run away' when outsiders approached them: 'Much remains to be done and I hope will be done, especially "population explosion".' Madras had seen terrific rains lately, and Janaki lamented the loss of a great number of plants in her garden at Maduravoyal.[6]

The Belated Padma Award

In 1977, when she was almost eighty, the Government of India finally decided to award Janaki a Padma Shri, the fourth-highest civilian award; a terribly belated recognition for the first Indian woman to be awarded a doctorate in the botanical sciences, the first woman to head a central government institution of science (the CBL of the BSI), the first such to be elected to scientific societies in India and abroad, the only Indian female presence at landmark international scientific conferences, and symposia even up until the 1960s, and perhaps the only practising Vavilovian evolutionary biologist and plant ecologist, male or female, in the country at this time, precociously aware of the need to conserve forests and genetic resources in general. It goes to show how little her science and its significance mattered to the powers that be; that she was a woman and a nomad didn't help either. The belated recognition is even more surprising given how close she was to Prime Minister Jawaharlal Nehru and V. K. Krishna Menon, in her capacity as a world-ranking expert in plant biology. Several male Indian botanists/agricultural scientists/geneticists (almost all of whom worked with her sometime or the other except perhaps M. S. Randhawa, and most with far less research experience or international exposure) had already been honoured with a Padma: T. S. Venkatraman (Padma Bhushan, 1956), Boshi Sen (Padma Bhushan, 1957), B. P. Pal (Padma Shri, 1958, Padma Bhushan, 1968), K. Ramiah (Padma Shri 1957, Padma Bhushan, 1970), N. Parthasarathy (Padma Shri, 1958), A. R. Gopal-Ayengar (Padma Shri, 1967), H. Santapau (Padma Shri, 1967), M. S. Randhawa (Padma Bhushan, 1972), M. S. Swaminathan (Padma Shri, 1967, Padma Bhushan, 1972) and T. S. Sadasivan (Padma Bhushan, 1974). Speaking of the few women Padma awardees prior to 1977, these were chiefly in recognition of their contribution to 'public affairs,' 'civil service,' 'literature and education,' 'arts' and 'social work'; a few went to the field of medicine (nursing chiefly, but one to a doctor), one to sports (Arati Saha, Padma Shri, 1960), and only two to scientists: Savitri Sahni (Padma Shri,

1969, as science administrator rather than for her own contribution to science), and Asima Chatterjee (Padma Bhushan, 1975). Perhaps, Janaki took consolation from Mary Poonen Lukose (1886–1976), from her home state of Kerala, who was the first woman Surgeon General in India, but was awarded the Padma Shri in 1975, a year before she died at the age of ninety!

The Emergency had just been officially lifted when Janaki took a flight from Madras to New Delhi in late March 1977, in time to attend a symposium on the 'Production and Utilisation of Vegetable Raw Materials for Steroid Hormones and Oral Contraceptives' (26–28 March 1977), and attending the Padma awards function. Her paper was on the *Solanum* species collected from the Paniyas of South India. The awards rehearsal was scheduled for 1 April; on the following day, she received her Padma Shri from the Acting President, B. D. Jatti, at an investiture ceremony held at Rashtrapati Bhavan. As a Padma awardee, she was allowed to bring three guests to witness the ceremony. Her niece Padma's husband, K. V. Padmanabhan (retired IFS Officer and former Indian Ambassador to Iran) accompanied her all the way from Madras, and they were joined by nephew Brigadier E. A. Ram Mohan (the late Vasudevan's son, residing at Pandara Park in Delhi) and grand-niece Soumini Ram Mohan. Elder sister Parvathi and youngest sister Devayani were her only siblings now alive (they had lost Raghavan in 1973), but were unable to make it, on account of ill health and other domestic issues.

Janaki would remain in Delhi for two more weeks after the investiture ceremony, as she had to attend official meetings in the city. She had been appointed a non-official member of the managing committee of the National Bureau of Plant Genetic Resources (NBPGR) of the ICAR headed by M. S. Swaminathan. A preliminary meeting had been fixed for 16 April, at which Swaminathan was expected to enunciate the aims and objectives of the NBPGR,[7] before the meeting of the committee scheduled for 7 May.[8] Janaki was also to attend a meeting of the Indian Society for Genetics and Plant Breeding. On 23 April, from Delhi, Janaki travelled to Jammu, permission for which had already been obtained from the Madras University. Her destination was of course the RRL and the medicinal plant farms where her *Solanum* and Cymbopogons were being commercially cultivated. Her next visit to Jammu would be in late 1979.

A Highly Mobile and Busy Life

On her return to Madras, Janaki was busy as ever. By April 1977, Jebhadas and Z. Abraham had been registered for their doctoral degrees with the University of Kerala, with Janaki as their supervisor. Every time Jebhadas went collecting in the South Travancore area, Janaki would send a note to C. A. Ninan of the University, to facilitate her student's research by obtaining a permit from the forest department and helping him identify the plants collected. 'Any help rendered by the forest department will be gratefully acknowledged in any publication,' Janaki assured Ninan.[9] Jebhadas was working chiefly with the Kochuvelan, Malakkuravan, Melappandarum and Kanikkar tribes of the region. In May, Janaki received a letter from the University of Madras qualifying her as a 'Supervisor for guiding research work of candidates leading to PhD degree.'[10] By this time, Vishwanathan had already been awarded a PhD (August 1976) by the University of Calicut, with Janaki as his guide.

Her knowledge, experience and exposure were a great asset and several sought her advice or consulted her on projects of cytosystematics or ethnobotany, but at times she was unable to cope with the unending demands on her time. Janaki for instance had promised the Jesuit botanist K. M. Mathew (of the Carnatic Flora Project, The Rapinat Herbarium) of Tiruchirappalli to review his draft of a technical programme related to an ICAR Research Project entitled 'Ethnobotany Among the Tribals (the Malayali tribes of the Eastern Ghats: Pacchaimalais, Kollimalais, Kalrayans and Shevroys),' but she had not yet found the time to do it. Mathew had even attempted to meet her during a visit to Madras, but she was away on a collecting tour that had taken her to Trivandrum, Palghat, Shoranur and Calicut.

She was also at this time focused on getting the Shoranur Sub-Station (Ashok) up and running. Her former student, Vishwanathan had been put in charge of it but being employed as Assistant Professor at the Rice Research Station in Pattambi, Palghat (not far from Shoranur though), he had very little time to spare.[11] He would nevertheless help Janaki deal with the nitty-gritty of running the Sub-Station, including on one occasion tackling a disgruntled caretaker, with the help of the Labour Officer. The caretaker had filed a petition in court demanding a large sum of money that he believed was due to him. There were fears that Ashok might be lost, but this was averted by Vishwanathan's timely intervention and diplomacy, even if it turned out to be somewhat of an expensive affair for Janaki. Vishwanathan would soon be replaced by one H. Abdullah. Abdullah would occasionally write to Janaki in Malayalam (letters were addressed to CAS, Madras), giving her news of Ashok, and at times complaining about his irregular wages. The practical issues of dealing with Ashok were sometimes overwhelming for one at her age.

Environmental Concerns

In 1976, widespread protests erupted against the Silent Valley Project. The story of the protests and Save the Silent Valley campaign against the hydro-electric project have become something of a legend today, invoked every time a threat to biodiversity is perceived, but absent from the narrative is Janaki's involvement with the campaign and the part she played in drumming up support against the project, as a pioneering woman cytogeneticist and the only female presence on the Committee on Science and Technology, Government of Kerala, and the Government of India. The Silent Valley issue would only reaffirm her inviolable belief in the ecological values of biodiversity, and the conservation of genetic resources and unique ecosystems.

The National Committee on Environment Planning and Coordination (NCEPC) was born thanks to the Stockholm Conference of 1972 on Human Environment organised by the United Nations, which led to nations adopting legislations relating to health and safety, and flora and fauna. India responded with many new special legislations, besides tweaking the already existing ones. As part of this process, Prime Minister Mrs Gandhi asked the NCEPC to create a task force to study the ecological issues relating to the Western Ghats. Accordingly, in April 1976, with Zafar Futehally (Vice-President of the World Wildlife Fund in India) as Chairman, and nineteen members, a task force was constituted to conduct an impact analysis on the planned hydel project in the Silent Valley. The task force concluded that the result of constructing a dam across the Kunthipuzha would be such that 'the last vestige of natural climax vegetation of the region and one of the last remaining in the country, will be lost to posterity' and that 'various adverse ecological consequences' would follow. Unequivocally it suggested abandoning the project and declaring the area as a Biosphere Reserve, for it was 'unjustifiable to sacrifice this unique environment for the sake of generating 120 MW of power.'[12]

However, the task force made the fatal error of providing a list of 17 'safeguards' to be followed in the event the Kerala Government was unable to withdraw from the project. The Government jumped at this escape clause and chose to proceed with the hydel project, after forming a monitoring committee to supervise the implementation of the safeguards, headed by the Trivandrum cytogeneticist, A. Abraham. The Kerala Government argued that only 10% of the area would be cleared for the project (they claimed that the area to be submerged by the dam was only 1,022 hectares, of which 150 hectares were *shola* forests), and the rest would be protected by implementing the stipulated safeguards. The Government also managed to publicise the findings of the Futehally task force even before the NCEPC, headed by the wheat geneticist B. P. Pal, had the time to study or comment on them; the end result was that the NCEPC was forced to simply tag along with the monitoring committee.

Making the Silent Valley a Global Issue

On 14 August 1977, Janaki sent Darlington a booklet on the Silent Valley, 'which is being sacrificed to provide the necessary power for Kerala'; perhaps this was the 'Report of the Task Force for the Ecological Planning for the Western Ghats' headed by Futehally (1977). She wanted it to become a global issue, with influential scientists on the international scene taking a stand and intervening on the issue, and for this purpose appealed to Darlington who had the reputation to be loud and hard-hitting when it came to issues of this nature. 'I did my best to save it at a meeting of the Science and Technology group of the Indian National Science Academy, of which I am a member,' she wrote. 'Please do what is in your power to help. Save this most ancient relic of evergreen forest in India if not the world.'[13] In two weeks, Janaki would make a dramatic announcement to Darlington: 'Dear Cyril, I am about to start a daring feat, I have made up my mind to make a chromosome survey of the Forest Trees of the Silent Valley, which is about to be made into a lake by letting in the waters of the River Kunthi.' She added, 'I was trying hard, as were some others to save the Valley and its ancient flora but "Big Business" in Kerala is I think going to have its own way.'[14] The forest department would however not give her access to the area for her study.

Illustrated Book on South Indian Medicinal Plants

By the end of the year, Janaki's proposal 'to bring out an illustrated book on South Indian Medicinal Plants (with special reference to Ethnobotany),' which was really about revisiting the *Hortus Malabaricus* from the standpoint of chromosomes, and that of decolonising botanico-medical knowledge of the Western Ghats, was sanctioned by the Department of Science and Technology (DST) on 27 December 1977; it was to end on 31 March 1981.

Scientists Mobilise Support

V. S. Vijayan of the Kerala Forest Research Institute, who analysed the impact of the hydro-electric projects on the environment and brought out a report, was reprimanded and the report hushed up. A zoologist from the University of Kerala, Sathish Chandran Nair, actually managed to visit the Silent Valley and returned with much visual documentation, which he then showed around tirelessly to spread the word against the project. Scientists associated with the Kerala Shastra Sahitya Parishad (KSSP) also played a major role in taking the message across the villages and towns of Kerala. As a result, the protest against the hydroelectric project in the Silent Valley gathered momentum, grabbing the attention of individuals like ornithologist Salim Ali, mathematician-ecologist Madhav Gadgil and scientist-administrator M. S. Swaminathan, and of institutions such as the Bombay Natural History Society and the Geological Survey of India besides the IUCN (International Union for the Conservation of Nature founded in 1948), the global authority on the status of the natural environment and the measures required to safeguard it. All of them, including the General Assembly of the IUCN, strongly recommended the conservation of the forest area that was Silent Valley. An ordinance was passed in 1978 (under the Janata government led by Morarji Desai) for the protection of 'the ecological balance in the Silent Valley Protected Area,' and this would be followed up by an act by the Kerala Assembly in March 1979. These however remained cosmetic measures.

In June that year, the Kerala Government began work on the project unfazed by the rising protests. Some members of the NCEPC believed in having a dialogue with the KSEB and the Kerala Government, suggesting alternatives to the Silent Valley hydro-electric project. One such person was N. Madhavan Nayar, Director of the ICAR's Central Plantation Crops Research Institution (CPCRI), Kasaragod (1977–82). In June 1978, Nayar sent a confidential letter to B. P. Pal, Chairman of NCEPC, about the matter. Being an Adviser to the DST headed by M. G. K. Menon, Janaki was also kept in the loop. A meeting was organised on the issue in New Delhi, attended by several scientists, including Janaki and Madhavan Nayar. On return to Kasaragod, Nayar called on the Minister and several senior engineers of the KSEB which was at this time just recovering from a paralysing 51-day strike by a section of its employees. Nayar found them emotionally charged and unable to view the issue objectively. They believed that a thermal station in north Kerala was a serious desideratum, whereas the Silent Valley project was located near the southernmost part of Malabar.

By virtue of the proposed dam's location in a narrow and deep gorge, the project was to be one of the cheapest hydro-electric power projects ever, given the amount of power it was estimated to generate from just a small quantity of water. The KSEB was convinced that 'human interference' in the region could be contained if the Government provided some legislative protection, but what they could not comprehend at all 'was the significance of an ecological niche' (the match of a species to a specific environmental condition). Their conviction was based on the success of the Sabarigiri Project, completed twenty years ago, the catchment area of which had been secured from any human intervention. However, on Nayar's probing, they admitted that this had been possible only because of its geographical situation, on the off-side of the Periyar Game Sanctuary. Nayar suggested an alternative to the Silent Valley Project: setting up a thermal station in Northern Kerala, to the north of Cannanore, where extensive vacant lands were available. The KSEB took to this idea wholeheartedly, for this would not only satisfy the need for an additional source of power for the state, but particularly one for Northern Kerala, thereby reducing expenses on high-power transmission lines, and the attendant power-shedding issues.

A very serious issue emerged when Nayar spoke in private to the engineer-officials: they said the urgency being shown to get the Silent Valley Project started was linked to the compulsion to provide employment to a great number of technicians and engineers, 'who would be rendered surplus with the impending completion of the Idukki Power Project.' The Silent Valley Project was also pushed forward, they said, for the benefit of the Hindustan Construction Company (HCC, founded by Seth Walchand Hirachand, which had almost singlehandedly completed the construction of the colossal Idukki hydro-electric project), who then would not have to spend large amounts of money in transporting back the heavy machinery. What was very unsettling was that even before the results of the tenders were out, the HCC had been assured of the contract. Nayar claimed he had heard a 'lot of "loose talk" that this firm [had] been lobbying very strongly among the legislators, top KSEB officials, and other public people for getting Government of India's clearance for starting the Silent Valley Project.'

He also observed that the original proposal had recommended an earthen dam for the Project, which being labour-intensive, would generate large-scale employment, but this had been changed in the final proposal to a cement-concrete one, which only firms like the HCC could execute. A few KSEB officials admitted that they would be in favour of a thermal station, even if the power generated was more expensive than from a hydel source, 'if this [was] in the larger national interest, and will confer national advantages like conservation of nature, proving additional employment in the area round the site of the thermal station.'[15]

Nayar had not yet met with the Chief Conservator of Forests and other officials. The report by the Kerala Forest Research Institute, on the faunal survey of the Silent Valley Project ('Impact of Hydroelectric Project on Wild Life'), as we had earlier noted had been kept under wraps so as not to displease the Government. This was despite the fact that KFRI was an autonomous institution. 'I have come out of the above discussions with a very disturbed mind as the whole thing raises certain bigger issues than the question of it being just a power project and some additional power for the State/country. It brings in elements of powerful lobbying by powerful vested interests like a construction firm and by the Board because it will ensure continuity of employment to some people overlooking all other considerations,' Nayar wrote to B. P. Pal. He wanted Pal, with the lofty position he held in the world of Indian science, to do something 'very urgently and strongly to save the area . . . If we can get some kind of an assurance from the Planning Commission . . . it should be possible for us to negotiate seriously with the Kerala Government.'[16]

Nayar also mentioned the work of botanist M. K. Prasad of the Government Arts and Science College, Calicut, who had been doing his best in his individual capacity, and through the KSSP to bring to the attention of the public the scale of the ecological disaster that would follow the project, if sanctioned. Incidentally, it was Prasad who had successfully put forward a resolution at the KSSP's meeting in Kottayam in 1978, to put a lid on the plan for a hydel project; Prasad would famously accompany Salim Ali to the Silent Valley, where they would spend three days as guests of the Forest Department. Nayar however believed that Prasad's efforts had not made a deep impact, and to make matters worse, the newspapers were yet to take a stand on the issue. He had himself spoken to Prasad about recruiting a cadre of speakers among college and school teachers, and sending appeals signed by a great number of citizens to the Prime Minister and to the Chief Minister of Kerala. However, he believed all this would be possible only if a thermal station could be 'given' to the State, 'Kerala being so poor' from the point of view of power-generation.[17]

The Central Government Intervenes

On 28 July 1979, Charan Singh replaced Morarji Desai as Prime Minister. He received a confidential report from the NCEPC, which stated that it did not recommend the project and that it was of the strong opinion that the Silent Valley should be preserved as a whole, for it was 'a precious reservoir of genetic diversity which had not been fully explored. It is one of those species-rich areas where little-known plant and other forms of life have survived for centuries in the wild. It is to gene pools like this that man will have to return to in the future to find new materials for agriculture, for the life-saving drugs and the many other needs which will arise.'[18] The influence of Janaki in preparing this report is unmistakeable, because the concern for genetic diversity was at the very core of her scientific practice, much more than any one of her Indian contemporaries at this time.

Unlike Desai who had given the green signal to the Kerala Government to proceed with the project, Charan Singh read the NCEPC report and instituted a central committee, headed by Swaminathan, to re-open the issue, putting the Kerala Government in a quandary. Swaminathan visited the Silent Valley in October that year; he was accompanied by Nalni Dhar Jayal from his Ministry and A. Abraham, Chairman of the Kerala State Committee on Science and Technology, besides a few others. By the end of the visit, Swaminathan was convinced of the need to form a National Rain Forest Biosphere Reserve, an area of 40,000 hectares (including Silent Valley and the adjoining forests of New Amarambalam, Kundas and Attapadi). This, he said 'can become a sanctuary for valuable genes in several medicinal and plantation crops such as pepper and cardamom' and a 'reservoir of useful genes in rice, conferring resistance to some major pests.'[19] Swaminathan also stressed on the need to find alternative sources of power and irrigation.

The Kerala Government meanwhile set up its own panel of expert environmentalists and scientists—men like B. K. Nayar,[20] who would use sophistry to manufacture 'scientific facts' about the Silent Valley in the Government's favour. Nayar for instance would claim that only 240 flowering plants, 'with no new or rare species at all' were to be found in the region, and that it could hardly be referred to as an undisturbed or virgin forest.[21]

Cytogenetics Survey of the Silent Valley Flora

When Janaki received from Swaminathan his report on the Silent Valley, she found a few pages missing in her copy. Requesting him for another set, Janaki hailed his intervention in the matter and made clear her intention of conducting a cytogenetic survey of the Valley's flora: 'I am very happy that you are taking action to preserve the flora of the Silent Valley. I have sent a scheme for financial assistance to Dr B. P. Pal from the Man and Biosphere Committee of the DST. The proposed project is entitled "Cytogenetic survey of the flora of the Silent Valley".'[22] She had turned eighty-two only three months earlier. Her intention of doing such a survey, we may recall, had already been communicated to Darlington. This project would be sanctioned, and the project was to begin on 22 October 1980.

Twin Projects

Janaki had by this time written to Dr (Mrs) Manju Sharma, Principal Scientific Officer of the DST, New Delhi seeking clarifications regarding her two projects. The ongoing one was 'The illustrated book on South Indian Medicinal Plants (with special reference to Ethnobotany)' sanctioned by the DST in December 1977; it was to end on 31 March 1981, but as some funds were remaining, Janaki had requested for a six-month extension of the project, without any additional burden to the DST. She was yet to receive a response from the DST on this. The second project, 'Cytogenetic survey of the flora of the Silent Valley' had been sanctioned (it was to begin on the 22 October 1980) but she had been given no details yet (despite a reminder), regarding her honorarium or fellowships for those who would be working on the project. Janaki hoped that Manju Sharma (who appears to have been an acquaintance) would sort out these issues for her at the earliest.[23]

The extension would soon be granted, and P. T. Kalaichelvan would be appointed Senior Research Fellow, in place of Z. Abraham, who had left having completed his PhD under Janaki's supervision. Determined that the Silent Valley be saved at all costs, and aware that she was fast running out of time, Janaki wrote to A. Abraham, Chairman of the Silent Valley Environmental Monitoring Committee (and Chairman of the State Committee for Science and Technology and Ex-officio Secretary to Government of Kerala, Department of Planning and Economic Affairs), requesting him to grant her permission to visit the Silent Valley. She did not forget to remind him that she was a member of the National Committee for the Man and Biosphere programme of the Government of India. Janaki explained to Abraham, who incidentally was on the side of the Kerala Government (like B. K. Nayar), that she wanted to study the genetic composition of the vegetation on the spot (in situ) and to collect live and herbarium specimens (and propagules) of the constituent elements of the flora.[24] She would not hear from Abraham.

Notes

1 DP: C. 109 (J. 116), E. K. Janaki Ammal to C. D. Darlington, letter dated 31 August 1977.
2 E. K. Janaki Ammal to the Registrar, University of Madras, letter dated 21 September 1976 and 12 October 1976, private collection.
3 Ibid.
4 DP: C109 (J. 116), E. K. Janaki Ammal to C. D. Darlington, letter dated 9 July 1976 [it should be 9 September 1976].
5 Ibid., letter dated 18 December 1976.
6 Ibid.
7 The National Bureau of Plant Introduction came to be known as NBPGR from January 1977. Its previous avatars were as follows: in 1961, The Division of Plant Introduction, in 1956, Plant Introduction and Exploration Organization, and 1946, it was the Plant Introduction Scheme of the Botany Division of the IARI (1946).
8 E. K. Janaki Ammal to M. S. Swaminathan, letter dated 13 April 1977, private collection
9 E. K. Janaki Ammal to C. A. Ninan, letter dated 5 May 1977, private collection.
10 M. E. Pannirselvam, Controller of Examinations, to E. K. Janaki Ammal, letter dated 2 May 1977, private collection.
11 In July 1978, the Kerala Agricultural University, Mannuthy, sent Janaki a letter that it had no objection in sparing the services of Vishwanathan, 'without prejudice to his normal work under this University'.
12 D'Monte, 'Storm over Silent Valley', p. 33.
13 DP: C. 109 (J. 116), E. K. Janaki Ammal to C. D. Darlington, letter dated 14 August 1977.
14 Ibid., letter dated 31 August 1977.
15 N. M. Nayar to B. P. Pal, letter dated 19 July 1978, private collection.
16 Ibid.
17 Ibid.
18 D'Monte, 'Storm over Silent Valley', pp. 48–49.
19 Ibid., p. 48.
20 Nayar, Silent Valley: An Ecological Hyperbole.
21 Manilal, 'B. K. Nayar (1927–2012)', p. 1219.
22 E. K. Janaki Ammal to M. S. Swaminathan, letter dated 29 February 1980, private collection.
23 E. K. Janaki Ammal to Manju Sharma, letter dated 8 January 1980, private collection.
24 E. K. Janaki Ammal to A. Abraham, letter dated 20 March 1980, private collection.

28

MADRAS–NILGIRIS

Hill Tribes and Secret Herbs

I cannot get away from *Saccharum*—it was my first love &
will be my last one too.

—E. K. Janaki Ammal (1977)[1]

The Family Planning Association of India (FPA) was established in 1949 to
promote sexual health and family planning. It had several branches across
the country, including in the Nilgiris.[2] The Chairman of the Nilgiris Branch
was one Marie Buck, who had been a physical education instructor at the
WCC, Madras during Janaki's time. In 1946, three years after her husband
died, Buck joined Simpson & Co. in Madras, which was part of the Amalgamations group of industries, as Welfare Director. It was at Simpson's that
she founded the first family planning advisory centre. Buck was extremely
well-connected in Madras; the American Ambassador to India, the Consul-
Generals, the lawyer Govind Swaminadhan, and of course Sivasailam
Anantharamakrishnan (founder of the Amalgamations Group, affectionately called 'J') were all her dear friends. She had also started a farm (called
the J Farm) in the neighbourhood of the Pallikarnai Marsh, about 50 km
from Madras, where she cultivated food crops, including vegetables, which
would be supplied to the Simpson canteen.

Buck would later move to Coonoor in the Nilgiris, where she would reside,
at 'Sanjiv,' a spacious bungalow on the Kotagiri-Ootacamund Road; Field
Marshal Sam Manekshaw was her neighbour. The FPA office was located
at Bedford Circle in the town. Marie Buck played an important part in the
field of family planning in India, but what this chapter reveals is that she
facilitated ethnobotanical researches in the Nilgiris, even if it was chiefly to
promote her own agendas. Referring to an article published in *The Hindu* (7
March 1976), 'Trees whose leaves prevent Blindness,' Buck wrote to Prime
Minister Indira Gandhi during the Emergency: 'If the source of supply could
be known Mr. B. Sivaram, Additional Secretary, United Planters Association
of Southern India (UPASI) would gladly arrange Plantations at various altitudes in the Nilgiri Tea Estates—And I in my own garden.' She also spoke in

DOI: 10.4324/9781003267089-28

her letter of the secret forest products used by the Nilgiri Hill Tribes: 'Our quiet Family Planning study since 1969 informs us of the methods but not yet the Forest Products formulae ensuring: Positive, harmless Control of Conception, harmless encouragement of Conception, Overnight, painless, no surgery Female Sterilization, and Sex Selection as Tribal strength requires (Dangerous?).' She spoke of the four Nilgiri tribes, and how they remained 'secure and comfortable within their chosen environment and upon their own terms.' A private California Foundation concerned about fertility, she said, had suggested a United Nations Task Force to work with her Branch of the FPA on indigenous methods of fertility control, and she wanted to enlist Mrs Gandhi's support for this project. The national and worldwide potential of this, Buck stated, was far beyond their 'amateur capacity' and they needed to involve scientists on their programme.[3]

Janaki Renews Contact

It was timely for Buck that in late March 1976 Janaki renewed contact with her, after several decades. Janaki updated Buck on her work over the past five years (1971–76) at CAS, University of Madras. 'I have always been interested in the Flora of the Nilgiris and its affinities with the Himalayan plants. I now wish to make a more detailed study of this affinity and its relationships with the Tribes of the Nilgiris,' Janaki wrote to her. It was this research problem that she had got Z. Abraham take up for his PhD project. 'Any help rendered towards this cause will be gratefully received. We shall discuss the ways and means when we meet—I hope in June,' Janaki told her.[4]

Buck agreed to do the needful and used the opportunity to bombard Janaki with the draft of a proposal she had prepared and invited her to the Nilgiris to work with her in the field of family planning. Buck very kindly introduced Abraham to her local informants, including one Dr S. Narasimhan (1917–78), the founder of the Nilgiris Adivasi Welfare Association; she also arranged for his accommodation at the Cherambady Guest House in the forest.

Intelligence on Secret Herbs

Buck could not wait to have Janaki visit her. She wrote in this very unusual telegraphic or staccato style: 'You may come soon. Really stupendous. Enclosed copy of mine to the Prime Minister. . . . To you, soon, a copy to Dr Karan Singh, GOI Minister for Family Planning. Why not your University Grant be supplemented by the Proposed U. N. Task Force if it comes to us. Mr Abraham [Z. Abraham] knows the Cherambady guest house.' Ahead of her arrival, Buck shared with Janaki much of the intelligence she had gathered about the use of herbal contraceptives by the hill tribes. 'We believe the Irula there have and use forest products ensuring Positive Control of Conception, Positive encouragement of Conception and Sex Selection,' she wrote. The 'Kurumbas below and to the west of Coonoor,' she claimed, knew herbs which could induce female sterilization overnight. She also mentioned to Janaki about the fungus 'Life juice,' which the 'Rock' Kurumbas used, and which sustained them, despite centuries of inter-marriage. Buck promised to find out more from the Tribal Welfare Board, including obtaining a list and location of all Nilgiri tribes: 'They say the Nilgiris has more and a greater variety than any other place anywhere.'[5]

Buck prepared an itinerary for Janaki: She was to drop by at Sanjiv, en route to Cherambady, spend a night with her and meet her dear friends Lynn Townsend and his Parsi wife, Seloo. Townsend, Buck claimed, had met with Irulas twenty-five years ago; he was a treasure trove of information. Buck also shared with Janaki the intelligence that Schering Corporation, the German pharmaceutical company,[6] had visited the Nilgiris recently, to gather information on the Kurumba herb that could induce overnight sterilization. Buck was keen that Janaki be part of her proposed project.[7] By this time, she had posted her research proposal to the Prime Minister (and to Karan Singh, the Minister for Health and Family Planning, whom Janaki knew from her time in Jammu). She wanted to know what Janaki thought about the proposal: 'You hope the PM will agree? How can she refuse?' Buck queried impatiently. 'You can provide her with your Botanical Knowledge.' Sometimes she addressed Janaki as the 'Genius' or as a Modern Marco Polo, given her frequent exploratory tours. Buck assured Janaki that it would all work out well for her, but wished 'her brother and sister could have lived to share—pride in [her].'[8]

In late April 1976 Buck sent her a sample of a tree mushroom, which her friends Dr and Mrs Vasudeva Rao from Nihung, Kotagiri, had brought to her. Buck introduced Mrs Vasudeva Rao as 'a precious mine of information,' who had tasted it [the mushroom] twice, but after the second time had ended up violently ill. The fungus, with a disagreeable odour, appeared upon dead Eucalyptus trees after a shower of rain, she explained, and was unique to the lower Kotagiri climate. Buck had also found two new informants recently; they would act as her guides, Buck informed Janaki.[9] She was irrationally optimistic that a large grant would be awarded for organising a Task Force and so did everything possible to help Janaki and her researchers find their way around, including sharing her contacts freely with them, only so that they would discover the secrets of 'the Kurumba overnight Female sterilization.'[10] The Irulas, Buck stated, were the medicine men for all tribes and the Kurumbas were the 'deeply feared sorcerers. . . . You alone can share their Secrets,' she wrote to Janaki. They were the 'first Psychologists, first Psychiatrists anywhere.' Buck also mentioned the 'stinging-nettle fabric' which was essential to all the tribes.

Janaki Visits the Nilgiris

In February 1977, Buck received news of Janaki's impending arrival on the hills, in the company of her former doctoral student at Jammu, Usha Zutshi. Buck would wait for them at the Coonoor railway station with her old Willis jeep which she still drove at ninety years of age! Janaki and Zutshi were to visit the Irulas on the following day.[11] They returned to Madras after a few days of touring the place and collecting specimens.

While Buck was in the process of redrafting her project proposal for submission to the Prime Minister, Janaki sent her a letter from Madras clearly stipulating that her name should not be included in any proposal requesting financial assistance because she was already the project leader of the UGC-funded project, 'Ethnobotanical survey of South Indian economic plants with special reference to Primitive cultivars.' Buck was terribly disappointed, but still replied with information on bamboo seeds, reputed to be a powerful aphrodisiac, being sold in the tea shops beyond Gudalur on each market day. This belief stemmed from the behaviour of forest rats, which were attracted to flowering bamboos; as soon as the rats feasted on the flowers, a great number of offspring were produced. Buck wished to know from Janaki whether it was worth asking the Forest Department to 'observe, Gather seed . . . Should the seeds be sent for testing?'[12]

Eileen Plans a Visit to Madras

Meanwhile, Janaki's long-time friend the geneticist Eileen Macfarlane was looking to spend some time in India. Though she expressed assent to Eileen, Janaki was not too happy about it. She would remark to Darlington: 'Eileen is doing her best to come to India. I do not know if it is the best thing to do—I am leading a quiet and peaceful life. I feel she will upset it.'[13] Eileen, who had lost her husband, James B. Macfarlane in 1974, was on the verge of retirement from the Automated Medical Services, Mansfield, Ohio, where she was working for Raymond Thabet (1975–77).

About this time, Kenneth L. Jones, a long-time faculty member of the Department of Botany, University of Michigan, had sent Janaki a copy of *The Harley Harris Bartlett Diaries (1926–1959)* (1971). 'You will enjoy it, not much about our era,' Eileen would comment on the *Diaries* in a letter to Janaki. 'Ken did not know your address. I asked Frieda [Blanchard was Jones' faculty colleague], so he phoned her and she gave him your address. She & Dorothy still live in the old Geddes Av. House, which is stuffed with memorabilia. She seldom goes out. An octagenarian?'[14] In a few months, Frieda Blanchard would be no more. While Janaki thought the *Diaries* contained 'mostly trivia,' Eileen felt, 'his simple human side brought back a lot of old memories.' Janaki's reply made Eileen very 'happy & home-sick for INDIA:'

> Thank you for inviting me to come on a visit. Your work sounds interesting & challenging & I'd like to come & help for a while. I have told Dr Thabet I'd probably return to Wheeling next September . . . Now I have something worthwhile to plan for, if I can come to Madras in October 1977, for a few months & do something useful with you again. Please don't be nervous about it, I can adjust easily & always liked Madras. If you can put me up fine, or find me a room. How do you get around? I am very healthy and do not eat much. *Let me know what you think soon & don't back out.* This gives me a goal in life. What would you like from U.S.A.? How about one of the new 'pocket' calculators? A book on Mammalian genetics?[15]

When Eileen received Janaki's reply, in time for her birthday, she was thrilled. In another letter to Janaki, Eileen claimed that India was her 'spiritual home' and that it would be wonderful to work with her 'dear old friend EKJ.' Janaki had asked Eileen for her CV and list of publications for submission to the Madras University, but given the terrible spell of winter in Ohio, delay was inevitable. In the meantime, Eileen suggested, that the University could look up (ironically) the *American Men of Science*, 'under Macfarlane' to know more about her. When the book was later renamed, the *American Men and Women of Science*, the highly independent Eileen objected to it rather surprisingly, saying 'Man embraces Woman unless the latter is unpleasant.'[16] Eileen felt she may have met Marie Buck before, but was not very sure. She was keen to know if the Bosotto Brothers bakery was still around in Madras and whether Janaki's laboratory was to the north of the city.[17] When Dr Thabet, the Lebanese doctor she worked for, expressed concern about any 'political upheavals' in India (given the state of Emergency) that might affect her, Eileen responded that in that case, Janaki would not have invited her. Moreover, 'after war-time in London' no unrest could affect her, she claimed, just as Janaki would.

Eileen was a go-getter and a border-worker like Janaki. She wrote to Malcolm Adiseshiah, the Vice-Chancellor of Madras University, introducing herself as a cytogeneticist and anthropologist. 'When I was in India some years back at the invitation of Dr Miss E. K. Janaki Ammal, I worked on the Blood-groups of tribals of Kerala and Assam and collected plants used by them. Now I [would] like to return to India and join Dr Janaki Ammal at the Centre for Advanced Study in Botany of your University to continue my life-long interest in tribal medicine and habitats. I am about to retire from my medical work as Cancer Cytologist for Automated Medical Services of

Ohio Inc. I am in very sound health to undertake serious research work,' wrote a seventy-eight-year-old Eileen.[18]

In April (1977), Eileen updated Janaki on her preparations for the trip to India. She had read that the elections had gone off peacefully, and hoped that it would not affect the Provinces. 'I was surprised that Mrs Gandhi was ousted. The U. S. Press was against her. They are a rotten lot, love to tear any achiever down. I have never been interested in politics. I prefer vegetation & look forward to learning the S. Indian Flora again. The little spring flowers are out here (28°F–40°F) *Sanguinaria, Hepatica, Trillium, Nivea, Cardamine* & *Claytonia*, do you remember them,' she asked Janaki, reminding her of Ann Arbor.[19] Eileen wrote again three weeks later, to send Janaki her flight-details. 'We will both look very different due to remote birthdays!' she joked. She warned Janaki, that she should not expect her to 'know anything much about the tribals & Indian Botany after all these years in other fields. I'll be the student interested in chromosome numbers, breaks et al. I actually never was in the Nilgiris & will enjoy it very much.'[20] Closer to her date of arrival, Eileen wrote again to put Janaki at ease; she had after all stayed three months in the city (when the QMC was closed, in 1933–34), without much discomfort, during her first ever visit to India. 'It should all work out for the best. We can have a pleasant reunion & I can learn about your work in Ethnobotany & assist your student, or Mrs Marie Buck,' she added. 'If anything happens & I am an embarrassment or nuisance I can always return,' she reassured Janaki. Eileen was 'eagerly looking forward' to being with Janaki again, 'to seeing changed Madras & Coonoor or Ooty for the first time.' This was a 'trial visit,' she hinted, meaning she was planning a return if the present trip went well.[21]

Eileen Arrives

Janaki had just recovered from a bad bout of influenza, when it was about time to receive Eileen at the airport. They were expected to travel together to the Nilgiris within a few days of Eileen's arrival, but the rainy season disrupted plans. Buck would not recommend it either: 'Please take no risk until quite strong again,' she warned Janaki. 'When you know your arrival date—a phonogram please . . . The white Family Planning Van will meet the Train and a most delighted proud and happy welcome will await you here. If the flu is still a problem directly to bed,' she wrote to Janaki with affection.[22] Buck suggested that Janaki might want to study three important groups of tribes: the Kurumbas, the 'Rock' Kurumbas and the Irulas; the tribes believed that their secrets shared would at once lead to punishment by 'total loss of all personal and tribal value and power,' she added. And in another letter, 'Rest—6000 feet altitude—cosmic rays—rest as long as needed. The Tribes have been here for long. Happiest welcome awaits you. With love.'[23]

Eileen arrived in early June 1977; Janaki and Z. Abraham met her at the Madras airport with the Department of Botany car, an 'Ambassador, made in India.' She was accommodated in 'an air-conditioned room at the International Guest House of the University of Madras, on the Marina, overlooking the Bay of Bengal.' As soon as Eileen settled into the Guest House, she intimated Darlington of her arrival: 'After two years of microscopic cancer detection for Dr Raymond Thabet, I was eager to see India again. An old friend of my University of Michigan days said I could come and help her with ethnobotany. So I returned to Wheeling, bought a round-trip ticket by Air India and arrived in Madras June.' She was obviously joking when she referred to Janaki as 'an old friend of my University of Michigan days' as if she was someone unfamiliar to Darlington. Eileen found Madras 'hot and humid' and so was very glad when Janaki took her away to 'Coonoor at 6000 ft altitude, in the solid granite Nilgiri Hills (Blue Mountains), among large Tea Estates.' There, Janaki introduced Eileen to Marie Buck, the 'oldest USA citizen East of Suez,' who was the Chairman of the Nilgiris Family Planning Branch, and wife of Harry C. Buck, the YMCA Physical Education Director at Nandanam, Madras, in the 1920s.[24]

Janaki returned to Madras as soon as she settled Eileen into Buck's bungalow, which would be her residence for the next three months. Very keen on knowing more about the botanical contraceptives used by the local hill tribes, Buck went out of the way to provide Eileen with a jeep and driver to move around and explore. She had also arranged with two young informants, who were developing a vegetable farm just above the cave of the Kurumbas, to act as guides. Eileen explored the hills and managed to go to Parkside to meet some 'Cave'/'Rock' Kurumbas. It being a lovely sunny day, she was hopeful of getting some good photographs too. She had also begun collecting plants, but was 'proceeding slowly' with a make-shift press; luckily for her, she found a retired tea-planter and hunter who helped her with it, and her herbarium specimens were ready in a matter of just 72 hours.[25]

'In India they drive on the left and petrol costs nearly $ 3 a gallon (Rupees 3.7 per litre),' Eileen would enlighten Darlington. She had begun work in the hills full-throttle. 'This method is usually kept a secret,' she explained, referring to the tribal practice of sterilization without surgery. On 4 July, she would attend the 'usual American Tea Party' to celebrate American Independence Day, at the Consul General's House at Adyar, Madras, having flown down all the way from Coimbatore just for the event. Buck was unable to leave Coonoor to attend the celebration. Janaki, who had still not recovered entirely from the flu, was away in Tellicherry at this time, and so would not attend. Buck in the meantime had written to Janaki that one Major Mottram, who lived in the place below Sanjiv, had stumbled upon an extensive area of lush wintergreen, full of 'oil herbs unknown to him.' He wanted this to be kept a secret, but was willing to share the information only with Buck, Janaki and Eileen, for fear that 'collectors might trample and destroy herbs even more valuable.' 'W.G. [wintergreen] is widespread occurrence. If you and Eileen wish to investigate—the Jeep is yours. Major Mottram, 2 days notice, could be your guide,' she wrote in her typical style.[26]

By the time Janaki returned to Madras from Tellicherry, Eileen had already returned to Coonoor; she had also collected the copy of 'Fyson's Flora of Madras' [perhaps it was J. S. Gamble's *Flora of the Presidency of Madras* (1915–21)] and the herbarium press that Janaki had left for her at the department. Major Mottram took Eileen in Buck's jeep to Doddabetta, 'into uncultivated country' where they found some interesting plants, including 'a white rose and a ground orchid—*Habenaria* . . . Fyson will be a help,' Eileen remarked to Janaki.[27] She occasionally visited the Public Library at Ootacamund, and scribbled her letters while seated there. In mid-July, she wrote to Janaki from an extremely cold Ooty; she had bought herself a second-hand woollen coat. The cold and wet weather had made it difficult for her to dry out her plant collection, among which were some interesting specimens of *Rosa leschnaultiana*; we may recall that the genus *Rosa* was the subject of Eileen's doctoral research. Eileen was planning a few meetings with retired planters, those who had employed tribals on their estates, such as a Mr and Mrs Neale; the Neales employed Irulas and Kurumbas, and lived near Masinagudi.[28]

Buck Disappointed

Janaki had about this time just returned from Delhi, after attending official meetings of the DST, Government of India. She had been so busy with her field-work and with meeting the official demands on her, that she had been unable to visit the Nilgiris for quite some time now. Meanwhile, Buck had begun to get restless and even unpleasant because she worried that nothing had emerged from all that she had been doing for Janaki and Eileen. Buck was hoping for 'more definite details,' which could be incorporated in the final proposal for the 'Million-dollar grant,' as Eileen would sarcastically put it. Unfortunately for Buck, the Central Government all on a sudden decided to stop funding for family planning, given the rising instances of misuse of funds. 'This is a big upset for Marie,' Eileen would comment to Janaki. She was having a difficult time, trying to get Buck understand that her visit was purely exploratory 'in order to determine the feasibility & best modus operanda' for a 'Complete Survey of the Plants & Primitive Tribes of the Nilgiris,' and for a secondary study of blood groups and female cervical cancer detection. Eileen reported to Janaki once again three days later. This time, she let out that Buck had had a complete volte-face and was now 'very happy' about her stay with her. The reason for this was that the ageing and somewhat eccentric Buck had turned hopeful that Janaki and Eileen would be able to befriend the tribals and 'steal' their secrets.[29]

Meanwhile, Eileen wished to push her blood group project forward and insisted that Jebhadas, Janaki's doctoral student prepare a complete bibliography of all the blood-group studies that had been done on the tribes of the Nilgiris and Kerala, since her own pioneering ones from the 1930s. 'We need this. I have never seen it, but various reports (journalistic) come out now & then. He must go to the Central Library (Index Medicus?) & get Librarians to help,' she suggested to Janaki.[30]

On 15 September, Buck organised a private party at Sanjiv in honour of the visiting American Ambassador Robert F. Goheen, his wife and youngest son Charley, a college student. Goheen's parents were close family friends of the Bucks. Buck wanted Janaki and Eileen to meet the Ambassador and tell him about their researches in the Nilgiris, but Janaki was not interested. Meanwhile, Buck had learnt from Eileen that Janaki had given away her *Solanum khasianum* collection to a multinational company, to produce contraceptives (there does not appear to be any truth in this however). That Buck was unhappy with giving it away to an American multinational is rather surprising; she felt it should have gone to an Indian pharmaceutical company like the Sarabhai Chemicals.

From Within the Family

Janaki was yet to finish reading Darlington's *The Evolution of Man and Society*. She told him that she was working on a book on the medicinal plants of South India, with illustrations and chromosome numbers. This ambitious project aimed to produce a modern or cytotaxonomical and anti-colonial version of the *Hortus Malabaricus*—if such a thing was at all possible, given that the *Hortus* was a 12-volume compendium covering more than 700 plants. Janaki had hired the services of her niece, E. K. Shantha (daughter of her half-brother E. K. Govindan, Diwan of Pudukottai in the early 1930s), who had retired as Principal of the Tellicherry College for Women, to make the drawings; she wanted some of the drawings in the *Hortus* reproduced in her book. In his previous letter to her, it appears that Darlington had asked what her full name was, and what the surname 'Ammal' stood for, a strange question to ask someone he had known intimately for more than four decades. 'You asked me for my full name,' Janaki would write, without any feeling of surprise: 'EDAVALETH KAKKAT JANAKI AMMAL. You see now why I do not write it—fully—In fact it is only the family or house name, used only in documents. EDAVALETH means—to the right and to the left. This is a twin house of my father in Mahe (French territory) (EDAM—right, VALAM—left). Our house name is just EDATHIL—"To the right". "AMMAL" is only a term of respect like Miss and has become part of my name.'[31]

Hybrid in Short Blade

Janaki would have to postpone a tour to Malabar she had planned for September (1977), because of an urgent call from the DST concerning a grant application she had made: her book-writing project on the medicinal plants of South India. She would have to travel to Delhi to deal with it. When she visited the Shoranur Sub-Station later as part of her tour, A. Mahadevan, the Director of CAS joined her to inspect its progress. Meanwhile, Janaki received a letter from T. V. Sreenivasan of the Sugarcane Station, giving her news that her *Saccharum* x *Zea* hybrid was in 'short blade': the ideal time to fix pollen mother cells for meiotic studies. She responded saying that she would start immediately for Coimbatore and hinted that she would like to rewrite the paper based on this, but with him alone. Janaki was excited that K. Mohan Naidu, a young sugarcane scientist, had managed to induce flowering in her hybrid; she had been unsuccessful with it at Trombay. 'I have much to do. Do you think I should write to Dr Jegathesan too as he is your head so to say? I shall say I have to solve one problem in my *Saccharum* hybrids. I have *Saccharum* x *Imperata* here. Perhaps you haven't got it—we must make something out of it. It was backcrossed to POJ 2725 & produced the wonder cane [the Stud-bull],' she wrote to Sreenivasan. 'Somehow I cannot get away from *Saccharum*—it was my first love & will be my last one too. As for *Saccharum* x *Zea*, Darlington is keen to join in but I am not agreeable. Please let me have your frank remarks,' she added.[32] The *Saccharum* x *Zea* was supremely her own and not even Darlington would be allowed a share of it. However, again nothing would come out of this.

Injured Again

Disaster struck in late November 1977, two months after Janaki's visit to Coimbatore. She had had a bad fall while on a tour in Kerala, and fractured her thigh bone (right or left is not definite) just when she was supposed to board a flight to Delhi to attend a meeting of the NBPGR (in the IARI campus, Pusa). She would have to undergo surgery at the 'University Hospital (which one, is not known).' K. L. Mehra of the NBPGR, who received her telegram about her inability to attend, responded with much concern: 'I was sad to learn from your telegram that your leg bone has been fractured. I pray to God for your early recovery and good health so that you continue to give your advice and guidance to several of us. We held the meeting and I shall shortly be sending you its proceedings.'[33] She would be out of action for almost four months, and when she began to walk again it was with a slight limp, which would stay with her for the rest of her life. A fragment of her thigh bone had to be removed and Janaki would save it in a little box, it is said, to be cremated with her body.

In the New Year, Janaki wrote to Darlington from her hospital bed. This was the first time she was sitting up after her surgery and her writing was very illegible as a result. 'I am writing to you from a bed in the University Hospital, where I was brought with a broken femur over a month ago! I am afraid breaking bones is become one of my past-times [sic]! This one happened while I was about to leave for Delhi from Kerala after visiting tribals. I am now able to sit up in bed & my first note comes to you. Lying in bed one has a chance to look back—and you have been very much in my thoughts.' She had by this time finished reading his magnum opus, *The Evolution of Man and Society* and had 'ordered' her students to consider it as a textbook, 'preliminary to their studies in ethnobotany.' Janaki had widened the scope of her research, becoming 'a Vavilovian, if I [can be] bold enough to say so,' she would remark. She was clearly in a very fragile state of mind, and looked back on her life, and remembered with gratitude Bartlett and her father. She remarked: 'It was H. H. Bartlett . . . and my father that gave me the zest for the study of man.' With visible pride, she told Darlington that she was now adviser to the Ministry of Science and Technology. 'Not that I have been able to do very much to shake up old ways but an attempt is being made.' The ethnobotanical garden at Maduravoyal contained a good collection of primitive cultivars of economic plants gathered from the wild, she told him, and that she had managed to encourage the department in introduction and preservation.

'I often wish I could meet you. Can't you come to India again? And may I have the privilege of nominating you as a Fellow of the Academy of Science, India INSA—it is our highest honour,' she asked with much affection. Janaki also shared with Darlington news of Eileen, without hiding her mixed feelings towards her: 'Eileen visited India 2 months ago. Mr Macfarlane is no more so I am wondering if she would find a fourth husband also in India. She said she may be old in years but physically still young.'[34]

Appeals to Sadasivan

Janaki would reply in the affirmative to a letter from T. S. Sadasivan requesting her to endorse the nomination of Krishna R. Surange (Director of the Birbal Sahni Institute of Paleobotany, Lucknow) for the Presidentship of the Botany Section of the Indian Science Congress for 1979. Incidentally, The Indian Society of Genetics and Plant Breeding had decided to hold a special meeting during the Congress to discuss the steps to be adopted to make the Congress of 1983 a success; the Representative Council of the International Genetics Federation at its meeting held in Moscow on the 26 August 1978 had resolved by majority vote to accept the invitation of the Indian Society to host the next Congress in New Delhi in 1983. Janaki used this opportunity to write to Sadasivan about her idea of proposing Darlington's name as a Foreign Fellow of INSA: 'I hope I shall have your support for the same. I have already got his permission to do so.'[35] Janaki also sent Sadasivan, who was at Lucknow at this time attending meetings, copies of her letters to the DST on the book-writing project on South Indian Medicinal Plants; she wanted him to follow it up and use his powers to influence the relevant authorities to sanction it. It appears that this helped, for we soon gather from Janaki's correspondence that the project was sanctioned and funds allotted. She was at this time at Edam still recovering from her surgery, and was yet to return to Maduravoyal.

Naming a Genus After Janaki

About this time, J. Joseph and V. Chandrasekaran of the BSI, Coimbatore, named a newly discovered plant, *Janakia arayalpatra* (an interesting 'tuberous low shrub,' found on the rocky hill slopes of Kurusumalai, near Bonaccord Estates on the Western side of the Agasthiyar Hills), in appreciation of Janaki's invaluable contributions to cytotaxonomy. Janaki's botanical tours to Malabar and the Nilgiris invariably took her to the BSI office and herbarium at Coimbatore, where Joseph was based. The epithet *arayalpatra* was a reference to the leaves, which resembled those of the *Ficus religiosa*. Joseph and Chandrasekaran had introduced the plant to the BSI garden at Coimbatore in 1975. It is important to note that this was the first and the only instance of a genus or species dedicated to Janaki, despite her huge contribution to the field of botanical science.[36] Of course, a variety of *Magnolia kobus* (*Magnolia kobus* var. 'Janaki Ammal') was named after her, but it was only a variety, and not a genus or species. It is a matter of fact that in general, very few plants of the world have been named after women; they are mostly dedications to men, scientists or otherwise.

Contacts Buck Again

Janaki was back in Maduravoyal by early March when she heard from Buck: 'My legs are not strong enough for normal walking. I shall certainly come there when I am fit enough,' Janaki replied. She thanked Buck for taking such good care of Eileen during her three-month stay with her. Janaki disclosed that she wished to gather as much information as possible on medicinal plants of the different regions of the Nilgiris for her project, and therefore was intending to send Jebhadas to the hills. 'I hope you will give him all the chances to meet the different Tribes. I hope it will be convenient for you to find him a place to stay at Y.M.C.A. or similar boarding home . . . I wish I could have come with him. Please give him all the knowledge you have stored which will be duly acknowledged in all the publications.'[37]

Eileen Returns and Tensions Rise

Eileen would return to the Nilgiris in April 1978, to carry out her researches with the help of Marie Buck. Janaki had not yet gone back to Maduravoyal. 'I am so fit and so old that I decided to take some more trips before

I disintegrate! Poor Janaki looks dreadful, half blind, bent & lame. She had another fractured femur in November & is probably retired now in Tellicherry. . . . She was very game and got around. Last year the Central Government gave her the title *Padmashree* for introducing high yielding Citronella from Burmah,' she would tell Darlington. This was however an inaccurate piece of information; Janaki had been honoured for her life-long contribution to the field of botanical science, rather than for any particular accomplishment. Incidentally, by July that year, Janaki would be fit enough to travel to New Delhi to interview applicants for the post of Joint Director for the BSI.

Eileen befriended a retired Tea Planter this time, who willingly acted as her interpreter in her conversations with the Irula 'Herb Doctor.' This was about the secret plant which had the ability to induce sterility in men and women permanently. Eileen believed she had almost managed to recognise the plant from the leaves and twigs the man had brought her. She would later have its identity confirmed by the BSI at Coimbatore. A prolific publisher, Eileen wasted no time to write to Darlington that 'she would 'like to send a short note on the plant' and 'a few things' she had learned from an Irula medicine-man. 'If I write it up could you send it to some little magazine? I do not want big drug Compys. to go & mess with the Hillman,' she remarked. Eileen thought it might interest 'some British semi-scientific magazine,' rather than a purely academic one. She also told him that she had met with a 'medical genius' in Coimbatore, a 'Tamil Brahmin' named K. S. Ram, who knew 'some wonderful remedies (no quack) including something to sterilize a woman for only 3 yrs.'[38]

Meanwhile, Marie Buck had written to Janaki. She was still hopeful that together they would be able to organise a trial of the 'harmless' contraceptive and the 'overnight Female Sterilization herb' the Kurumbas used. 'You are so busy, but I hope you can help us,' Buck appealed to her. She had sent Janaki a parcel containing some powder of the bark of a tree which somebody had given her, accompanied by a note: 'They have not revealed the secret,' but this 'really may be a secret leaf for research.'[39] When Janaki got wind of Eileen's plan to publish on the Irula herb, she was livid. She complained to Darlington that it was 'very naughty' of Eileen to have thought of publishing on the medicinal plants of Nilgiris, when Z. Abraham (Janaki's student) was working on it for his doctoral thesis. 'However, we have lots more to do here and she can have a slice of our bread,' an angry Janaki commented.[40]

Opening More Fronts

The DST sanctioned the creation of a Sub-Station at Shoranur in late May 1978, and this was the ideal time to begin planting operations in Ashok. Janaki also wanted a herbarium of medicinal plants set up in the laboratory attached to the Station. With these in mind, she wrote to the Registrar of the University of Madras, perhaps from Edam where she was recovering from her surgery, requesting him to release some funds: 'The place requires *pukka* fencing and minor repairs so as to make it secure from pilferage of our valuable collection of medicinal plants. The building in this property also requires white-washing and furnishing. May I therefore request you to immediately sanction an advance of Rs 3000/- for this purpose.'[41] The funds would take a couple of months to be released. Meanwhile, Janaki contacted the Calicut University botanist, K. S. Manilal, about a botanical artist who could work on her illustrated book project, though she had been getting 'the sketches done by someone at a local (Madras) College' (and also by her niece Shantha). Manilal assured her that he would be happy to help in whatever manner possible.[42]

Janaki's research publications in the 1970s had dwindled in number, and were mostly in the capacity of co-author.[43] The bulk of her energies were being directed towards attending national-level committees, and meetings of scientific societies, presenting public lectures, guiding doctoral students, and developing the gardens at Maduravoyal and Shoranur. Manilal wished to offer a collection of 'some rare and important plants' to the ethnobotanical gardens at these two places; Janaki immediately sought the permission of the Registrar of the Madras University to send Jebhadas, the senior research fellow, accompanied by field-man P. Gopinathan,[44] to collect these from the Calicut University. The plan was to combine this with a bit of collecting in and around Shoranur, Wayanad (Paniya settlements of Vythiri and Sultan Battery) and Attapadi (where Mudugas and Irulas were concentrated). Incidentally, Manilal's work on the *Hortus Malabaricus* had by now been three years in the making (in October) and as he was not expecting the UGC to give him an extension, he was more than willing to offer the services of Narayana Vaidyar (whose inputs were being used for his *Hortus* project) for the Shoranur Sub-Station, if Janaki needed him.[45] It does not however appear that she employed him; her chief source of information remained the aboriginal tribes of the Western Ghats. Manilal was also organising a symposium in connection with the 300th anniversary of the *Hortus* at Calicut University and was keen that Janaki contribute a paper to it, which it appears she did. Incidentally, at this time the title of her DST-supported project recorded a slight change; it had now become ' "An Illustrated Book" on South Indian Medicinal Plants (with special reference to Ethnobotany).'

Hybrid in Short Blade Again

In mid-1978, Janaki would hear that her *Saccharum* x *Zea* hybrid was in 'short blade' again. She wanted to make the best use of the opportunity and demanded that Darlington return her paper, which he had rejected; she was hopeful of making new discoveries this time, backed by solid evidence.[46] In less than a week, Janaki was at Coimbatore, fixing her material and closely studying it under the microscope; she found that the B chromosomes (the supernumerary chromosomes) had increased in number owing to years of vegetative growth.

Communicates with Darlington

Janaki was also now well enough to travel to where the Irulas lived, the *Marutho Male* ('Medicine-Hill') in the Nilgiris. The Irulas usually collected medicinal plants and sold them to 'native physicians,' Janaki would enlighten Darlington. She had collected live plants to cultivate at Madura-voyal, to be subjected later to cytological and chemical analysis. She also shared with Darlington her discovery of an interesting *Solanum* hybrid species, a cross between *S. incanum* and *S. indicum*, both used by the tribals as contraceptives.[47]

Janaki was very sorry to hear that Darlington's house had been burgled and that the traditional brass lamp she had gifted him several years ago (in Trivandrum in 1933), and of which he was very fond, had been robbed. 'If there is anyone coming to England, I will be happy to replace the brass lamp, which was stolen,' she assured him. She had just received 'a personal invitation' to attend the World Forestry Conference to be held in Jakarta. She was however not in any position to go, she told him.[48] The sight of Darlington's letter, in reply to hers, pleased her immensely. 'I am always happy when I see your handwriting on an envelope addressed to me,' she would remark. In another letter, she gave him a brief account of the history of the lamp which had been stolen from his house; the story had to do with how the Malabar Namboothiris married to maintain a pure line, an aspect she had discussed in an unpublished paper on man and society in India (which incidentally was also discussed in her Wenner Gren paper):

> By the way it is the same sort of lamp mentioned in the bible, which the foolish Virgins forgot to fill with oil! This type of lamp is found only in Malabar, and I have a feeling it must have come with the early Syrian Xian Fathers (followers of St Thomas the Disciple of Jesus). Not many homes possess this lamp. It was used in ceremonial occasions to welcome an honoured guest like the Nambudiris, who take Nair women as 'bedfellows'. Only the eldest son of a . . . Nambudiri is allowed to marry. The others cohabit with Nairs, who consider it a great privilege too. Since matriarchy [matriliny] is the rule in Malabar—it matters little who your father is. It is the Mother's family, which counts for prestige. You will ask what happens to the Nambudiri women when their men folk marry outside. They remain like nuns or they were adopted by castes lower than themselves—in a sense they were outcasted.

Many families in Malabar claim their descent from such Nambudiri women (genetic[ally] improved female seed).[49]

Janaki was keen to have copies of Darlington's recent books and offprints of his papers, she wrote to him, with the intention of reviewing them for *Current Science, Journal of the Botanical Society* or *The Indian Journal of Genetics and Plant Breeding*. 'I am too poor to order them for myself from England. With my salaams, Yours always, Janaki,' she signed off with much affection.[50] Interestingly enough, she had never published a review essay before, but obviously felt Darlington (or rather the John Innes camp) deserved to be better known in India, certainly better than Gates (the King's College camp). However, like several of her writing projects, this would also be shelved. In a few weeks, she would hear from Margaret, Darlington's former wife, giving her news of the Darlington children and grandchildren, which made her very happy.

In October 1978, Janaki attended meetings of the Man and Biosphere (MAB) committee of the DST, 'which is trying to preserve the Flora & Fauna of specially interesting regions.' Eileen would express happiness at the Indian Government's initiatives in making 'good nature preserves,' for she had been deeply upset to see the 'deforestation & planting of Australian trees . . . in the Nilgiris.'[51] Janaki had shared her anxiety with Darlington over the 'destruction of forests in India in the recent years for the cultivation of commercial crops like tea, coffee, rubber and even potatoes in the Nilgiri Hills! Result—Floods!!' Reiterating that she was working on a book on the medicinal plants of South India, she said that 'a lot of original information from the Tribals' was now available. 'Ethnobotany is becoming the fashion, or shall I say the "Rage" among botanists,' Janaki would remark; we might recall that she had observed just the opposite a few years earlier—that ethnobotany was unfashionable and therefore did not attract funds. '*Costus speciosus* is the newest plant for promotion of contraceptives,' she confided. The plant grew in the hills by the side of the Ghat roads in Malabar. Janaki had still not been able to spot another of those 'quaint brass lamps . . . Perhaps I will find one in Kerala,' she reassured Darlington.[52]

An Untiring Collector

Janaki left for Malabar in November that year; she would turn 81 that month. She was to stop by at Shoranur, to drop off the plants meant for Ashok, but for some reason decided to continue her journey to Tellicherry. She wrote to Z. Abraham from Edam after a couple of days, and let him know that she had been collecting on the outskirts of Tellicherry, 'along the Mahe Road.' She wanted to go to Wayanad as the weather was looking good, but it was a handicap that Gopi, her assistant and field-man, was not around. She gave Abraham a set of instructions: 'I want you to *somehow* send Gopi to Tellicherry to accompany me as well as to pack up fruit trees for Ashok. . . . When he comes and I hope it will be the very next day after you get this letter—ask him to bring the large basket with the handle, which is in his room (I think). It is shaped like this . . . [drawing] He will know it. I used to use it for taking cats. In it he must bring a large amount of papaya seedlings and wild plantain seedlings and as many seedlings of Curry Leaf plants *Murraya koenigii*, as it will hold—the larger the better. Besides the above I want six seedlings of Cocoa. Will leave for Manantoddy as soon as Gopi arrives in Telly taking my niece [perhaps E. K. Shantha] with me as companion (we will stop two days at least). Please redirect all letters to Tellicherry. This is best time for collecting in Kerala—Many grasses & annuals are in flower. You will have to pay Gopi's train fare . . . I'll go to Shoranur only after Wynaad.'[53]

She would again write to Abraham the very next day, with more instructions for him. She wanted Gopi to bring (besides papayas, wild bananas and cocoa) nutmegs (if available), and a lot of tetraploid curry leaf plants. Janaki also disclosed to Abraham her plan to start an arboretum on top of the hill (the hilly part of the Edam homestead), 'as the first step to send the tenant away.' Gopi was asked to bring two dozen labels for naming the trees. 'I will make a large label here—'Arboreta of Medicinal Trees' or 'Hortus Malabaricus.' She had decided she would go to Silent Valley and Wayanad only after Gopi arrived; 'so please send him soon,' she wrote impatiently. Meanwhile, she awaited Vishwanathan's arrival at Edam.[54] Janaki wanted Wayanad to be made a Subsidiary Station in North Kerala for her project. Incidentally, she was to attend a seminar at the Sugarcane Breeding Institute, Coimbatore but decided against it. She would visit the Institute however on her return journey to Madras and also look up Ashok at Shoranur.

News of Marie Buck

By the end of 1978, Buck would leave Sanjiv, to reside in Golfston, Kotagiri. Her health had begun to suffer badly, and the family planning funds had ceased. Buck's friends P. Vasudeva Rao and wife appealed to Janaki in late December to make a short visit to Buck: 'My husband and I feel that this is the time she needs help and will appreciate the visits of good friends like you—who will understand her in her present confused state of mind.' The American Consul General of Madras was making arrangements to find a female companion to stay with her, they said, but Buck was not happy with having a stranger in her house. 'The house that she is living now is a large one, and there will be accommodation for you to stay with her,' Mrs Rao wrote, suggesting that Janaki give her company for some time.[55] This was however not possible, given Janaki's various official commitments and her own fragile state of health. In fact, Eileen had been writing to Janaki with much concern, requesting her to pay attention to her diet, to keep it 'well-balanced, including plenty of Vitamins (lots of Vit. C) so that she did not 'get run down.'[56]

More Bad News

In December (1978), Janaki would hear from Josephine Fuller, former Registrar of the Royal Holloway College, London, that the British mycologist Margaret Madge (Rita Madge), who had trained under her at Wisley, had died on the 21st of the month: 'I found her in the morning peacefully sitting in her favourite chair clearly having fallen asleep and not woken up again. . . . I know you and Rita had been friends for many years . . . many many times I have heard her speak of "Janaki" with love and affection . . . I am sorry to give you this sad news,' Fuller wrote. In fact, it so happened that Janaki had written to Rita only a couple of weeks before her death and the letter had been a pleasant surprise to her.[57]

Reflections on Ethnobotany

Late that month, Janaki travelled to Meerut to attend the First Botanical Conference organised by the Indian Botanical Society. She had by now become an expert, on the indigenous tribes of the Western Ghats, and the plants used by them; her paper at the conference was titled 'Ethnobotany—Past and Present.' Janaki's thesis was that the Aryans had obtained the knowledge of the use of herbs from the proto-Dravidians, or 'the flat-nosed, short, wide-faced Kadars, Pulayas, the Irulas and Kanikkars of South India'; it was this which constituted the ethnobotany of the past, she stated categorically. The Samhitas and the Ayurveda treatises, 'which are the foundations of Indian's heritage in medical-lore,' she said, acknowledged their debt to the 'dark men' of the forest. 'Today we still have the aboriginal tribes living in the forest. This knowledge of the uses of plants is often kept secret and passed on from father to son. It is to discover these hidden and secret uses of the flora of our land that ethnobotany has become an important part of our investigation,' Janaki stated. Plants which could restrict fertility were most important to India, she believed, given the 'population explosion' the country was witnessing. Janaki's students Z. Abraham and Jebhadas had accompanied her to the conference. While Abraham's paper discussed ethnobotanical aspects relating to the tribes of the Nilgiris, chiefly, Todas, Kotas and Irulas, Jebhadas's dealt with the ethnobotany of the Kanikkars of South India. Among others who presented papers on ethnobotany was Manilal from Calicut University, on the rices of Malabar. Incidentally, it was at this conference that Madhav Gadgil (with V. D. Vartak) presented a paper on the Sacred Groves of Maharashtra.[58]

'Ethnobotany: Past, Present and Future' was also the subject of the third M. O. P. Aiyengar Memorial Lecture at the Madras University delivered by

Janaki (in early 1979?). Janaki's tribute to Aiyengar was expressed in her opening line: 'As an Alumn[us] of this University and one of the first among woman students to work under the inspiring guid[ance of] Prof. Iyengar [sic], I feel still more honoured to face this gathering.'[59]

Eileen on Pantheism

Meanwhile, Darlington had put Eileen in touch with the well-known American ethnobotanist, Richard E. Schultes, considered the 'father of ethnobotany,' for publication of her paper on what the Irula 'herb doctor' had shared with her. Eileen in turn sent Darlington a book 'about the weird side of Hindu theology—a mixture of metaphysics and magic,'[60] which she thought would interest him. She believed Janaki was 'into that to some extent & has a Sai Baba whose photos (or pictures) shed Holy ashes with which she smears her forehead. Her friend 'the medical genius' had told her, claimed Eileen, that 'it was a lot of non-sense.' Like some of her other observations about Janaki, this one too appears to be a misrepresentation of facts. Despite having lived in India for a good amount of time, and having had Janaki for a friend for so long, Eileen did not have a great opinion about the Indian way of life. 'Indians mostly are very chauvinistic & expect to convert the world as you'll see in Pagal Baba's book. It is all beyond me. I am repelled by pantheism,' she remarked.[61] It is no wonder that Janaki's relationship with her was a very complex and volatile one.

A Monk-Like Existence

By this time, Govindamma, a Tamil woman from the Maduravoyal village, had become Janaki's constant companion at the Field Station. She would brush Janaki's long tresses and tie them into a neat knot (as it had become rather tedious for Janaki to do this herself), cook for her and maintain her modest dwelling. Wherever she went, she would take Govindamma along, including to Edam. Janaki's was a spartan existence; she had stopped wearing silk saris (in saffron/yellow) several years before, replacing them with cotton ones when it dawned on her that she was wearing a fabric that involved the killing of thousands of silkworms. Her vegetarianism and the conscious choice of not wearing silk were part of her *ahimsa* way of life, no doubt inspired by Gandhi. The only thing she possessed was her library, and her great stash of correspondence; the books would be housed in wooden cupboards with glass doors and the letters that she held very close to her heart locked up in another. One end of the simple long room at the Field Station, was set apart as her private space, marked simply by a cot and a mosquito net over it. The other end was used by her students in the daytime; this was where she discussed projects with them. It is said that she kept a

517

cow, and named its calf Freddy; it would be Govindamma's duty to take care of them. Like in Jammu, Janaki kept several cats at Maduravoyal, and to each she would assign a name; Lakshman and Sita were her dear ones. The cats however were not mere pets, they were for her, subjects of genetic investigations.

In addition, her students, in particular Z. Abraham whom she mostly relied on, made life easier for her by chaperoning her to railway stations and airports, buying her medicines, aerogrammes and provisions, and typing out her official letters. In Abraham's absence, Jebhadas would become her personal assistant. Earlier, Jebhadas's imported typewriter was used to type out official letters, but later a machine was purchased out of the UGC grant, for the department's common use. Janaki would occasionally visit the ancient temple at Thiruverkadu (literally meaning 'forest of holy herbs and roots,' a name that resonated with her project on medicinal plants), not far from Maduravoyal. Abraham on one occasion accompanied her to the temple of Thirunelli in Wayanad, situated in the midst of beautiful forests, during one of their collecting excursions in North Malabar. He recollected how he had to physically support her as she climbed the several steps leading to the sanctum.

Revisiting the Hortus Malabaricus

Between April 1978 and March 1979, Janaki and her team worked hard towards producing a modern version of the *Hortus Malabaricus*, 'an up-to-date book, well illustrated, and with short precise descriptions of each species for identification in the field and with medicinal properties, the different diseases for which the plant was used and the mode of application and thus to harness the local natural resources to ward off diseases and to promote scientific investigation with a view to re-establishing and resuscitating the indigenous system of medical practice with special reference to Ethnobotany.' If the original *Hortus Malabaricus* from the 1670s was based on information gathered through the brokers of knowledge such as the Ezhava (Itti Achuden) and Konkani brahman physicians (Ranga Bhatt, Vinayaka Bhatt and Appu Bhat) of Dutch Malabar (Cochin), the modern *Hortus* envisioned by Janaki three centuries later, was an anti-colonial compendium of medicinal plants of the Western Ghats (chiefly Malabar) reordered from the chromosomal point of view, but based on information collected in situ from the forest dwellers (tribals), who she believed were the primary interlocutors of indigenous medical knowledge. Janaki had often spoken of the urgent need to record the knowledge of forest flora that had come down through generations of tribal people, which was in danger of being lost; this ethno-medical project was really a response to that. In a way, it was also her homage to her father's ancestors, the famed Oracheri *Gurunathanmar* of Chokli, Tellicherry, well-known as herbal healers and astrologers.

In all, Janaki's team had undertaken fourteen field trips to collect medicinal plants and data connected with them (such as locality, habit, habitat and phenological details), from across parts of Kanyakumari, Tinnevelly, Coimbatore, Shoranur, Palghat, Olavakot, Tellicherry, Nedumpoyil, Wayanad, Cannanore, Manantoddy, Calicut, Kottayam and the Nilgiris. They collected live plants to be grown in the two ethnobotanic gardens, of Maduravoyal (240 accession numbers added) and Ashok at Shoranur (56 accession numbers added). Besides, voucher specimens of these plants (586 field numbers) had been processed into an herbarium of South Indian Medicinal Plants at Maduravoyal. Attempts were also made to obtain information on the folk-uses of medicinal and edible plants by interacting with the Nayadi tribe at Kalliayankulam Government Nayadi Colony (Olavakot), Velliyadu and Cherukattupuram near Vaniankulam (Palghat district) and Pillayam Kulambu in Shoranur. With the help of Janaki's niece, E. K. Shantha, fifty illustrations (recognition sketches) had been completed. Some of these plants were also cytologically investigated, including the *Costus speciosus*, *Aloe vera*, *Furcraea gigantea* and *Pancratium triflorum*.

Shantha's daughter E. C. Vasanthakumari, Principal of the Sri Narayan Guru College, Chelannur, Calicut, assisted Janaki occasionally in collecting medicinal plants. Janaki's team visited the Botanical Garden of the Calicut University and the indigenous medicinal garden of the Directorate of Indian Medicine, Madras, to add to information. Further, to confirm the identities of the plants collected and to obtain information on the distribution of these, several visits were made to the Herbarium of the BSI, Coimbatore, and one visit each to the Herbarium of medicinal Plants at the College of Indian Medicine, Palayamkottai, and the Herbarium of the Forest Research Institute and College, Dehra Dun.

Cytology Laboratory at Ashok

In April 1979, Janaki wrote to Vishwanathan that she was 'seriously thinking of establishing a cytology laboratory at Ashok, Shoranur.' She had also received the sanction of the DST to appoint a part-time research associate to work on the book-writing project; she wanted Vishwanathan to accept the post. She wished to appoint her present doctoral student, Nagendra Prasad, to take charge of the Shoranur Sub-Station (Ashok), but only under Vishwanathan's 'supervision both for his PhD thesis and for the book-writing project.' Prasad had been sent by Manilal from the Calicut University and Janaki had employed him on her project a year before; he had registered for a PhD at CAS under her supervision.

Meanwhile, A. Mahadevan, Director of CAS, planned a visit to the Shoranur Sub-Station sometime in the first week of May for his regular round of inspection. Janaki wanted to get there at least a couple of days ahead, with Nagendra Prasad in tow carrying the necessary laboratory equipment.[62] She had already placed a request with Mahadevan for equipment to set up the 'little' cytology laboratory, including a research microscope with illuminator attached, a camera lucida attachment—prism type, an incubator with thermostat, a refrigerator and an old microscope 'for rough work.' By this time, Janaki had employed a new laboratory-cum-field attendant, M. Robert. After spending a few days at Shoranur setting up the laboratory and tending to the garden, Janaki and her team toured Tellicherry, Nedumpoyil (Mahadevan would accompany them to these two places), Calicut and Wayanad for botanical collection.

While she was on tour, Janaki received a letter from M. S. Swaminathan, now Principal Secretary to the Ministry of Agriculture and Irrigation, Government of India, inviting her, given her wide experience and knowledge of the subject, to a meeting on 'the question of developing an All-India coordinated research project on the collection, conservation, evaluation and improvement of bamboo, neem, suitable leguminous shrubs, and indigenous plants which are of value as industrial raw material, fodder, feed, fuel and bio-fertilizer.' A group had been constituted for the purpose under the chairmanship of S. K. Seth (Retired Inspector General of Forests), and the members included besides Janaki and T. S. Mahabale, S. D. N. Tewari (Conservator-in-Chief, Madhya Pradesh), S. Kedharnath and R. C. Ghosh (both from the Forest Research Institute) and the Director of the BSI, S. K. Jain.

Saving the Silent Valley

The Silent Valley controversy would remain one of Janaki's greatest concerns during this period. In the first week of July 1979, she would write to V. Ramalingaswami, President of INSA, reiterating the need to protect the Silent Valley come what may. In response to her letter, INSA agreed to constitute a panel of experts (of no more than four or five members) to advise the Kerala Government regarding the steps to be taken 'either to preserve the Silent Valley or minimise the damage to the same.' Ramalingaswami suggested that C. V. Subramanian of CAS, at this time on a Nehru Fellowship, also be requested to serve on the panel. INSA had already discussed the formation of the panel with B. P. Pal and H. Y. Mohan Ram but wanted Janaki to suggest suitable names for the panel, 'experts either in forest management or ecology of hill streams or well-versed in flood control.'[63]

Towards the end of that month, Janaki wrote to the Conservator of Forests, Palghat District, saying that she wished to make a botanical collection in the Silent Valley region and would like to avail of a jeep on 8 August, and the services of a forest-guide. Strategically, she mentioned that she was a member of the State Committee on Science and Technology (Government of Kerala) and that she was working on a book-writing project on medicinal plants sponsored by the DST. Janaki also provided a Palghat address to the Forest authorities: c/o A. K. Raghavan, Reghukulam, Palghat[64];

Dr Raghavan was the husband of Janaki's niece Tara (Cousalya's elder daughter). Janaki was not allowed access to the Valley. The Forest Department claimed there was 'some rock-blasting going on in the area,' making it unsafe for anyone to venture there at this time. Janaki was terribly disappointed, and even somewhat annoyed.

A meeting of the INSA panel (with H. Y. Mohan Ram as Chairman and seven other scientists, including N. M. Nayar and C. K. Varshney) to discuss the Silent Valley Project was fixed for 10 August (1979) at the UBL (Chepauk), Madras. Janaki attended and there is no doubt she narrated the incident of her being denied entry to the Silent Valley, on the pretext of the 'rock-blasting to make an approach road.' As a consequence of the meeting, INSA decided to 'send a small team to the project area to ascertain for itself . . . the impact of the construction programme on vegetation and wildlife in the area.' She had just received from M. K. Prasad a recently published report on the Silent Valley prepared by some members of the KSSP (M. K. Prasad, M. P. Parameswaran, V. K. Damodaran, K. N. Syamasundaran Nair and K. P. Kannan), titled 'The Silent Valley Hydro-Electric Project—A Techno-economic & Socio-Political Assessment' (1979).

Five days after the INSA panel meeting, Janaki wrote to friend B. P. Pal once again, about the need to do something to protect the flora and fauna in the fast-dwindling tropical rain forests of India, especially those in the Western Ghats. Pal replied that the 'NCEPC [was] seized of the problems and is alive to the importance and urgency of the issue. The National Committee, in its wish to preserve unique and endangered flora and fauna, has launched a number of programmes and activities, one of which is the identification, designation and management of biosphere reserves, but being an advisory Committee of the Government of India . . . its powers are limited . . . the solution to such problems will ultimately lie in increased awareness among the public regarding these vital issues.'[65]

Janaki would not give up. She wrote next to her friend S. K. Jain, Director of BSI saying that she had attempted to visit the Silent Valley but had been denied entry by the Forest Department. She conveyed to Jain how she was 'anxious' to make a chromosome survey of the plants of the Valley 'before they [became] exterminated' and that materials for this would be fixed in situ. She also intended to make an extensive collection of live plants from the area, which would be preserved at the ethnobotanical and medicinal garden, Shoranur for further studies.[66] On the same day, she posted a letter to the Divisional Forest Officer, Palghat. Janaki called to his attention her earlier attempt at entering the Silent Valley, and how it was cancelled. 'As an elder botanist of this University I am very much interested in the evolution of floras especially of the Malabar. I hope you will help me to living specimens of saplings of the trees of the Silent Valley so that I can grow them in suitable places in Kerala, probably in Shoranur for analysis of their genetic composition and evolution. As these trees are liable to be cut down during

521

the course of the Silent Valley Hydro-electric Project, it is important that we have a representative collection available for botanists for future studies,' she appealed to the Forest Officer.[67]

K. V. Sankaran Nair, the Divisional Forest Officer replied condescendingly after twenty long days (on the basis of advice given him perhaps) that he was 'glad to note that she was very much interested in the evolution of floras especially of Malabar,' but that she was under the wrong impression that 'trees from the whole Silent Valley forest' would be cut down to make way for the Project. Toeing the government line Nair stated that the 'clearance for the Project [would] be only [be] about 10 % of the Silent Valley Forest.' 'If you want to collect large number of specimens, I request you to obtain prior sanction of Kerala Government,' he advised, obviously passing the buck.[68] Janaki was livid and decided to give Sankaran Nair a piece of her mind. She wrote to him that she was 'surprised to find . . . a Forest Officer favouring the de-forestation of the Silent Valley. As Biologists, we should be interested to preserve the total population of the Silent Valley which has been existing for millions of years. The opinion of the Scientists all over the world is, that the Valley should remain untouched. I hope you will co-operate in our efforts in preserving the total population from extinction.'[69]

Project on the Flora of the Silent Valley

It was all very frustrating, but irrespective of setbacks Janaki decided to prepare a research proposal for a 'Cytogenetics Survey of the Flora of the Silent Valley' to send to the DST for consideration. Virtually, no work had been done so far on the cytology of plants in the Valley, given the difficulty in accessing the place, 'clothed with dense impenetrable vegetation sheltering wild animals.' Only a few had made collections, but even that was at best patchy and from only the margins of the forests. 'The Palghat gap in Western Ghats, the Nilgiri Plateau, the very heavy rainfall in the valley, its impenetrable nature and its continuous history of several million years of evolution has considerably influenced the rich vegetation and it is perhaps the last remains of the evergreen biota of the tropics,' Janaki noted in her proposal. At the time of applying for a grant, she was drawing an honorarium of Rs 1,250/- from the DST (she had no other income), while Z. Abraham and P. Nagendra Prasad, her co-investigators, were receiving monthly fellowships of Rs 500/- each under the project 'To bring out an illustrated book on South Indian Medicinal Plants with special reference to Ethnobotany,' which was to end on 31 March 1981. In early July (1979), E. K. Shantha, who had been assisting Janaki on the project was appointed research associate, receiving an honorarium of Rs 400/- monthly.

'I am very sad to get news of two deaths—your brother's and Pio's [Koller]—perhaps my turn will be next,' she remarked bleakly to Darlington in August 1979. She told him that she was still 'fighting a losing battle with the Kerala Government over the building of a dam across a river to form a lake of the famous "Silent Valley", the last remnant of the Tropical Forests of Kerala.' Janaki explained that she was an adviser to the Ministry of Science and Technology of India (formed in May 1971) but the Department of Forestry was controlled by the State and not the Centre, and that the State 'was determined to have their own way.' 'The result will be disastrous to the Flora and Fauna of the region and perhaps to the whole of S. India. I am sending you a booklet about this [this was perhaps the Kerala Shastra Sahitya Parishad report, "A Techno-economic and socio-political assessment of the Silent Valley Hydroelectric Project" (1979)]. I want you to make it an international problem in Biology,' she appealed to Darlington once again. 'The Kerala Government is out to make money by exterminating the rare species of flora and fauna as well as that are native to the Silent Valley,' she remarked angrily.[70]

On His Magnus Opus and the Silent Valley

In one of his earlier letters, it appears that Darlington had enquired when she had attended the symposium 'Man's Role in Changing the Face of the Earth.' Perhaps he was preparing a biographical note on her, and also the reason why he had earlier asked for an expansion of her initials in her name. Janaki of course had replied that it was held in Princeton in 1955, organised by the Wenner Gren Foundation for Anthropological Research and that she was the lone participant from India. Janaki added that she was indebted to her teacher, the botanist-anthropologist Bartlett, who had also attended the conference, for recommending her name. She also remembered that the Wenner Gren Foundation had 'royally' treated the seventy participants who had attended. In a couple of weeks, Janaki would also post him her copy of the symposium proceedings, in which her paper on subsistence economy had been published. 'I have no use for the volume,' she would remark.[71]

In the same letter, referring to his book, she remarked in typical style: 'I think it is a jolly good production and quite readable too unlike some of your other writings.' She equally enjoyed reading his responses to the criticisms raised. Not only was the volume splendid, she was sure he had inherited his father's memory. For his amusement she had signed this letter in full: EDAVALETH KAKKAT JANAKI AMMAL, reminded of his earlier letter asking to explain her name. Obviously missing him, Janaki remarked that sometimes she wished 'India was not so far away from England'; she wondered whether she would see him before she departed and added, 'It is a longing I cannot get rid of!' Quickly, her thoughts would revert to the pressing issue of the Silent Valley. 'I want you to help to save "Silent Valley",' she appealed to him again. She mentioned that she had been making an attempt to visit the 'unique forest' with permission from the Kerala Forest Department, but at the last hour had been told she could not because 'there was blasting of rocks being done near the Valley and it would not be a safe place to go!' She told him that 'Dr B. P. Pal [was] doing the best to save the Valley. With an FRS after his name perhaps he will be able to do more than I can. I am not going to be silent. In fact, I am arranging to start a Cytological Survey of the Forest Trees they are cut[ting] and also grow seedlings in our Ethnobotanical Garden at Shoranur, Kerala. We need your good wishes!'.[72]

In another letter to him, she queried: 'Will you accept an invitation from the University?' Janaki wanted Darlington to see the Silent Valley before it disappeared under the waters. 'Please help to save the Valley—Kerala Marxists are determined to construct a dam across the Valley,' she appealed repeatedly, as if in a frenzy.[73] She had in fact planned it all; had spoken to A. Mahadevan, the Director of CAS, who was insistent that Darlington visit as the guest of Madras University. Janaki also thought it was an auspicious time for a visit, as the university was going to celebrate its centenary. She wanted Darlington to present 'a lecture or two' if he was willing. 'I hope you will not slash India too much. We are doing our best in spite of many problems,' she warned.

In early October 1979, after posting her letter to Darlington, Janaki boarded the Tamil Nadu Express for Delhi in connection with 'literature collection' for her DST project on an illustrated book on South Indian medicinal plants. She spent about a couple of months reading at the libraries of CSIR, INSA and IARI (Pusa). She would not omit to mention the Silent Valley issue again in her next letter to him sent from Delhi: '"The Silent Valley" is engaging my attention. I hope you received the literature about it [that] I sent you. Your visit will add to our efforts to keep it from being destroyed. I have not yet visited the forest, which is going to be destroyed when the Valley is converted into a lake! I am making a note of the flora of the Valley to see the chromosome situation of it.' Janaki hoped Gwendolen would accompany him to India.[74]

Janaki would be thoroughly disappointed when she received his letter three weeks later, telling her that he had cancelled his trip. 'We were looking forward to honouring a "great scientist",' she would comment in deep sadness. At least she had good news to convey to him: that the Kerala Government had been 'forced to give up destroying the Silent Valley.' She had still not been able to visit the place; but it being the rainy season was full of leeches, and she dreaded 'them more than Tigers.' 'As the oldest Forest of India if not the world we have sent up a project to the India Govt. to make a genetic study of the trees. We will bring some of the plants and grow them in our Ethnobotanical Garden—my ethnobotanical garden at Shoranur, Kerala, which has a climate similar to the Valley—being only 40 miles or so away,' she explained to Darlington. Janaki now wanted to transform Ashok into a microcosm of the Silent Valley flora, especially, the trees.

She had planned a trip to Jammu (RRL) and Dehra Dun (Forest Research Institute) for 'seeds of the genera of plants of the Valley,' she told Darlington.

'I need the trees so that I can make a beginning right away.' Janaki reminded him about her being an adviser to the Ministry of Science and Technology, by virtue of which she had a 'say in many matters of scientific value.' Janaki had travelled to Delhi in early December 1979, where she had met with 'good friends' B. P. Pal and M. S. Swaminathan, who were with the Ministry of Science and Technology. As a non-official adviser to the Ministry, she was often called up to Delhi for meetings. 'Our Minister is Dr Swaminathan—well-known scientist and a good friend of this University [Madras] and myself,' she told him.[75] Janaki had not got that quite correct, for Swaminathan was Director General of the ICAR, and Principal Secretary in the Ministry of Agriculture and Irrigation at this time, and not a Minister. Perhaps it was just a slip of her pen, or it had to do with her age, but there is no doubt she was becoming increasingly forgetful.

Heartbroken

Janaki travelled to Jammu in late December—early January (1980) braving the extreme cold. On her return journey, she came down with a bad case of flu, leaving her completely drained of energy and low in spirits. She would also not hear from Darlington for several months at a stretch; she had not received replies to her letters. Finding this very unusual, in early July, she decided to write to him nevertheless: 'I hope you are keeping fit and busy with another book. I enjoyed reading the last one—in spite of certain disturbing parts.' What exactly she found disturbing in *The Evolution of Man and Society* is not definite, but it might have been the parts on race; it was after all for its racist conclusions that the book was considered controversial. The emotional intensity of her relationship with him had hardly waned; it was as intense as it was in the 1930s. She wrote to him: 'When shall I see you—Will I before I die—I long to be with you. Yours ever, Janaki.'[76] Little did she know that he would be gone first, in fact in less than nine months of writing this letter. Her letter also conveyed some bad news; she had lost her youngest sister, Devayani (in February). 'She was the prop of Edathil House more than I could be. She had a sudden heart attack early morning,' she informed Darlington; in fact, he had himself suffered an attack only a few months before, a second one, but Janaki was unaware of this.[77]

She was in Central India (for what purpose, is not clear) when Devayani passed away. Niece Shantha's daughter Vasanthakumari would inform her about it after attending the funeral: 'Everything went off well but all our sympathies are with Parvathiedathi [sister Parvathi] who is left alone,' she would remark. Parvathi would however have nieces, E. U. Shantha and her sister Savithri (who mostly resided at Edam) to give her company.[78] Despite her physical distress (having recently broken her arm), E. U. Shantha would write to enquire if Janaki was well, and if she was 'taking multivitamins daily and barley water,' and on how Govindamma and her cats were faring.

At Last, Silent Valley Is Saved

On 12 January 1980, Indira Gandhi (who had been swept back to power in the General Elections earlier that month) had visited Kerala to campaign for the State Assembly elections. When asked by the press reporters about the Silent Valley issue, Mrs Gandhi stated in no unclear terms that while all must be done to improve the power situation in the state, 'the world is getting more and more conscious that all such developments should keep in view ecosystem or ecology because destroying or disturbing it will leave long-term adverse effects . . . It is worthwhile to see whether we can get some benefits without destroying these forests.'[79] By mid-July, Mrs Gandhi had written a third letter to the Kerala Chief Minister E. K. Nayanar with regard to the Silent Valley Project, saying that while she appreciated the reasons for his anxiety to see the project through, it was important to do what was necessary to protect the environment. She was convinced of this from the preliminary scientific reports that had reached her from scientists of the BSI, the Birbal Sahni Institute of Palaeontology, the National Bureau of Plant Genetic Resources, Lucknow, and the CAS and UBL, Madras—the last two chiefly represented by Janaki. Mrs Gandhi concluded that 'from the point of view of ecology and importance of biosphere reserves, it would not be desirable to approve the Silent Valley Hydro-Electric Project.' In fact, she made it clear that she would like him to begin work on the 'development of the Silent Valley National Park without further delay.'[80] Eventually, on 26 December 1980, the Kerala Government declared the Silent Valley a National Park.

The Man and Biosphere Committee of the DST (Government of India), in view of the fact that the Silent Valley was to be kept untouched, expressed interest in Janaki's proposal to study cytogenetically the flora of the ancient forest. As for her, she believed that this would lend greater significance to the work done at the Maduravoyal Field Research Laboratory over the past nine years under her direction. She would suggest to the Vice-Chancellor of the Madras University that a department of Ethnobotany and Cytology be established at the earliest at Maduravoyal. Researches done under the two schemes—the proposed cytogenetic survey of the flora of Silent Valley and the ongoing project of bringing out an illustrated book on South Indian medicinal plants—she said, would contribute to the making of an active school of research on ethnobotany and cytology in Madras. Janaki referred to the 100 dominant species of the Silent Valley flora, listed in the KSSP report, most of which were medicinal in value and would be the focus of her proposed cytogenetic study. The study aimed to map the genetic changes that had occurred in the course of isolation, from the chromosomal point of view.

More Collecting in Malabar

It would be late March 1980 before Janaki next visited Edam. She was once again on a collecting tour in Malabar, in the company of her research fellows, chiefly in and around Tellicherry and Wayanad. At Manantoddy (Wayanad), while getting off a bus (she insisted on taking the bus, recalls Z. Abraham), Janaki dislocated her left knee. She would spend the next ten days resting at Edam, massaging her injured knee with medicated oils. On her return to Maduravoyal, she sent for her niece E. K. Shantha to help her process the material and all the information collected until then. Shantha had by now been appointed full-time research fellow on the project. Janaki wrote to her niece with much affection: 'I shall make your stay here as comfortable as possible. You know, Maduravoyal is a rustic village. You can be my guest and stay with me in my room. If possible, kindly bring one mosquito net with you. I shall send Mr. Abraham to the Central Railway Station to receive you on the 10th [August 1980] . . . Please bring with you all the available notes on Medicinal Plants.'[81] Shantha had already sent her a collection of medicinal plants (collected by her daughter, Vasanthakumari) in the new year (1980), based on those mentioned and described in the ayurvedic text, *Ashtangahridaya*.[82]

One Final Push

Four months later, Janaki posted an urgent letter to Shantha reminding her that their project was to end on 31 March 1981. She was frantic because the final report of the DST project had to be submitted in April. 'In order to complete the book, I want you to come to Madras during the first week of January (1981) as soon as possible. Please bring with you all your manuscripts, notes and the books on medicinal plants,' Janaki instructed her.[83] Shantha had compiled information on about 1,100 species of plants used medicinally, providing the botanical name, family, names in English, Malayalam, Tamil, Sanskrit and Hindi (wherever possible), a short description enabling identification, habitat and medicinal uses. The information had been gathered from R. N. Chopra's (with S. L. Nayar and I. C. Chopra) *Glossary of Indian Medicinal Plants* (1956), J. F. Dastur's *Medicinal Plants of India and Pakistan* (1970), *Wealth of India: India's Raw Material Resources* (an encyclopaedic series, 1948–1976),[84] *Ashtanga Nikhandu* (Malayalam), and Malayalam health magazines such as *Arogya Bandhu*, *Vaidya Kaumudi* and *Vaidya Bharathi*. Finally, the report on the project, 'To bring out an illustrated book on South Indian Medicinal Plants (with special reference to Ethnobotany)' was submitted to the DST on schedule in April 1981.

Figure 28.1 Janaki and Eileen Macfarlane, with doctoral students at the Ethnobotanical Garden, Maduravoyal, University of Madras. 1970s.

Source: Courtesy of Dr. Z. Abraham.

Notes

1 E. K. Janaki Ammal to T. V. Sreenivasan, letter dated 29 September 1977, private collection.
2 Incidentally, the FPA was affiliated to the International Planned Parenthood Federation founded in November 1952 in Bombay by the American birth-control
activist Margaret Sanger (considered a supporter of eugenics) and Lady Rama
Rau (Dhanvanti Rama Rau, 1893–1987) of Madras; the organization had its
headquarters in London.
3 Marie Buck to Indira Gandhi, letter dated 9 March 1976, private collection.
4 E. K. Janaki Ammal to Marie Buck, letter dated 13 April 1976, private collection.
5 Marie Buck to E. K. Janaki Ammal, letter dated 1 April 1976, private collection.
6 The company later merged with the American firm of Plough Inc.; before the
merger, it had exported diphtheria medicines to America and opened a branch
in New York in 1929. Until the end of World War II, the amalgamated company
thrived on the sales of a sex hormone.
7 Marie Buck to E. K. Janaki Ammal, letter dated 1 April 1976, private collection.
8 Ibid., letter dated 7 April 1976, private collection.
9 Marie Buck to E. K. Janaki Ammal, letter dated 26 April 1976, private collection.
10 Ibid., letter dated 16 October 1976, private collection.

11 E. K. Janaki Ammal to Marie Buck, letter dated 4 February 1977, private collection.

12 Marie Buck to E. K. Janaki Ammal, letter dated 25 March 1977, private collection.

13 DP: C. 109 (J. 116), E. K. Janaki Ammal to C. D. Darlington, letter dated 10 March 1977.

14 Eileen Macfarlane to E. K. Janaki Ammal, 12 December 1976, private collection.

15 Ibid.

16 Ibid., letter dated 28 January 1977, private collection.

17 Ibid., letter dated 16 January 1977, private collection.

18 Eileen W. Macfarlane to Malcolm Adiseshiah, letter dated (perhaps March) 1977, private collection.

19 Eileen W. Macfarlane to E. K. Janaki Ammal, letter dated 10 April 1977, private collection.

20 Ibid., letter dated 28 April 1977, private collection.

21 Ibid., letter dated 26 May 1977, private collection.

22 Marie Buck to E. K. Janaki Ammal, letter dated 3 June 1977, private collection.

23 Ibid, letter dated 9 June 1977, private collection.

24 DP: C. 111 (J. 151), Eileen Erlanson to C. D. Darlington, undated letter, but April—June, 1977.

25 Eileen W. Macfarlane to E. K. Janaki Ammal, letter dated 28 June 1977, private collection.

26 Marie Buck to E. K. Janaki Ammal, letter dated 4 July 1977, private collection.

27 Eileen Macfarlane to E. K. Janaki Ammal, letter dated 7 July 1977, private collection.

28 Ibid., letter dated 16 July 1977, private collection.

29 Ibid., letter dated 20 July 1977, private collection.

30 Ibid., letter dated 17 July 1977, private collection.

31 DP: C. 109 (J. 116), E. K. Janaki Ammal to C. D. Darlington, letter dated 31 August 1977

32 E. K. Janaki Ammal to T. V. Sreenivasan, letter dated 29 September 1977, private collection.

33 K. L. Mehra to E. K. Janaki Ammal, letter dated 24 November 1977, private collection.

34 DP: C. 109 (J. 116), E. K. Janaki Ammal to C. D. Darlington, undated (but January 1978).

35 E. K. Janaki Ammal to T. S. Sadasivan, letter dated 11 February 1978, private collection.

36 Joseph and Chandrasekaran, 'Janakia arayalpatra—A New Genus and Species of Periplocaceae from Kerala, South India'. In more recent times, a world-renowned South-Indian rose breeder, S. Viraraghavan of Kodaikanal, named a rose variety that he created, after Janaki Ammal. See 'A Rose Named E. K. Janaki Ammal', The Telegraph, 6 June 2019.

37 E. K. Janaki Ammal to Marie Buck, letter dated 9 March 1978, private collection.

38 Ibid., letter dated 25 April 1978, private collection.

39 Marie Buck to E. K. Janaki Ammal, letter dated 22 May 1978, private collection.

40 DP: C 109 (J. 116), E. K. Janaki Ammal to C. D. Darlington, letter dated 14 August 1978.

41 E. K. Janaki Ammal to the Registrar, University of Madras, letter dated 31 May 1978, private collection.

42 K. S. Manilal to E. K. Janaki Ammal, letter dated 27 June 1978, private collection.

43 In 1978, Janaki's paper on the aromatic grasses of India, co-authored with her Jammu student B. K. Gupta (at this time teaching at the D. A. V. Post-Graduate College, Dehra Dun), was published in the *Indian Journal of Forestry*.

44 Gopinathan had initially come to work as a cook at the university department, before he was spotted and appointed by Janaki as field-man on the UGC project.

45 It would take another three decades of tireless work before his English translation of the *Hortus Malabaricus* (from the original Latin) would be published by the University of Kerala (2003); a Malayalam translation would appear in 2008. Incidentally, in 2003, Manilal would receive the E. K. Janaki Ammal National Award for Taxonomy.

46 DP: C 109 (J. 116), E. K. Janaki Ammal to C. D. Darlington, letter dated 20 July 1978.

47 Ibid., letter dated 14 August 1978.

48 Ibid.

49 Ibid., letter dated 21 October 1978.

50 Ibid.

51 Eileen W. Macfarlane to E. K. Janaki Ammal, letter dated 4 December 1978, private collection.

52 DP: C 109 (J. 116), E. K. Janaki Ammal to C. D. Darlington, letter dated 19 November 1978.

53 E. K. Janaki Ammal to Z. Abraham, letter dated 22 November 1978, private collection.

54 Ibid., letter dated 27 November 1978 (wrongly dated by Janaki; it should have been 22 November 1978).

55 Mrs P. Vasudeva Rao to E. K. Janaki Ammal, letter dated 30 December 1978, private collection.

56 Eileen W. Macfarlane to E. K. Janaki Ammal, letter dated 4 December 1978, private collection.

57 J. Fuller to E. K. Janaki Ammal, letter dated 29 December 1978, private collection.

58 Puri and Murty (eds.), *The Journal of the Indian Botanical Society*, Supplement, Vol. 57, December 1978, pp. 60–66.

59 Unpublished manuscript in private collection.

60 Pagal Baba, *Temple of the Phallic King: Yogis, Swamis, Saints and Avataras*, New York: Simon and Schuster, 1973.

61 DP: C. 111 (J. 151), Eileen Erlanson to C. D. Darlington, letter dated 11 January 1979.

62 E. K. Janaki Ammal to T. V. Vishwanathan, letter dated 25 April 1979, private collection.

63 J. N. Nanda, Executive Secretary, INSA to E. K. Janaki Ammal, letter dated 10 July 1979, private collection.

64 E. K. Janaki Ammal to The Conservator of Forests, Palghat District, letter dated 31 July 1979, private collection.

65 B. P. Pal to E. K. Janaki Ammal, letter dated 21 August 1979, private collection.

66 E. K. Janaki Ammal to S. K. Jain, letter dated 22 August 1979, private collection.

67 E. K. Janaki Ammal to The Divisional Forest Officer, letter dated 22 August 1979, private collection.

68 K. V. Sankaran Nair to E. K. Janaki Ammal, letter dated 12 September 1979, private collection.

69 E. K. Janaki Ammal to K. V. Sankaran Nair, letter dated 28 September 1979, private collection.

70 DP: C 109 (J. 116), E. K. Janaki Ammal to C. D. Darlington, letter dated 11 August 1979.

71 Ibid.
72 DP: C 109 (J. 116), E. K. Janaki Ammal to C. D. Darlington, letter dated 24 August 1979.
73 Ibid., letter dated 4 October 1979.
74 Ibid., letter dated 9 November 1979.
75 Ibid., letter dated 10 December 1979.
76 DP: C 109 (J. 116), E. K. Janaki Ammal to C. D. Darlington, letter dated 4 July 1980.
77 Ibid.
78 E. C. Vasanthakumari to E. K. Janaki Ammal, letter dated 7 February 1980, private collection.
79 Ramesh, *Indira Gandhi: A Life in Nature*, p. 269.
80 Ibid., p. 272.
81 E. K. Janaki Ammal to E. K. Shantha, letter dated 25 July 1980, private collection.
82 Ibid., letter dated 8 January 1980, private collection
83 E. K. Janaki Ammal to E. K. Shantha, letter dated 16 December 1980, private collection.
84 This was the revised and expanded version of George Watt's *Dictionary of Economic Products* (1889–1893), published by the CSIR.

29

THE FINAL SALAAMS

When shall I see you—Will I before I die—I long to be with you.

—E. K. Janaki Ammal to C. D. Darlington, July 1980[1]

Janaki would continue to protect Edam fiercely, refusing to sell it as long as she was alive. In fact, she wanted to be cremated in the Edam *paramba*. In September 1980, Janaki would send her field-man Gopinathan to Edam, to 'renovate the garden' there. She wrote to sister Parvathi, 'As far as I know he is honest and trust-worthy. I will be coming there before the end of this month and I will discuss the matters. I hope you will feed him.'[2] In just a few months' time, niece E. U. Shantha would pass away, leaving the family distraught beyond words. Janaki, who was in Tellicherry at this time, attended the death ceremonies.

She was unaware at this time that Darlington had passed away, on 26 March 1981; he had taken ill and had been admitted to the John Radcliffe Hospital, Oxford, where he died soon after. Janaki would learn of his death only six months later, when Oliver Darlington managed to track down the address of Janaki's niece in Madras, with some help from Eileen, and conveyed the sad news:

> I knew Janaki when I was a child and she lived with my family in London during the war. I have some sad news for Janaki. I had written to her old address and have not had any acknowledgement so I fear she may have moved or may be unwell. . . . My news concerns my father, Cyril Darlington, who was a close colleague and no doubt wrote several papers together. My father was taken ill in March and was taken to hospital where he died shortly afterwards. All my family were of course very shaken by this but mercifully his illness was very brief and he was 77 when he died . . . I would be very glad if you would pass this news, or my letter, to Janaki, and I would be glad also to have some news of her. She figured very

DOI: 10.4324/9781003267089-29

large in my childhood years and she told me, when she visited us in 1955, that she measured her time in Britain by my age. I was born roughly the same time as she arrived here.[3]

Darlington's death marked the end of an intense, inter-continental and five-decade-long emotional and intellectual relationship between a brilliant even if controversial English scientist, from a small cotton town in Lancashire, and a pioneering Indian woman scientist from Tellicherry on the Malabar coast, older to him by six years—an unusual tale by any standards. It is not difficult to imagine the state of mind Janaki was in when she received the terribly belated news of Darlington's death. She was already on the threshold of dementia by this time.

A Fading Presence

Janaki carried on with her work at Maduravoyal, irrespective of age-related illnesses, which were not many or even serious, except for the progressing dementia and her longstanding fibrositis. All her basic needs were taken care of by Govindamma. Her nieces in the city visited regularly, and sometimes took her to their homes for short stays. She was however happiest in her room at Maduravoyal (a name that resonated with the sweetness of her much-loved *Saccharum*) amidst her cats and her ethnobotanical garden—her hermitage as she called it. She would continue with her researches, but with much greater reliance on her doctoral students, or rather student—at this time only Nagendra Prasad. Her focus would continue to be on the cytogenetic survey of the Silent Valley flora, although she would never make it to the Silent Valley.

Prasad would instead make a botanical collection from the Valley in the company of Manilal, his former teacher at the Calicut University; with Janaki as co-author, he would publish at least one paper, on the chromosome numbers of ten species collected from the place, the floristic composition of which was complex and little understood. Their study revealed that speciation was progressing at a very fast rate in this ecosystem, and if protected as a biosphere (as it must, Janaki reiterated), it could be used as a 'control' for reference and experimentation, as if the Valley was a large laboratory.[4] Another publication , with her as co-author, was a note on the chromosome count of *Centella asiatica*, collected from the Silent Valley. This was a medicinal plant belonging to the family Apiaceae, used in Ayurvedic preparations as a hair tonic. They found it to be a diploid (the primitive stage of development), confirming once again Janaki's surmise that given their isolation here, most species would be diploids.[5]

Prasad (with Janaki) had discovered an alternative source for diosgenin, an alkaloid that went into the making of steroidal hormones used in the manufacture of anabolic agents, sex hormones and oral contraceptives. A great percentage of the world supply of diosgenin came from the rhizomes of the *Dioscorea* species, but in their paper, they showed how the rhizomes, roots, stems, roots and leaves of *Costus malortieanus* could be profitably used as a suitable alternative. The occurrence of diosgenin in *Costus speciosus* had already been reported, and it was this that had led them to look for the alkaloid in a related species.[6] Their next stage of investigation was the relationship between polyploidy and the diosgenin content in different parts of the *Costus speciosus*, which would contribute to developing suitable agropractices. They investigated the distribution of diosgenin at different ploidy levels (diploid, triploid and tetraploid), at different stages of growth (vegetative, flowering and fruiting), and in the various parts of the plant during its life-cycle (root, rhizome, stem and leaf). The plant was collected from different locations: Nedumpoyil (a diploid), Kakkayam (a triploid) and Jammu (a tetraploid), and their clones planted at the Field Research Laboratory, Maduravoyal. Contrary to expectations, of the three races studied, they found the diosgenin content highest in the diploid, followed by the triploid and the tetraploid. They also observed that diosgenin synthesis started soon after sprouting, to gradually increase until it reached a maximum at the stage when it was about to flower.[7] Janaki and Prasad followed this up with a survey for germplasm evaluation of the *Costus speciosus*; a total of thirty populations of this species were surveyed in Tamil Nadu (Kanyakumari and Madurai) and Kerala (Trivandrum, Tellicherry, Nedumpoyil, Idukki, Kottayam, Shoranur, Calicut University, Kakkayam and Wayanad), besides a collection from Jammu. The plants were grown experimentally at Maduravoyal under uniform conditions (soil and environment). While the triploids were the most robust, the diploids had the greatest amount of diosgenin content.[8]

Edam in Its Final Ebb

Janaki's niece, Savithri, residing at Edam, wrote to Janaki in early 1983 to let her know that Parvathi (Janaki's elder sister) had indeed received her monthly remittance; she complained that her aunt was refusing to sit down to write a letter acknowledging it. The 91-year-old Parvathi was constantly laid up, Savithri said, so much so the doctor had to be summoned. Dr V. K. Keshavan, who examined her, prescribed some medicines and 'Glucose D.' This had revived her somewhat, but Savithri continued to be anxious for her. She was also very concerned about Janaki and hoped she was coping alright. 'We heard that there is water scarcity at Madras & Schools & Colleges will be closed by 31st March. . . . When will you be coming this side?' she enquired with unease.[9]

Exactly a year later, in early January 1984, Parvathi passed away. At eighty-six, Janaki had survived every one of her siblings, as well as Darlington. Late that month she fell seriously ill. Her nieces were summoned, and they removed her to Dr Mehta's Hospital on McNichols Road, Chetpet.

Govindamma stayed by her side day in and day out, while Janaki's anxious family took turns to visit her. Meanwhile, her dear student Z. Abraham (who was now based in Lucknow, first at the Drug Research Institute and later at CIMPO) was in Madras visiting his wife and their newborn. Learning of Janaki's hospitalisation, he rushed to see her. Govindamma could not hold back tears on seeing Abraham. Janaki was being tube-fed. She could barely open her eyes, was mostly unconscious and struggling to breathe, an extremely sorry sight for her loved ones. She would remain in this condition for more than a week and on 7 February 1984, at ten in the night, she would be gone for ever.

Her body was taken to her niece's house on Harrington Road in the city and kept there briefly for relatives, friends and colleagues to pay their last respects. It would be consigned to flames at the Nungambakkam crematorium, alas not at Edam as was her wish. That grand house and its legacy, which she so fiercely protected until the end of her life, would be gone too, marking the end of a saga, of an exceptional life in science. It is a matter of shame that no memorial stands in her name anywhere, not even in her hometown of Tellicherry, despite her being India's foremost woman scientist ever!

Figure 29.1 Janaki with her doctoral students at the Maduravoyal Field Station, University of Madras. Early 1980s.

Source: Courtesy of Dr. Z. Abraham.

Notes

1 DP: C 109 (J. 116), E. K. Janaki Ammal to C. D. Darlington, letter dated 4 July 1980.
2 E. K. Janaki Ammal to E. K. Parvathi, letter dated 7 September 1980, private collection.
3 Oliver Darlington to Janaki's niece, Uma Ramachandran, letter dated 19 September 1981, private collection.
4 Prasad and Janaki Ammal, 'Chromosome number report of some plants from Silent Valley', 1985.
5 Ibid., 'Chromosome count of Centella asiatica (Linn.) Urban.', 1985.
6 Ibid., 'Costus malortieanus H. Wendl, a new source for diosgenin', 1983.
7 Janaki Ammal and Prasad, 'Relationship between polyploidy and diosgenin content in different parts of Costus speciosus', 1984.
8 Ibid., 'Germplasm evaluation of Costus speciosus in relation to diosgenic content', 1984.
9 E. U. Savithri to E. K. Janaki Ammal, letter dated 3 March 1983, private collection.

EPILOGUE
Portrait of a Nomad Woman Scientist

[M]y life is too much of a failure to write about.
—E. K. Janaki Ammal (1952)[1]

Movement was central to Janaki's life. Modelled around such key fluid figures of modernity as the nomad, the exile, the pilgrim, and the *bhikshu*,
or the wandering religious mendicant or vagabond, her evolving feminist
subjectivity was heavily marked by her geographical wanderings in pursuit
of science—a life that turned out to be an unlikely pilgrimage.[2] Janaki was
a curious pilgrim; not only was mobility a way of being for her, but also
eminently a way of making knowledge. She was a border-crosser *non pareil*,
who chose self-expatriation as a deliberate strategy to do science. Danger,
tragedy and chaos, she found alluring, even as she set her heart on creating order. It was as if an ever-present death wish, a spartan existence and
a parlous life-style were crucial to her in achieving scientific goals, which
made even wartime England, with the bombing and danger to life, a great
stimulant to creative work.

Janaki's life in science is in fact a classic demonstration of how separation or isolation and serendipity of the transitional or liminal space, height-
ened productivity. When she described herself as a pilgrim of science, it
was not the self-flagellating pilgrim of history that she was referring to,
but the modern, adventurous and curious pilgrim, to whom pilgrimage or
circumambulation offered a means of creative bonding and imagining innovative futures, outside of the purview of the state. It was a journey towards
the 'unknown and the uncertain, enabling the self to develop and progress,
while the crossing back to the familiar reinterpret[ed] the movement into a
life-course.'[3] Shaped by the tension between inertia and mobility, chaos and
order, security and freedom, solitude and companionship 'and the fantasies
of an independent free-floating existence,' Janaki's life was a rendezvous
with nomad science.[4]

Sciences are diverse and very importantly there are, and have been, diverse
ways of doing science than those the Nobel Prize Committee or some similar

537

authority is willing to recognise, and these ways demand equal attention if not more. It will still be science worthy of contemplation and scrutiny if small, and not 'big' science, if field-centred (or located in the border zone of the laboratory and the field), and not 'lab-centred,' if slow and not 'fast' science, if it is science in the making and not 'finished' science, and if cross disciplinary and unifying (even embracing the humanities), and not confined to imagined isolated silos of disciplines.

In their treatise on wandering subjectivities, *A Thousand Plateaus*, the French philosopher Gilles Deleuze and his radical psychoanalyst collaborator, Félix Guattari referred in general to two categories of sciences: the 'major,' 'royal' or 'state' science, and the 'minor,' 'nomad' or 'ambulant' science.[5] While state science is a centric (drawing everything towards the centre), formalised and finalised kind of top-down science, not only solid, cold and rigid (not unlike the state itself), privileging a fixed order of things, but also fast, and sometimes even hasty, nomad science is an eccentric (moving away from the centre), imaginative and creative scientific practice, and therefore necessarily slow, exploring problems that are transitional and fluid and focused on working from matter upwards—also very importantly—one that is punctuated by moments of wonderment. The present study yields among other things to a science in action analysis (of science in the making as against finished science), inspired by the constructivist turn in the philosophy of science, which, explored cultures of science and technology and demonstrated how scientific facts and objects are constructed.[6]

However, in the place of a team of (male) scientists doing science in a privileged site such as an European laboratory (of microbiology for instance) the focus here is on the twentieth-century career of a highly mobile Asian woman scientist, which spanned continents; the chief protagonist is one who did science 'in-between'—in the border zone of the lab and the field—and importantly while trapped between 'a rock and a hard place, between [what] nourishes and inspires and the State that imposes.'[7]

This biography offers a unique opportunity to consider the unlikely discipline of cytogenetics as nomad science among other things.[8] Unlikely, because the state (both in the colonial and post-colonial contexts) was the chief employer of cytogeneticists (at plant breeding and agricultural stations; universities, much less so), with its goal of reducing the gap between research and application, and speedily at that. How does 'nomad science,' necessarily slow, then become an appropriate description of Janaki's practice, when she chiefly operated within such regimented settings? Using Deleuzian lenses, combined with insights from border studies, including those that investigate liminality and mobility in the modern world, and in the lab-field border zone, we make a modest attempt to shine light on those nomadic tendencies that animated Janaki's life in science. The exercise is not an exhaustive one nor aimed at 'fixing' her as a nomadic individual (she is identified as nomadic in a given context)[9] but only attempts to provide an outline of what one might develop more fully elsewhere. In fact, it is a thrilling thought to imagine Deleuze, the philosopher, in the 1940s (a student at this time) crossing paths with our Asian woman cytogeneticist somewhere on a boulevard, unbeknownst to one another, at a time when they were both residents of Paris!

This biography adds a further dimension to the Deleuzian idea by arguing that nomad science is by definition slow science, and that Janaki's nomadism, of which this study is a clear demonstration, manifested in at least four ways—in-betweenness, cross disciplinarity, opening lines of flight and adopting a problem-based model of doing science. The following sections consider each of these briefly.

In-betweenness

The site of Janaki's scientific practice was the border zone between what the state imposed on her and her own personal research goals. The 'in-between has . . . consistency and enjoys both an autonomy and a direction of its own. The life of the nomad is the intermezzo,' observed Deleuze and Guattari.[10] The intermezzo or the interlude signifies transition and mobility, and it is in such liminal spaces that multiplicity or heterogeneity thrives,[11] and where nomadic individuals enjoy the freedom to test out new ideas and undertake creative experimental work, without being cowed down by the tyranny of hierarchy, or even patriarchy. However, borders are ambiguous and unstable, rendering life in the border zone—the bridge that connects 'what is' and 'what can or will be'—precarious and volatile, filled with pain, anxiety and even humiliation. Also, the independence experienced in the liminal zone is short-lived, because the state is forever looking to enslave the nomad into achieving the goals set by it, and very importantly, economically and in the shortest possible time. It somehow manages eventually to subdue the nomad's originality and creative inventiveness. One can cite several instances from Janaki's life to illustrate this, but her Coimbatore interlude in the early part of her career (in the 1930s) is a particularly telling one.

As earlier noted, the nomad is not only a way of being but also thinking. Janaki devoted her first three years as sugarcane-geneticist at the Imperial Sugarcane Breeding Station in Coimbatore to making scientifically controlled crosses, besides studying their chromosomal behaviour at meiosis, both equally challenging and time-consuming processes—this was slow science in the making. While the first task was primarily in aid of the state, but provided her with an exceptional flow of research material, the second was undertaken solely to fulfil her personal research goals. The choice of hybrids for the sugarcane expert T. S. Venkatraman depended on their uses to the state and the sugar industry in particular, in contrast to Janaki for whom these were all epistemic things with the potential to throw light on the evolutionary pathways of the cultivated cane. Venkatraman believed that such academic work was better suited to the university rather than a breeding station, where it could only be treated as a minor activity. State science, Deleuze enlightens us, constantly imposes itself on the processes of minor science, allowing them to exist only in the 'capacity of "technologies" or "applied science".'[12] As the nomad scientist is constantly pushing back the boundaries of knowledge, some of these boundaries become attractive to state science; Venkatraman, for instance, never hesitated to acknowledge the importance of Janaki's work, because her ingenious crosses (especially her intergeneric ones), could potentially sire a highly economic cane. The case in point being her *Saccharum-Imperata* hybrid, later discovered to be a highly desirable 'stud bull.' However, rather than attribute this achievement to Janaki, Venkatraman would unfairly identify the hybrid as parthenogenetic (meaning of 'virgin birth,' or born without a male parent) and erase all traces of her role as its breeder.

State science represented the dominant model current in the period, especially so in colonial (and post-colonial) settings. In such settings, there was unimaginable pressure on scientists to produce practical results speedily, while maintaining stringent economy at all times. At one point Janaki was overwhelmed by the tyrannising presence of several male actors of state science at the Breeding Station—these included the Coimbatore economic biologists (such men as Venkatraman), who thought cytogenetical research was beyond the scope of the Station, as well as visiting scientists like John Russell of the Rothamsted Experimental Station, who assumed she was perhaps getting on to some unproductive path, and Ruggles Gates of King's College, London, who cast suspicion on the authenticity of her amazing intergeneric *Saccharum-Zea* cross. And as for Darlington, focused as he was on extracting constants out of variables in the nascent field of cytogenetics, his practice fitted hand-in-glove with the Deleuzian description of state or royal science. While he was willing to hail the work of the rice expert K. Ramiah and his form of applied science, Darlington was unwilling to grant Janaki anything more than an ability to engage in elementary exploration and play second fiddle to the geneticist. He obviously assigned a superior status to his own top-down practice centred at the laboratory (of one of the world's leading centres of research on cytogenetics), which involved reducing data to construct theories or systems (submitting everything to laws), and an inferior one to Janaki's ambulant but nevertheless rigorous, and undoubtedly slow science—working on one specific example after another, and generating a series of local singularities, before synthesising or federating (as the French philosopher Michel Serres would have it)[13] them to construct the global picture—while, located at an imperial breeding station and in the border zone of the field and the laboratory, outside of theory. Janaki's mode of practice was procedural (walking one step at a time), cumulative, and combinatorial; she adopted an ecological/encyclopaedic approach to making scientific knowledge. It involved making a great series of local, small-scale and singular discoveries, slowly and progressively over a life-time of wandering—carving out fragments of an ever changing, and many hued imaginary evolutionary jigsaw puzzle of planetary scale. This was slow science in the making, and an unfinished and indeterminate pursuit by definition, and which went against the grain of state science.[14]

State science, Deleuze reminds us, only takes what it can appropriate of nomad or ambulant science, with the rest either repressed, treated as noise, or simply denied scientific status.[15] The fate of Janaki's invention, the *Saccharum-Zea* hybrid, which she aptly referred to as the 'Cinderella of the Sugarcane Station' is substantial evidence of this marginalisation. It would take her four decades, a painfully long time (when the hybrid flowered almost magically, for the first time ever not unlike in a fairy tale) to convince the authorities of state science that her hybrid was indeed genuine.

While the in-between enabled women in particular to craft a creative life in science, it was not without challenges in institutions of the state apparatus, where hierarchy and patriarchy were rife. When Janaki accepted Nehru's offer in 1952, of the post of special duty officer in charge of the reorganisation of the BSI, she knew it was going to be a very challenging task, not only because it was an onerous one, but because she would be faced with much resistance in executing her unorthodox and disrupting idea of 'contaminating' taxonomy with cytology. The 'fresh, undigested, bitter taste of newness'[16] of her ideas was too much for the new-born Indian state to palate, which instead preferred to heed the advice of the Kew Gardens (a centre of royal science, as far as plant taxonomy was concerned), rather than a nomad scientist like herself. Change and innovation are after all abominable to the state, which desires quick results and demands that status quo is maintained at all times. In essence, the in-between is a veritable fount of fascinating nomad tales of self-creating individuals, of which Janaki's life in science is a shining example of one punctuated by war, death, loss, love and creativity.

Cross-disciplinarity

Just as the state and the nomad mutually constitute each other and coexist, so do the laboratory and the field, which are not entirely different or always separate cultures of doing science; in other words, one is inconceivable without the other. Importantly, the location of their meeting is not a line or a boundary but a broad frontier of vigorous interaction, exchange and intermingling, and with no clear advantage to either. One of the characteristic traits of the lab-field border is movement, and movement involves time for those who occupy the border zone.[17]

Cytogeneticists like Janaki demanded very little by way of capital investment in their science; it was small-scale science, as against "big science", funded by the state. All it required was a powerful microscope, with a camera lucida attached to it (used for making drawings) and chemicals that would go into the preparation of high-quality slides. It however required great expertise and experience to 'entrap' a slice of life between two pieces of glass and bring the chromosomes into sharp focus, and even more to make accurate observations on the basis of these. Janaki had a way with this, and it went a long way in making her a respectable cytologist across the world.[18] It was however an extremely slow science. Janaki's border practice involved collecting material (sometimes involving long stretches of time, and in places involving much travel) in the field, preparing slides in the laboratory, and spending innumerable hours of study under the microscope—studying chromosomes for their shape, number and of anything extraordinary that might be happening, besides undertaking experiments in the field, both to improve plants through hybridisation and polyploidy (breeding and introduction) and for taxonomic ends (experimental taxonomy).

A border career demanded a 'synthesis of sympathy and intellect, observation and experiment, spontaneity and control, laboratory and field,'[19] qualities Janaki possessed and techniques she employed to great benefit in her practice. She was a border person with a border career, corralling the best features of traditional natural history (thus connecting her to the colonial naturalists of the East India Company by an unbroken thread) and experimental lab biology in a creative synthesis. Also, 'the biological side of the problem [was] only one side. It [was] quite as much a problem of history, in archaeology, in anthropology, in nutrition, in sociology.'[20] Janaki was among those few who creatively united in their scientific practice diverse knowledge streams, including the humanities, especially anthropology. Her method was a reflection of her rhizomatic epistemology: it was fluid, adaptive and synthetic, and aimed at inventing new ways of ordering or classifying transitional flora. In fact, she lived multiple lives: as a cytogeneticist, a field biologist, a plant geographer, a palaeobotanist, an evolutionary systematist, an experimental breeder, an ethnobotanist, a historian and ethnographer, and not least a naturalist, traveller-explorer and plant ecologist, identities that might easily be identified with the blueprint of a border culture. It is difficult to name even one among her Indian (male) geneticist contemporaries (almost all did royal science), who adopted such a cross-disciplinary methodology in their researches. Janaki's was a deep and thick or many-layered kind of science, the result of careful exploration of the local, complex and singular connections between myriad fields of knowledge—a practice that was genuinely cross-disciplinary.

Opening Lines of Flight

Janaki's everyday life at the breeding station was a dramatic unfolding of science in action, developing in unexpected ways. She was euphoric when she succeeded in creating new hybrids—her epiphanic moments—especially the challenging intergeneric ones, and overwhelmed with excitement when she discovered unusual chromosomal behaviour, but also overcome with misery and frustration, when for instance, stray goats gobbled up a precious plant, reducing months of labour to naught, or when she was not permitted to pub-lish her research findings, or simply when there was nothing "happening". After an initial high at the Coimbatore Breeding Station, when an exciting new world of research had opened out to her, and she was able to contribute a thing or two to sugarcane breeding and cytotaxonomy, Janaki grew increas-ingly frustrated with the state's attempts to discipline and direct her actions. She had already performed the escape act once before, after a particularly short interlude at Trivandrum where she was a professor at a men's college, because she found the city's conservatism and patriarchy not to her taste, but more importantly, because her research had to be put on the back-burner. Janaki was willing to trade any amount of money for the freedom to wander into the forests, along the *holzwege*, away from regimented paths and mimetic grooves and accomplish something original by way of research in cytogenet-ics; this was in fact how she justified her very existence. She was morally com-pelled as it were to move away from the state, in her constant effort to be a nomad—to find time, space and quiet to address her personal research goals.

Janaki was a border-worker frequently changing the locale or the terri-tory of her practice, forcing the biographer to redraw her nomadic cartogra-phy continually. As someone who pushed back the boundaries of knowledge (in the field of plant cytogenetics), her life shines light on the vexing ques-tion of what motivates scientists to move. Darlington, we may recall, com-mented on one occasion that her only drawback from an employer's point of view was that she would never stay anywhere very long. The nomadic is a tendency towards deterritorialisation, a 'movement by which "one" leaves the territory,' or what has been termed a 'line of flight' by Deleuze and Guattari.[21] Janaki's life in science was moulded by her frequent escapes, the lines of flight (*ligne de fuite*, which could connote fleeing or leaking), from controlling or repressive environments. Her life might be read as a series of interludes, transitions, and anti-mimetic moves, for the nomad values free-dom and the journey rather than the destination. Her rhizomatic and femi-nist subjectivity was one of radical rootlessness (at home nowhere) and of multiple belongings (at home everywhere); at one point in her life after the experience of a deadly war, she felt the whole world had become her home.

Having said that, it was not the actual act of travelling that defined the nomad, as much as 'the subversion of set conventions,' or the undermin-ing of the power and authority of the state.[22] Janaki was also however a literal nomad, traversing the fringes of state science institutions across three continents, and frequently opening lines of flight to break free from mental confinement, and explore novel liberatory and creative trajectories, even when these were only leaps of faith; without a sense of danger, life was not worth living as far as she was concerned. Importantly, by being nomadic,

Janaki the scientist was also challenging the traditional view of the nomad as embodied in the male persona.[23] Her nomadic wandering was punctuated by occasional acts of disappearance—of becoming invisible to the state—in her case, by retreating to the dense forests of the Western Ghats, or the Himalaya to collect wild plants and gather knowledge of their uses from the original dwellers of these ecosystems; she was after all strongly vocal about the need to conserve these 'hideaways,' or sanctuaries, and privilege the indigenous over the colonial from the point of view of knowledge. Janaki's project was one of botanical modernisation and decolonisation, one which fitted hand in glove with her ontological and epistemological nomadism.

When Janaki became tired of the strictures of the Coimbatore Breeding Station—its unscientific and controlling atmosphere, she began desperately looking for a mode of escape, which sometimes appeared before her in the form of epistemic communities such as science congresses, the international genetics conference at Edinburgh (1939) being a case in point. She reimagined herself at the outbreak of war as a scientist-refugee in the manner of several other scientists at this time, only that hers was a self-imposed exile. Britain became her refuge, from an imperial breeding station where patriarchy and pseudoscience were omnipresent; also, she found Coimbatore too 'Brahmanical' for her taste (in contrast to cosmopolitan Madras). Janaki—an independent woman, a border-crossing plant cytogeneticist, involved in a necessarily slow science such as evolutionary biology or cytogeography—felt India (both colonial and post-colonial) with its focus on state-led fast science, that further bolstered gender and other inequalities, had nothing to offer her by way of facilities, a stimulating atmosphere, or research career.

She was also highly mobile during the war years, shuttling between the John Innes and the Kew Gardens, besides the Cambridge School of Agriculture, as part of her researches. She received just enough by way of maintenance, and London was a dangerous place to be in, with all the bombing and destruction to life, and yet she bloomed like a desert flower amidst the thorns. The nomad couldn't be happier, and her science stood to gain. Janaki's exile in Britain lasted nine years, during which time she suffered the ravages of war and extreme loneliness, but it was also the period that saw her become highly productive and evolve into a world-ranking cytogeneticist. Her brief return to India in 1948, only months after India's independence, turned out to be depressing in the extreme for her, because state science had taken hold of the country like never before and it seemed as if all research had ended in the newly independent nation. This time, her escape was to the Nepal Himalaya, the first woman to go on a solitary exploration in the region in the name of science. An important aspect of Janaki's nomadic subjectivity was the persistent temptation to die (the death drive), a drive to escape life, hardwired into it. Her deep-seated death wish, her tendency to embrace tragedy rather than evade it, in fact, helped her break free of oppressive, unstimulating or even seemingly impossible situations in extraordinary ways. Janaki was also one among a new generation of naturalists, deeply inspired by accounts of heroism, adventure, fortitude and tenacity; her private library, we know, was filled with books by explorers, men and women who transgressed boundaries and lived adventurous lives, to which this biography is proof.

Working as a cytologist at the RHS Wisley Garden was yet another interlude of doing science in-between. Her work was very rewarding, enabling her to move in the upper echelons of the intellectual and cultured society in Britain, but it did not provide her with the ambience required for research. It gave her access to new research material, but the realisation that she had ended up as a chromosome-counter and a maker of tetraploids, rather than a scientist, impelled her to break free and reach for the horizons and its promise of open unregimented spaces. This took her across the borders to Paris, where she researched on the Camellia supported by a grant, which also gave her the opportunity to travel in new lands and learn a thing or two about aromatic flora. But she soon found herself grounded for want of funds; this time however no amount of networking would help her find the much-needed support to pursue the Camellia to its many homes across the continents. The rest of her life as an in-between would be a tale of 'separation and entanglement of living here and remembering/desiring another place,' especially Ann Arbor.[24]

Forging national and international connections was integral to the nomad scientist. Amiable, focused, articulate and tenacious, besides being a cosmopolite with a feisty and defiant sense of humour, Janaki made allies easily. Several such linkages and connections resulted in close friendships across class, race gender and culture—forming heterogeneous, non-hierarchical and dynamic constellations—which helped her immensely; these included such people as Mary Agnes Chase, H. H. Bartlett, N. L. Bor and Eileen Macfarlane, who held positions of authority within institutions of the state apparatus (except Eileen) and were themselves nomadic in temperament, besides those eminent figures of royal science like Darlington and J. B. S. Haldane, men and women of social position such as Hilda Seligman, F. C. Stern, Eva Forncrook and Ralph Peer, and several non-scientists including members of her own family—whom she could rely on to execute her escapes or to join with, combine powers and innovate.

Adopting a Problem-Based Model

The problem of the origin of cultivated plants 'is an adventure in apparent chaos,' noted Edgar Anderson.[25] Chaos is however seductive to the nomad; rather than something to run away from, it is an opportunity to be creative. A genuine understanding of the problem of evolutionary relationships of plants, both in the past, and the latent genetic qualities to be explored in the future, depended on a thorough and detailed knowledge of the evolutionary pattern of each individual group, that is the evolutionary steps traversed before a species came into being, and it was a bird's eye view of this complex process that the *Chromosome Atlas* provided.

The authors of the *Atlas* represented differing approaches to science, but complemented each other very well: while Janaki the nomad scientist worked out and accumulated chromosome numbers for the hybrids, polyploids and wild relatives of cultivated plants, caught up in the fluid processes of becoming, Darlington the royal scientist provided an overview and analysis of these, which was an exercise in reduction. Rather than work from essences, Janaki encountered the problem and 'followed' it to where it took her, generating much empirical data in the process. Her projects were about assimilation, compilation and synthesis, rather than analysis, and (very importantly) always works in progress on the much tangled and dynamic history of relationship between man and plants. Very tellingly, she referred to the alternative, self-organised places she created for research at Shoranur and Maduravoyal as a garden and hermitage respectively, allowing solitude, rather than as an 'Institute' or 'Centre,' terms associated with hierarchical structures of state science.

Janaki's scientific contributions may be organised under six heads: cytotaxonomy, experimental breeding and plant introduction, cytogeography, origin and diffusion of farming practices, genecology and ethnobotany. She did not invent or use a model organism in her researches in the manner of her contemporaries (most of who clearly were representatives of royal science)—for instance, the *Crepis* of Babcock and Stebbins, the *Oenothera* of the Michigan botanists (including Janaki's doctoral supervisor B. M. Davis and head of the department H. H. Bartlett), the *Tradescantia* or corn of Edgar Anderson, the maize of McClintock, the *Nicotiana* of Kostoff and so on. Nor did she use a theorematic model in her researches, which involved boiling data down to construct axioms or extract universal constants based on a notion of fixed essences.

The model she followed instead was a problem-based one: she was focused on concrete problems, on mapping changes, transmutations, affinities, similitudes or relations, and celebrating singularities, or the singular conditions under which something new was produced. Janaki worked on an extensive genera of flowering plants, adding a new group each time she moved to a new location, so much so that by the end of the 1970s, she had investigated a whole range of temperate and tropical flowering plants, both the domesticated and their wild counterparts, and all those hybrids or entities in-between, from the point of their origin and distribution: from economic plants like the sugarcane and other grasses, ornamentals like the *Magnolia*, medicinal plants like the *Solanum*, aromatics like the *Cymbopogon*, and finally to the primitive cultivars of the humid forests of the Western Ghats. It is worth mentioning here that rather than expend energy fighting the Kerala government over the Silent Valley issue, she chose the path of least resistance, and creatively floated the idea of undertaking a pioneering project on the cytogenetics of the flora of these undisturbed forests.

The ontological similarities between the human and non-human (plants in the present case) elements of the assemblages that Janaki was part of are particularly striking. Her plants were nomads, hybrids and polyploids (displaying a multiplicity of things at once), not unlike herself, and equally possessed the power of metamorphosis; they were fluid or transitional entities, with the potential to escape (like breaking free from sterility), akin to her lines of flight, to a higher ploidy for example, and thereby evolve, albeit with some help from a colchicine-like toxin and the hands of a skilful breeder. The heterogeneous alliances she made with people and plants, the entanglements, the assemblages or multiplicities, amplified in complexity and performative power, especially through creative re-combinations; this was especially true of her colchicining work at Wisley, which linked her to several young cytologists (mostly women) and also to non-scientists, chiefly nursery-men and gentlemen horticulturists. Also, plants that once earned her attention, would never end up as abandoned workings, but would instead become part of her growing assemblage. Every time an opportunity turned up, and turn up they did, by serendipity and/or science, these would be revisited and new connections and meanings extracted. Top most among them were the nightshades, beginning with the *Nicandra physalodes* of her doctoral research, the eggplant and later the *Solanum khasianum*, and then the grasses, in particular, the *Saccharum* (her first and last love, she would call it), but there were also those like the *Magnolia*, *Camellia* or the *Buddleia*, or the *Vinca rosea*, which thrived in the harsh monazite sands of South Travancore.

Janaki had several epiphanic moments in her career, chiefly to do with her success in creating tetraploids using colchicine. In a plant breeder's career, the manifestation of a plant's tetraploid marked an inflection point in the history of its cultivation, the polyploid being fertile and true-breeding. For the cytologist-breeder it was a thing of joy, as the plant he or she had so expertly created was equipped to pursue an independent course of evolution. The nomad scientist, Deleuze and Guattari state, must 'follow' the flow of matter (or material),[26] or follow a problem through to where it might lead, much like how Janaki pursued (or wished to pursue) the sugarcane and the *Camellia*, among several other plant genera to their near and distant, and many homes. Adding the ethnographic tool to her practice, by invoking layered and comparative narratives of plant-man relationships played out in forests for instance, was one of the many ways by which she resisted reductionism or the quantitative processing of royal science and its practitioners.

Janaki lived a double life at the Wisley Garden: making tetraploids and running chromosome counts (of ornamentals such as the *Rhododendron* and *Magnolia*) for members of the Royal Horticultural Society (RHS) by day and doing independent research on the cytogeography of flowering plants by night, a new disciplinary trajectory that she was pioneering. By the end of her tenure at Wisley (1951), she had collected enough chromosome counts of a range of plants and produced several new polyploids, besides publishing journal articles, but this was still at the level of undisciplined wandering. Quiet reflection, a stimulating research atmosphere and time were crucial if the larger map of the link between polyploidy and plant migration was to be delineated, but this was not to be had, and the reason she wanted to take flight again.

A return to India was ineluctable, but when she had the opportunity to head the Central Botanical Laboratory there, she cleverly included her cyto-geographical project among the aims of the institution, in the hope that she could continue with this line of research with the help of a small team of dedicated workers, and eventually publish a path-breaking thesis on the subject, complete with distribution maps of ecological border zones. Janaki's practice involved decoding the invisible chromosomal songlines, where the axes of time and place intersected, and held the secrets to how plants evolved.[27] With its aim of mapping evolution across landscapes on a plan-etary scale, of complex assemblages comprising interrelated forms, Janaki's cytogeographical practice was grounded in archipelagic thinking, a view of the world as a scatter, albeit interconnected, rather than of consolidated forms. By adopting such an approach, she had moved beyond the 'centric' ideas articulated in the *Atlas*, and had placed Vavilov, her muse and a royal figure of science, on his head as it were. Running a centre of calculation from his institute(s), Vavilov assigned world flora to different 'centres of origin,' a reterritorialising centripetal arrangement, while Janaki was sug-gesting in its place an eccentric, fluid and centrifugal or deterritorialising alternative through her cytogeographical method. A construct of cytologi-cal, geographical, historical and cultural dimensions, it was a way of prob-ing the entangled histories of human and plant migration across time and space, in terms of dynamic relations and processes, rather than as something static or fixed and emanating from a centre. In short, Janaki's cytogeogra-phy was nomad science in the making, privileging the metamorphic, the unstable or transitional over firmly delimited identities. Her ultimate goal was to weave a deep, thick or many-layered cultural history of nature, and slowly, step by step, using chromosomes as the warp and the weft.

That Janaki was unable to explore fully this original line of thinking, of chromosomal phytogeography, made her deeply uneasy and unhappy, so much so that she felt her life had ended a failure and was not worth fussing about. This biography however demonstrates quite the opposite, that her life is a blazing testament to intellectual integrity, tenacity and creativity in the face of upheavals and adversities, to transgressing boundaries and reinventing oneself incessantly, but slowly, and fundamentally, to cultivating 'the habit of freedom'[28]—to become who one is, while giving back to 'human reason,' its original churning and vigour.[29]

Notes

1 BHL, UoM: University Herbarium Records (University of Michigan) H. H. Bartlett series, Box 11, E. K. Janaki Ammal to H. H. Bartlett, letter dated 15 June 1952.
2 For a recent review of literature on key figures in the production of modern mobility, see, Salazar, 'Theorizing mobility through concepts and figures'.
3 Beckstead, 'Liminality in acculturation and pilgrimage: when movement becomes meaningful', p. 384, 392.
4 Engebrigtsen, 'Key figure of mobility: The nomad', p. 43; also, Peters, 'Exile, nomadism, and diaspora: the stakes of mobility in the Western canon'.
5 Deleuze and Guattari, *A Thousand Plateaus*, pp. 387–467.
6 Latour, *Science in Action: How to Follow Scientists and Engineers Through Society*.
7 Deleuze and Guattari, *A Thousand Plateaus*, p. 400.
8 For the case of cybernetics, see Pickering, 'Cybernetics as nomad science' and his *The Cybernetic Brain: Sketches of Another Future*.
9 Stengers, *Cosmopolitics II*, p. 364.
10 Deleuze and Guattari, *A Thousand Plateaus*, p. 419.
11 Deleuze and Parnet, *Dialogues*, p. viii.
12 Deleuze and Guattari, *A Thousand Plateaus*, p. 411.
13 For a guided insight into Serres's thinking, see Watkin, *Figures of Thought*.
14 https://christopherwatkin.com/2019/09/29/michel-serres-book-excerpt-serresal-gorithmic-universal/ accessed 1 August 2022
15 Ibid., p. 400.
16 French novelist Michel Tournier on his friend and fellow-student Deleuze at the Lycee Carnot in the early 1940s. Tournier, *The Wind Spirit: An Autobiography*, p. 128.
17 Kohler, *Landscapes and Labscapes: Exploring the Lab-Field Border in Biology*, pp. 11–19.
18 Many hundreds of her slides were stored at Edam, but unfortunately not a single one survives today.
19 Kohler, *Landscapes and Labscapes*, p. 34.
20 Anderson, *Plants, Man and Life*, p. 108.
21 Deleuze and Guattari, *A Thousand Plateaus*, p. 559.
22 Braidotti, *Nomadic Subjects: Embodiment and Sexual Difference in Contemporary Feminist Theory*, p. 8.
23 Wandering women saints, to whom spirituality provided the means to freedom and self-actualisation, are however not altogether unknown in Hindu mythology or Indian culture; examples being the naked saint Akka Mahādēvi, the skeletal one, Karaikal Ammaiyār of medieval South India, and the legendary poet Avvaiyar, none of who were monastic institution builders in the manner of their male contemporaries. See, Ramaswamy, 'Rebels–Conformists? Women Saints in Medieval South India'.
24 Clifford, *Routes: Travel and Transition in the Late Twentieth Century*, p. 255.
25 Anderson, *Plants, Man and Life*, p. 207.
26 Deleuze and Guattari, *A Thousand Plateaus*, pp. 410–411.

27 Songlines or dreaming tracks of the Australian aborigines narrate the story about the creation of the lands and the seas by their ancestors; it refers to a kind of cultural memory of the land that they lived on. The travels of these ancestors across landscapes and over time, linking important sites and locations made up the songlines. The English writer Bruce Chatwin described them in his book, *Songlines* (1987) as 'the labyrinth of invisible pathways which meander all over Australia'.

28 The phrase 'habit of freedom' is from Virginia Woolf, *A Room of One's Own*, p. 171.

29 Bachelard, *L'Engagement Rationaliste*, p. 7.

ARCHIVAL SOURCES

United States of America

The Bancroft Library, University of California, Berkeley

    Sauer Papers: Carl Sauer-E. K. Janaki Ammal correspondence

University of Michigan

    Bentley Historical Library

        Barbour Scholarships for Oriental Women Papers and
        Minutes of the Barbour Scholarships (1920s–1940s)
        Blanchard Family Papers c. 1835–c. 2000)
        M. U. Botanical Gardens (Correspondence series)
        University Herbarium Records (University of Michigan) H. H. Bartlett series
        Harley Harris Bartlett Papers 1909–1960
        E. K. Janaki Ammal (Alumni File)
        The Michigan Alumnus
        The Michigan Alumnus Quarterly Review
        The Michiganensian
        The Michigan Daily

    Buhr Building (Dissertations)

        E. K. Janaki Ammal, "Chromosome Studies in *Nicandra physalodes* (L.) Gaertn" unpublished doctoral dissertation, 1931

Huntington Library, San Marino, California

    Mss Peer, Box 2, Folder 6

Mount Holyoke College, Massachusetts

    Eleanor Mason letters
    Dorothy Elizabeth Williams letters

Alma Grace Stokey letters
Photograph of E. K. Janaki Ammal, with Checha T. George, c. 1923

Oakridge Nuclear Laboratory, Tennessee

ORNL-The News (1955)
Letter from Russell Hilf (class of 1954), Professor Emeritus, Department of Biochemistry/Biophysics, School of Medicine and Dentistry, University of Rochester
Photographs of E. K. Janaki Ammal at the ORNL

Smithsonian Institution Archives, Washington

Record Unit 229, United States National Museum, Division of Grasses, Folder 42, Box 4, E. K. Janaki Ammal—Mary Agnes Chase correspondence (1920s–1930s)

US National Herbarium

Janaki Ammal's Collections

Great Britain

Bodleian Library, Oxford

Darlington Papers: correspondence with E. K. Janaki Ammal, E. W. Macfarlane, Pio Koller and Boshi Sen (1931–1980); and Darlington's pocket diaries (1931–1951)

British Library, London

Asylum Press Almanac, Madras, 1892
Private Papers: Louise Carolina Maria Ouwerkerk (1931–1933)
India Office Records: L/P & J/11/1 (passports)
Principal's Journal, The Women's Christian College, Madras (1920s)
Annual Report of the Basel Mission, 1840–42

The Madras Mail (1895, 1924)
The Times of India (1948–1951)

The John Innes Horticultural Institution, Norwich

Record of Work of the John Innes Horticultural Institution, 1910–1948
Visitors Register (1931–1948)
Staff file for E. K. Janaki Ammal

Linnean Society, London

Certificate of Recommendation for Election to Fellowship CR/144 (1931), CR/166 (1953) and LL/10

Rothamsted Experimental Station, Harpenden

E. John Russell Papers: Diaries of his travels in India (1936)

Royal Botanic Garden, Edinburgh (Library & Archives)

B. L. Burtt correspondence, letter dated 12 October 1974

Royal Botanic Gardens, Kew (Library & Archives)

Director's Correspondence (1883)
E. J. Salisbury, W. B. Turrill, N. L. Bor, and F. C. Stern and Lady Stern
    correspondence
Printed Notebooks, Herbarium E. K. Janaki Ammal (1930s–1940s)
Purchase and Exchange of Plants Vol. I
Correspondence with Janaki Ammal (1945–1970)
I. H. Burkill, Miscellaneous correspondence
Director George Taylor's Memorandum to Russell, April 1962
E. K. Janaki Ammal, Herbarium Records

Royal Geographical Society, London

E. K. Janaki Ammal, Fellowship Certificate

Royal Horticultural Society, Lindley Library, London

E. A. Bowles correspondence
RHS/Minutes/WY/WY Garden Advisory Committee, Box No. 5, Wis-
    ley Advisory Committee Minute, 1945–1947
Visitor's Book, Cytology Department, RHS Wisley, maintained by E.
    K. Janaki Ammal from 1 October 1946–30 September 1951
E. K. Janaki Ammal, Herbarium Records

Royal Horticultural Society Garden, Wisley

Photographs, Annual Reports and the live collection of Magnolias on
    Battleston Hill

Wellcome Library, London (online access)

J. B. S. Haldane Papers
Hans Grüneberg Papers

India

Birbal Sahni Institute of Palaeobotany, Lucknow

Second Silver Jubilee Lecture on 'Plants and Man' by E. K. Janaki
    Ammal (1972)

Botanical Survey of India, Herbarium and Library (Sibpur)

Official file on the Reorganization of the BSI by E. K. Janaki Ammal, June 1953

Centre for Advanced Study in Botany, University of Madras

Janaki Herbarium, Botany Field Research Laboratory, Maduravoyal

University of Madras, Senate House Library

Madras University Calendars (1890–1940)

National Archives of India, New Delhi

Files (Government of India and Residency) on the Grow More Food campaign

Secret Collection, Government of India, Ministry of EA & CR C. A. Branch: Proposed expedition to Nepal by Dr E. K. Janaki Ammal (1948)

Regional Research Laboratory, Jammu

Annual reports and photographs (1950s–1960s)

The Hindu, Archives, Chennai

1940s

The Madras Mail

1895, 1924

The Times of India, Bombay

1934, 1948, 1949, 1950, 1955

Women's Christian College, Madras (Chennai)

Staff Register (1920s) and Principal's journals

A Note: In addition to the primary sources (mostly unpublished) listed above, back volumes of relevant journals of science from the late nineteenth and early twentieth centuries were perused, including the complete run of *Current Science, Proceedings of the Indian Academy of Sciences (Part B), The Indian Journal of Genetics and Plant Breeding* and *Science and Culture*. Oral history sources, in the form of interviews with Janaki's descendants, colleagues and students also went a long way in writing this biography.

E. K. JANAKI AMMAL'S
PUBLICATIONS

1931. "A Polyploid Eggplant, Solanum melongena Linn.," *Papers of the Michigan Academy of Science, Arts and Letters* 15: 81–83.

1932. a.—"Chromosome Studies in *Nicandra physalodes*," *La Cellule* 4: 49–110.
b.———(with C. D. Darlington), "The Origin and Behaviour of Chiasmata. I. Diploid and Tetraploid tulipa," *Botanical Gazette* 93, no. 3 (May): 296–312.
c. Abstract—"Polyploidy in Solanum melongena Linn." *Proceedings of the Nineteenth Indian Science Congress* (Bangalore), 313. Bangalore: Indian Science Congress.

1933. "The Chromosome number of Cleome viscosa," *Current Science* 1, no. 10 (April): 328.

1934. "Polyploidy in Solanum melongena Linn.," *Cytologia* 5: 453–459.

1936. a.—with T. S. N. Singh. "Cytogenetic analysis of Saccharum spontaneum L. Chromosome studies in Indian forms," *Indian Journal for Agricultural Sciences* 6: 1–8.
b. Ibid. "Cytogenetic analysis of Saccharum spontaneum L. 2. A type from Burma," *Indian Journal for Agricultural Sciences* 6: 9–10.
c. Ibid. "A preliminary note on a new Saccharum-Sorghum hybrid," *Indian Journal for Agricultural Sciences* 6: 1105–1106.

1937. a. Abstract—"Chromosome Studies in Saccharum arundinaceum L." *Proceedings of the Twenty Fourth Indian Science Congress* [Botany], 268. Hyderabad: Indian Science Congress.
b. Abstract—"Inheritance of habit in Saccharum spontaneum L." *Proceedings of the Indian Science Congress* [Agriculture], 365. Hyderabad: Indian Science Congress.
c. Abstract—"Tetrasomic inheritance in two Saccharum officinarum and Saccharum spontaneum hybrids." *Proceedings of the Indian Science Congress* [Agriculture], 365. Hyderabad: Indian Science Congress.
d. "Sugarcane-Sorghum Hybrids in Wild State," *Current Science* 6, no. 7 (November): 235.

1938. a. Abstract—"Chromosome behaviour in S. spontaneum x Sorghum durra hybrids, Studies in S. arundinaceum." *Proceedings of the Twenty Fifth Indian Science Congress*, 143. Calcutta: Indian Science Congress.

b. "Chromosome numbers in sugarcane x bamboo hybrids," *Nature* 141, no. 3577 (21 May): 925.

c. "A Saccharum-Zea cross," *Nature* 142, no. 3596 (1 October): 618–619.

d. Abstract—"The species concept in the light of cytology and genetics." *Proceedings of the Twenty Fifth Indian Science Congress*, 205–208. Calcutta: Indian Science Congress.

1939. a. "Triplo-Polyploidy in Saccharum spontaneum L.," *Current Science* 8, no. 2 (February): 74–76 [first presented at the 7th International Genetics Congress, 1939, Edinburgh].

b. "Supernumerary Chromosomes in Parasorghum," *Current Science* 8. no. 5 (May): 210–211.

1940. a. "Chromosome Numbers in Sclerostachya fusca," *Nature* 145, no. 3673 (23 March): 464

b. "Chromosome Diminution in a Plant," *Nature* 146, no. 3713 (28 December): 839–840.

1941. a. "Intergeneric Hybrids of Saccharum I-III," *Journal of Genetics* 41: 217–253.

"Intergeneric Hybrids of Saccharum I," *Journal of Genetics* 41: 217–230.

"Intergeneric Hybrids of Saccharum II, Saccharum x Imperata," *Journal of Genetics* 41: 231–242.

"Intergeneric Hybrids of Saccharum III, Saccharum x Zea," *Journal of Genetics* 41: 243–253.

b. "The Breakdown of Meiosis in a Male-Sterile Saccharum," *Annals of Botany* 5: 83–88.

1942. "Intergeneric Hybrids of Saccharum IV, Saccharum x Narenga," *Journal of Genetics* 44: 23–32.

1945. a.—with C. D. Darlington, "Adaptive iso-chromosomes in Nicandra," *Annals of Botany* 9, 1945: 267–281.

b.—with C. D. Darlington. *Chromosome Atlas of Cultivated Plants*. London: George Allen & Unwin Ltd.

1948. "The Origin of the Black Mulberry," *Journal of the Royal Horticultural Society* 73, part 4 (April): 117–120.

1949.—with A. P. Wylie, "Chromosome numbers of cultivated Narcissi." *RHS Daffodils & Tulip Year Book* 15, 33–40. Royal Horticultural Society.

1950. a. "Polyploidy in the Genus Rhododendron." *Rhododendron Year Book* [RHS] 5, 92–98. Royal Horticultural Society.

b.—with I. C. Enoch and Margery Bridgwater, "Chromosome numbers in Species of Rhododendron." *Rhododendron Year Book* [RHS] 5, 78–91. Royal Horticultural Society.

c.—"A triploid Kniphofia," *Journal of the Royal Horticultural Society* 75: 23–26.

1951. a. "Chromosomes and Horticulture," *Journal of the Royal Horticultural Society* 76, part 7 (July): 236–239.

b. "Chromosomes and the Evolution of Garden Philadelphus," *Journal of the Royal Horticultural Society* 76, part 8 (August): 269–275.

c. "The Chromosome History of Cultivated Nerines," *Journal of the Royal Horticultural Society* 76, part 10 (October): 365–371.

d.—with Margery Bridgwater, "Chromosome numbers in Hybrid Nerines," *Journal of the Royal Horticultural Society* 76, part 10 (October): 372–75.

e. Abstract—"Polyploidy as a factor of active speciation in South-East Asia." In *Annual Meeting of the British Association for the Advancement of Science*, Edinburgh.

1952. a. "Chromosome Relationships in Cultivated Species of Camellia." *The American Camellia Year Book*, 106–114. Gainesville, Florida: American Camellia Society.

b. "The Race History of Magnolias," *The Indian Journal of Genetics and Plant Breeding* (IJGPB) 12: 82–92.

c.—with Richard Seligman "Notes on the Occurrence of Chromosome Races in Dianthus monspessulanus in Northern Italy," *Journal of the Royal Horticultural Society* 77, part 6 (June): 221–223.

d. "The Story of Primula malacoides," *Journal of the Royal Horticultural Society* 77: 287–290.

e.—with Barbara Saunders, "Chromosome Numbers in Species of Lonicera," *Kew Bulletin* 7, no. 4: 539–541.

1953. a. "Chromosomes and the Species Problem in the Genus Viburnum," *Current Science* 22, no. 1 (January): 4–6.

b. "A New Interspecific Citrus Hybrid," *Current Science* 22, no. 6 (June): 178–179.

1954. a. "The Botanical Survey of India: a retrospect," *Science and Culture* 19: 322–328.

b. "The Cyto-Geography of the Genus Buddleia in Asia," *Science and Culture* 19: 578–581.

c. "The Scope and Functions of the Reorganized Botanical Survey of India," *Science and Culture* 20: 275–280.

1956. a. "Introduction to the Subsistence Economy of India." In *Man's Role in Changing the Face of the Earth*, edited by William L. Thomas, 324–335. Chicago & London: The University of Chicago Press. 2 vols.

b. "Genetic Homeostasis," (Review article), *Current Science* 23, no. 12 (December): 415.

1958. a. "Report on the Humid Regions of South Asia." In *Problems of Humid Tropical Regions* (Kandy, Sri Lanka, 22–24 March 1956), 43–51. Paris: UNESCO. https://unesdoc.unesco.org/ark:/48223/pf0000017765 accessed 1 December 2018.

b. "An Introduction to the Genetical Analysis of the Humid Forest Flora of South Asia." In *Symposium on the Study of Tropical Vegetation* (Kandy, Sri Lanka, 19–21 March 1956), 137–39. Paris: UNESCO. https://unesdoc.unesco.org/ark:/48223/pf0000019459?posInSet=1&queryId=N-EXPLORE-e8a5e4d2-6aba-440f-a2ab-fe196613968b accessed 3 December 2018.

c.—with R. S. Sundara Raghavan. "Polyploidy and vitamin C in Emblica officinalis Gaertn," *Proceedings of the Indian Academy of Sciences* 47, no 5 (May): 312–314.

d. "Iso-chromosomes and the origin of triploidy in hybrids between old and new world species of Philadelphus," *Proceedings of the Indian Academy of Sciences* 48, no. 5 (November): 251–258.

1959. a. Abstract—"Distribution Patterns of some Indian Plants." In *Proceedings of the Forty Sixth Session, Indian Science Congress* [Botany], 112. New Delhi.

b. "Central Botanical Laboratory," *Bulletin of the Botanical Survey of India* 1, no. 1: 33–37.

c. "A cyto-systematic survey of Bambusae I. The slender bamboos of Asia and S. America," *Bulletin of the Botanical Survey of India* 1: 78–84.

d. "Cyto-geography of Some Indian Plants," *Current Science* 28, no. 2 (February): 55.

1960. a. "The Effect of the Himalayan Uplift on the Genetic Composition of the Flora of Asia," *Journal of the Indian Botanical Society* 39: 327–334.

b. "The Genetic Pattern in the Distribution of Some Indian Plants," *Memoirs of the Indian Botanical Society*, Memoir 3: 1–4.

1962. a. "Genetics and Man in India," Presidential Address, Indian Society of Genetics and Plant Breeding, *IJGPB* 22, no. 2: 103–107.

b.—with S. N. Sobti, "The Origin of the Jammu Mint," *Current Science* 31, no. 9 (September): 387–388.

c. "Tetraploidy in Rauvolfia serpentina Benth," *Current Science* 31, no. 12 (December): 520–521.

d.—with S. N. Sobti, "Polyploidy in the Genus Terminalia," *Science and Culture* 28, no. 8: 378–379.

e.—with S. N. Sobti, "Chromosome Relationship in Calendula Species," *Proceedings of the Indian Academy of Sciences, Section B* 55, no 3: 128–130.

f.—with S. D. Singh, "Induced tetraploidy in Dioscorea deltoidea Wall," *Proceedings of the Indian Academy of Sciences, Section B* 56, no. 6 (December): 329–331.

1963. a. "The significance of polyploidy in the improvement of medicinal plants," *Bulletin of the Regional Research Laboratory, Jammu* 1: 85–86.

b.—with H. P. Bezbaruah, "Induced tetraploidy in Catharantus roseus," *Proceedings of the Indian Academy of Sciences* 57, no. 6 (June): 339–341.

1964. a.—with S. N. Sobti and K. L. Handa, "The interrelationship between polyploidy, altitude and chemical composition in Acorus calamus," *Current Science* 33, no. 13 (August): 500.

b.—with J. N. Baruah and R. Rao. *Some Vegetable Tanning Materials of Assam.* New Delhi: CSIR.

1966. a.—with S. D. Singh and R. D. Dhar, "On the occurrence of natural population of Rauwolfia serpentina in Jammu," *Current Science* 35, no. 5 (5 March): 132.

b.—with B. K. Gupta, "Oil content in relation to. polyploidy in Cymbopogon," *Proceedings of the Indian Academy of Sciences Section B* 64, no. 6 (December): 334–335.

1967. a.—with B. L. Kaul, "Cytomorphological Studies in autotetraploid Asparagus officinalis L.," *Proceedings of the Indian academy of Sciences Section B* 65, no. 1: 1–9.

b.—with R. D. Dhar, "Heredity versus Environment in the production of alkaloids in Rauvolfia serpentina," *Proceedings of the Indian Academy of Sciences Section B* 65, no. 3 (March): 103–105.

c. "Genetical improvement of Datura innoxia and D. mete," *Annual Report* (for 1967), Regional Research Laboratory, Jammu.

1968.—with B. L. Kaul, and Usha Zutshi, "Transference of solasodine from Solanum incanum to S. melongena." *Annual Report* (for 1968), Regional Research Laboratory, Jammu.

1969. a.—with B. K. Gupta, "The role of vivipary in the Genus Cymbopogon," *Current Science* 38, no. 2: 576.

b.—with S. N. Khosla, "Breaking the barrier to polyploidy in the genus Eucalyptus," *Proceedings of the Indian Academy of Sciences* 70, no. 5 (November): 248–249.

1970. a.—with V. Abraham, "On the occurrence of supernumerary chromosomes in the Genus Apluda," *Current Science* 39, no. 14 (20 July): 330.

b.—with Usha Zutshi, "Tetraploidy and alkaloid content in Datura metel Linn.," *Proceedings of the Indian Academy of Sciences* 71, no. 1 (January): 1–3.

1971. a.—with Bharati Bhatt, "Induced tetraploidy in Solanum khasianum Clarke," *Proceedings of the Indian Academy of Sciences* 74, no. 2 (August): 99–101.

b.—with Bharati Bhatt, "Observations on the Glyco-alkaloid content of diploid and tetraploid Solanum khasianum Clarke," *Current Science* 40, no. 3 (5 February): 70–71.

c.—with T.V. Sreenivasan, "Observations on the cytology of the Madras Mint," *Current Science* 40, no. 20 (22 October): 544–545.

1972.—with D. Jegathesan and T. V. Sreenivasan, "Further Studies in Saccharum x Zea Hybrid I. Mitotic Studies," *Heredity* 28, no. 1: 141–142

1974. a.—with T. V. Vishwanathan, "A High Alkaloid Containing Race of Solanum incanum Linn., collected from the Paniyas of Kerala," *Current Science* 43, no. 12 (June): 378.
b. "Plants and Man." *Second Silver Jubilee Lecture* (delivered in 1971), 1–6. Lucknow: Birbal Sahni Institute of Palaeontology.

1977. "India's wealth in medicinal and aromatic plants—Its exploitation and improvement." In C. K. Atal and B.M. Kapur (eds.) *Cultivation and Utilization of Medicinal and Aromatic Plants* [Companion volume to *Cultivation and utilization of medicinal plants*] 468–71. Jammu Tawi: Regional Research Laboratory.

1978. a.—with B. K. Gupta, "The aromatic grasses of India: an appraisal," *Indian Journal of Forestry* 1, no, 1: 19–21.
b. "Ethnobotany—Past and present," *Journal of the Indian Botanical Society* [Supplement] 57: 60–61.
c.—with A. W. Jebhadas, "Ethnobotany of Kanikkara of South India," *Journal of the Indian Botanical Society* 57: 66–67.

1983.—with P. Nagendra Prasad, "Costus malortieanus H. Wendl, a new source for diosgenin," *Current Science* 52, no. 17 (September): 825–826.

1984. a.—with P. Nagendra Prasad, "Relationship between polyploidy and diosgenin content in different parts of Costus speciosus (Koen.) Sm.", *Current Science* 53, no. 11 (5 June): 601–602.
b.—with P. Nagendra Prasad, "Ethnobotanical findings on Costus speciosus (Koen.) Sm. among the Kanikkars of Tamil Nadu," *Journal of Economic and Taxonomic Botany* 5: 129–134.
c.—with P. Nagendra Prasad, "Germplasm evaluation of Costus speciosus in relation to diosgenin content," *Indian Journal of Forestry* 7, no. 2: 144–149.

1985. a.—with P. Nagendra Prasad, "Chromosome count of Centella asiatica (Linn.) Urban," *Current Science* 54, no. 14 (20 July): 706–707.
b.—with P. Nagendra Prasad, "Chromosome number reports of some plants from Silent Valley: 1," *Indian Journal of Forestry* 8, no 3: 205–207.

1987. a.—with P. Nagendra Prasad, "Karyology of Costus malortieanus H. Wendl," *Journal of Indian Botanical Society* 66, nos. 1 & 2: 151–153.
b.—with P. Nagendra Prasad and A. W. Jebadhas, "Medicinal plants used by the Kanikkars of South Kerala," *Journal of Economic and Taxonomic Botany* 11, no. 1: 149–155.

GENERAL BIBLIOGRAPHY

Abraham, A. 1939. "Chromosome Structure and the Mechanisms of Mitosis and Meiosis," *Annals of Botany* 3, no. 11: 545–568.

Abraham, J. 2009. "The Stain of White: Liaisons, White Men, Memories and White Men as Relatives," *Men and Masculinities* 9, no. 2: 131–151.

Agha, S. 1933. *Some Aspects of the Education of Women in the United Provinces*. Allahabad: Indian Press.

Aiyappan, A. 1936. "Blood Groups of the Paniyans of the Wynaad Plateau," *Man* 36: 191.

Allen, G. E. 1978. *Thomas Hunt Morgan, the Man and His Science*. Princeton, NJ: Princeton University Press.

Allen, G. E. 1992. "Julian Huxley and the Eugenical View of Human Evolution." In *Julian Huxley: Biologist and Statesman of Science*, edited by C. K. Waters and A. Van Helden, 193–222. Houston, TX: Rice University Press.

Alleva, L. 2006. "Taking Time to Savour the Rewards of Slow Science," *Nature* 443: 271.

Anandhi, S. 2000. "Reproductive Bodies and Regulated Sexuality." In *A Question of Silence? The Sexual Economies of Modern India*, edited by M. E. John and Janaki Nair, 139–166. London and New York: Zed Books (Kali for Women, 1998).

Anderson, E. 1952. *Plants, Man and Life*. Boston, MA: Little, Brown and Company.

Anderson, Robert S. "Empire's setting sun? Patrick Blackett and military and scientific development of India," *Economic and Political Weekly* 36, no. 39 (September 29–October 5, 2001): 3703–20.

Anon. 1895. "Influenza in England," *Evening Post*, LIX, no. 46, 3 February.

Anon. 1925. "The Botanical Garden: A Fascinating Place," *Michigan Alumnus* 32, no. 3: 64.

Anon. 1942. "Science and International Politics," *Current Science* 11, no. 5: 177–182.

Anon. 1943. "The Way and Spirit of Science," *Current Science* 12, no. 10: 267–268.

Anon. 1944. "Freedom for the Scientist," *Current Science* 13, no. 11: 291.

Anon. 1947a. "Review of the Work of the Royal Botanic Gardens, Kew, during 1946," *Kew Bulletin* 2, no. 1: 1–7.

Anon. 1947b. "The Indian Science Congress," *Current Science* 16, no. 1: 1–2.

Anon. 1950. "Review of the Work of the Royal Botanic Gardens, Kew, during 1949." *Kew Bulletin* 5, no. 1: 1–23.

Anon. 1952. "Tracer Techniques in Agriculture," *Current Science* 21, no. 5: 119.

Anon. 1954. "Radiation Genetics," *Current Science* 23, no. 12: 391.

Anon. 1980. *University Botany Laboratory: Golden Jubilee 1980 Yearbook* (Centre for Advanced Study in Botany). Madras: University of Madras.

Anon. 1992a. "Radioisotopes and Health: Trace of Hope," *Oak Ridge National Laboratory Review* 25, nos. 3–4: 36–38.

Anon. 1992b. "Alexander Hollaender: A Radiant Biologist," *Oak Ridge National Laboratory Review* 25, nos. 3–4: 39–41.

Arnold, M. H. 1991. "Joseph Burtt Hutchinson 21 March 1902–16 January 1988," *Biographical Memoirs of Fellows of the Royal Society* 37: 278–297.

Atkinson, L. R. 1968. "Alma Stokey," *American Fern Journal* 58, no. 4: 145–152.

Bachelard, G. 1972. *L'Engagement rationaliste*. Paris: Presses Universitaires de France.

Bailey Ogilvie, M. 2019. *For the Birds, American Ornithologist Margaret Morse Nice*. Norman, OK: University of Oklahoma Press.

Banerji, M. L. 1963. "Outline of Nepal Phytogeography," *Vegetatio* 11, nos. 5–6: 288–296.

Barahona, A. 2015. "Transnational Science and Collaborative Networks. The case of Genetics and Radiobiology in Mexico," *Dynamis* 35, no. 2: 333–358.

Barahona, A. 2020. "Women and the Workplace. Collaborative Networks of Women Geneticists in the 1960s and early 1970s," *Perspectives on Science* 28, no. 2: 201–220.

Barber, C. A. 1927. *Tropical Agriculture Research in the Empire with Special Reference to Cacao, Sugarcane, Cotton and Palm*. London: HMSO.

Barndt, K. and C. M. Sinopoli (eds.). 2017. *Object Lessons, the Formation of Knowledge: The University of Michigan Museums, Libraries, & Collections 1817–2017*. Ann Arbor, MI: University of Michigan.

Bartlett, H. H. and K. L. Jones. 1975. *The Harley Harris Bartlett Diaries (1926–1959)*. Ann Arbor, MI: K. L. Jones.

Beckstead, Z. 2010. "Liminality in Acculturation and Pilgrimage: When Movement Becomes Meaningful," *Culture & Psychology* 16, no. 3: 383–393.

Bell, G. D. H. 1986. "Frank Leonard Engledow 20 August 1890–3 July 1985," *Biographical Memoirs of Fellows of the Royal Society* 32: 189–219.

Belling, J. 1926. "The Iron-aceto-carmine Method of Fixing and Staining Tissues," *Biological Bulletin* 50: 160–163.

Belling, J. and A. Blakeslee. 1924. "The Configurations and Sizes of the Chromosomes in the Trivalents of 25-Chromosome *Daturas*," *Proceedings of the National Academy of Sciences of the United States of America* 10: 116–120.

Beteille, A. 2003. "The Social Character of the Indian Middle Class." In *Middle Class Values in India and Western Europe*, edited by Imtiaz Ahmad and Helmut Reifeld, 73–85. New Delhi: Social Science Press.

Bhagavathi Kutty Amma, P. R. and T. Ekambaram. 1940. "Sugarcane x Bamboo Hybrids," *Journal of the Indian Botanical Society* 18: 209–229.

Biagioli, M. and J. Riskin (eds.). 2012. *Nature Engaged: Science in Practice from the Renaissance to the Present*. London: Palgrave.

Biffen, R. H. 1904. "Experiments with Wheat and Barley Hybrids Illustration Illustrating Mendel's Laws of Heredity," *Journal of the Royal Agricultural Society of England* 65: 337–345.

Biffen, R. H. 1905. "Mendel's Laws of Inheritance and Wheat Breeding," *Journal of Agricultural Science* 1: 4–48.

Biffen, R. H. 1906. "The Application of Mendel's Laws of Inheritance to Breeding Problems," *Journal of the Royal Agricultural Society of England* 67: 46–63.

Bonavia, E. 1886. "On the Probable Wild Source of the Whole Group of Cultivated True Limes (*Citrus acida* Rox., *C. medica* var. Acida of Brandis, Hooker and Alph. De Candolle)," *Botanical Journal of the Linnean Society* 22, no. 145: 213–218.

Bonavia, E. 1890. *The Cultivated Oranges and Lemons of India and Ceylon.* London: W. H. Allen.

Bonneuil, C. 2019. "Seeing Nature as a 'Universal Store of Genes': How Biological Diversity Became 'Genetic Resources,' 1890–1940," *Studies in History and Philosophy of Biology & Biomedical Sciences* 75: 1–14.

Bor, N. L. 1938. "The Vegetation of the Nilgiris," *The Indian Forester* 64, no. 10: 600–609.

Bor, N. L. 1940. "Three New Genera of Indian Grasses," *The Indian Forester* 66, no. 5: 267–272.

Bor, N. L. 1947. "Tropical Ecology and Research," presented at the Fifth British Empire Forestry Conference, Great Britain, 1–6.

Borges, J. L. 1962. "An Approach to Al-Mutasim." In *Fictions.* New York: Grove Press.

Borthakur, A. and P. Singh. 2013. "History of Agricultural Research in India," *Current Science* 105, no. 5: 2.

Bowler, P. J. 1989. *The Mendelian Revolution: The Emergence of Hereditarian Concepts in Modern Science and Society.* London: The Athlone Press.

Braidotti, R. 1994. *Nomadic Subjects: Embodiment and Sexual Difference in Contemporary Feminist Theory.* New York: Columbia University Press.

Bremer, G. 1925. "Cytology of the Sugarcane," *Genetica* 7: 293–322.

Brown, A. 2007. *J. D. Bernal: The Sage of Science.* Oxford: Oxford University Press.

Bulmer, M. 1998. "Galton's Law of Ancestral Heredity," *Heredity* 81, no. 5: 579–585.

Burkill, I. H. 1960. "The Organography and Evolution of Dioscoreaceae, the Family of the Yams," *Journal of the Linnean Society of London* 56, no. 367: 319–412.

Burns, W. 1941. "Some Ideas and Opportunities for Plant Geneticists in India," *The Indian Journal of Genetics and Plant Breeding (IJGPB)* 1: 2–3.

Burns, W. 1943. "The Teaching of Plant Genetics," *IJGPB* 3: 1–6.

Campos, L. 2010. "Mutant Sexuality: The Private Life of a Plant." In *Making Mutations: Objects, Practices and Contexts*, edited by L. Campos and A. von Schwerin, 49–70. Preprint 393, Berlin: Max-Planck-Institut für Wissenschaftsgeschichte (eprint).

Capasso, F., T. S. Gaginella, G. Grandolini and A. A. Izzo. 2003. "Active Principles." In *Phytotherapy: A Quick Reference to Herbal Medicine*, 31–44. Berlin and Heidelberg: Springer-Verlag.

Carl Rufus, W. 1942. "Twenty-five Years of the Barbour Scholarships," *Michigan Alumnus Quarterly Review* 49, no. 11: 15.

Carlson, E. A. 1966. *The Gene: A Critical History.* Philadelphia, PA: Saunders.

Carnaghan, C. 2012. "Jean Rasmussen," *Asian Affairs* 43, no. 3: 546–547.

Champion, H. G. 1936. "A Preliminary Survey of the Forest Types of Indian and Burma," *Indian Forest Records* 1: 1–286.

Chandrasekharan, S. N. and S. V. Parthasarathy. 1948. *Cytogenetics and Plant Breeding*. Madras: P. Varadachary & Co.

Chatwin, B. 1987. *Songlines*. New York: Viking.

Chiarelli, B. 1961. "The Chromosomes of Orang-Utan (*Pongo pygmaeus*)," *Nature* 192, no. 4799: 285.

Chopra, R. N. and L. D. Kapoor. 1952. "Some Botanical Aspects of Kashmir," *Palaeobotanist* 1: 115–119.

Chowdhury, I. 2016. *Growing the Tree of Science: Homi Bhabha and the Tata Institute of Fundamental Research*. New Delhi: Oxford University Press.

Clarke, A. E. and J. H. Fujimura (eds.). 1992. *The Right Tools for the Job. At Work in Twentieth-Century Life Sciences*. Princeton, New Jersey: Princeton University Press.

Clausen, R. E. and T. H. Goodspeed. 1923. "Inheritance in Nicotiana Tabacum III. The Occurrence of Two Periclinal Chimeras," *Genetics* 8, no. 2: 97–105.

Clifford, J. 1997. *Routes: Travel and Transition in the Late Twentieth Century*. Cambridge, MA: Harvard University Press.

Comfort, N. 2003. *The Tangled Field: Barbara McClintock's Search for the Patterns of Genetic Control*. Cambridge, MA: Harvard University Press (first edition, 2001).

Coon, E. M. 1924. "The Women's Christian College," *The 90th Annual Report of the American Madura Mission, South India*. Yale Divinity Library Collection: 59–60. https://hdl.handle.net/10079/digcoll/4866117 accessed 16 October 2018.

Crow, J. F. 1994. "Hitoshi Kihara, Japan's Pioneer Scientist," *Genetics* (Genetics Society of America) 137, no. 4: 891–894.

Crow, J. F. 2001. "Plant Breeding Giants: Burbank, the Artist; Vavilov, the Scientist," *Genetics* 158, no. 4: 1391–1395.

Curry, H. A. 2010. "Making Marigolds: Colchicine, Mutation Breeding, and Ornamental Horticulture, 1937–1950." In *Making Mutations: Objects, Practices and Contexts*, edited by L. Campos and A. von Schwerin, 259–284. Preprint 393, Berlin: Max-Planck-Institut für Wissenschaftsgeschichte.

Curry, H. A. 2016. "Atoms in Agriculture: A Study of Scientific Innovation Between Technological Systems," *Historical Studies in the Natural Sciences* 46, no. 2: 119–153.

Damodaran, V. 2013. "Gender, Race and Science in Twentieth-Century India: E. K. Janaki Ammal and the History of Science," *History of Science* 51, no. 3: 283–307.

Damodaran, V. 2017. "Janaki Ammal, C. D. Darlington and J. B. S. Haldane: Scientific Encounters at the End of Empire," *Journal of Genetics* 96, no. 5: 827–836.

Darlington, C. D. 1930. "A Cytological Demonstration of 'Genetic' Crossing-Over (Hyacinthus)," *Proceedings of the Royal Society of London* (Series B) 107, no. 748: 50–59.

Darlington, C. D. 1932. "The Origin and Behaviour of Chiasmata: VI Hyacinthus Amethystinus," *Biological Bulletin* 63, no. 3: 368–372.

Darlington, C. D. 1938. "Structure of Chromosome," *Nature* 141: 371–372.

Darlington, C. D. 1939. "Cytology," *Nature* 144: 816–817.

Darlington, C. D. and L. F. La Cour. 1942. *The Handling of Chromosomes*. London: George Allen & Unwin Ltd.

Darlington, C. D. 1947a. "A Revolution in Russian Science," *Discovery* 8: 40–43 and *Heredity* 38: 143–148.

Darlington, C. D. 1947b. "The Retreat from Science in Soviet Russia," *Nineteenth Century and After* 142: 157–168.

Darlington, C. D. 1948. *The Conflict of Science and Society* (Conway Memorial Lecture). London: Watts & Co.

Darlington, C. D. and K. Mather. 1949. *The Elements of Genetics*. London: Allen and Unwin.

Darlington, C. D. 1969. *The Evolution of Man and Society*. New York: Simon and Schuster.

Darlington, C. D. 1981. "Genetics and Plant Breeding, 1910–80," *Philosophical Transactions of the Royal Society of London* (Series B) 292, no. 1062: 401–405.

Darwin, C. R. 1868. *The Variation of Plants and Animals Under Domestication*. London: John Murray.

Darwin, F. and E. H. Acton. 1894. *Practical Physiology of Plants*. Cambridge: Cambridge University Press.

Dasgupta, S. 2000. *Jagdish Chandra Bose and the Indian Response to Western Science*. New Delhi: Oxford University Press.

Datta, R. M. 1935. "Indian Women in Science," *Modern Review* 57, no. 3: 279.

Davis, B. M. 1911. "Some Hybrids of *Oenothera biennis* and *O. grandiflora* that resemble *O. Lamarckiana*," *The American Naturalist* 45, no. 532: 193–233.

de Candolle, A. 1959 (1886). *Origin of Cultivated Plants*. London: Hafner Publishing Co.

Deleuze, G. and F. Guattari. 1987a. *A Thousand Plateaus: Capitalism and Schizophrenia* (Brian Massumi, trans.). Minneapolis, MN: University of Minnesota Press (first edition 1980).

Deleuze, G. and C. Parnet. 1987. *Dialogues*. London and New York: The Athlone Press (French, 1977).

Desmond, R. 1994. *Dictionary of British and Irish Botanists and Horticulturists*. London: Taylor & Francis and the Natural History Museum.

D'Monte, D. 1985. "Storm Over Silent Valley." In *Temple or Tombs? Industry Versus Environment: Three Controversies*, edited by D. D'Monte, 29–59. New Delhi: Centre for Science and Environment.

Dobzhansky, T. 1950. "Evolution in the Tropics," *American Scientist* 38, no. 2: 209–221.

Doctor, G. 2016. "Celebrating Janaki Ammal, Botanist and Passionate Wanderer of Many Worlds," *The Wire*, 6 November. https://thewire.in/science/janaki-ammal-magnolia-edathil accessed 15 December 2016.

Dronamraju, K. 2010. "J. B. S. Haldane's Last Years: His Life and Work in India," *Genetics* 185, no. 1: 5–10.

Dronamraju, K. 2017. *Popularizing Science: The Life and Work of J. B. S. Haldane*. Oxford: Oxford University Press.

Dunn, L. C. 1965. *A Short History of Genetics*. New York: McGraw Hill.

Eigsti, O. J. and P. Dustin. 1955. *Colchicine in Agriculture, Medicine, Biology and Chemistry*. Ames: Iowa State College Press.

Endersby, J. 2009. "Lumpers and Splitters: Darwin, Hooker, and the Search for Order," *Science* 326, no. 5959: 1496–1499.

Engebrigtsen, A. I. 2017. "Key Figure of Mobility: The Nomad," *Social Anthropology* (Special Issue: Key Figures of Mobility) 25, no. 1: 42–54.

Engledow, F. L. and K. Ramiah. 1930. "Investigations on yield in cereals. VII. Study of Development and Yield of Wheat Based Upon Varietal Comparison," *The Journal of Agricultural Science* 20, no. 2: 265–344.

Fairchild, D. 1938. *The World Was My Garden*. London: C. Schreibner's Sons.

Ferry, G. 2000. *Dorothy Hodgkin: A Life*. New York: Cold Spring Harbour Laboratory Press (London: Granta Publications, 1998).

Ferry, G. 2014. "Telling Stories or Making History?" In *Writing About Lives in Science*, edited by P. Govani and Z. Franceschi, 55–63. Göttingen: V & R Unipress.

Fisher, R. 1930. *The Genetical Theory of Natural Selection*. Oxford: The Clarendon Press.

Forbes, G. 2004. *Women in Modern India. The New Cambridge History of India*. IV.2. Cambridge, UK: Cambridge University Press.

Forbes, J. 1834. *Oriental Memoirs: A Narrative of Seventeen Years Residence in India* (2 vols). London: Richard Bentley (first edition 1813).

Fowler, C. and P. Mooney. 1990. *Shattering: Food, Politics, and the Loss of Genetic Diversity*. Tuczon, AZ: University of Arizona Press.

Fraser Roberts, J. A. 1964. "Reginald Ruggles Gates, 1882–1962," *Biographical Memoirs of Fellows of the Royal Society* 10: 83–106.

Fyson, P. F. 1912. *A Botany for India*. Madras: Christian Literature Society for India.

Fyson, P. F. 1932. *The Flora of the South Indian Hill Stations: Ootacamund, Coonoor, Kotagiri, Kodaikanal, Yercaud and the Country Round* (3 vols). Madras: Government Press.

Gamble, J. S. 1957 [1915–21]. *Flora of the Presidency of Madras*. 3 vols. Calcutta: Botanical Survey of India.

Ganeshaiah, K. N., R. Vasudeva and R. Uma Shanker. 2009. "In Search of *Sanjeevani*", *Current Science* 97, no. 4: 484–489.

Gardiner, J. 2006. "The "Rhododendron" Horticultural Society: The Symbiotic Relationship Between the Royal Horticultural Society and the Genus Rhododendron," *The Azalean* 28, no. 4: 83–88.

Gates, R. R. 1936. "Recent Progress in Blood Group Investigations," *Genetica* 18: 47–65; also see his "Blood groupings and racial classification." In *Proceedings of the 25th Indian Science Congress, 1938*, Part 4: 39–40.

Gates, R. R. 1937. "Double Structure of Chromosomes," *Nature* 140: 1013–1014.

Gates, R. R. 1938. "The Jubilee Meeting of the Indian Science Congress," *Science* 87, no. 2260: 357–359.

Gates, R. R. with S. V. Mensinkai. 1938. "Double Structure of Chromosomes," *Nature* 141, no. 2: 607.

Godbole, R. and R. Ramaswamy. (eds.). 2008. *Lilavati's Daughters: The Women Scientists of India*. Bangalore: Indian Academy of Sciences.

Golinski, J. 2005 [1998]. *Making Natural Knowledge: Constructivism and the History of Science*. Chicago, IL: Chicago University Press.

Gopal, S. 2004 [1989]. *Jawaharlal Nehru: A Biography*. New Delhi: Oxford University Press.

Gopal-Ayengar, A. R. and E. V. Cowdry. 1947. "Deoxyribose Nucleic Acid from Isolated Chromosome Threads in Experimental Epidermal Methylcholanthrene Carcinogenesis in Mice," *Cancer Research* 7, no. 1: 1–8.

Gould, S. J. 2002. "No Science without Fancy, No Art without Facts: The Lepidoptery of Vladimir Nabokov." In *I Have Landed*, 29–53 (Part II: Disciplinary Connections: Scientific Slouching across a Misconceived Divide). Cambridge, MA: Harvard University Press.

Govani, P. and Z. Franceschi. 2014. *Writing About Lives in Science: (Auto)Biography, Gender, and Genre.* Göttingen: V & R Unipress.

Grüneberg, H. 1946. "Review of *Chromosome Atlas of Cultivated Plants*, London, 1945," *The Eugenics Review* 38, no. 2: 93–94.

Gundert, H. 1983. *Tagebuch aus Malabar 1837–1859.* Ulm: Kommissionsverlag J. F. Steinkopf Verlag.

Gupta, B. K. 1970a. "Studies in the Genus *Cymbopogon* spreng–III. Contribution to Cytogeography of some Indian Species of *Cymbopogon*," *Proceedings of the Indian Academy of Sciences B* 71, no. 1: 4–8.

Gupta, B. K. 1970b. "Study in the Genus Cymbopogon Spreng–IV. A New Species of Cymbopogon from Ladakh," *Proceedings of the Indian Academy of Sciences B* 71, no. 1: 9–12.

Hacking, I. 1999. *Social Construction of What?* Cambridge, MA: Harvard University Press.

Haldane, J. B. S. 1938. "Forty Years of Genetics." In *Background to Modern Science*, edited by J. Needham and W. Pagel, 225. Cambridge: Cambridge University Press.

Haldane, J. B. S. 1959. "The Theory of Natural Selection Today," *Nature* 183, no. 4663: 710–713.

Harman, O. S. 2003. "The British and American Reaction to Lysenko and the Soviet Conception of Science," *Journal of the History of Biology* 36, no. 2: 309–352.

Harman, O. S. 2004. *The Man Who Invented the Chromosome: A Life of Cyril Darlington.* Cambridge, MA and London: Havard University Press.

Harvey, J. D. 1997. "Almost a Man of Genius." In *Clemence Royer, Feminism, and 19th Century Science.* New Brunswick, NJ: Rutgers University Press.

Haycraft, J. B. 1900. *Darwinism and Race Progress.* London: S. Sonnenschein; New York: Scribner.

Hellman, G. J. 1950. "Curator getting around," https://www.newyorker.com/magazine/1950/08/26/curator-getting-around accessed 13 August 2020

Heidegger, M. 1962. *Being and Time*, (trans. J. Macquarrie and E. Robinson). London: SCM Press (published in German in 1927).

Heidegger, M. 2002. *Off the Beaten Track* (trans. and edited by J. Young and K. Haynes). Cambridge: Cambridge University Press (published in German in 1950).

Henson, P. M. 2002. "Invading Arcadia: Women Scientists in the Field in Latin America, 1900–1950," *The Americas* 58, no. 4: 577–600.

Henson, P. M. 2003. " 'What Holds the Earth Together': Agnes Chase and American Agrostology," *Journal of the History of Biology* 36: 437–460.

Holmes, S. J. 1936. *Human Genetics and Its Social Import.* New York: McGraw-Hill Book Co.

Hubbard, C. E. 1975. "Norman Loftus Bor (1893–1972)," *Kew Bulletin* 30, no. 1: 1–10.

Hughes, A. 1959. *A History of Cytology*. London: Abelard-Schuman.

Indian Science Congress. 1921. *Madras Handbook 1922*. Madras: Madras Diocesan Press.

Indian Science Congress. 1938. *Proceedings of the Twenty-Fifth Indian Science Congress* (Silver Jubilee Session). Calcutta: Royal Asiatic Society of Bengal.

Indian Society of Genetics and Plant Breeding. 1957. "Genetics and Plant Breeding in South Asia: International Symposium," *Indian Journal of Genetics and Plant Breeding* 17, no. 2.

Indian Society of Genetics and Plant Breeding and S. Ramanujam. 1966. "The Impact of Mendelism on Agriculture, Biology and Medicine: Proceedings of the 3rd International Symposium Organized in February 1965 to Celebrate the Centenary of Mendelism and Silver Jubilee of the Society," *Indian Journal of Genetics and Plant Breeding* 26A.

Innes, C. A. and F. B. Evans. 1951. *Madras District Gazetteers*, vol. 1. Madras: Malabar Government Press.

Jacob, K. T. 1941. "Cytological Studies in the Genus *Sesbania*,'" *Bibliographia Genetica* 8: 225–300.

Jerdon, T. C. 1847. *Illustrations of Indian Ornithology*. Madras: P. R. Hunt, American Mission Press.

Jerdon, T. C. 1851. "A Catalogue of Species of Ants Found in Southern India," *The Madras Journal of Literature and Science* 17: 103–127.

Jerdon, T. C. 1862–64. *Birds of India*. Calcutta: Military Orphan Press.

Jeyabaskaran, R. and P. S. Lyla. 1996. "Krusadai Island, the Biologist's Paradise," *Seshaiyana* 4, no. 1: 63–72.

Jorgensen, C. A. and M. B. Crane. 1927. "Formation and Morphology of Solanum Chimaeras," *Journal of Genetics* 18, no. 2: 247–273.

Joseph, J. and V. Chandrasekaran. 1978. "*Janakia arayalpatra*—A New Genus and Species of *Periplocaceae* from Kerala, South India," *Journal of the Indian Botanical Society* 57: 308–312.

Kadam, B. S. 1951. "The Light of Agricultural Research in India," *The IJGPB* 11, no. 2: 127–135.

Karunakaran, E. K. and R. A. Pai. 1967. "E Trisomy Syndrome in a Dyzygotic Twin," *Indian Paediatrics* 4, no. 3: 145–149.

Kass, L. B. 2003. "Records and Recollections: A New Look at Barbara McClintock, Nobel Prize-Winning Geneticist," *Genetics* 164: 1251–1260.

Kedharnath, S. 1988. "Edavaleth Kakkat Janaki Ammal (1897–1984)," *Biographical Memoirs of Fellows of the Indian National Science Academy* 13: 90–101.

Keller, E. F. 1983. *A Feeling for the Organism: The Life and Work of Barbara McClintock*. New York and San Francisco, CA: W. H. Freeman and Company.

Kemp, N. D. 2002. *A Merciful Release: The History of the British Euthanasia Movement*. Manchester: Manchester University Press.

Kevles, D. J. 1985. *In the Name of Eugenics: Genetics and the Uses of Human Heredity*. Berkeley and Los Angeles, CA: University of California Press.

Kleinman, K. 1999. "His Own Synthesis: Corn, Edgar Anderson, and Evolutionary Theory in the 1940s," *Journal of the History of Biology* 32: 293–320.

Knight, H. 1954. *Food Administration in India, 1939–47*. Stanford, CA: Stanford University Press.

Kohler, R. 1994. *Lords of the Fly: Drosophila Genetics and the Experimental Life*. Chicago, IL: University of Chicago Press.

Kohler, R. 2002. *Landscapes and Labscapes: Exploring the Lab-Field Border in Biology*. Chicago, IL: The University of Chicago Press.

Koshy, T. K. 1933. "Chromosome Studies in Allium I the Somatic Chromosomes," *Journal of the Royal Microscopical Society* 53: 299–318.

Kraft, A. 2004. "Pragmatism, Patronage and Politics in English Biology: The Rise and Fall of Economic Biology 1904–1920," *Journal of the History of Biology* 37: 213–258.

Krementsov, N. 2005. *International Science Between the World Wars: The Case of Genetics*. Oxon and New York: Routledge.

Krishna, V. V. 2011. "Organization of Industrial Research: The Early History of CSIR, 1937–1947." In *Science and Modern India: An Institutional History, c. 1784–1947*, edited by U. Das Gupta (History of Science, Philosophy, and Culture in Indian Civilization v. 15, pt. 4), 157–184. New Delhi: Pearson Education.

Krishnan, E. K. 1933. "Control of Lantana: A Suggestion," *Indian Forester* 59, no. 11: 713.

Krishnan, E. K. 1939a. "Prickly Pear for Gum," *Indian Forester* 65, no. 7: 441.

Krishnan, E. K. 1939b. "Forest Fires," *Indian Forester* 65, no. 9: 567–572.

Krishnan, E. K. 1940. "Evergreens in Cudappah," *Indian Forester* 66, no. 8: 482–485.

Krishnan, E. K. 1953. "Forest Reminiscence: Mostly about Snakes," *Indian Forester* 79, no. 11: 614–616.

Krishnaswamy, N. and V. S. Raman. 1949. "A Note on the Chromosome Numbers of Some Economic Plants in India," *Current Science* 18, no. 10: 376–377.

Krishnaswamy, N. and G. N. Rangaswami Ayyangar. 1935a. "Chromosome Numbers in *Sesbania grandiflora* PERS. The Agathi Plant," *Current Science* 3, no. 12: 488.

Krishnaswamy, N. and G. N. Rangaswami Ayyangar. 1935b. "Chromosome Numbers in *Cajanus indicus* SPRENC," *Current Science* 3, no. 12: 614–615.

Kulkarni, C. G. 1927a. "Inheritance Studies of White-Capping in Yellow Dent Maize," *Papers of the Michigan Academy of Science, Arts and Letters* 6: 253–273.

Kulkarni, C. G. 1927b. "The Date of the Bhagvadgita," *Papers of the Michigan Academy of Science, Arts and Letters* 7: 251–268.

Kulkarni, C. G. 1929. "Meiosis in Pollen Mother Cells of Strains of *Oenothera pratincola* Bartlett," *Botanical Gazette* 87: 238–258.

Kumar, N. (ed.). 2009. *Women and Science in India: A Reader*. New Delhi: Oxford University Press.

Kurup, K. K. N. 1988. *Modern Kerala: Studies in Social and Agrarian Relations*. New Delhi: Mittal Publications.

La Cour, L. 1929. "New Fixatives for Plant Cytology", *Nature* 124: 127.

La Cour, L. 1931. "Improvements in Everyday Technique in Plant Cytology", *Journal of the Royal Microscopical Society* 51, no. 2: 119–126.

Latour, B. 1987. *Science in Action: How to Follow Scientists and Engineers Through Society*. Cambridge, MA: Harvard University Press.

Lesley, J. W. and M. M. Lesley. 1929. "Chromosome Fragmentation and Mutation in Tomato," *Genetics* 14: 321–337.

Levine, M. M. 2005. *Defining Women's Scientific Enterprise: Mount Holyoke Faculty and the Rise of American Science.* Lebanon, NH: University Press of New England.

Lewis, D. 1983. "Cyril Dean Darlington 1903–1981," *Biographical Memoirs of Fellows of the Royal Society* 29: 113–157.

Lewis, D. 1986. "Leonard Francis La Cour 1907–1984," *Biographical Memoirs of Fellows of the Royal Society* 32: 357–375.

Lewis, D. 1992. "Kenneth Mather 1911–1990," *Biographical Memoirs of Fellows of the Royal Society* 38: 249–266.

Loskutov, I. G. 1999. *Vavilov and His Institute: A History of the World Collection of Plant Genetic Resources in Russia.* Rome: International Plant Genetic Resources Institute.

MacArthur, R. H. and E. O. Wilson. 1967. *The Theory of Island Biogeography.* Princeton, NJ: Princeton University Press.

Macfarlane, E. W. 1936. "Preliminary Note on the Blood Groups of Some Cochin Castes and Tribes," *Current Science* 4, no. 9: 653–654.

Macfarlane, E. W. 1937a. "Supplementary Note on the Blood Groups of the Aborigines of Bihar," *Current Science* 6, no. 4: 284.

Macfarlane, E. W. 1937b. "Review of S. J. Holmes' Human Genetics and Its Social Import," *Current Science* 5, no. 9: 494–495.

Macfarlane, E. W. 1938. "Blood Group Distribution in India with Special Reference to Bengal," *Journal of Genetics* 36: 225–237.

Macfarlane, E. W. 1940. "Blood Grouping in the Deccan and Eastern Ghats," *Journal of the Royal Asiatic Society of Bengal* 6: 39–49.

Macfarlane, E. W. 1941. "Tibetan and Bhotia Blood Group Distribution," *Journal of the Royal Asiatic Society of Bengal* 8: 180–181.

Macfarlane, E. W. and S. S. Sarkar. 1941. "Blood Groups in India," *American Journal of Physical Anthropology* 28, no. 4: 397–410.

Maeda, T. 1928. "The Spiral Structure of Chromosomes in the Sweet Pea (Lathyrus odoratus, Linn.)," *The Botanical Magazine* XLII, no. 496: 191–195.

Maheshwari, N. 2004. "Remembering Panchanan Maheshwari – An Eminent Botanist of the Twentieth Century," *Current Science* 87, no. 12: 1756–1760.

Maheshwari, P. 1952. "Lysenko's Latest Discovery-the Conversion of Wheat into Rye, Barley and Oats," *Nature* 170: 66–68.

Maheshwari, R. and A. Raman. 2014. "The Knight of Sugar Industry: T. S. Venkatraman (1884–1963)," *Current Science* 106, no. 8: 1146–1149.

Malabar Marriage Commission. *Report of the Malabar Marriage Commission with Enclosures and Appendices.* 1891. Madras: Lawrence Asylum Press.

Manilal, K. S. 1988. *Flora of Silent Valley.* Calicut: Mathrubhumi.

Manilal, K. S. 2003. *Van Rheede's Hortus Malabaricus. English Edition, with Annotations and Modern Botanical Nomenclature* (12 Vols). Trivandrum: University of Kerala.

Manilal, K. S. 2012. "B. K. Nayar (1927–2012)," *Current Science* 103, no. 10: 1219.

Manton, I. 1953. "The Cytological Evolution of the Fern Flora of Ceylon," *Symposia of the Society for Experimental Biology* 7: 174–185.

Martínez-San Miguel, Y. and M. Stephens. (eds.). 2020. *Contemporary Archipelagic Thinking: Towards New Comparative Methodologies and Disciplinary Formations.* New York: Rowman & Littlefield Publishers.

Mather, K. 1943. "Polygenic Inheritance and Natural Selection," *Biological Reviews* 18, no. 1: 32–64.

Mathews, J. K. 1990. "The Legacy of Charles W. Ranson," *International Bulletin of Missionary Research* 14: 108–112.

Mayr, E. 1982. *The Growth of Biological Thought: Diversity, Evolution and Inheritance*. Cambridge, MA: Harvard University Press.

McClintock, B. 1929. "A Method for Making Aceto-carmine Smear Permanent," *Stain Technology* 4: 53–56.

McDougall, E. 1926. *A Missionary College at Madras: Women's Christian College, Madras 1915-25*. Madras: Diocesan Press.

McGucken, W. 1984. *Scientists, Society and State: The Social Relations of Science Movement in Great Britain 1931-1947*. Columbus, OH: Ohio State University Press.

McOuat, G. 2017. "J. B. S. Haldane's Passage to India: Reconfiguring Science," *Journal of Genetics* 95, no. 5: 845–852.

Mehra, G. 2007. *Nearer Heaven than Earth: The Life and Times of Boshi Sen and Gertrude Emerson Sen*. New Delhi: Rupa.

Mendel, G. 1866. "Versuche über Pflanzen-Hybriden" published in the *Proceedings of the Natural History Society of Brünn*. For an English translation, see Druery, C. T and W. Bateson. 1901. "Experiments in Plant Hybridization," *Journal of the Royal Horticultural Society* 26: 1–32.

Michener, D. and A. Reznicek. 2017. "Harley Harris Bartlett (1886–1960): Visionary Botanical Scholar." In *Object Lessons, The Formation of Knowledge: The University of Michigan Museums, Libraries, & Collections 1817–2017*, edited by K. Barndt and C. M. Sinopoli, 168–172. Ann Arbor, MI: University of Michigan.

Morgan, T. H. 1910. "Chromosomes and Heredity," *The American Naturalist* 44: 449–496.

Muldal, S. and D. H. Valentine. 1950. "Cytology, Genetics and Classification," *Nature* 166: 769–771.

Mundon, A. 2003. "Renaissance and Social Change in Malabar—A Study with Special Reference to Ananda Samajam, Siddha Samajam and Atma Vidya Sangham," Unpublished Thesis, Department of History, University of Calicut. http://hdl.handle.net/10603/50479 accessed 13 April 2014.

Myers, W. M. 1960. "Some Limitations of Radiation Genetics and Plant Breeding," *IJGBP* 20, no. 2: 89–92.

Nayar, B. K. 1980. *Silent Valley: An Ecological Hyperbole*. Trivandrum: Parisara Samrakshana Asoothran Samithi.

Newland, A. R., J. Perree and R. Schaap. 2001. *Koson Ohara—Amsterdam Rijksmuseum, "Crows, Cranes & Camellias. The Natural World of Ohara Koson 1877–1945. Japanese Prints from the Jan Perree Collection"*. Leiden: Hotei Publishing.

Newton, W. C. F. and C. D. Darlington. 1929. "Meiosis in Polyploids, I. Triploid and Pentaploid *Tulipa*," *Journal of Genetics* 21: 1–15.

Nietzsche, F. 1997. *Untimely Meditations* (R. J. Hollingdale, trans. and edited by D. Breazeale). Cambridge: Cambridge University Press (first published 1876).

Ogilvie, M, J. Harvey and M. Rossiter. 2014. *The Biographical Dictionary of Women in Science: Pioneering Lives from Ancient Times to the Mid-20th Century* (2 Vols). London and New York: Routledge (First published 2000).

Oommen, M. P. and A. I. Vogel. 1930. "Syntheses of Cyclic Compounds. Part VII. The Stereoisomeric βγ-Diphenyladipic Acids," *Journal of Chemical Society* 21: 48–54.

Osvald, H. and E. Aberg. (eds.). 1950. *Proceedings of the Seventh International Botanical Congress.* Stockholm: Almquist & Wiksell.

Ouwerkerk, L. 2011. *No Elephants for the Maharaja: Social and Political Change in Travancore 1921–1947* (edited and introduced by D. Kooiman). New Delhi: Manohar Publications (first published 1994).

Padmasini, K. 1947. "Antiseptic Culture of Ragi Seedlings- and Their Response to Vitamins," *Current Science* 16: 227–228.

Pai, M. K. and G. McKendrik. 1911. "The Rate of Multiplication of Micro-Organisms: A Mathematical Study," *Proceedings of the Royal Society of Edinburgh* 31: 649–655.

Pal, B. P. 1948. "Some Aspects of Wheat Improvement in India", *IJGPB* 8, Nos. 1 and 2: 59–66.

Pal, B. P. and S. Ramanujam. 1944. "Plant Breeding and Genetics at the Imperial Agricultural Research Institute, New Delhi," *IJGPB* 4: 43–53.

Panje, R. R. 1933. "*Saccharum spontaneum* Linn., A Comparative Study of the forms grown at the Imperial Sugarcane Breeding Station, Coimbatore," *Indian Journal of Agricultural Sciences* 3, no. 6: 1013–1044.

Parthasarathy, N. 1941. "An Indian Source for Colchicine," *Current Science* 10, no. 10: 446.

Parthasarathy, N. 1946. "Chromosome Numbers in Bambusae", *Current Science* 15, no. 8: 233–234.

Parthasarathy, N. 1948. "Origin of Noble Sugar-Canes (*Saccharum officinarum*)", *Nature* 161: 608.

Penny, F. E. 1908. *On the Coromandel Coast.* London: Smith, Elder & Co.

Perkins, J. H. 1997. *Geopolitics and the Green Revolution: Wheat, Genes, and the Cold War.* Oxford and New York: Oxford University Press.

Peters, J. D. 2006. "Exile, Nomadism, and Diaspora: The Stakes of Mobility in the Western Canon." In *Spaces of Visual Culture*, edited by J. Morra and M. Smith, 141–160. London: Routledge.

Pickering, A. (ed.). 1992. *Science as Practice and Culture.* Chicago, IL: University of Chicago Press.

Pickering, A. 2010a. "Cybernetics as Nomad Science." In *Deleuzian Intersections: Science, Technology, Anthropology*, edited by C. B. Jensen and K. Rödje, 155–162. New York and Oxford: Berghahn Books.

Pickering, A. 2010b. *The Cybernetic Brain: Sketches of Another Future.* Chicago, IL and London: The University of Chicago Press.

Pickersgill, B. 1977. "Taxonomy and the Origin and Evolution of Cultivated Plants in the New World", *Nature* 268: 591–595.

Pillay, A. P. 1931. *Welfare Problems in Rural India.* Bombay: Taraporewala & Sons.

Pinsdorf, K. 1929. *Relations between Argentina and Brazil.* Stanford, CA: Stanford University Press.

Piramal, G. 1998. *Business Legends.* New Delhi: Penguin Books.

Pistorius, R. 1997. *Scientist, Plants and Politics: A History of the Plant Genetic Resources Movement.* Rome: International Plant Genetic Resources Institute.

Portin, P. and A. Wilkins. 2017. "The Evolving Definition of the term "Gene"," *Genetics* 205, no. 4: 1353–1364.

Prashad, B. 1938. *The Progress of Science in India in the Past Twenty-Five Years*. Calcutta: Indian Science Congress Association (Silver Jubilee).

Pripas-Kapit, S. R. 2015. "Educating Women Physicians of the World: International Students of the Woman's Medical College of Pennsylvania, 1883–1911", doctoral dissertation, University of California, Los Angeles. https://escholarship.org/uc/item/9gh5b9j1 accessed 5 June 2019.

Pugh, J. 2013. "Island Movements: Thinking with the Archipelago," *Island Studies Journal* 8, no. 1: 9–24.

Punnett, R. C. 1941. *Proceedings of the Seventh International Congress of Genetics: Edinburgh, Scotland, 23–30 August 1939*. Cambridge: The University Press (issued as a Supplementary volume to the *Journal of Genetics*).

Puri, V. and Y. S. Murty. (eds.). 1978. "First Botanical Conference, Indian Botanical Society: Abstracts," *The Journal of the Indian Botanical Society* 57 (Supplement).

Pycior, H. M., N. G. Slack and P. G. Abir-Am. (eds.). 1996. *Creative Couples in the Sciences*. New Brunswick, NJ: Rutgers University Press.

Raghavan, T. S. 1952. "Sugarcane x Bamboo Hybrids," *Nature* 170: 329–330.

Rajbhandari, K. R. 1976. "History of Botanical Exploration in Nepal," *Journal of the Bombay Natural History Society* 73: 468–481.

Ramamurti, V. 2006. Anekal Ramaswamiengar Gopal-Ayengar (1 January 1909–09 September 1992)," *Biographical Memoirs of Fellows of the INSA* 30: 17–27.

Ramanujam, S. 1944. "Genetical Research as Applied to Plant Breeding in Post-War India," *Current Science* 13, no. 3: 63–65.

Ramaswamy, V. 1992. "Rebels—Conformists? Women Saints in Medieval South India," *Anthropos* 87, H. 1./3.: 133–146.

Ramesh, J. 2017. *Indira Gandhi: A Life in Nature*. London, New York, Sydney, Toronto and New Delhi: Simon & Schuster.

Ramesh, J. 2020. *Chequered Brilliance: The Many Lives of V. K. Krishna Menon*. Haryana: Penguin Random House India.

Ramiah, K. 1934. "Rice Research in Madras," *Current Science* 3, no. 1: 36.

Ramiah, K. 1937. *Rice in Madras, a Popular Handbook*. Madras: Government Press.

Ranson, C. W. 1938. *A City in Transition: Studies in the Social Life of Madras*. Madras: The Christian Literature Society of India.

Rao, C. H. 1915 [1916]. *The Indian Biographical Dictionary*. Madras: Pillar & co.

Rao, V. 2015. "J. B. S. Haldane, an Indian Scientist of British Origin," *Current Science* 109, no. 3: 634–638.

Ray, B. (ed.). 2005. *Women of India: Colonial and Post-Colonial Periods* (Series: History of Science, Philosophy and Culture in Indian Civilization, Gen. Ed. D. P. Chattopadhyaya, Vol. IX, Part 3). New Delhi: Sage Publications.

Reddy, B. M., T. S. Vasalu and K. C. Malhotra. 1993. "Physical Anthropology as a Science: A Commentary," *Current Science* 64, no. 1: 16–21.

Reddy, M. 1930. *My Experience as a Legislator*. Madras: Current Thought Press.

Revathi Amma, C. K. 1977. *Sahasrapoornima*. Kottayam: Sahitya Pravartaka Sahakarana Sangham.

Rheinberger, H. J. 1997. *Towards a History of Epistemic Things: Synthesizing Proteins in the Test Tube*. Stanford, CA: Stanford University Press.

Richharia, R. H. 1960. "Origins of Cultivated Rices," *IJGPB* 20, no. 1: 1–14.

Richmond, M. 2010. "Women in Mutation Studies: The Role of Gender in the Methods, Practices, and Results of Early Twentieth Century Genetics." In

Making Mutations: Objects, Practices and Contexts, edited by L. Campos and A. von Schwerin, 11–48. Preprint 393, Berlin: Max-Planck-Institut für Wissenschaftsgeschichte.

Richmond, M. 2015. "Women as Mendelians and Geneticists," *Science and Education,* 24, nos. 1–2: 125–150.

Richmond, M. 2017. "Women as Public Scientists in the Atomic Age: Rachel Carson, Charlotte Auerbach, and Genetics," *Historical Studies in the Natural Sciences* 47, no. 3: 349–388.

Richmond, M. 2018. "Women in the Historiography of Biology." In *Historiography in Biology, Historiographies of Science, vol. 1,* edited by M. Dietrich, M. Borrello and O. Harman. New York and Cham: Springer.

Richmond, M. 2020. "South American Field Work, Cytogenetic Knowledge: The cytogenetic research program of Sally Hughes-Schrader and Franz Schrader" in "Heredity and Evolution in an Ibero-American Context," ed. Ana Barahona and Marsha Richmond. *Special Issue: Perspectives on Science* 28, no. 2: 127–169.

Robinson, G. 1979. *A Prelude to Genetics: Theories of a Material Substance of Heredity, Darwin to Weismann.* Lawrence, KS: Coronado Press.

Roll-Hansen, N. 2005. "The Lysenko Effect: Undermining the Autonomy of Science," *Endeavour* 29, no. 4: 143–147.

Rossiter, M. W. 1982. *Women Scientists in America, Struggles and Strategies to 1940.* Baltimore, MD: Johns Hopkins University Press (paperback 1984).

Rossiter, M. W. 1995. *Women Scientists in America: Before Affirmative Action, 1940–1972.* Baltimore, MD: Johns Hopkins University Press.

Roxburgh, W. 1832. *Flora Indica or Description of Indian Plants* (Vols. 2 and 3). London: Printed for William Thacker.

Rudwick, M. 1996. "Geological Travel and Theoretical Innovation: The Role of 'Liminal' Experience," *Social Studies of Science* 26: 143–169.

Rümke, C. L. 1934. "Saccharum-Erianthus Bastaarden," *Archief voor de suikerindustrie in Nederlandsch-Indië,* 211–63.

Russell, E. J. 1937. *Report on the Work of the Imperial Council of Agricultural Research in Applying Science to Crop Production in India.* Delhi: The Manager of Publications.

Salazar, N. M. 2017. "Theorizing Mobility Through Concepts and Figures," *Social Anthropology* (Special issue: Key figures of mobility) 25, no. 1: 5–12.

Sampath, S. and K. Ramanathan. 1949. "Chromosome Numbers in Indian Economic Plants- III," *Current Science* 18, no. 11: 408–409.

Santapau, H. 1954. "The Revival of the Botanical Survey of India," *Science and Culture* 19, no. 8: 377–381.

Santapau, H. 1960. "I. H. Burkill in India," *Garden's Bulletin (Singapore)* 17, Series 4i: 341–349.

Sapp, J. 1983. "The Struggle for Authority in the Field of Heredity, 1900–1932: New Perspectives in the Rise of Genetics," *Journal of the History of Biology* 16, no. 3: 311–342.

Sauer, C. O. 1938. "Theme of Plant and Animal Destruction in Economic History." *Journal of Farm Economics* 20, no. 1: 765–775.

Sax, K. 1931. "The Smear Technique in Plant Cytology," *Stain Technology* 6: 117–122.

Schaffer, G. 2007. "Scientific Racism Again?': Reginald Gates, 'the Mankind Quarterly' and the Question of 'Race' in Science after the Second World War," *Journal of American Studies* 41, no. 2: 253–278.

Schiebinger, L. 2014. "Following the Story." In *Writing About Lives in Science*, edited by P. Govani and Z. Franceschi, 43–54. Göttingen: V & R Unipress.

Schneider, W. H. 1995. "Blood Group Research: Great Britain, France and the United States between the World Wars," *American Journal of Physical Anthropology* 38, issue supplement S2: 87–114.

Schneider, W. H. 1996. "The History of Research on Blood Group Genetics: Initial Discovery and Diffusion," *History and Philosophy of the Life Sciences* 18: 277–303.

Searle, G. R. 1979. "Eugenics and Politics in Britain in the 1930s," *Annals of Science* 36: 159–169.

Segal, L. 1933. *Modern Russia: The Land of Planning*. London: Industrial Credits & Services.

Sen, B. 1940. "Vernalization," *Indian Farming* 1: 55–59.

Sengupta, J. C. 1959. "Botanical Survey of India: Its Past, Present and Future," *Bulletin of the Botanical Survey of India* 1, no. 1: 9–29.

Shackman, G. 2001. "The Botanical Garden on Iroquois," *Ann Arbor Observer*, May. aaobserver.aadl.org/aaobserver/15253 accessed 8 August 2019.

Shah, A. 2007. *Vikram Sarabhai (1919–1971): The Renaissance Man of Indian Science*. Haryana: Penguin Random House India.

Shamdasani, S. (ed.). 1996. "Introduction." In *The Psychology of Kundalini Yoga: Notes of the Seminar Given in 1932*, edited by C. G. Jung. London: Routledge.

Singh, R. and S. C. Roy. 2018. *A Jewel Unearthed: Bibha Chowdhuri*. Düren, Germany: Shaker Verlag gmbH.

Sinopoli, C. M. 2017. "Walter N. Koelz (1895–1989) and Rup Chand (1902–94): Partners in Asian Collecting." In *Object Lessons, The Formation of Knowledge: The University of Michigan Museums, Libraries, & Collections 1817–2017*, edited by K. Barndt and C. M. Sinopoli, 178–186. Ann Arbor, MI: University of Michigan.

Sircar, S. M. 1939. "Some Aspects of Vernalization," *Science and Culture* 4: 438–442.

Sismondo, S. 2010. *An Introduction to Science and Technology Studies*. Oxford: Blackwell (first edition 2004).

Smocovitis, V. B. 1992. "Unifying Biology: The Evolutionary Synthesis and Evolutionary Biology," *Journal of the History of Biology* 25, no. 1: 1–65.

Söderqvist, T. 2011. "The Seven Sisters: Subgenres of 'Bioi' of Contemporary Life Scientists," *Journal of the History of Biology* 44, no. 4: 633–650.

Soyfer, V. N. 2003. "Tragic History of the VII International Congress of Genetics," *Genetics* 165: 1–9.

Stebbins, G. L. 1950. *Variation and Evolution in Plants*. New York: Columbia University Press.

Stebbins, G. L. 1958. *Ernest Brown Babcock 1877–1954*. Washington, DC: National Academy of Sciences.

Stebbins, G. L. 1971. *Chromosome Evolution in Higher Plants*. London: Edward Arnold Ltd.

Steere, W. C. 1931. "A New and Rapid Method for Making Permanent Acetocarmine Smears," *Stain Technology* 6: 107–111.

Stengers, I. 2011. *Cosmopolitics II* (R. Bononno, trans.). Minneapolis, MN: University of Minnesota Press.

Stengers, I. 2018. *Another Science is Possible: A Manifesto for Slow Science.* Cambridge, UK: Polity Press.

Štrbáňová, S. 2016. *Holding Hands with Bacteria: The Life and Work of Marjory Stephenson.* Berlin and Heidelberg: Springer Verlag.

Strutevant, A. H. 1965. *A History of Genetics.* New York: Harper and Row.

Subramanian, C. V. 1991a. "Plant Physiologist and Humanist: A Birth Centenary Tribute to Prankrushna Parija," *Current Science* 60, no. 6: 380–382.

Subramanian, C. V. 1991b. "Professor Birbal Sahni—the Man and His Message," *Current Science* 61, nos. 9–10: 564–569.

Subramanian, C. V. 2007. "Edavaleth Kakkat Janaki Ammal," *Resonance* 12, no. 6: 4–9.

Sundara Raghavan, R. 1957. "Chromosome Numbers in Indian Medicinal Plants," *Proceedings of the Indian Academy of Sciences-Section B* 45, no. 6: 294–298.

Sundara Raghavan, R. 1958a. "A Chromosome Survey of Indian Dioscoreas (Communicated by E. K. Janaki Ammal)," *Proceedings of the Indian Academy of Sciences-Section B* 48, no. 1: 59–63.

Sundara Raghavan, R. 1958b. "Chromosome Numbers in Indian Medicinal Plants–II," *Proceedings of the Indian Academy of Sciences-Section B* 47, no. 6: 352–358.

Sundara Raghavan, R. 1959a. "A Note on Some South Indian Species of the Genus *Dioscorea*," *Current Science* 28, no. 8: 337–338.

Sundara Raghavan, R. 1959b. "Chromosome Numbers in Indian Medicinal Plants–III," *Proceedings of the Indian Academy of Sciences-Section B* 49, no. 4: 294–298.

Sur, A. 2011. *Dispersed Radiance: Caste, Gender, and Modern Science in India.* New Delhi: Navayana Publishing.

Swaminathan, M. S. 1988. "100th Birth Anniversary of Academician N. I. Vavilov," *Indian Journal of Plant Genetic Resources* 1, nos. 1–2: 1–5.

Swaminathan, M. S. 1996. "Benjamin Peary Pal," *Biographical Memoirs of Fellows of the Royal Society* 42: 266–274.

Taylor, W. R. 1924. "The Smear Method of Plant Cytology," *Botanical Gazette* 78: 236–238.

Taylor, W. R. 1966. "Records of Asian and Western Pacific Marine Algae, Particularly Algae from Indonesia and the Philippines," *Pacific Science* 20: 342–359.

Thomas, R. and T. S. Venkatraman. 1930. "Sugarcane—Sorghum Hybrids," *Agricultural Journal of India* 25: 164.

Thomas, W. L. (ed.). 1956. *Man's Role in Changing the Face of the Earth* (2 vols). Chicago, IL and London: The University of Chicago Press.

Thuljarama Rao, J. 1963. "Tiruvadi Sambasiva Venkatraman," *Biographical Memoirs of the Indian National Science Academy* 11. www.insaindia.res.in/BM/BM11_8612.pdf accessed 9 November 2016.

Thurston, E. 1894. *Notes on Tours Along the Malabar Coast.* Madras Government Museum Bulletin Series 1, no. 2. Madras: Government Press.

Thurston, E. 1897. *Eurasians of Madras and Malabar.* Madras Government Museum Bulletin Series 2, no. 2. Madras: Government Press.

Thurston, E. 1900. *Notes on Some of the People of Malabar.* Madras Government Museum Bulletin Series 3, no. 1. Madras: Government Press.

Thurston, E. 1903. *Some Marriage Customs in South India.* Madras Government Museum Bulletin Series 4, no. 3. Madras: Government Press.

Thurston, E. 1909. *Castes and Tribes of South India* (7 Vols., Vol. 7). Madras: Government Press.

Tilman, H. W. 1952. *Nepal Himalaya*. Cambridge: Cambridge University Press.

Tournier, M. 1988. *The Wind Spirit: An Autobiography* (trans. of Le Vent Paraclet, Paris: Editions Gallimard, 1977). Boston, MA: Beacon Press.

Trotsky, L. 1932 (1930). *History of the Russian Revolution*. 3 vols (trans. into English by Max Eastman). New York: Simon and Schuster.

Turesson, G. 1993. "The Scope and Import of Genecology," *Hereditas* 4, no. 1–2: 172.

UNESCO. 1956. *Study of Tropical Vegetation: Proceedings of the Kandy Symposium* (Kandy, Sri Lanka, 19–21 March 1956). Paris: UNESCO.

UNESCO. 1958. *Problems of Humid Tropical Regions* (Kandy, Sri Lanka, 22–24 March 1956). Paris: UNESCO.

Vavilov, N. I. 1997. *Five Continents* (published in English, trans. of the Russian version published in 1987). Rome: The International Plant Genetic Resources Institute.

Venkatraman, T. S. 1924. "Simple Contrivances for Studying Root Development in Agricultural Crops," *Agricultural Journal of India* 19: 509.

Venkatraman, T. S. 1927 [1925]. "Sugarcane Breeding in India." In *Imperial Botanical Conference* [July 7–16, 1924]: *Report of Proceedings*, edited by F. T. Brooks, 57. Cambridge: Cambridge University Press.

Venkatraman, T. S. 1927. "Sugarcane Breeding, Indications of Inheritance," *Memoirs of the Department of Agriculture in India* 14, no. 3: 113–129.

Venkatraman, T. S. 1938. "Presidential Address: Hybridization in and with the Genus Saccharum (Its Scientific and Economic Aspects" (Section: Agriculture). In *Proceedings of the Twenty-Fifth Indian Science Congress*. Calcutta, (Silver Jubilee Session), 267–84," *Current Science* 6 (Supplement: Indian Science Congress Silver Jubilee Session, Calcutta, 1938) no. 8: 425–427.

Venkatraman, T. S. 1942. "Message of the Sugarcane," *IJGPB* 2: 3–10.

Venkatraman, T. S. and R. Thomas. 1930. "Brief Note on Sugarcane-*Sorghum* Hybrids," *Proceedings of the International Society of Sugarcane Technologists* 4, Bulletin no. 67: 1–8.

Vishwanathan, T. V. 1973. "A New Source of Glyco-Alkaloid *Solanum trilobatum* Linn., and Its Tetraploid Derivative," *Current Science* 42, no. 22: 805.

Wagner, W. H. 1970. "Biosystematics and Evolutionary Noise," *Taxon* 19, no. 2: 146–151.

Wagstaff, J. 2009. "Dr Marion Wood: Early Life and Career," *Devon Newsletter* (National Council for the Conservation of Plants and Gardens) (Summer): 19–22.

Watkin, C. 2020. *Michel Serres: Figures of Thought*. Edinburgh: Edinburgh University Press.

Whyte, R. O. and P. S. Hudson. 1933. *Vernalization, or Lyssenko's Method for the Pre-Treatment of Seed*. Aberystwyth: Imperial Bureaux of Plant Genetics.

Wilks, W. (ed.). 1907. *Report of the 3rd International Conference on Genetics, Hybridisation and General Plant-Breeding*. London: Royal Horticultural Society.

Williams, M. 1987. "Sauer and 'Man's Role in Changing the Face of the Earth'," *Geographical Review* 77, no. 2: 218–231.

Woolf, V. 1935 [1929]. *A Room of One's Own*. London: Hogarth Press.

Woolf, V. 1938. *Three Guineas*. London: Hogarth Press.

Yasui, K. 1933. "Ethyl Alcohol as a Fixation for Smear Methods," *Cytologia* 5: 140.

Yule, H. and A. C. Burnell. 1903. *Hobson-Jobson: A Glossary of Colloquial Anglo-Indian Words and Phrases, and of Kindred Terms, Etymological, Historical, Geographical and Discursive*, new edition, edited by W. Crooke. London: Murray.

Zimmer, D. 1998. *A Guide to Nabokov's Butterflies and Moths*. Hamburg: D. E. Zimmer.

Zutshi, U. 1966. "The Transference of Solasodine from *Solanum incanum* to S. Melongena," *Current Science* 35, no. 17: 439.

Zutshi, U. 1967a. "Interspecific Hybrids in Solanum- I. *Solanum indicum* Linn. and *Solanum incanum* Linn.," *Proceedings of the Indian Academy of Sciences Section B* 65, no. 3: 111–113.

Zutshi, U. 1967b. "Variation in the Morphology and Solasodine Content of Two Races of *Solanum incanum* Linn.,' *Proceedings of the Indian Academy of Sciences Section B* 65, no. 3: 108–110.

INDEX

Alstonia scholaris 410
Amalgamations Group (J Farm),
    Madras 499
Amaranth 278–279, 285
Amaryllis 278
Amorphophallus 151
amphiploid(s) 197, 238
Amsterdam 101, 156–157, 159; Sixth
    International Botanical Congress
    (1935) 157–159
Anacardiaceae 443
Anamalai Hills 378, 380–381, 397, 399
Anderson, Edgar 63, 111, 121, 239,
    244–245, 308, 310, 375–376,
    381, 390, 545–546, 549–550;
    Introgressive Hybridization (1949)
    376; *Plants, Man and Life* (1952)
    121, 376
Anderson, E. G. 44, 94
Andropogoneae 111, 174, 185, 197,
    220, 278, 453
angiosperm(s) (flowering plants) 70,
    324, 406, 442, 444, 468; woody 250,
    254, 261, 310
anthropology xxxi, 48, 52–53, 55, 61,
    112, 114, 142, 170, 177, 181–182,
    202, 376–377, 386, 424–425,
    427–428, 435, 480, 542
anthropologist(s) 113, 159, 167, 183,
    202, 210, 377, 385, 398, 422–423,
    427–428, 454, 503, 523
Antibes, Provence (France) 340
apes 469
Apluda 453, 456
Apocynaceae 395, 472
Appleton, E. V. 360
Arabia(n) 160, 380, 406, 427
Aralam (Indo-Russian Farm) 470
Arber, A. 111
archipelagic thinking xviii,
    xxxvi–xxxvii, 330, 379, 394–395,
    398, 401, 406, 408, 548
Archipelago xxxvi–xxxvii, 109, 133,
    330, 379, 396
Argikar, G. P. 422
Arnold Arboretum (Harvard University)
    63, 37
aromatic plants 151, 437, 480–482,
    486, 530, 545–546
Artemisia 430–431
Artocarpus 382, 398
Arundo 298

Asana, J. J. 159
Asana, R. D. 309, 403
Ashby, E. 315
Ashtangahridaya 527
Ashworth, D. 247
asparagus 250–251, 262, 289, 319,
    334, 450
Assam 30, 161, 174, 185, 196, 201,
    220–221, 251, 263, 274, 296,
    306–307, 321, 330, 342, 353, 358,
    363, 381, 382, 393–394, 396–398,
    401, 402, 409, 417, 421, 460,
    466, 469, 478, 482, 503; Regional
    Research Laboratory (RRL), Jorhat
    417, 421
Association of Economic Biologists *see*
    Coimbatore
Atkinson, G. 251
atomic bomb 232, 371
atomic energy 370–371, 387, 393, 407,
    410
Atomic Energy Committee (Bombay)
    369; Indian Atomic Energy
    Commission 313, 372
Atomic Energy Establishment (later
    Bhabha Atomic Research Centre,
    BARC, Trombay) 390, 452; Medical
    and Biology Division 412, 452
atomic radiation 370
Attapadi forests (Western Ghats) 468, 496
Australia(n) 214, 247, 319, 346, 379,
    380, 392, 403, 406, 408, 460, 466,
    550
autoecology *see* ecology
autopolyploid 237
autopolyploidy 241, 444
Avdulov, N. P. 298
ayurveda 4, 350, 469, 516, 527
ayurvedic 433
Ayyangar, G. N. Rangaswamy
    134–135, 230

Babcock, E. B. 72–73, 82, 123, 546
Bachelard, G. 550
backcrossed 205, 461, 508
Badagas 465
Bai, Sunanda, K. 283
Baker, Laurie 462
Baker, W. 228
bamboo(s) 111, 170–171, 174, 189,
    197, 201, 203, 246, 277, 298, 409,
    458, 476–477, 502, 520

Goodspeed T. H. 79, 157
Gopal-Ayengar A. R. 252, 296, 322, 371, 410, 412, 422, 439, 449, 452, 456, 458, 489
Gopinath, P. 441
Gould, N. K. 248, 262, 271, 300
Gourou, Pierre 377
Govindamma 517–518, 526, 533, 535
Govindarajalu, E. 472
Govindaswamy, S. 422
Gowri Amma, K. R. 404
gram 282, 382
Gramineae (Poaceae) 111, 138, 179, 185, 213, 229, 276–277, 355, 409, 482
grapes 187
grass(es) 49–51, 68, 70, 103, 107–108, 111, 117, 120, 133, 138–139, 146, 148, 156–157, 170, 173–174, 176, 184–185, 188, 190–191, 193–198, 201–202, 204–205, 215, 217–222, 229–230, 237, 242, 246–247, 251, 259, 269, 277–278, 284, 285, 298, 306, 322, 334, 340, 354–355, 357, 386, 401, 409, 429, 453, 468, 474–476, 482, 515, 530, 546–547; grassland 63, 276, 465, 468 (see also shola)
grasshoppers 75–76
Gray, Asa 60, 62–63, 328, 339, 378
grazing 378, 398
'great traveller(s)' 260–261
Green, D. E. 248
Green, S. 467
Green Revolution xxxiii, 406, 414, 463
Gregoirè, V. 112, 157
Gregory, R. 113, 194, 296, 311
Griffith, W. 352
Grow More Food campaign 302–305, 307, 314, 339, 378, 383, 397; R. K. Narayan, All about Rice (1950) 314
Grüneberg, H. 243, 245, 411, 414, 452
Guattari, F. 538–539, 542, 547, 549, 550
Guha, B. S. 424
Gulag 234
Gundert, H. 4, 17
Gupta, B. K. 433–434, 450, 530
Gupte, Miss. K. G. 283
Gustafson, F. 61
Gustafsson, Åke 402
Gwyer, M. 282
gymnosperms 324, 444, 468

Haddon, A. C. 113
Haeckel, E. 206
Haldane, J. B. S. xxxiii, 86–88, 92–93, 113, 121, 148, 182, 207, 209, 225, 233, 243, 251–252, 257, 270, 311, 325, 347, 376, 387, 391, 405, 411, 414, 420, 422–427, 434–435, 441, 443, 450, 454, 545; Journal of Genetics 27, 219, 252, 424; Indian citizenship 425; limerick on Gayre 424; population genetics (beanbag genetics) 87; protest against USIS 425
Hall, D. 175, 211
Hamerton, J. 257
Hanger, F. 321–322, 332
Hannyngton, John Child 6–8, 15, 17–18, 255; John Caulfield Hannyngton 7
Hanuman 423, 471
Harlan, J. 153, 234
Harland, S. 207, 308
Harman, O.S. 71, 82, 101–102, 121, 165, 167, 244, 435, 451, 456, 524
Harrow, R. L. 247–248
Hartmann, M. 206
Harvey, G. 293, 405, 524
heat and cold shocks 242
Heidegger, M. xxxvii, 204
Helmericks, Constance 253
hemp 187, 239, 382
Henderson, D. M. 476
Herrick, A. C. 319
Hertwig, R. 73, 206
Herzog, M. 374
heterochromatic 241
heterochromatin 242, 245
heterogeneity (multiplicity) 539
heterogeneous (connections, populations, practices, races) xxxii, xxxv, 233, 240, 545, 547
heteroploid(s) 157
heterosis 174, 180, 479; see also hybrid vigour
hexaploidy 441–443
Heyerdahl, T. 440
high altitude agriculture 429–430, 436–437, 440
Hill, A. V. 204, 313
Hill, A. W. 193, 219
Himalaya(n) xviii, 189, 196, 201, 243, 250, 261, 263, 272–277, 279, 288, 295, 299, 321, 339–340, 349,

Botanical Congress (1935) 157–159; an anthropological curiosity 48; anti-Brahman sentiments 108, 217, 265, 544; applying to the University of Ceylon 253; 'Ashok' (ethnobotanical/medicinal garden, Shoranur) 412–413, 439, 462, 470, 472, 491, 511, 514–515, 519, 524; attic flat, Wisley 306; 'August 15th 1947' 272, 280; Bengal politics 347; Birbal Sahni Medal (1960 421; birth 8 (birth certificate 8, 16); Board of Studies for Natural Science 46, 51, 105, 115, 148, 298; 'born wanderer' 168, 194; botanising family 108; botany as a familial enterprise 58; Buddhist 24, 426; cats 215, 438, 515, 518, 526, 533; 'Chromosome Catalogue of Medicinal Plants' 459–460; 'chromosome counter' 256, 325, 545; 'Cinderella of the Sugarcane Station' 184, 447, 541; collecting marine algae 89, 109–110, 126, 136, 142; concerns about population explosion 114, 342, 381, 383, 384, 402, 405, 423, 468, 481, 489, 516; connections and similitudes 379, 401–402, 407–408, 482, 542, 547; 'contaminate' taxonomy with cytology 347, 541; cytogeography 261, 302, 309, 321, 330, 332, 334–335, 339, 346, 351, 358, 361, 379, 402, 404–406, 409, 415, 421, 438, 446, 448–449, 475, 544, 546–548; Dabbler in Radioisotope Procedures 372; death 534–535; death wish 191, 195, 215, 223, 537, 544; 'Dictionary of Medicinal Plants' (*see also* 'Chromosome Catalogue of Medicinal Plants') 449, 460; Director of Agriculture (India) 299, 302, 383; Doctor of Laws (1956) 373, 387–388; drafting a science policy for Kerala xvii, 462, 466, 474, 477; early pioneer of genetic conservation 241, 381, 463, 496; Edam (Edathil House, Tellicherry) 8, 10, 12–14, 21, 32, 34–36, 47–48, 52, 57–58, 103, 139, 142, 168–170, 185–186, 190–191, 215, 255, 284–286, 316, 341, 345, 364, 399, 404, 413, 426, 440, 445, 446, 448, 462–463, 465,

470, 478, 487, 507, 510, 512, 514–515, 517, 525, 527, 532, 534–535, 549; Edavaleth House (Mahe) 6, 8, 426, 463, 507; eggplant (*Solanum*) 45, 51, 93–94, 102, 127–128, 130, 382, 547; Emeritus Professor (J&K University 1964) 437, 445–446, 448; eugenics 44, 113–114, 159, 384, 423; falling out with Darlington over the *Atlas* 322–323, 358–359; Fellow of the Indian Academy of Sciences (1934) 148; Fellow of the Indian National Science Academy (1957) 403; Fellow of the Indian Society for Genetics and Plant Breeding (1975) 477; Fellow of the Linnean Society (1931) 104, 120; Fellow of the Royal Geographical Society (1950) 138, 294, 302, 316; fibrositis 341, 533; field notebook/s 153, 173, 201–202, 204, 270, 284–285, 288, 305, 438; 'Flora of Assam' 421; 'Flora of Chingleput District' 51; 'Flora of Malabar' 49; 'Flora of the Nilgiris' 221, 500; 'Flora of South India' 51, 103, 109; Gandhi 19, 24, 487, 517 (meeting Gandhi 142); Ganjam explorations 53, 55; garden at Capper House 19, 22; "Genetical Analysis of Humid Forest Flora of South Asia" (1956) 393–396; "Genetics and Man in India" (1961) 427, 435; 'great restlessness' 152, 168, 195, 403; 'gypsy life' 105; 'hermit' 189, 256; 'hermitage' 457, 533, 546; 'homesick' 20, 292 ('homesick for England' 401, 403–404); 'Hortus Malabaricus', Institute in Malabar 415, 431, 515; 'I do not like Congressmen' 307; India as an 'anthropological museum' 52; INSA project on ethnobotany 464–466, 473–474, 477; Institut National Agronomique, Paris 325, 377; interest in anthropology xxxi, 48, 52–53, 55, 61, 114, 142, 376–377, 386, 398, 425, 427–428, 480, 542; Jammu Symposium (1976) 481–482; *Janakia arayalpathra* 510, 529; *kathu kutthal* (Janaki's ear-piercing ceremony) 9; lantern slides 259, 299; longing for silence and

For Product Safety Concerns and Information please contact our EU
representative GPSR@taylorandfrancis.com
Taylor & Francis Verlag GmbH, Kaufingerstraße 24, 80331 München, Germany

www.ingramcontent.com/pod-product-compliance
Lightning Source LLC
Chambersburg PA
CBHW070711220326
41598CB00026B/3689